D0767781

©2000 Algrove Publishing Limited
ALL RIGHTS RESERVED.
No part of this book may be reproduced in any form, including photocopying, without permission in writing from the publishers, except by a reviewer who may quote brief passages in a magazine or newspaper or on radio or television.

Algrove Publishing Limited
36 Mill Street, P.O. Box 1238
Almonte, Ontario, Canada K0A 1A0

Telephone: (613) 256-0350
Fax: (613) 256-0360
Email: sales@algrove.com

National Library of Canada Cataloguing in Publication Data

Hiscox, Gardner D. (Gardner Dexter), 1822?-1908.
1800 mechanical movements and devices.

(Classic reprint series)
Reprint of Mechanical movements : powers and devices, by Gardner D. Hiscox,
12th ed., published: New York : N.W. Henley, 1911, with new cover title added.
On cover: Originally published in 1899.
Includes index.
ISBN 1-894572-18-1

1. Mechanical movements. I. Title. II. Title: Eighteen hundred mechanical movements and devices. III. Series: Classic reprint series (Almonte, Ont.)

TJ181.H58 2000 621.8 C00-901337-7

Printed in Canada
#5-3-04

Publisher's Note

Although some of the information in this book (such as the extensive material on steam power) is of limited use to the current designer of mechanical devices, the sections on gearing, power transfer and the control of motion are alone worth the price of the book to a craftsman. We were tempted to reduce the size of the book by eliminating some of the more arcane information, but decided to be faithful to the original. The thought of eliminating any of the engravings to save a bit of paper seemed like bad economics in the long run. After looking through the book you will see why we were loath to make it a more "economical" production.

Leonard G. Lee, Publisher
Ottawa
December, 1999

MECHANICAL MOVEMENTS
POWERS AND DEVICES

CONTAINS:

AN ILLUSTRATED DESCRIPTION OF MECHANICAL MOVEMENTS AND DEVICES USED IN CONSTRUCTIVE AND OPERATIVE MACHINERY AND THE MECHANICAL ARTS, BEING PRACTICALLY A MECHANICAL DICTIONARY, COMMENCING WITH A RUDIMENTARY DESCRIPTION OF THE EARLY KNOWN MECHANICAL POWERS AND DETAILING THE VARIOUS MOTIONS, APPLIANCES AND INVENTIONS USED IN THE MECHANICAL ARTS TO THE PRESENT TIME

BY

GARDNER D. HISCOX, M. E.

Author of "Gas, Gasoline and Oil Engines," "Compressed Air," etc., etc

Illustrated by Eighteen Hundred Engravings

ESPECIALLY MADE FOR THIS BOOK

Twelfth Edition

NEW YORK:

THE NORMAN W. HENLEY PUBLISHING CO.

132 NASSAU STREET

1911

Copyrighted 1911
BY THE NORMAN W. HENLEY PUBLISHING CO.

Copyrighted, 1899 and 1903
BY NORMAN W. HENLEY & CO.

MACGOWAN & SLIPPER
PRINTERS
30 BEEKMAN ST., NEW YORK, N. Y.

PREFACE

The need for an illustrated and condensed work of reference for the inventor, the mechanical student, the artisan, and the workingman with the ambition of an inquiring mind, has become not only apparent to teachers of mechanics, but a real want among all who are interested in mechanical thought and work.

It is an interest the growth of which has been greatly encouraged by the rapid development of the inventive and mechanical arts during the past half century.

The increasing inquiries from inventors and mechanics, in regard to the principles and facts in constructive and operative mechanics have induced the author to gather such illustrations as have been found available on the subject of mechanical motions, devices, and appliances, and to place them in a form for ready reference with only sufficient text to explain the general principles of construction and operation, and as a partial exhibit of the mechanical forms in general use, with a view to place the largest amount of illustrated information within the limited means of the humblest seeker after mechanical knowledge.

The field of illustrated mechanics seems almost unlimited, and with the present effort the author has endeavored partially to fill a void and thus to help the inquirer in ideal and practical mechanics, in the true line of research.

Mechanical details can best be presented to the mind by diagrams or illustrated forms, and this has been generally acknowledged to be the quickest and most satisfactory method of conveying the exact conditions of mechanical action and construction.

Pictures convey to the inquiring mind by instantaneous comparison what detailed description by its successive presentation of ideas and relational facts fail to do; hence a work that appeals directly to the eye with illustrations and short attached descriptions, it is hoped, will become the means of an acceptable form of mechanical education that appeals to modern wants for the encouragement of inventive thought, through the study of illustrations and descriptions of the leading known principles and facts in constructive art.

The designing of the details of mechanical motion, devices, and appliances for specific purposes is an endless theme in the constructive mind, and if we may be allowed to judge from the vast number of applications for patents, of which there have been over a million in the United States alone, and of which over six hundred thousand have been granted in consideration of their novelty and utility, the run of mechanical thought seems to have become a vast river in the progress of modern civilization.

To bring into illustrated detail all the known forms and elements of construction is not within the limit of a human life; but to explore the borders of inventive design through the works that have passed into record has been the principal aim of the author of this book.

GARDNER D. HISCOX.

PREFACE TO TWELFTH EDITION.

The success of the previous editions of this work warrants the issue of this edition in enlarged and improved form, in which more than one hundred and fifty up-to-date mechanical movements and devices have been added, making it a most useful book of reference for those engaged in mechanical studies and pursuits, notably inventors and designers of machinery, in fact, for all who are interested in mechanics and its devices.

GARDNER D. HISCOX.

RECENTLY PUBLISHED

Mechanical Appliances

Mechanical Movements and Novelties of Construction

BY GARDNER D. HISCOX, M. E.

This is a complete work and a Continuation as a SECOND VOLUME of the author's work "Mechanical Movements, Powers and Devices."

The encouraging manner in which the first volume of "Mechanical Movements" was received has caused the publication of a second volume which unlike the first volume, which is more elementary in character, contains illustrations and descriptions of many combinations of motions and of mechanical devices and appliances found in different lines of machinery. The illustrations are of more complex machinery, are larger, with more extended descriptions. Some of the chapters included, are: Steam Power Appliances; Hydraulic Power Appliances; Road and Vehicle Devices; Horological Time Devices; Textile and Manufacturing Devices; Gearing and Gear Motion; Drafting Devices; Explosive Motor Power and Appliances; Mill and Factory Appliances; Perpetual Motion, etc. Each device being shown by a line drawing with a description showing its working parts and the method of operation. 396 pages, 1,000 specially made illustrations. Price, $2.50

☞ *Copies of this book will be sent prepaid on receipt of price.*

THE NORMAN W. HENLEY PUBLISHING CO.
132 Nassau Street New York City

CONTENTS.

SECTION I.

THE MECHANICAL POWERS.

SECTION II.

TRANSMISSION OF POWER.

SECTION III.

MEASUREMENT OF POWER.

SPEED, PRESSURE, WEIGHT, NUMBERS, QUANTITIES, AND APPLIANCES.

SECTION IV.

STEAM POWER.

BOILERS AND ADJUNCTS, ENGINES, VALVES AND VALVE GEAR, PARALLEL MOTION GEAR, GOVERNORS AND ENGINE DEVICES, ROTARY ENGINES, OSCILLATING ENGINES.

SECTION V.

STEAM APPLIANCES.

SECTION VI.

MOTIVE POWER.

SECTION VII.

HYDRAULIC POWER AND DEVICES.

SECTION VIII.

AIR POWER APPLIANCES.

SECTION IX.

ELECTRIC POWER AND CONSTRUCTION.

GENERATORS, MOTORS, WIRING, CONTROLLING AND MEASURING, LIGHTING, ELECTRIC FURNACES, FANS, SEARCHLIGHT, AND ELECTRIC APPLIANCES.

SECTION X.

NAVIGATION AND ROADS.

VESSELS, SAILS, ROPE KNOTS, PADDLE WHEELS, PROPELLERS, ROAD SCRAPERS AND ROLLERS, VEHICLES, MOTOR CARRIAGES, TRICYCLES, BICYCLES, AND MOTOR ADJUNCTS.

SECTION XI.

GEARING.

SECTION XII.

MOTION AND DEVICES CONTROLLING MOTION.

SECTION XIII.

HOROLOGICAL.

SECTION XIV.

MINING.

SECTION XV.

MILL AND FACTORY APPLIANCES.

SECTION XVI.

CONSTRUCTION AND DEVICES.

SECTION XVII.

DRAUGHTING DEVICES.

SECTION XVIII.

MISCELLANEOUS DEVICES.

ADDITIONS TO THE TWELFTH EDITION.

Section I.

THE MECHANICAL POWERS.

Forces and the Measure of their Work.

Force may be said to be the cause of motion and power in mechanics. Its condition may be static or dynamic; in the latter condition it becomes the means for the practical application of motion in the various forms of mechanical devices. Its statical condition is illustrated in the strains sustained in the material of construction and suspension.

The first and simple form of static force may be illustrated in the column, in the various positions in which it may be used for resistance of any kind; although in machinery, it may in itself become a moving body under stress. Static force may be represented by a column supporting weight; a beam under compressive strain; a body of water retained in a mill dam, steam pressure in a boiler, compressed air or liquefied gases, and a suspended weight; a coiled spring or anything that is under pressure without motion. The principal expressions for static force are compression, tension, and torsion, or their combinations. The resolution of forces is the geometrical relation and value of two or more forces acting upon a single point from different directions, or of a single force acting against several points of resistance.

The terms of resolution may be directional, static, or dynamic.

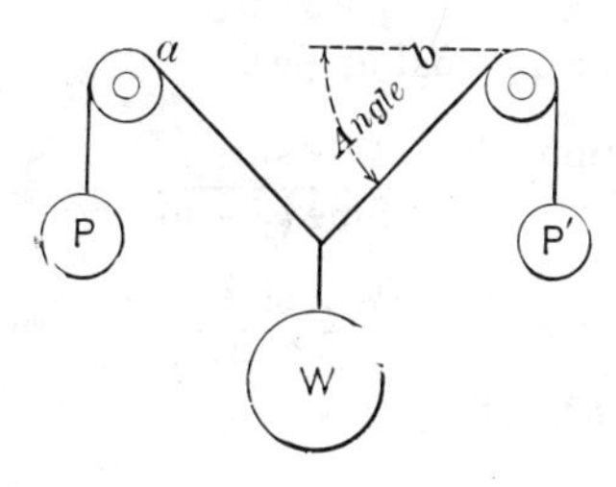

3. THE RESOLUTION OF SUSPENSION—in which W represents a force or the weight of gravitation, and P, P′ the resisting power or equivalent weights. Solution:

$$P \text{ and } P' = \frac{\text{half the weight}}{\text{Sine of angle of depression } a \text{ or } b}$$

when the angles are equal.

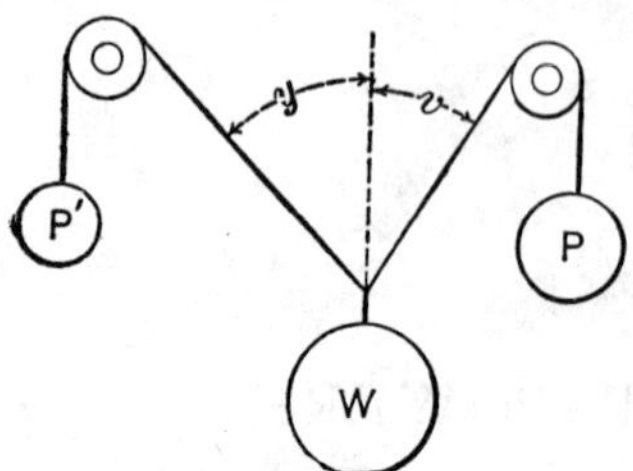

4. For unequal angles the forces vary as the sines of the angles from the vertical, respectively.

Solution: $P = \frac{W \times \text{sine } y}{\text{Sine }(y+v)}$

$P' = \frac{W \times \text{sine } v}{\text{Sine }(y+v)}$

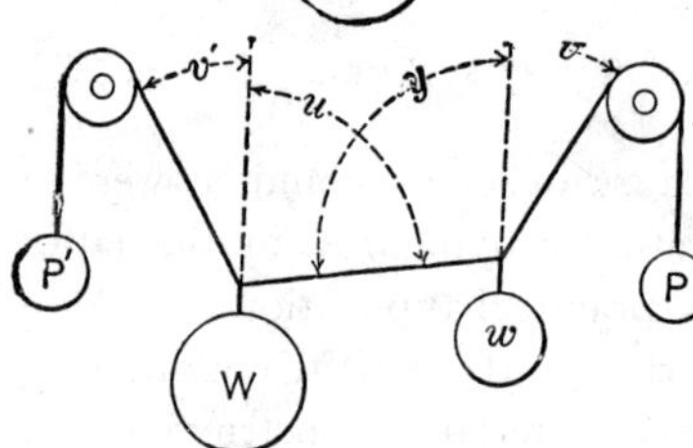

5. In a combination of forces the resolution involves the sines of the varying angles.

$P = \frac{w \times \text{sine } y}{\text{Sine}(y+v)}$, $P' = \frac{W \times \text{sine } v'}{\text{Sine}(u+v')}$

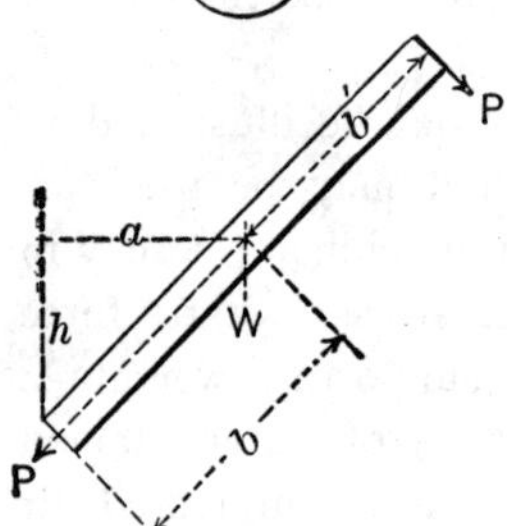

6. The forces in the direction of P and P′ in which the weight of a beam inclined and resting upon a point at P′ = W, at the centre of gravity.

$P' = \frac{W \times a}{b'}$ $P = \frac{W \times a}{b}$

The longitudinal thrust of struts or braces is the same as for tensional strains inversely, only that the weight of timbers or heavy materials should be considered separately, as shown further on.

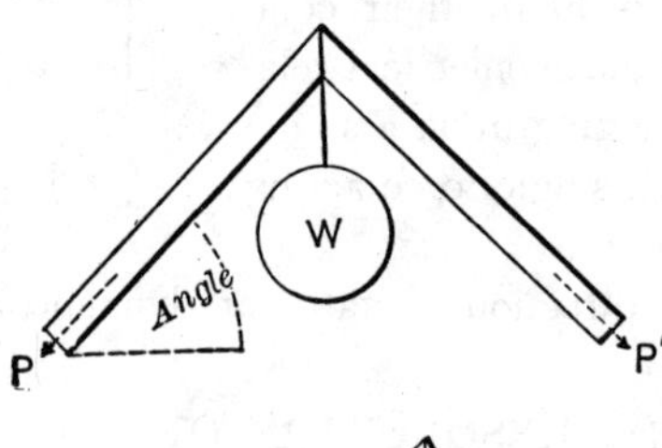

7. Where the members are of the same length and at equal angles.

$P \text{ and } P' = \frac{\text{half the weigh}}{\text{Sine of the angle } a \text{ or } b.}$

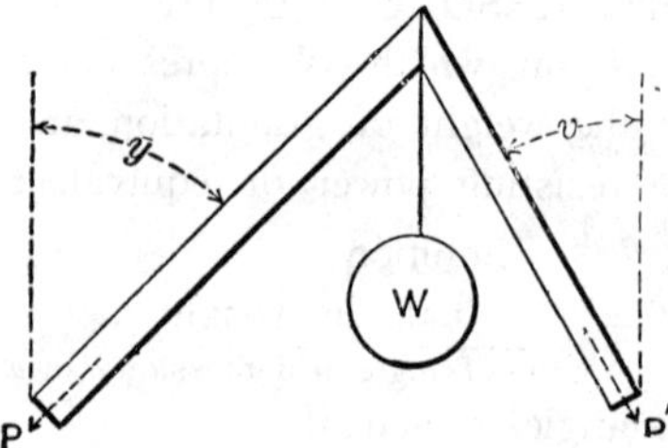

8. For unequal angles.

Solution: $P = \frac{W \times \text{sine } y}{\text{Sine }(y+v)}$

$P' = \frac{W \times \text{sine } v}{\text{Sine }(y+v)}$

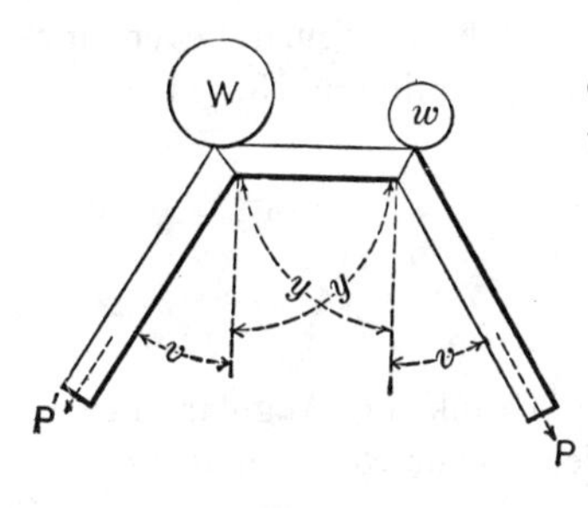

9. With truss beams carrying unequal weights the formulæ for end thrust are for equal angles.

$$P = \frac{w \times \text{sine } y}{\text{Sine } (y+v)}$$

$$P' = \frac{W \times \text{sine } y}{\text{Sine } (y+v)}$$

For unequal angles, the formula is the same as in No. 8.

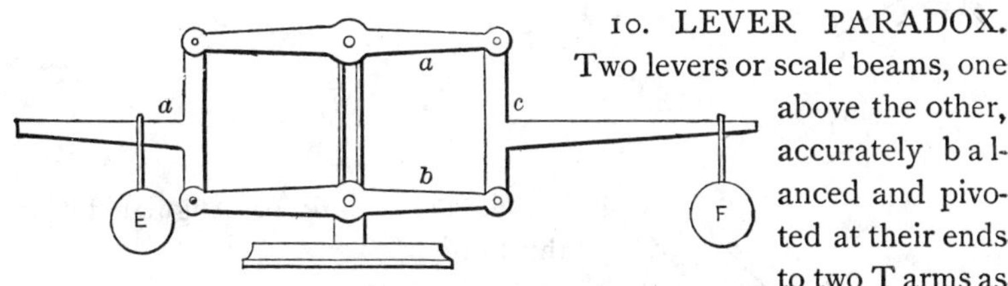

10. LEVER PARADOX. Two levers or scale beams, one above the other, accurately balanced and pivoted at their ends to two T arms as shown in the cut, may have equal weights hung at various distances on the arms, and they will be balanced on the centre line and at any angle above or below the centre line. A nut for amateurs to crack.

THE LEVER AND ITS POWER.

The weight of lever is not considered.

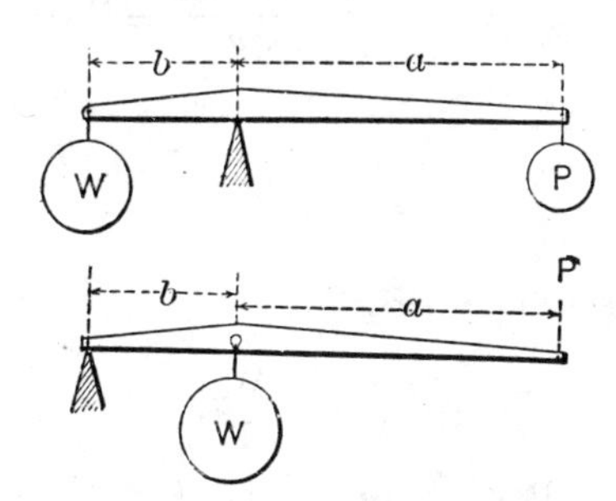

11. First order.

$$\frac{a}{b} = \frac{W}{P}, \quad \frac{P \times a}{b} = W. \quad \frac{W \times b}{a} = P$$

12. Second order.

$$\frac{b+a}{b} = \frac{W}{b}, \quad \frac{P \times (b+a)}{b} = W,$$

$$\frac{W \times b}{(b+a)} = P$$

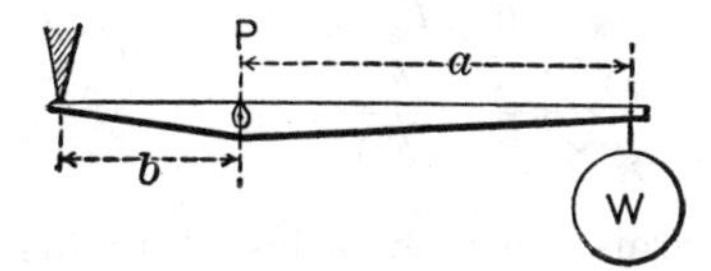

13. Third order.

$$\frac{b + a}{b} = \frac{P}{W}, \quad \frac{P \times b}{a + b} = W$$

$$\frac{W \times (b + a)}{b} = P$$

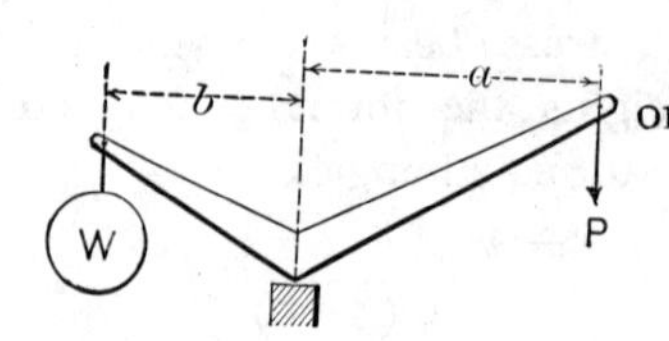

14. Bell Crank or Angular Lever, first order. Same notation as No. 11.

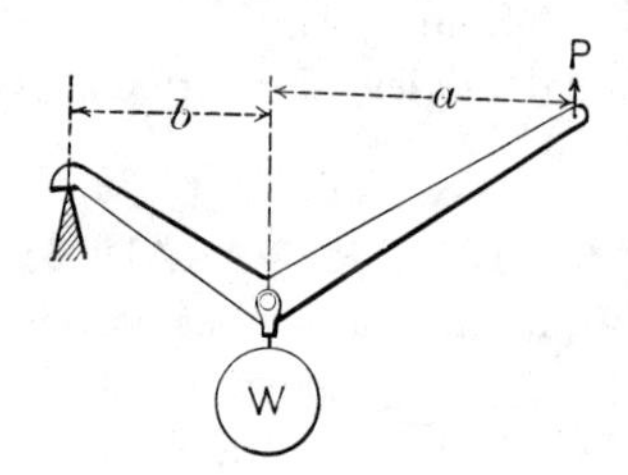

15. Bell Crank or Angular Lever, second order. Same Notation as No. 12.

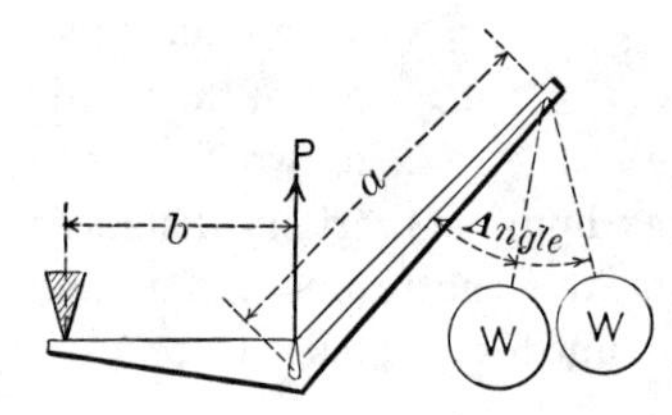

16. Bell Crank or Angular Lever, third order.

$$\frac{W \times \left(b + (a \times \text{cosine of angle})\right)}{b} = P$$

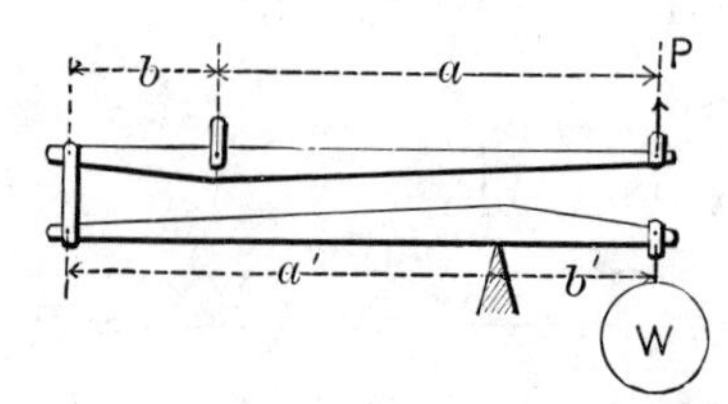

17. Compound Lever, first order.

$$\frac{a \times a'}{b \times b'} = \frac{W}{P}, \quad \frac{P \times a \times a'}{b \times b'} = W,$$

$$\frac{W \times b \times b'}{a \times a'} = P$$

18. Compound Lever, first and second orders.

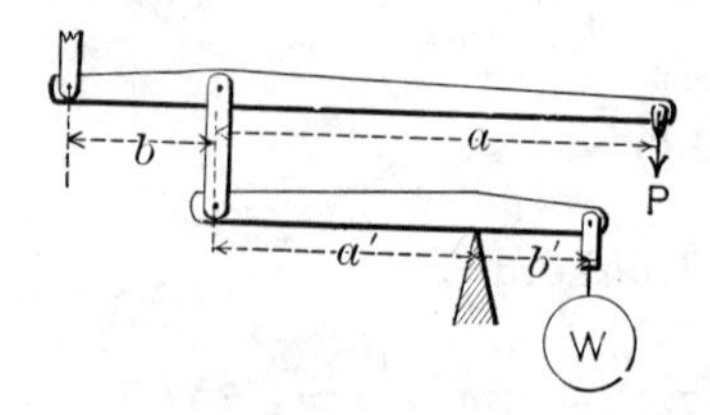

$$\frac{a}{b} \times \frac{a' \times b'}{b'} = \frac{W}{P},$$

$$\frac{P \times a \times (a' + b')}{b \times b'} = W$$

$$\frac{W \times b \times b'}{a \times (a' + b')}$$

The differential weight of lever arms must be adjusted to the proper factor for accurate computation.

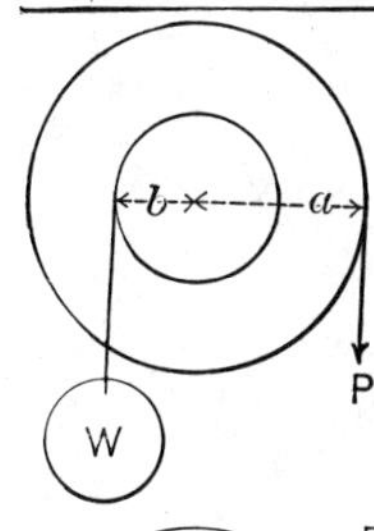

19. Revolving Lever, first order.

$$W = \frac{P \times a}{b}$$

$$P = \frac{W \times b}{a}$$

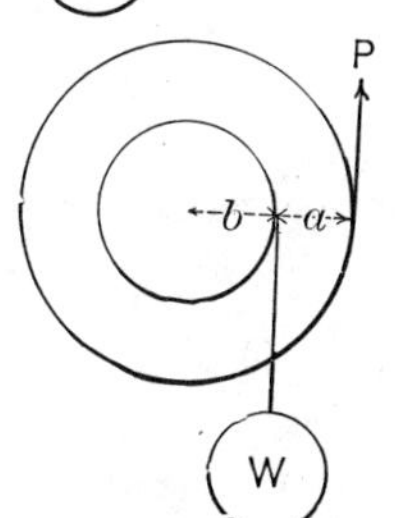

20. Revolving Lever, second order.

$$W = \frac{P \times (a + b)}{b},$$

$$P = \frac{W \times b}{a + b}$$

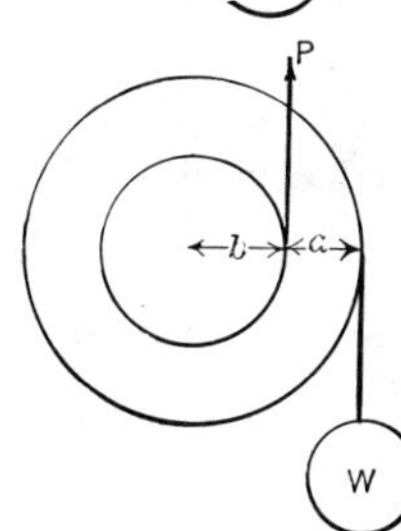

21. Revolving Lever, third order.

$$W = \frac{P \times b}{a + b},$$

$$P = \frac{W \times (a + b)}{b}$$

THE INCLINED PLANE.

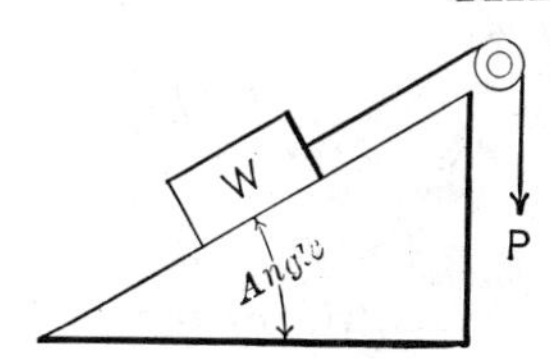

22. Weight sliding on inclined surface.

(W × sine of angle) + friction = P.

$$\frac{P}{\text{Sine of angle}} - \text{friction} = W.$$

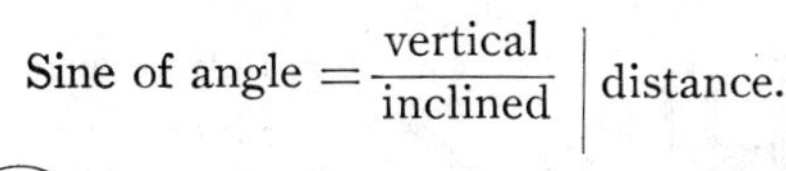

$$\text{Sine of angle} = \frac{\text{vertical}}{\text{inclined}} \Big| \text{ distance.}$$

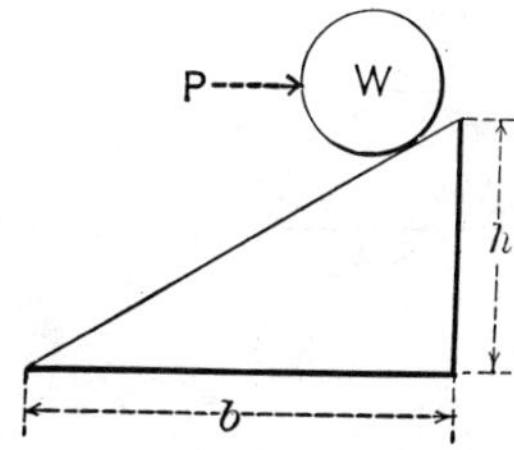

23. Rolling weight by horizontal push.

$$P = \frac{W \times h}{b}, \quad W = \frac{P \times b}{h}$$

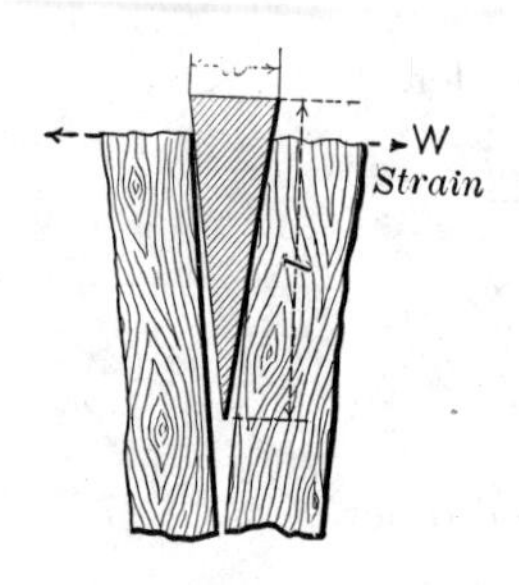

24. THE WEDGE.

$$\text{Strain} = \frac{\text{force of blow} \times l}{w}$$

l, length of wedge.
w, width of wedge.

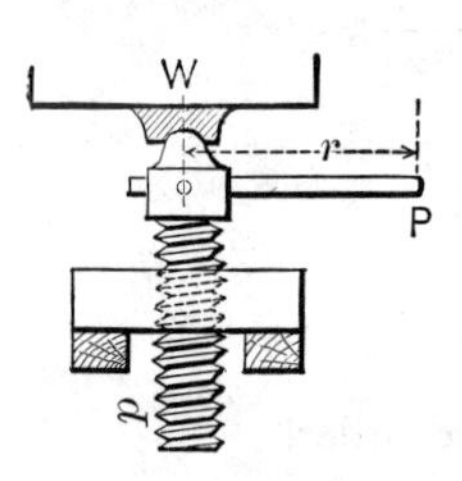

25. THE SCREW. All measures in equal units or inches.

$$W = \frac{P \times (2 \times r \times 3.1416)}{\text{Pitch of screw}}$$

$$P = \frac{W \times \text{pitch of screw}}{2 \times r \times 3.1416}$$

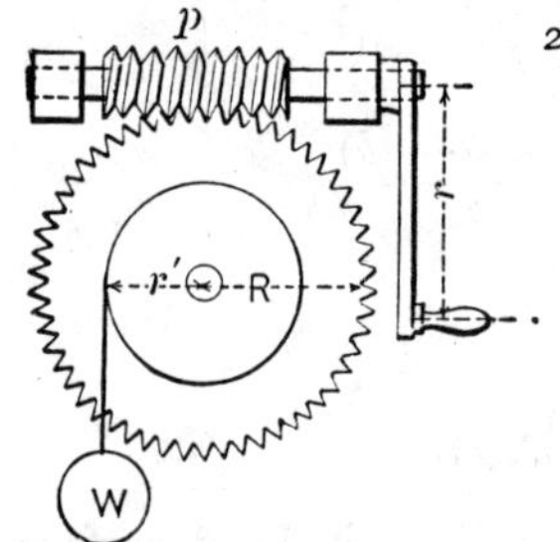

26. WORM GEAR or ENDLESS SCREW.

P = power.
r = length of crank.
R = radius of pitch line of gear.
p = pitch of screw.
r' = radius of winding drum.

$$W = \frac{P \times r \times 6.28 \times R}{p \times r'}$$

$$P = \frac{W \times p \times r}{6.28 \times r \times R}, \quad \frac{W}{2} \text{ if screw is double-thread.}$$

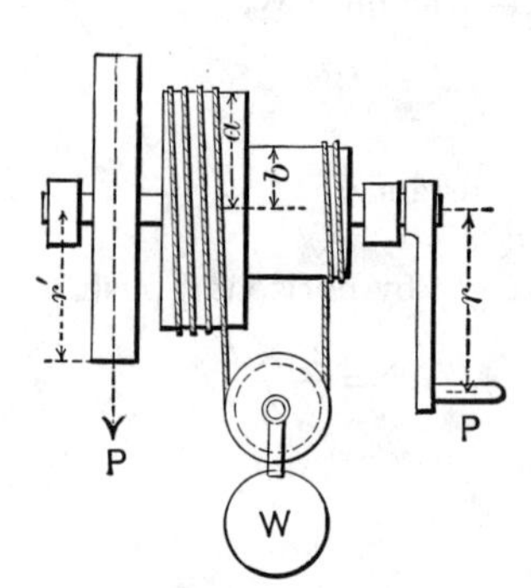

27. CHINESE WHEEL, or differential axle, with crank or pulley.

a = radius large drum.
b = radius small drum.

$$W = \frac{P \times r \times 2}{a - b}$$

$$P = \frac{W \times (a - b)}{r \times 2}$$

TACKLE BLOCKS.

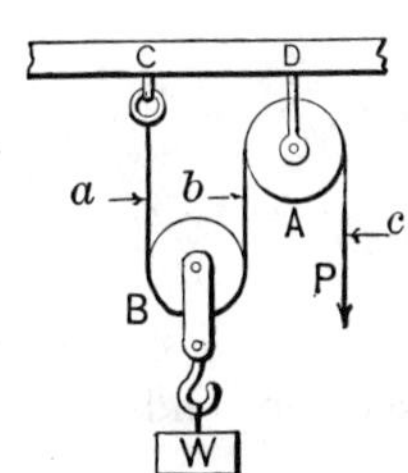

28. Two single sheaves. a, b, c are of equal strain. $a + b = W$. Sheave A only transfers the direction of P.

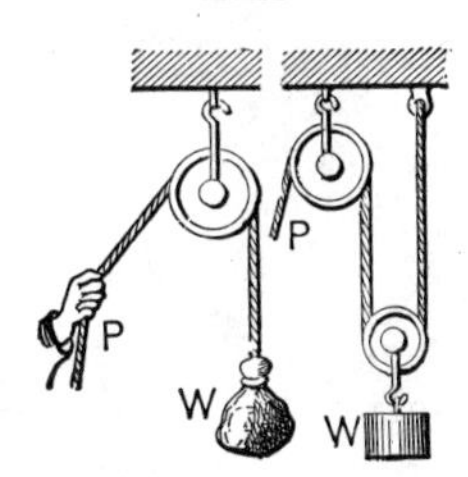

29. Simple sheave block.

$$P = W.$$

30. Two single sheave blocks—upper one fixed, lower movable.

$$P = \frac{W}{2}$$

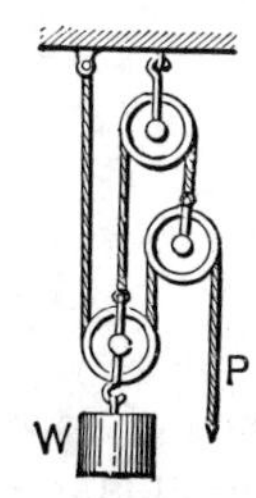

31. Three single sheave blocks—one block fixed, two blocks movable.

$$P = \frac{W}{4}. \quad W = P \times 4.$$

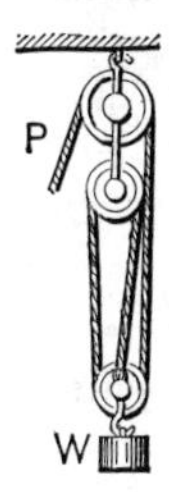

32. Three single sheave blocks, consisting of two fixed blocks and one movable block.

$$\text{Power: } P = \frac{W}{3}. \quad W = P \times 3.$$

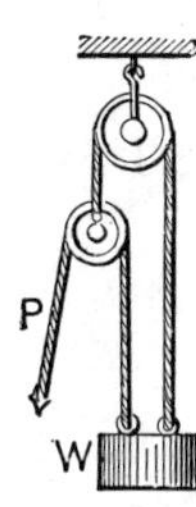

33. **One fixed sheave block, one movable sheave block.**

$$P = \frac{W}{3}. \quad W = P \times 3.$$

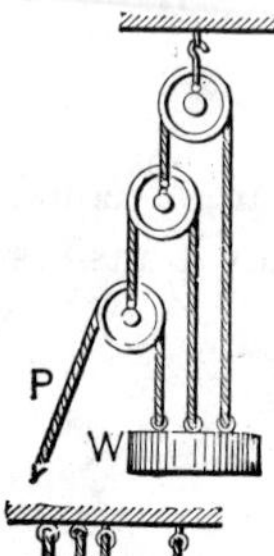

34. One fixed sheave block, two movable fixed blocks.

$$P = \frac{W}{7}. \quad W = P \times 7.$$

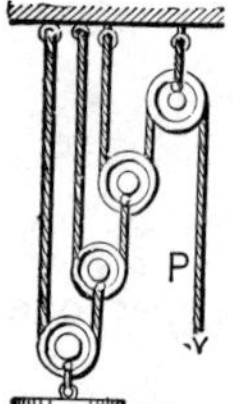

35. One fixed pulley block, three fixed rope ends.

$$P = \frac{W}{8}$$

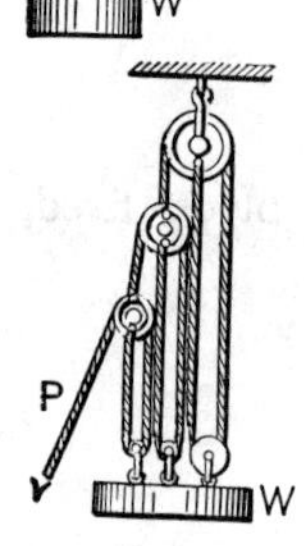

36. Multiple sheave blocks, all single.

$$P = \frac{W}{26}. \quad W = P \times 26.$$

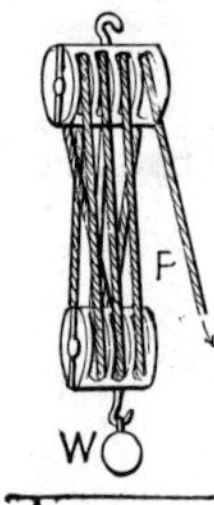

37. Four and three sheave blocks, with end of rope fixed to top block. Four sheave block fixed, three sheave block movable.

$$P = \frac{W}{6}. \quad W = P \times 6.$$

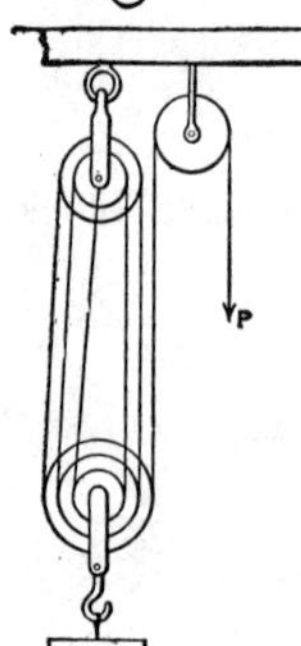

38. Roving of a three and two sheave pair of blocks, with a draw block fixed above.

$$P = \frac{W}{6}. \quad W = P \times 6.$$

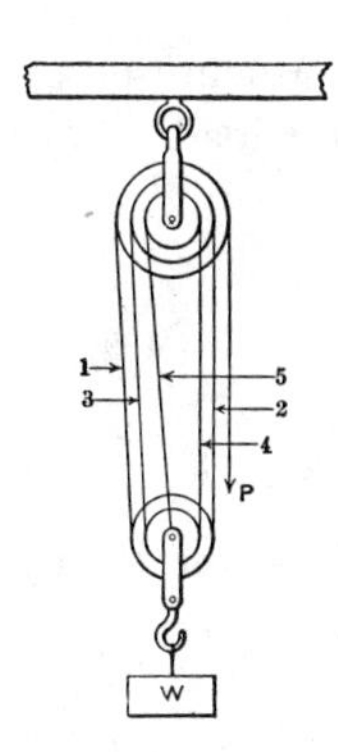

39. Roving of a two and three sheave pair of blocks, with end of rope fixed to lower block.

$$P = \frac{W}{5}. \quad W = P \times 5.$$

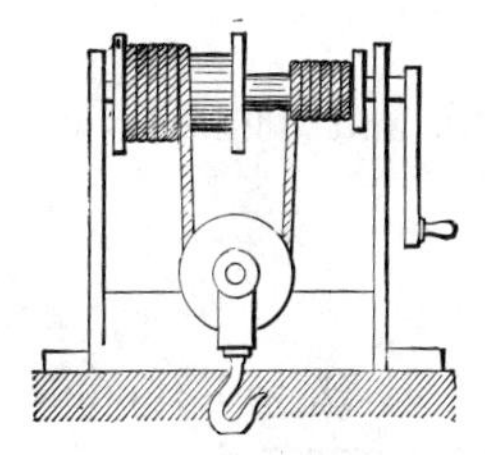

40. CHINESE WINDLASS.—The sheave and hook rises equal to one-half the difference in the circumference of the barrels for each turn of the crank. See No. 27 for the power.

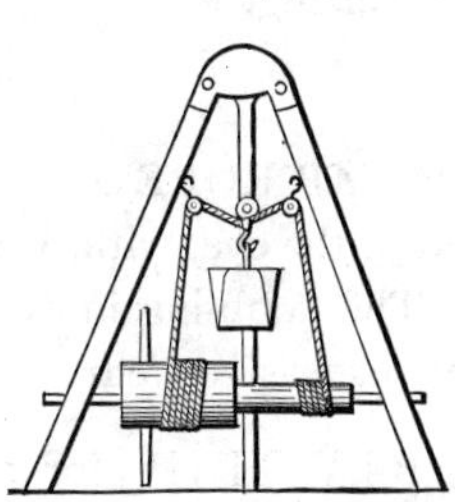

41. CHINESE SHAFT DERRICK.—The sheaves suspended from the upper part of the derrick legs allows the bucket to be raised above the mouth of the shaft or pit by the differential windlass.

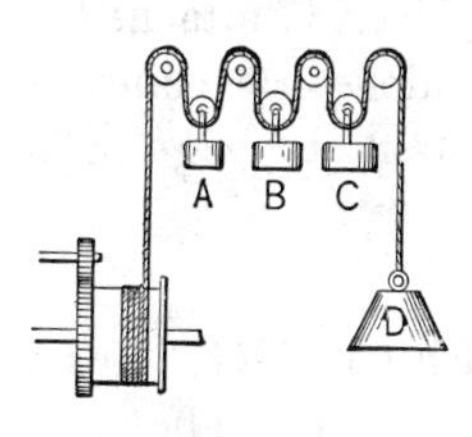

42. COMPOUND WEIGHT MOTOR, for a limited fall. The power is only equal to one-half of one of the weights. The time of falling and distance equals three times the time and distance of one weight.

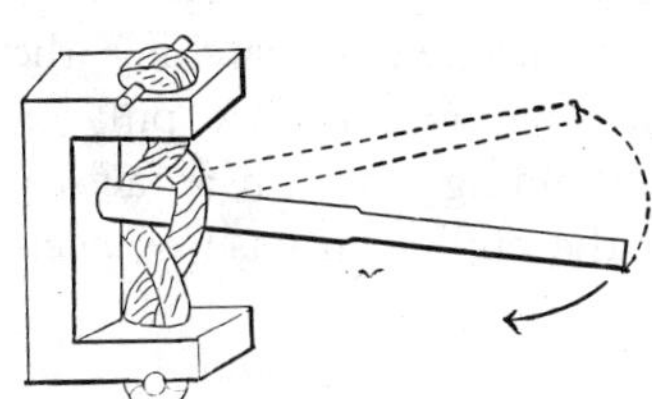

43. ROPE TWIST LEVER, for a temporary pull, or as a clamping device.

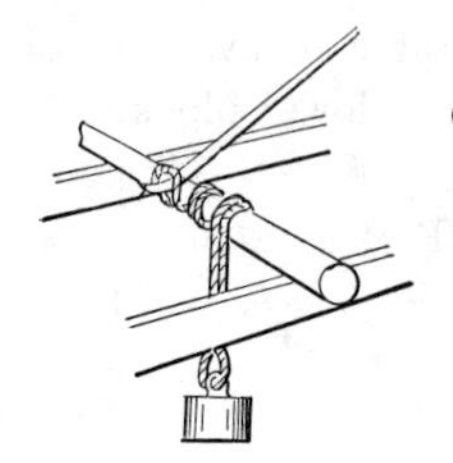

44. SPANISH WINDLASS.—Much used on over-truck frames for suspending the load.

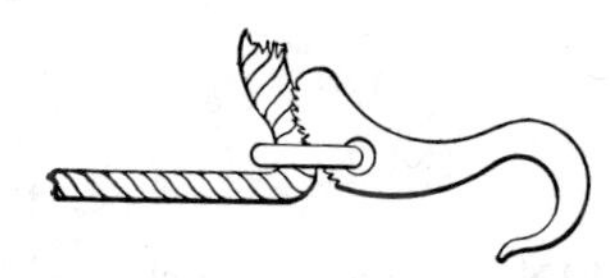

45. ROPE GRIP HOOK—for taking a temporary bite on a hawser.

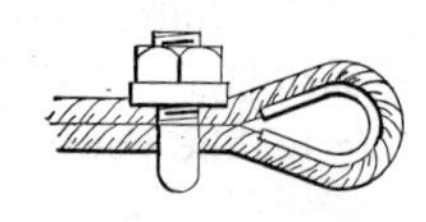

46. GUY ROPE CLIP and Thimble—for wire rope.

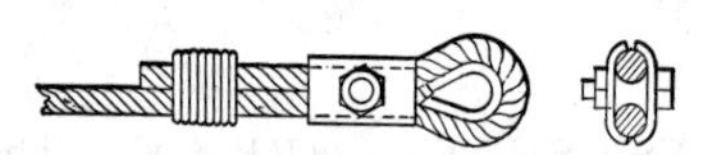

47. ROPE END, with thimble, clip, and yarn seizing.

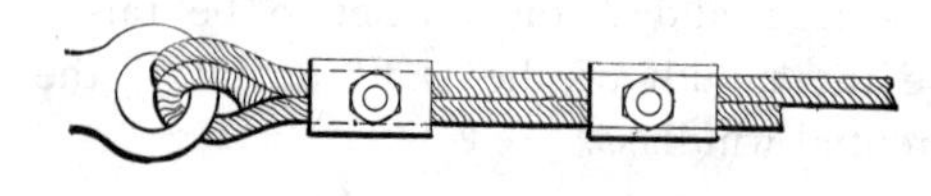

48. HEMP ROPE END, doubled in the eye, with two clips. The doubling in eye prevents excessive wear.

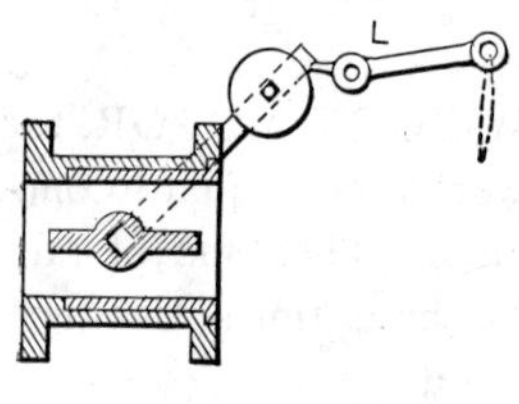

48*a*. LEVER SAFETY TRIP, for a throttle valve. The lever L attached to a lanyard extended along the lines of machinery enables instant stoppage of an engine in case of accident.

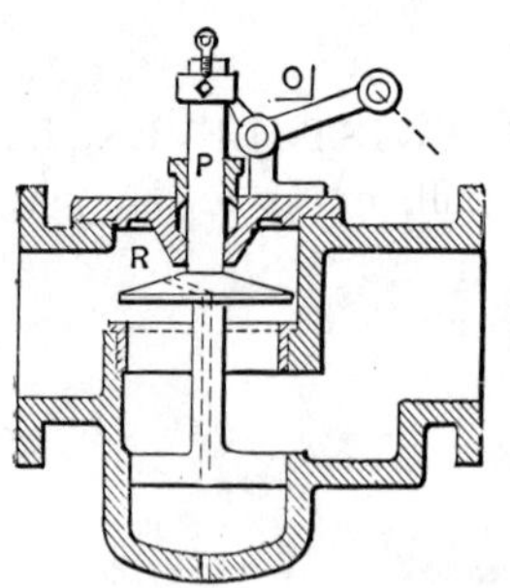

48*b*. LEVER SAFETY TRIP, for a balanced disk throttle valve. The lever O holds the valve open by catching the shoulder of the spindle P. A pull on the lanyard extending through a factory quickly stops the engine in case of accident.

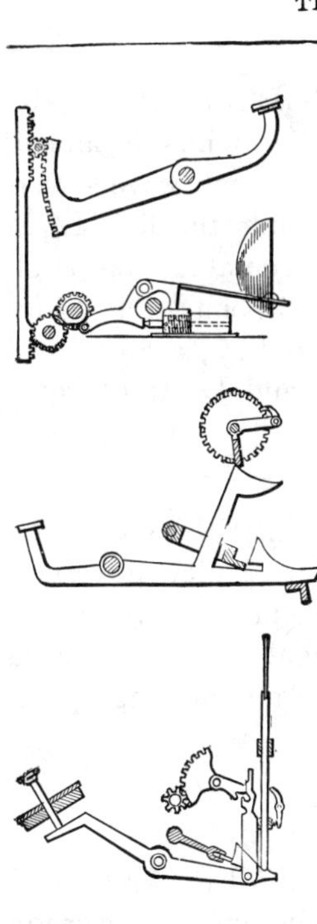

48*c*. A LEVER SECTOR operating the bell and indicator in a cash register. A sector on the lever moves the vertical rack and with it the pinions and striking pawl.

48*d*. LEVER AND RATCHET mechanism for a cash register. The pawl on a bell crank lever is operated directly from an arm on the finger lever.

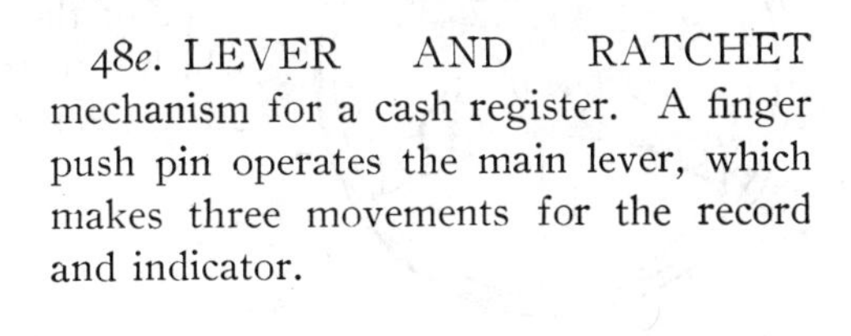

48*e*. LEVER AND RATCHET mechanism for a cash register. A finger push pin operates the main lever, which makes three movements for the record and indicator.

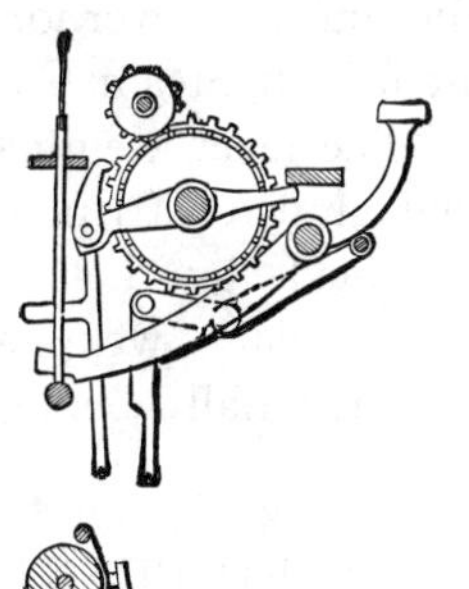

48*f*. LEVER MOVEMENT of a cash register. Through a single movement of the finger lever, three different movements are made, including the raising of the index number.

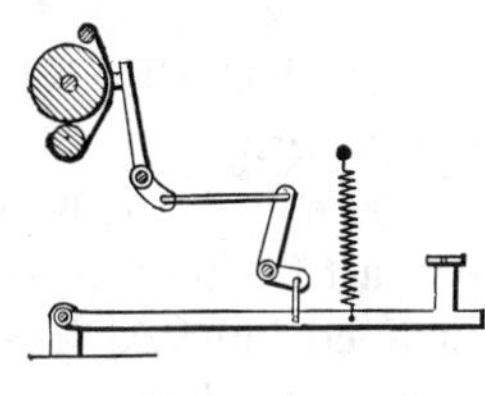

48*g*. LEVER ACTION in a typewriting machine. A main lever with finger stud operates the type lever through a bell crank and links.

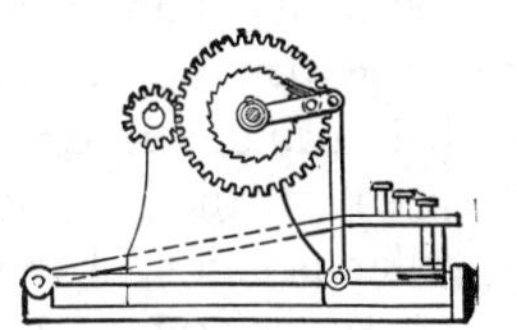

48*h*. LEVER ACTION in a typewriting machine. The long lever and finger stud is linked to a ratchet lever concentric with the type line barrel. One touch of the finger stud for close lines and two touches for open lines.

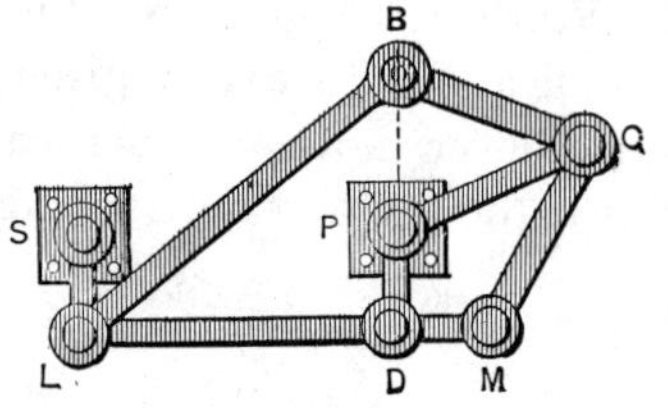

48*i*. STRAIGHT LINE LINKAGE.—With the joints S and P fixed the joint B will have a vertical motion while the link L, M will have a horizontal motion parallel with the fixed points S, P. Links P, C, C, M, and B, C, are of equal length. L, B and L, M of equal lengths, as are the short links, S, L and P, D.

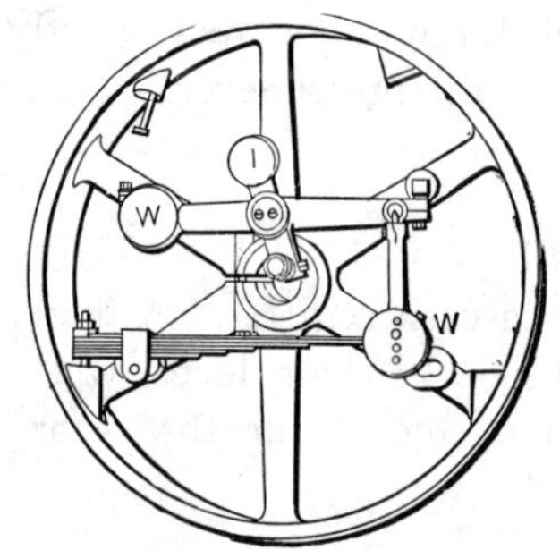

48*j*. THE LEVER AND ITS OFFICE in the pulley governor. Type of the Shepherd governor, in which centrifugal force and inertia are combined for regulating speed.

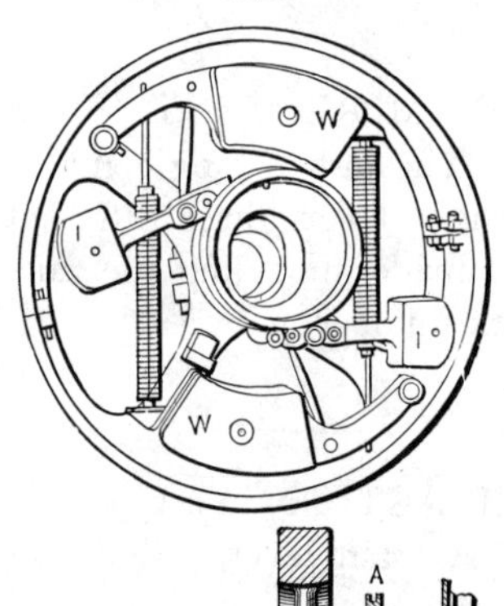

48*k*. THE LEVER AND ITS OFFICE in the pulley governor. Type of the Fitchburgh Steam Engine Company. The lever weights W, W are thrown out by centrifugal force and restrained by helical springs. The auxiliary weights I, I are moved tangentially by inertia.

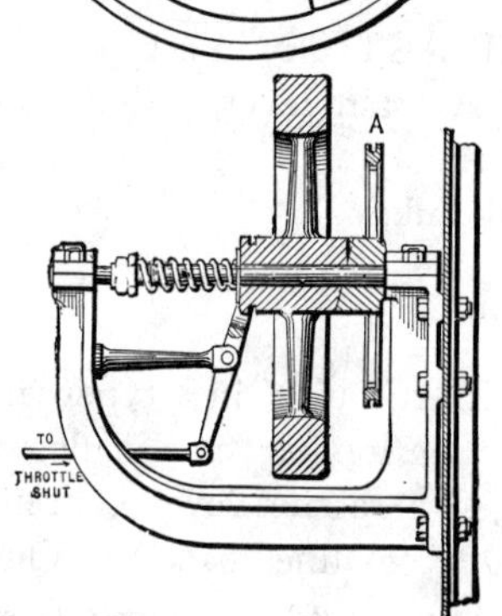

48*l*. THE INCLINED PLANE in a marine governor. The sprocket wheel A and inclined hub are fast on the shaft. The inertia wheel and its inclined hub are free on the shaft with its hub face pressed against the driving wheel hub by the coiled spring. Irregularity in the speed of the engine changes the angular position of the hub planes and so operates the throttle lever. See No. 1591.

Section II.

TRANSMISSION OF POWER.

ROPES, BELTS, FRICTION GEAR, SPUR, BEVEL, AND SCREW GEAR, ETC.

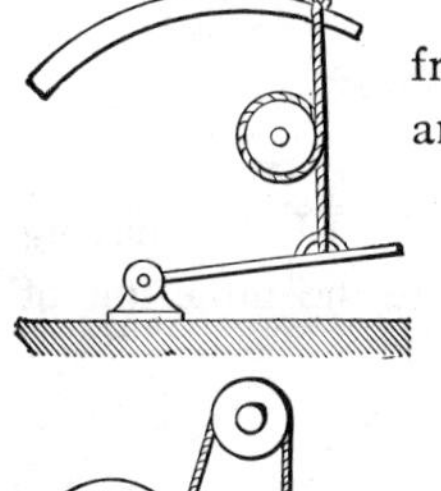

49. ALTERNATING CIRCULAR MOTION from the curvilinear motion of a treadle. The ancient lathe motion.

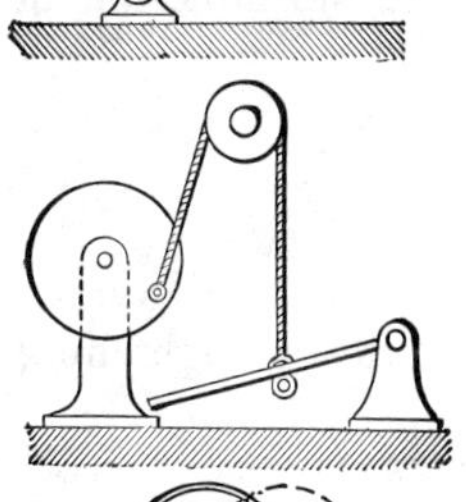

50. CIRCULAR MOTION from curvilinear motion of a treadle through a cord and pulley.

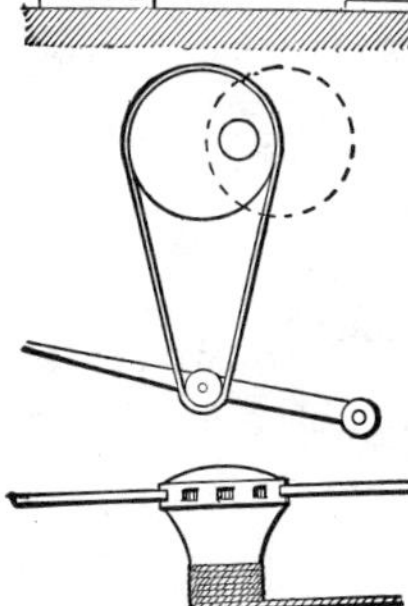

51. ECCENTRIC CRANK and Treadle.—A belt over the eccentric and a roller in the treadle. The equivalent of a crank.

52. CAPSTAN, OR VERTICAL WINDLASS.—The pawl falling in the circular rack at the bottom locks the windlass. The rope should always wind on the bottom and slip upward.

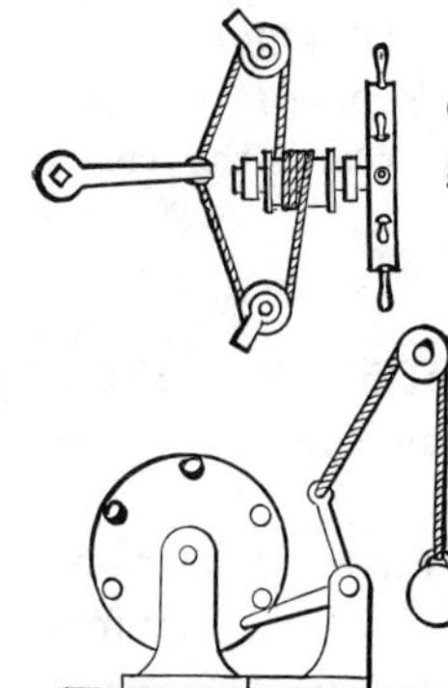

53. STEERING GEAR.—A hand wheel and drum on a shaft, carrying a rope rove through guide pulleys and attached to the tiller.

54. JUMPING MOTION given to a weight, or other body, by a pin wheel and bell-crank lever.

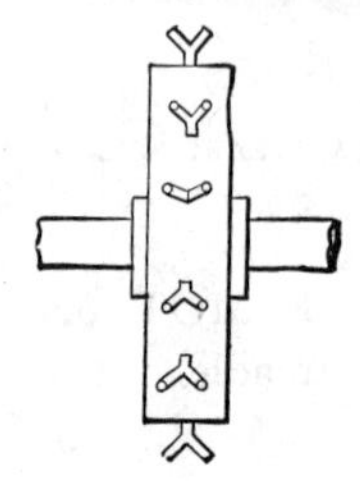

55. ROPE SPROCKET WHEEL, several modifications of which are in use in old-style hoists.

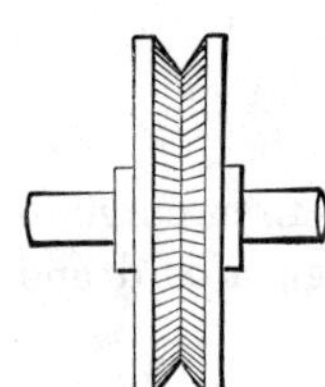

56. *V*–GROOVED ROPE PULLEY, having corrugated groove faces to increase the adhesion of the rope.

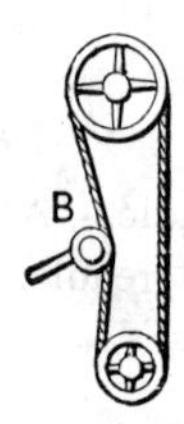

57. ROPE TRANSMISSION, with a tightening pulley, B.

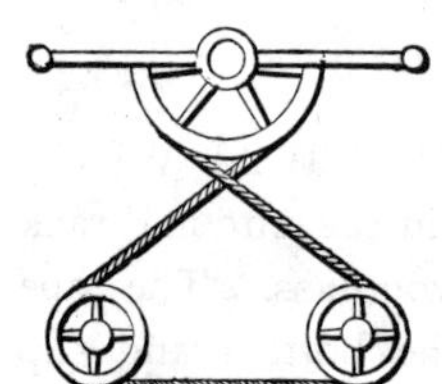

58. VIBRATORY MOTION to two shafts, transmitted from the rocking of a lever arm and sector.

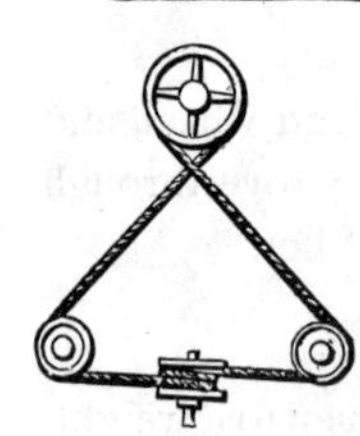

59. TRANSMISSION BY ROPE to a shaft at right angles to the driving-shaft. The guide sheaves give direction to the rope over the curve face of the driven pulley, the rope slipping towards the centre of the driven pulley.

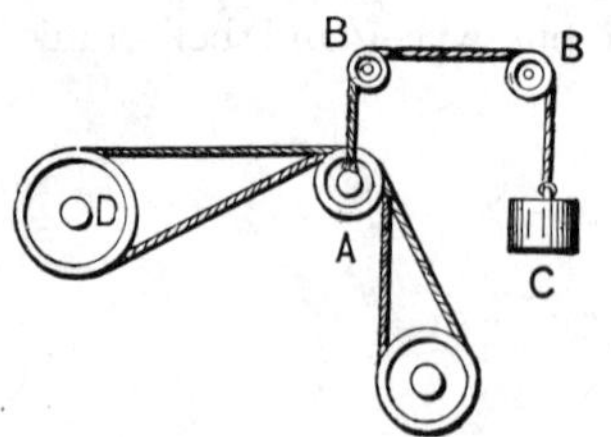

60. TRANSMISSION BY ROPE to a portable drill or swing saw.

D, driving sheave.

A, double loose sheaves in a frame, suspended by weight C attached by rope over sheaves, B, B. C, counter weight.

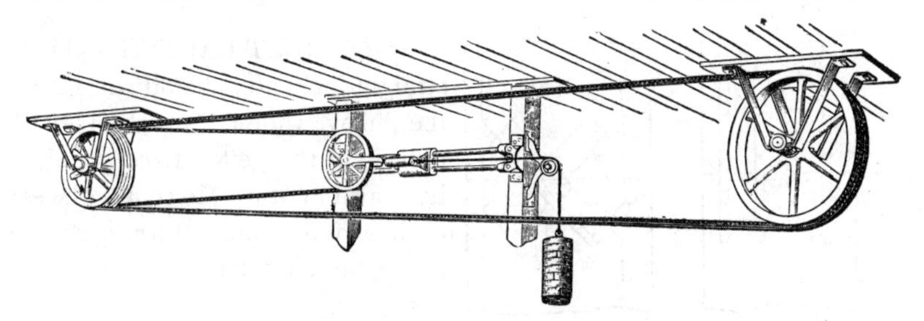

61. HORIZONTAL ROPE TRANSMISSION, with tension slide and weight.

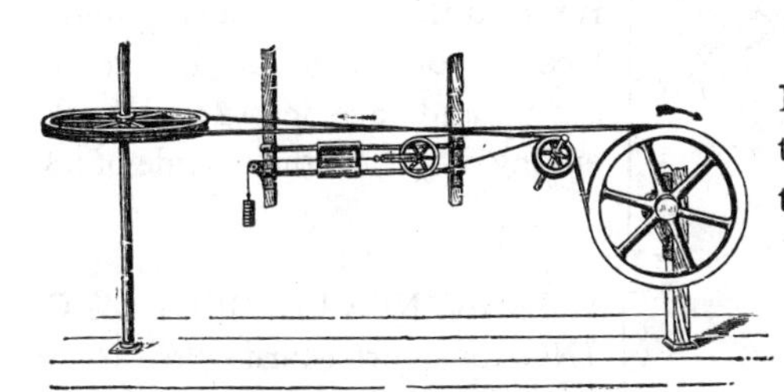

62. ROPE TRANSMISSION from vertical to horizontal shaft, with tension slide and weight.

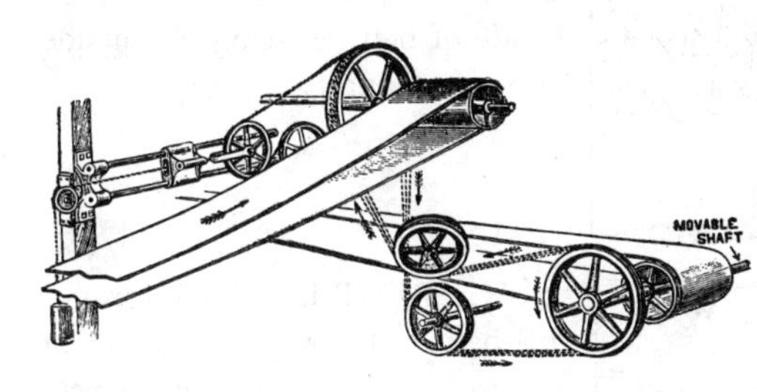

63. ROPE TRANSMISSION to a movable shaft at right angles from the driving-shaft, with tension slide and weight.

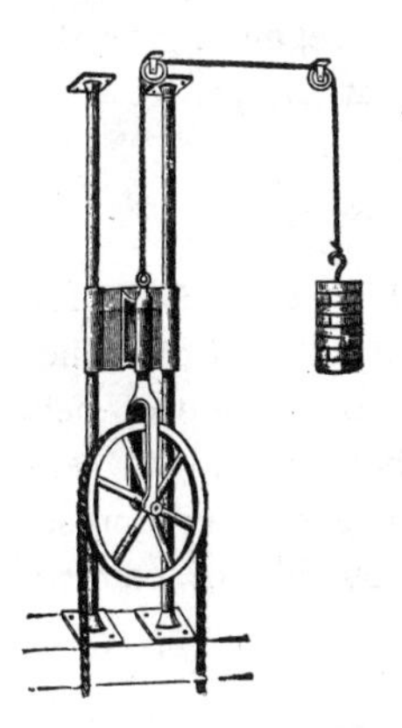

64. VERTICAL TENSION CARRIAGE, with slides and pulley guide.

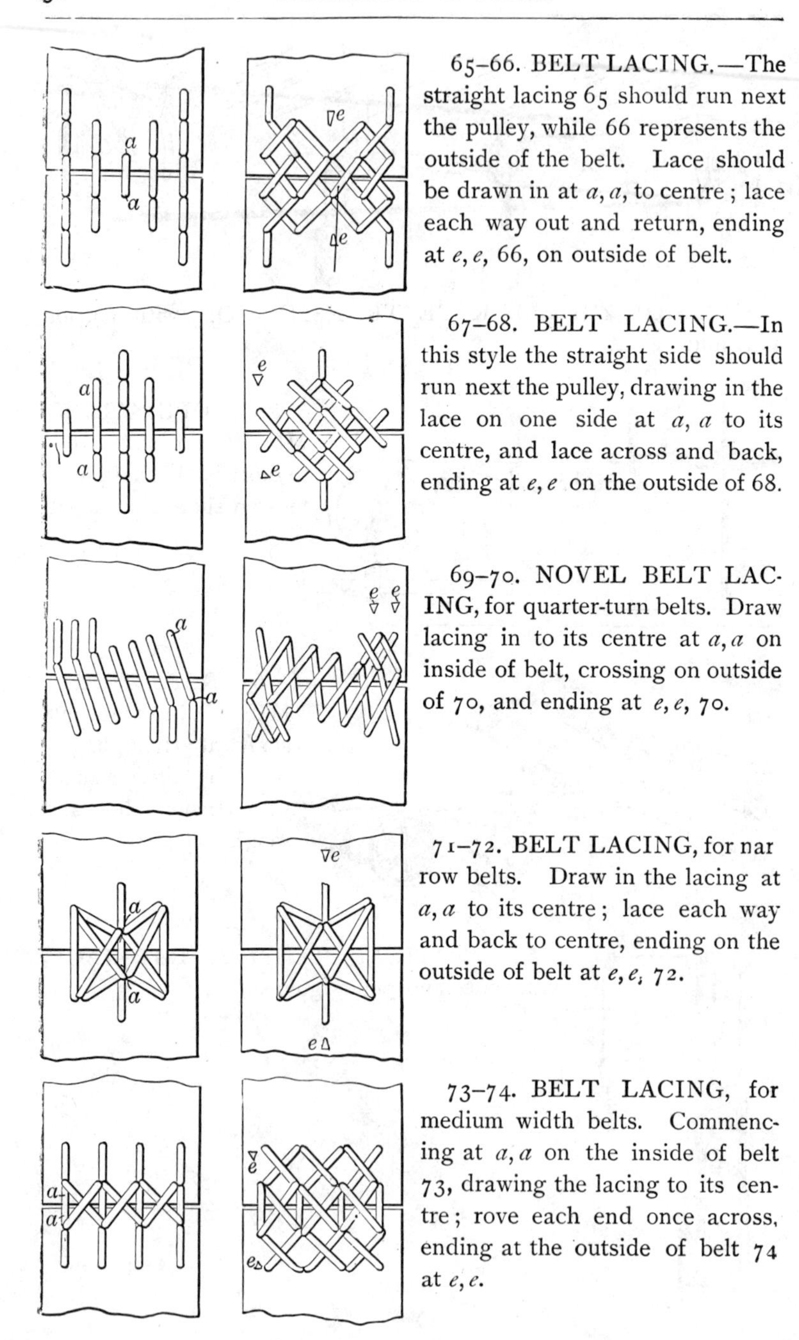

65–66. BELT LACING.—The straight lacing 65 should run next the pulley, while 66 represents the outside of the belt. Lace should be drawn in at *a*, *a*, to centre; lace each way out and return, ending at *e*, *e*, 66, on outside of belt.

67–68. BELT LACING.—In this style the straight side should run next the pulley, drawing in the lace on one side at *a*, *a* to its centre, and lace across and back, ending at *e*, *e* on the outside of 68.

69–70. NOVEL BELT LACING, for quarter-turn belts. Draw lacing in to its centre at *a*, *a* on inside of belt, crossing on outside of 70, and ending at *e*, *e*, 70.

71–72. BELT LACING, for nar row belts. Draw in the lacing at *a*, *a* to its centre; lace each way and back to centre, ending on the outside of belt at *e*, *e*, 72.

73–74. BELT LACING, for medium width belts. Commencing at *a*, *a* on the inside of belt 73, drawing the lacing to its centre; rove each end once across, ending at the outside of belt 74 at *e*, *e*.

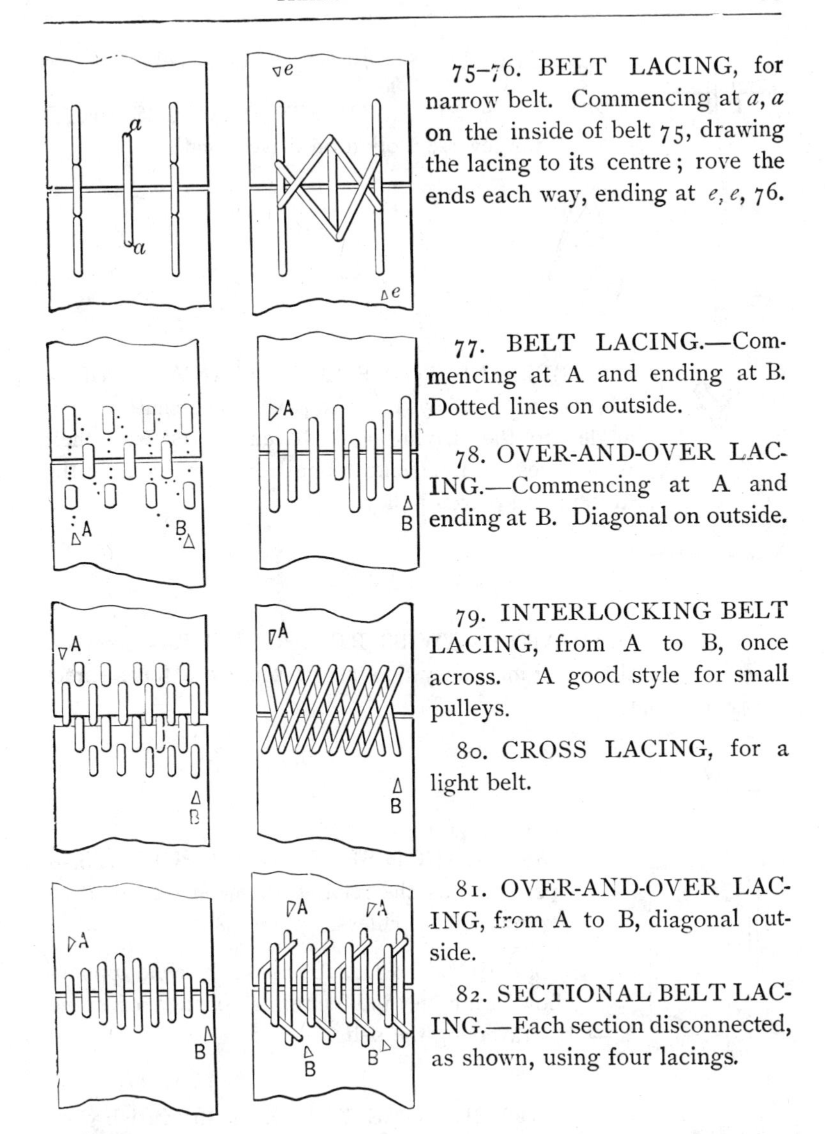

75–76. BELT LACING, for narrow belt. Commencing at *a*, *a* on the inside of belt 75, drawing the lacing to its centre; rove the ends each way, ending at *e*, *e*, 76.

77. BELT LACING.—Commencing at A and ending at B. Dotted lines on outside.

78. OVER-AND-OVER LACING.—Commencing at A and ending at B. Diagonal on outside.

79. INTERLOCKING BELT LACING, from A to B, once across. A good style for small pulleys.

80. CROSS LACING, for a light belt.

81. OVER-AND-OVER LACING, from A to B, diagonal outside.

82. SECTIONAL BELT LACING.—Each section disconnected, as shown, using four lacings.

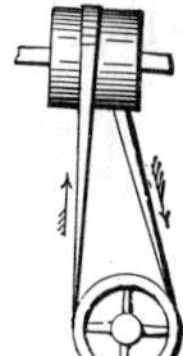

83. QUARTER TWIST BELT.—The arrows show the direction the belt should run.

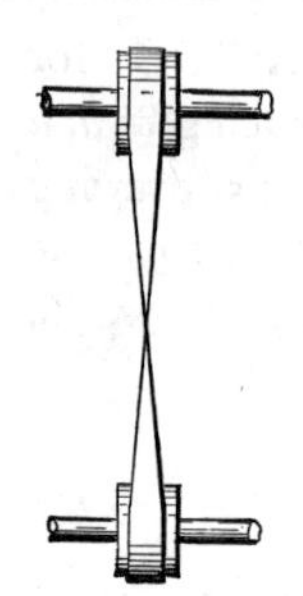

84. FULL TWIST BELT, or cross belt.

85. FULL TWIST OR CROSS BELT, for reverse motion on driven shaft.

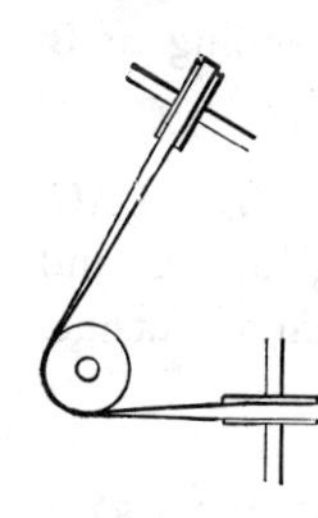

86. BELTING TO A SHAFT AT ANY ANGLE.—The two idler pulleys must be placed on a shaft at right angles to the driving and driven shafts, with their peripheries at the central line from centres from the driving and driven pulleys.

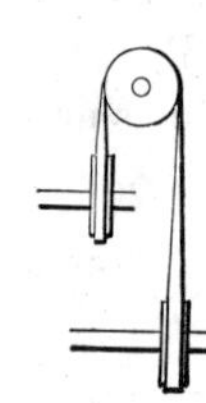

87. QUARTER TWIST RETURN BELT.—A method used for belting pulleys on shafts too close for a direct belt.

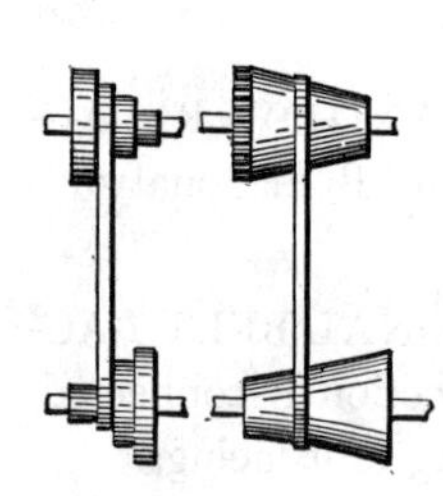

88. CHANGE SPEED STEP PULLEYS.—Speeds are as the relative diameters of the driving and driven pulleys.

89. CONE PULLEYS.—The cone pulleys allow of minute and continual change of speed by traversing the belt.

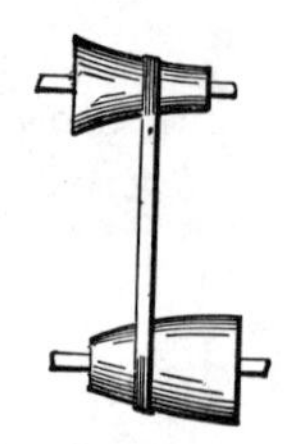

90. CURVED CONE PULLEYS, for variably increasing or decreasing speed by traversing the belt.

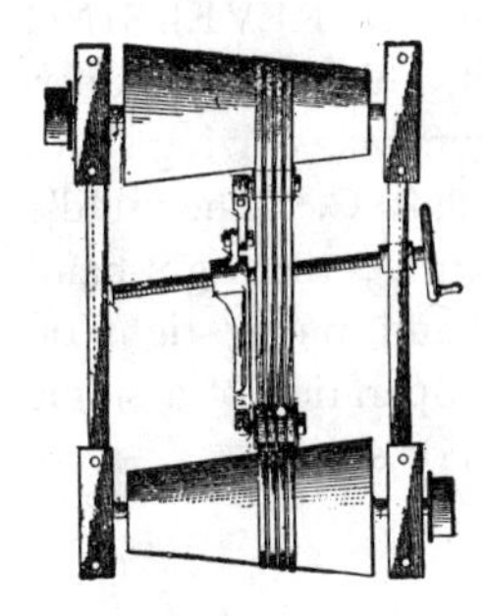

91. SHIFTING DEVICE FOR CONE PULLEYS.—Made efficient by a division of the proper belt width into a number of narrow belts, kept in place by webs on the belt tighteners, which are moved forward and backward by a carrier nut and screw shaft. This arrangement gives more power for a given width than with a single belt, and with less wear. It equalizes the stress on the belts by the set-up of the guide pulleys as tighteners.

Patent of P. D. HARTON, Philadelphia, Pa.

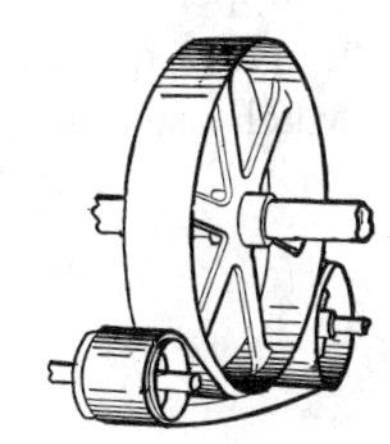

92. BELT TRANSMISSION, for short belt and close connection. The belt is wrapped close to and pressed against the driven pulley by a tightening pulley. For electric motor power or the driving of generators.

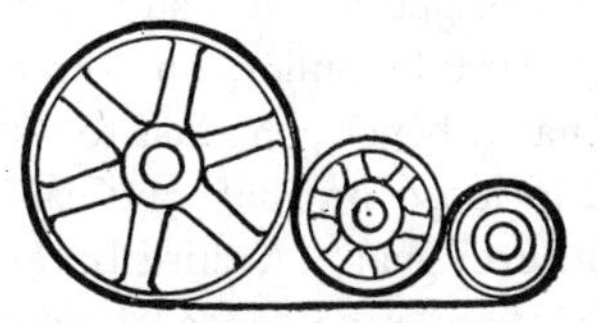

93. BELT TRANSMISSION OF POWER, at close range. A combination of friction gear increased by belt contact of the driving or driven pulley with a light intermediate pulley gives an additional belt pressure, with small belt strain on the slack side. It eliminates vibration of belt.

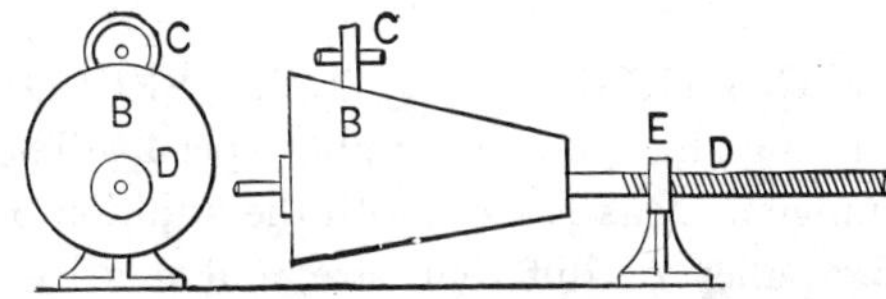

94. VARIABLE TRANSMISSION OF MOTION, from an eccentric conical pulley to a friction pulley. The riding pulley C traverses the cone, which moves forward or backward by the rotation of the screw in the nut stud E, producing a progressive variable motion in the pulley C, increasing or decreasing as the cone rotates forward or backward.

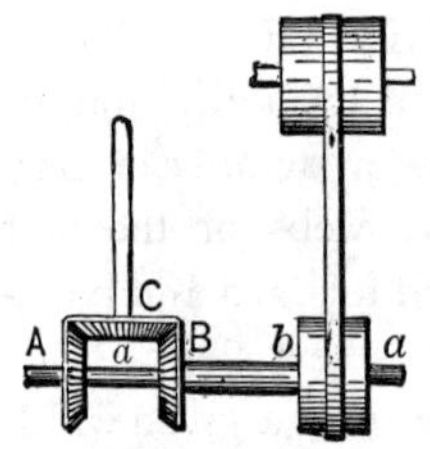

95. STOP, DRIVING, AND REVERSING MOTION with a single belt, which may be operated either way: from the drum on a driving shaft, or from the bevel gear on shaft C. The middle pulley being loose on shaft *a*, the right-hand pulley tight on shaft *a*, left-hand pulley tight on the hollow shaft B, *b*. The operation of a single shipper changes the motions or stops.

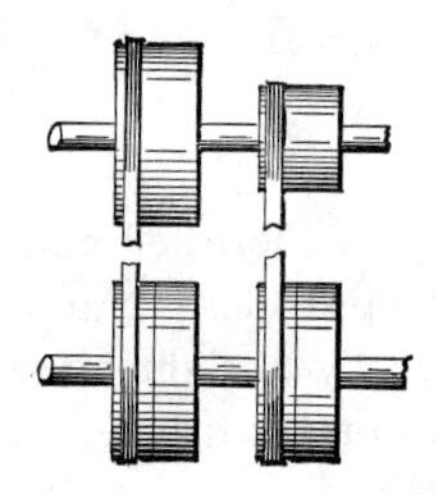

96. TWO SPEED PULLEYS AND BELTS.—Two pair of tight and loose pulleys on lower shaft, unequal broad tight pulleys on upper shaft. By crossing the belt from one of the pair a quick return speed may be obtained. Much used on tapping-machines and planers.

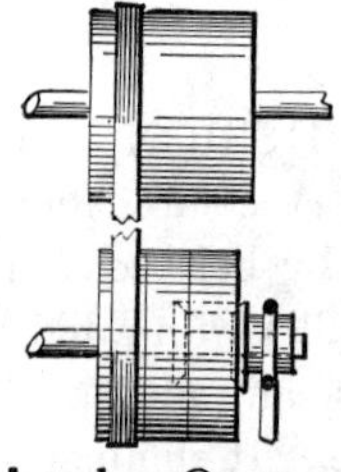

97. PULLEYS, COMBINED WITH A DIFFERENTIAL GEAR for two speeds, and stop-belt shown on loose pulley. Middle pulley on lower shaft is fast to shaft, and has a bevel gear fast to its hub. Pulley on the right is loose on shaft and carries, transversely, another bevel gear. A third bevel gear runs loose upon the shaft and is held by a friction band. On moving the belt to the middle pulley an ordinary motion is obtained; to the right-hand pulley a double speed is obtained.

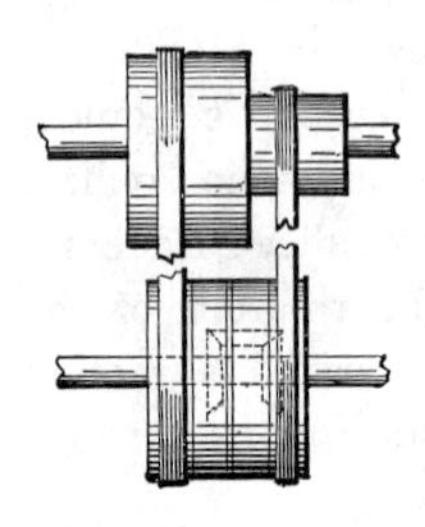

98. TRANSMISSION OF TWO SPEEDS from a driving shaft, one a variable speed. The same arrangement as No. 97, with the addition of a driving pulley of different size, and a driven pulley attached to the friction gear on the lower shaft. The right-hand belt shifts to the next pulley and may be straight or cross, making a variety of motions to the lower shaft.

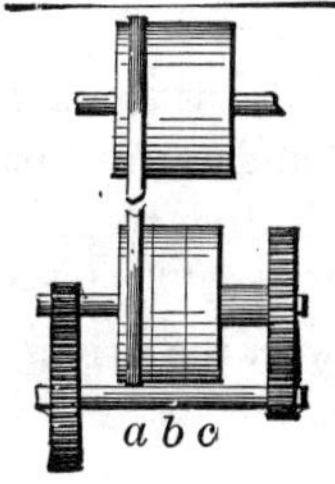

99. TWO SPEED GEAR from belt pulleys and one hollow shaft. A solid shaft with loose pulley (*a*) and fast pulley (*b*), fast pulley (*c*) on hollow shaft carrying large driving gear at the right.

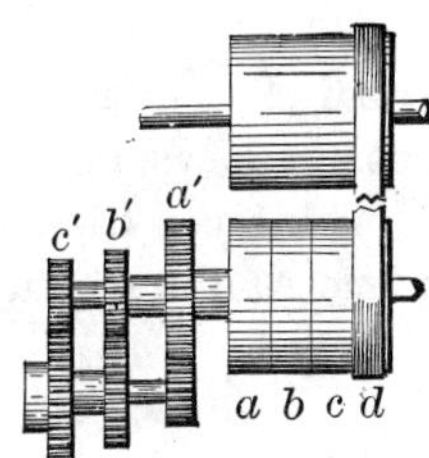

100. VARIABLE SPEED OR CONE GEARING.

a, tight pulley on outside hollow shaft.
b, tight pulley on inside hollow shaft.
c, tight pulley on inner or solid shaft.
d, loose pulley on solid shaft.
a′ b′ c′, differential spur gears for three speeds.

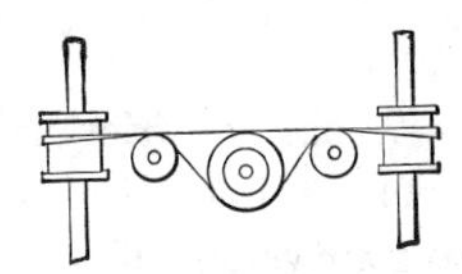

101. TRANSMISSION OF POWER from a horizontal shaft to two vertical spindles. A single belt, with two idlers, for tightening and directing the half twist of the belt.

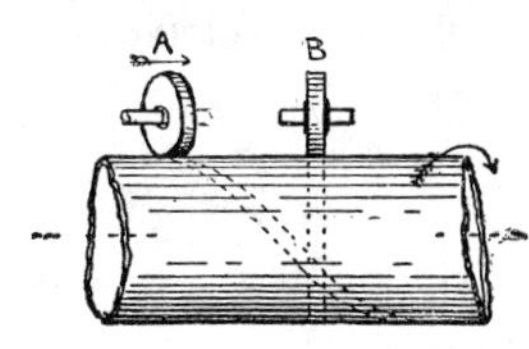

102. FRICTIONAL RECTILINEAR MOTION, from the angular position of a sheave or pulley rolling on a revolving barrel or long cylinder. A, forward motion; B, stop. The principle of the "Judson" railway propulsion. Efficiency was increased by the use of a small truck with several roller pulleys.

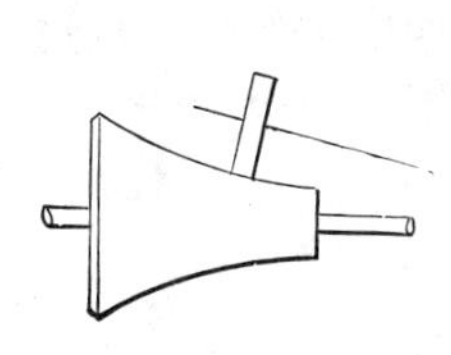

103. VARIABLE ROTARY MOTION from a friction pulley traversing a concave conical drum. The speed ratio of the traversing pulley is a variable one.

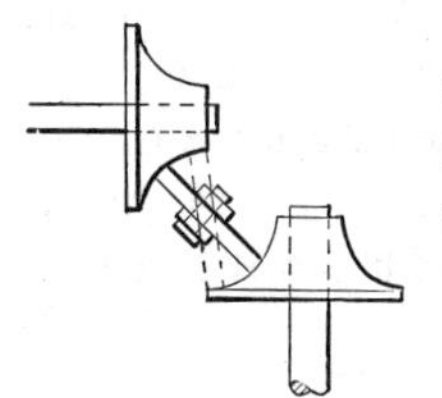

104. VARIABLE MOTION to a right-angled shaft, by curved cone friction pulleys with intermediate swinging pulley. A sewing-machine or other light power movement.

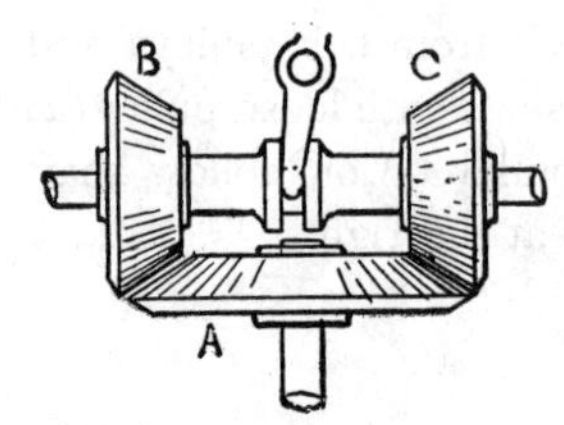

105. FRICTION GEAR, traversing motion. A, the driver. B and C are fast on the clutch sleeve which is free to slide on a feather on the driven shaft. The lever brings B or C in contact with the driving cone A for reversing.

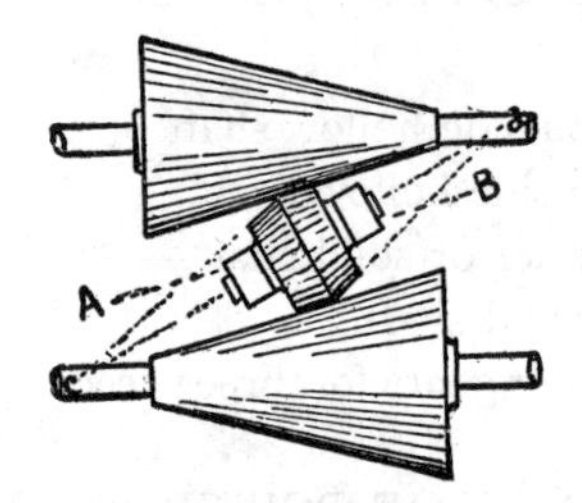

106. FRICTION GEAR.—Variable speed from a pair of cone pulleys, one of which is the driver. A double-faced friction pinion is moved on the line A, B in contact with both cones.

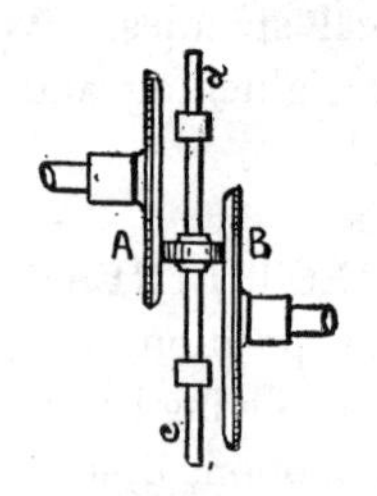

107. FRICTION GEAR.—A pair of friction discs A, B on parallel shafts out of line, with a traverse friction pinion on a transverse spindle *c, d* will give a great range of speed velocities.

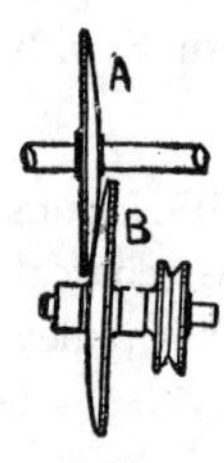

108. FRICTION GEAR.—Variable speed from a rocking shaft and convex discs. "Wright's" driving device for sewing-machines. A is the driving shaft with convex disc. B is a band shaft that swivels by the foot pedal and kept taut or released at its different positions.

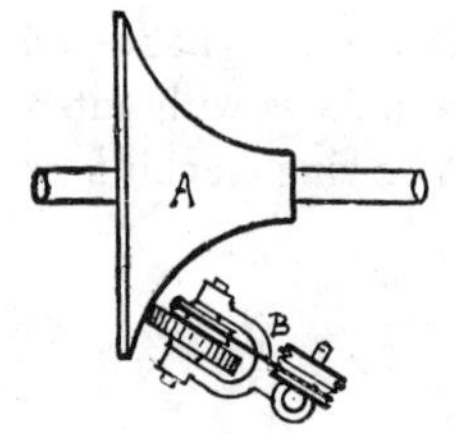

109. TRANSMISSION OF VARIABLE SPEED, for sewing-machines. A, driving concave cone. B, swivelling yoke carrying a friction pulley, with a band running a pair of pulleys at the swivel, one of which drives the sewing machine.

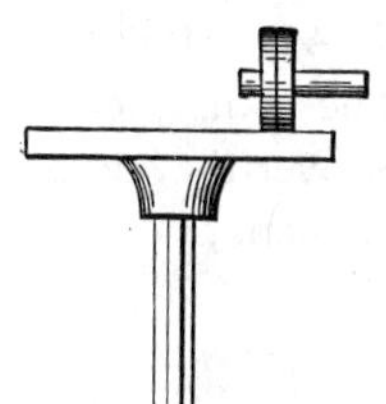

110. FRICTION GEAR, with variable speed by traversing a pulley to or from the centre of the face of a disc wheel. Leather or rubber facing for wheel and pulley makes best working condition.

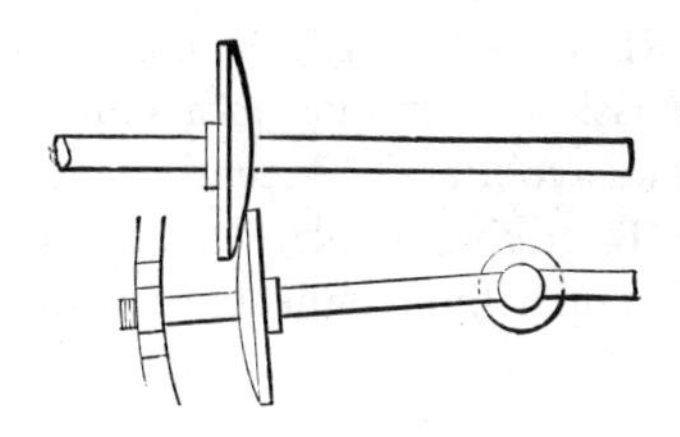

111. VARIABLE SPEED GEAR forsewing-machines, "Wright's" model. The upper shaft is the driver, the lower shaft carrying the band pulley, swivels by the foot for variable speed.

112. TRANSMISSION OF ROTARY MOTION to an oblique shaft by rolling contact of drums with concave faces.

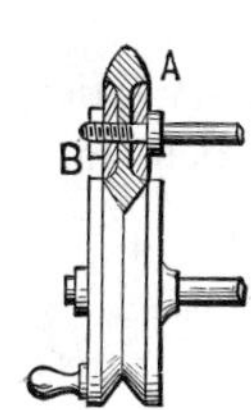

113. COMBINATION FRICTION GEAR, "Howlett's Patent." A rubber disc clamped between metal washers.

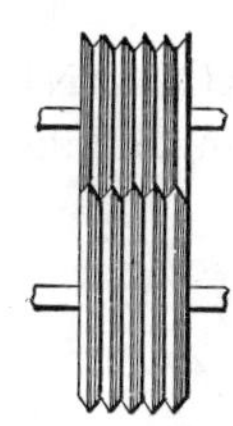

114. GROOVED FRICTION GEARING.—The loss of power by friction increases with the size and depth of the grooves. Friction increases inversely as the angles of the grooves.

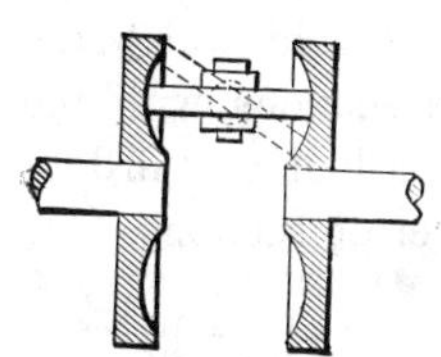

115. VARIABLE MOTION to a shaft in line by curved-faced discs, with a swinging pulley pivoted central to the curves on the face of the discs.

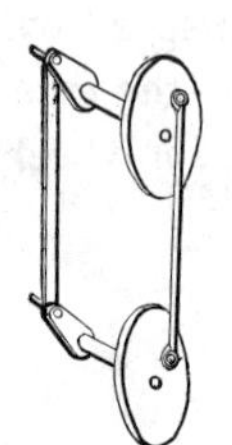

116. TRANSMISSION OF CIRCULAR MOTION by right-angled cranks on each shaft. The pair of crank connections carry the right-angled cranks over the centre. The principle of the locomotive wheel connections.

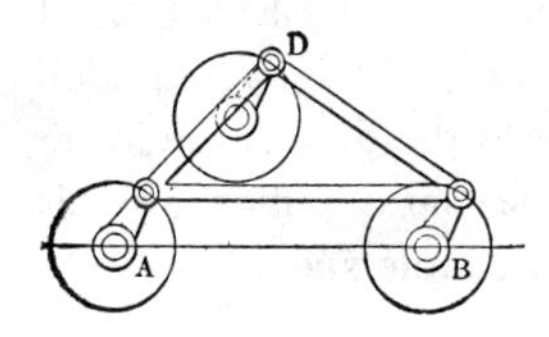

117. THREE CRANK LINK connection for transmission of motion to a crank by direct link to avoid a dead centre. A. driven crank; B, driving crank; D, a relief crank with triangular link connections with cranks A and B.

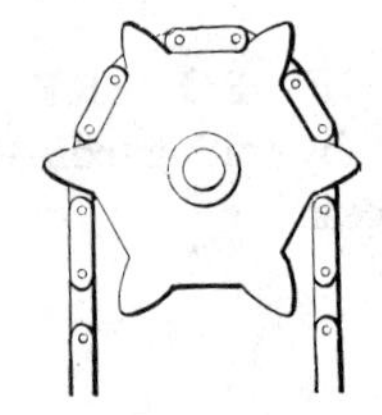

118. SPROCKET WHEEL AND CHAIN.—Pitch radius is at the centre of the rivets, with a slight clearance for easy running.

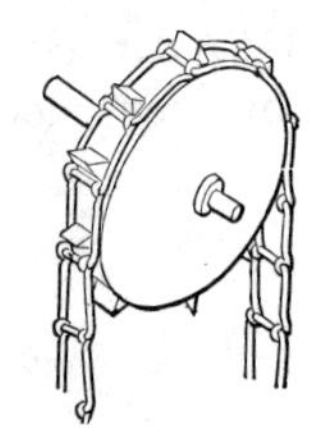

119. LINK BELT AND PULLEY.—A variety of hook link forms are in use for link belt transmission.

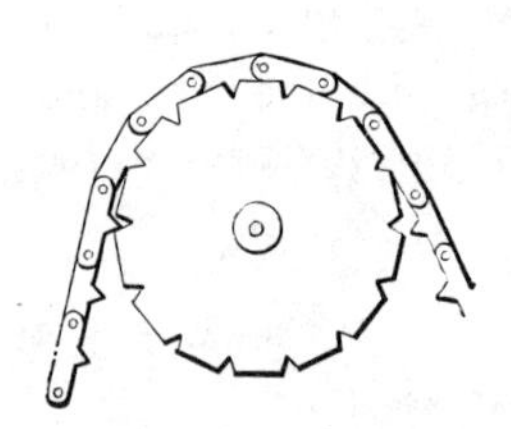

120. TOOTHED LINK CHAIN AND PULLEY, alternating double and single links.

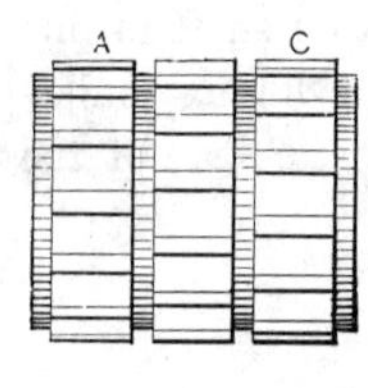

121. STEP GEAR.—A spur gear in which the face is divided into two or more sections, with the teeth of each section set forward a half or third of the pitch, according to the number of sections.

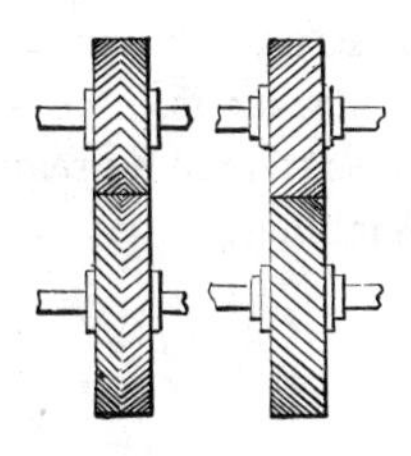

122. V-TOOTHED GEARING.—The obliquity of the teeth from the centre of the face neutralizes the longitudinal thrust of plain oblique teeth, as shown in the next pair.

123. OBLIQUE TOOTH GEAR.—A smooth running gear, with slight longitudinal thrust due to the inclined tooth surfaces.

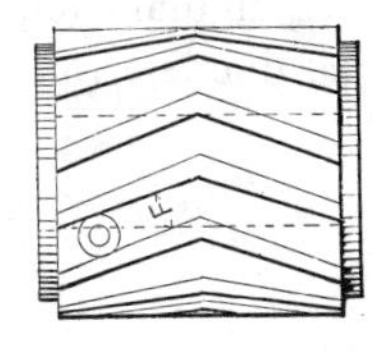

124. V-TOOTHED GEAR.—The teeth of which are usually inclined from the centre lines of the face equal to the amount of the pitch at the outer ends.

125. SPLIT SPUR GEAR, showing method of bolting on to the shaft of a trolley car.

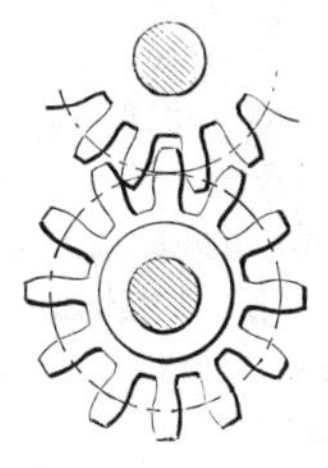

126. STAR WHEEL GEAR, for wringing machines, mangles, etc. Allows a variable mesh to the teeth.

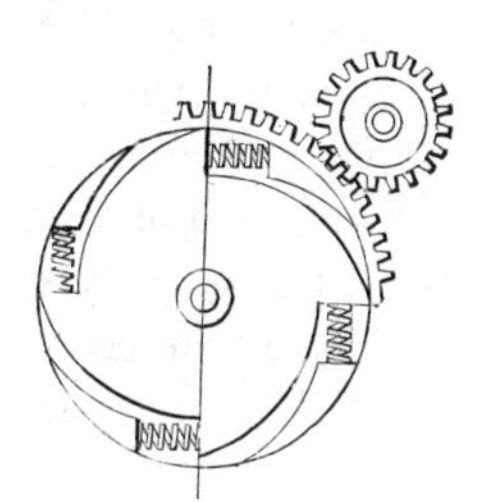

127. ELASTIC SPUR GEAR, to prevent back lash. The gear runs loose on the shaft; the ratchet-wheel is fast on the shaft. Compression springs are inserted between the shoulders of the gear and cam ratchet wheel.

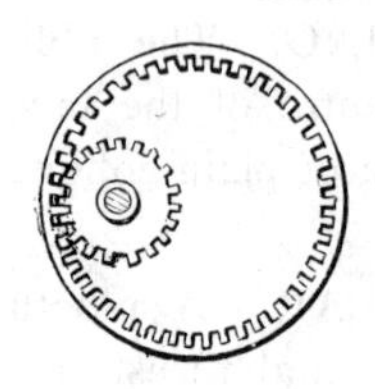

128. INTERNAL SPUR GEAR and Pinion.—In this style of gearing more tooth surface is in contact than with outside teeth; it has less wear and great power. Much used in hoisting machines.

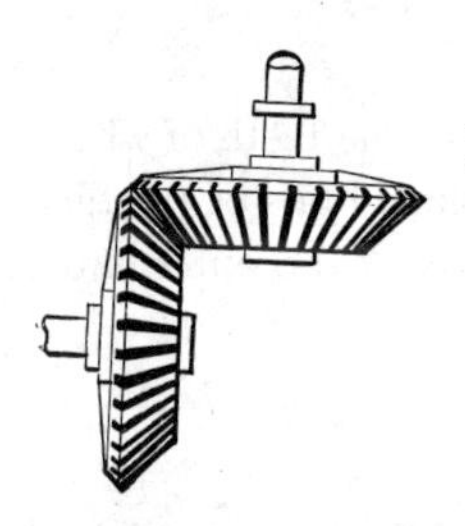

129. BEVEL GEARS, when of equal diameter. MITER GEARS, when of unequal diameter.

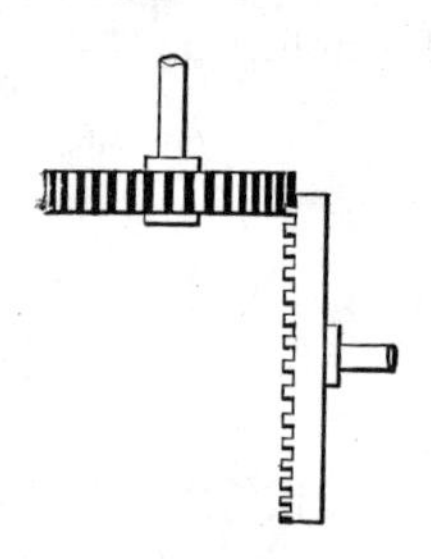

130. CROWN WHEEL geared with a spur wheel. Used for light work. A very old device.

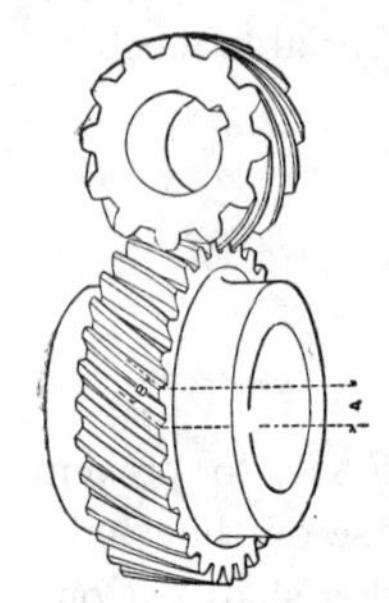

131. SPIRAL GEARING.—The velocity ratio of spiral gears cannot be determined by direct comparison of pitch diameters, as in spur gearing, but must be found from the angles of the spiral in each gear. Thus if the spiral angles of two gears are the same the velocity ratio will be inversely as the pitch diameters; but if the spiral angles are not equal, the number of teeth per inch of pitch diameter will vary. In any case the velocity ratio will depend upon the number of teeth and their spiral angle, as expressed in the following proportion: v, the velocity of the small gear is to V, the velocity of the large gear, as D, the pitch diameter of the larger, × by the cosine of its spiral angle, is to d, the pitch diameter of the smaller, × by the cosine of its spiral angle.

132. OBLIQUE SPUR AND BEVEL GEAR.—An oblique tooth spur gear and an oblique bevel gear, operating shafts running at an angular position.

133. OBLIQUE BEVEL GEAR on shafts at right angles and crossing out of axical plane.

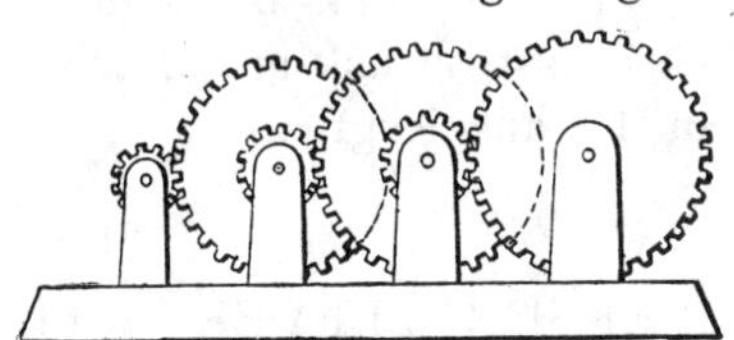

134. GEAR TRAIN—Solution for increased speed: Divide the multiple of the number of teeth in the driving gears by the multiple of the number of teeth in the driven pinions, or the multiple of each pair separately may be multiplied by the multiple of the next pair. For decreasing speed, divide the ratios.

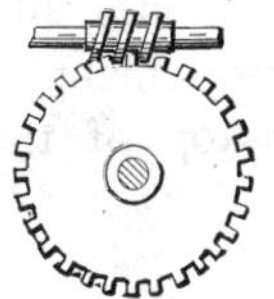

135. WORM GEAR.—With single thread the revolutions of the screw equal the number of teeth in the spur wheel for its revolution.

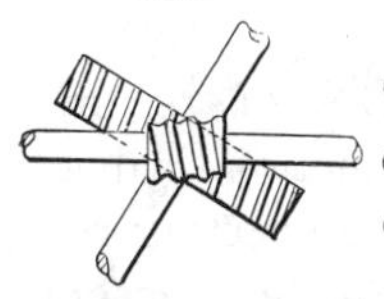

136. SKEW WORM AND WHEEL GEAR.—The angle of the teeth on this spur wheel must be equal to the angle of the screw shaft, less the angle of the screw at the pitch lines of both.

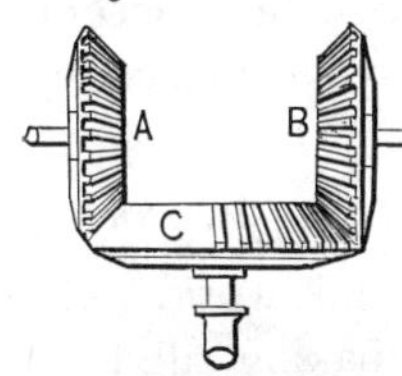

137. UNIFORM INTERMITTENT MOTION in opposite directions. The blank sector in the bevel wheel driver C interrupts the motion of A and B alternately.

138. VARIABLE SPEED BEVEL GEAR.—A bicycle novelty. One revolution of A gives two revolutions of B. A is an elliptic bevel gear central on the shaft. B is an elliptic bevel gear of one-half the number of teeth of A and revolves on one of its elliptic centres. The cranks are set opposite to the short diameter of the driving gear A, giving greater power to the tread and quicker motion at the neutral points of the crank.

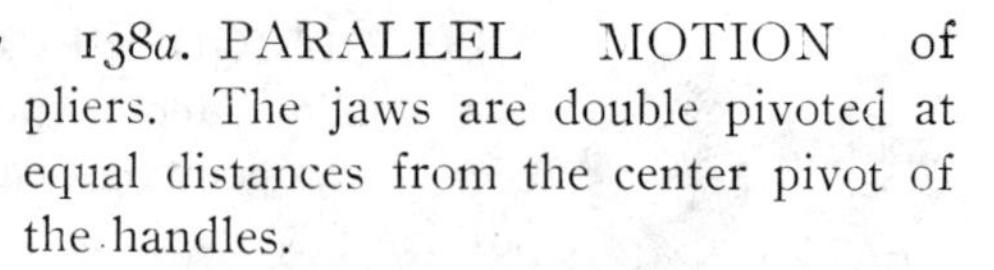

138*a*. PARALLEL MOTION of pliers. The jaws are double pivoted at equal distances from the center pivot of the handles.

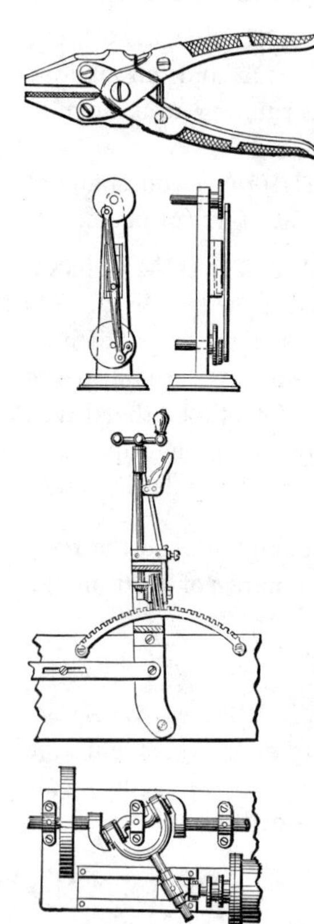

138*b*. TRANSMISSION OF CIRCULAR MOTION by link and sliding block. Block is fast on link at half distance for equal crank lengths.

138*c*. REVERSING LEVER with rack sector and worm gear. The worm wheel is lifted from the sector for large movements by the small latch lift and snaps back while a small movement is made by the handle at the top of the lever.

138*d*. TRANSMISSION of reciprocating motion into rotary motion by diagonal crank pins and yoke connecting rod. A sliding swivel on the cross head accommodates the swing of the yoke connecting rod.

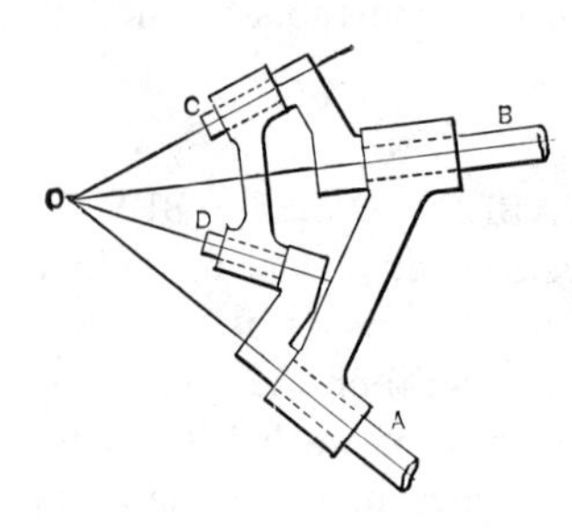

138*e*. LINK TRANSMISSION of shaft motion, or conic link work. The principle consists in making all lines of motion meet at a common center O. Cranks of equal length and also distances of rotating bearings of equal lengths and equal distances from O. Fixed shaft bearings also at equal distances from O.

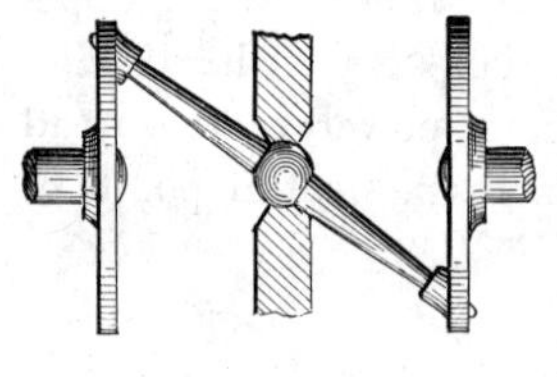

138*f*. GYRATING LEVER TRANSMISSION.—The lever swinging on its socket at its center and journaled at the edge of the disks will transmit power from one shaft to another in the same line.

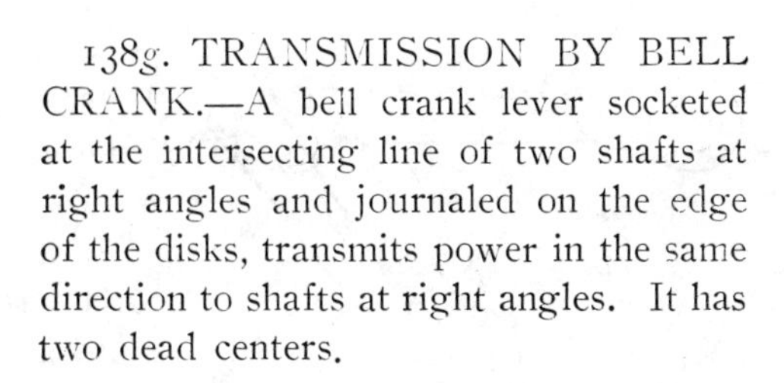

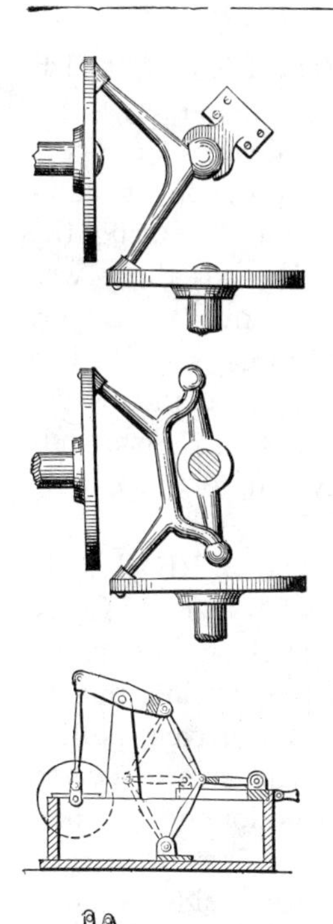

138*g*. TRANSMISSION BY BELL CRANK.—A bell crank lever socketed at the intersecting line of two shafts at right angles and journaled on the edge of the disks, transmits power in the same direction to shafts at right angles. It has two dead centers.

138*h*. GAMBREL JOINT LINKAGE for transmission at right angles, in which the dead centers of the bell crank linkage are avoided. The twisting motion at the dead center will be taken by the center bearing yoke.

138*i*. TWO REVOLUTIONS FOR ONE STROKE.—The toggle links passing their center line to the position shown by the dotted lines makes a second revolution of the wheel.

138*j*. EQUALIZING THRUST by cross links, not a true parallel motion. Pump rod is strained by the cross connection.

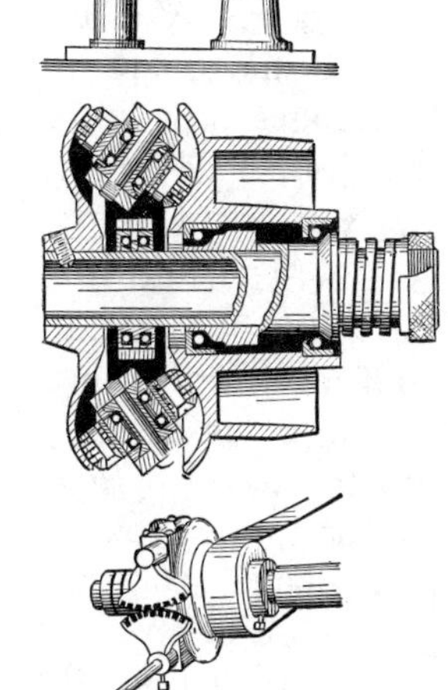

138*k*. SPEED CHANGING PULLEY.—The principle of action as shown in Fig. 115. The frame of the transfer pulleys is fixed and the change of angle made by the two sector gears and handle is shown in the second cut. All parts run on ball bearings.

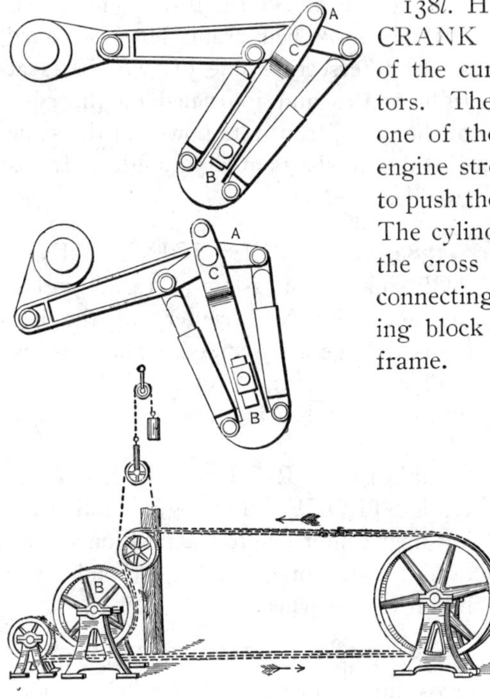

138*l*. HELPING THE CRANK over the center. One of the curious devices of inventors. The compression of air in one of the cylinders during the engine stroke is made the power to push the crank over the center. The cylinder frame is pivoted to the cross head at A and to the connecting rod at C. B is a sliding block pivoted to the engine frame.

138*m*. REVERSE MOTION DRIVE.—A being the driving pulley, B a driven pulley, will have a reverse motion by the belt running on the near side guided by the two idler pulleys.

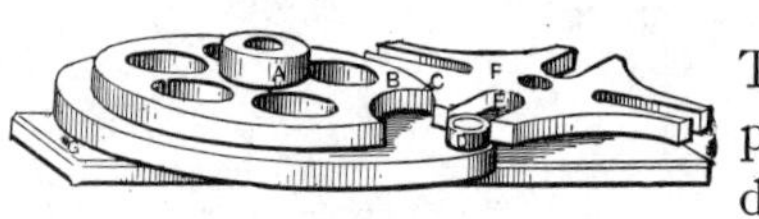

138*n*. INTERMITTENT TRANSMISSION power by spur gear. A is the driver. When B and C are together, gear F is locked. When pin roll D engages with E the driven gear F will revolve ¼ turn, more or less, as designed.

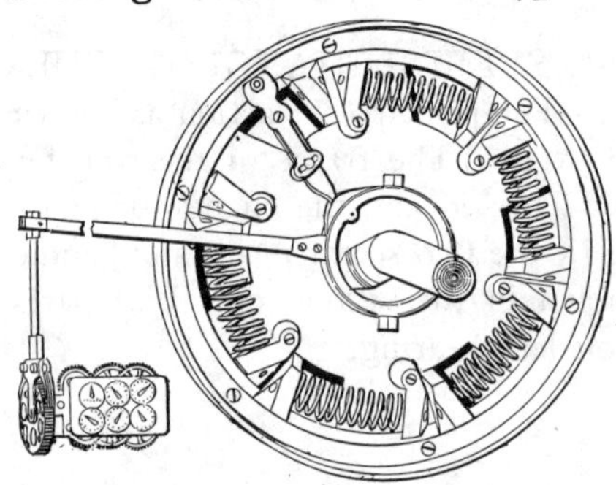

138*o*. A CONTINUOUS REGISTERING DYNAMOMETER.—Two flanged hubs on a shaft with a loose pulley between to receive or give power. The springs abut on the shaft flange and displace the loose pulley. An eccentric displaced by the power pull acts upon a recording dial by a lever.

Section III.

MEASUREMENT OF POWER.

SPEED, PRESSURE, WEIGHT, NUMBER, QUANTITIES, AND APPLIANCES.

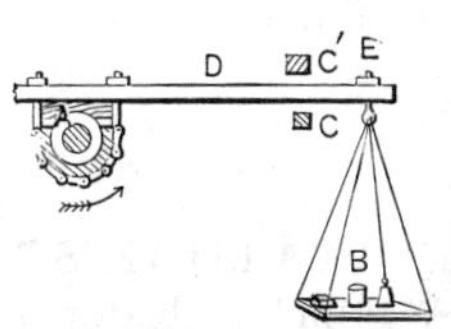

139. PRONY BRAKE, for the measurement of power. A is power shaft and pulley, enclosed in friction blocks and strap; D, the lever; C′, C, stops to control excessive movement of the lever; B, weights to balance the friction of the pulley, which should be tightened by the strap nuts until its full power at the required speed is balanced by the weight put upon the scale platform.

THE PRONY BRAKE.

RULE.—Diameter of pulley in feet × 3.1416 × revolutions per minute = speed of periphery of pulley per minute. The lever is of the third order. Its length from centre of shaft to the eye holding the weight, divided by the radius of the pulley, all in feet, or decimals of a foot = the leverage. Then the *pounds* weight × by the *leverage* and by the speed = the *foot-pounds*, which divided by 33,000 = the *horse-power*. Weight of lever at E when loose on the pulley should be deducted from the weights put on platform.

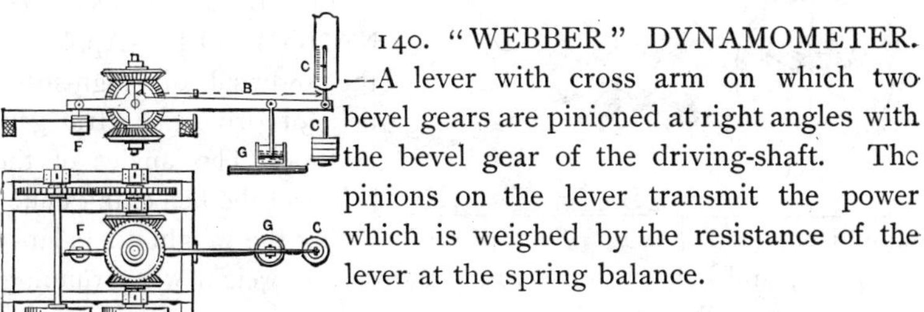

140. "WEBBER" DYNAMOMETER.—A lever with cross arm on which two bevel gears are pinioned at right angles with the bevel gear of the driving-shaft. The pinions on the lever transmit the power which is weighed by the resistance of the lever at the spring balance.

The H. P. indicated is:

$$\frac{B \times 6.2832 \times R \times W}{33{,}000} = \text{H. P.}$$

B = radius of the lever in feet. R = revolutions per minute.
W = weight on the scale.

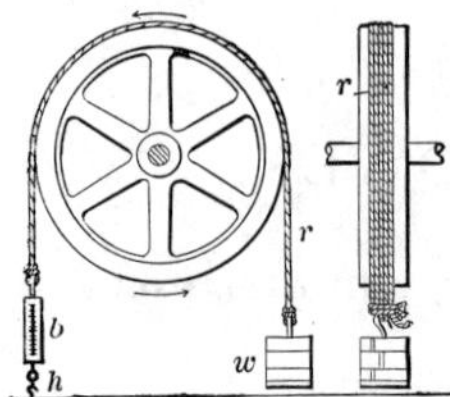

141. MEASUREMENT OF POWER.—The Rope Brake. Several ropes over a pulley gathered in a knot, to which is hung a specific weight less than the range of the spring scale attached to the other end. The pounds of relief from the stated weight by the motion of the pulley, multiplied by the velocity of the periphery of the pulley in feet per minute, gives the foot-pounds power per minute.

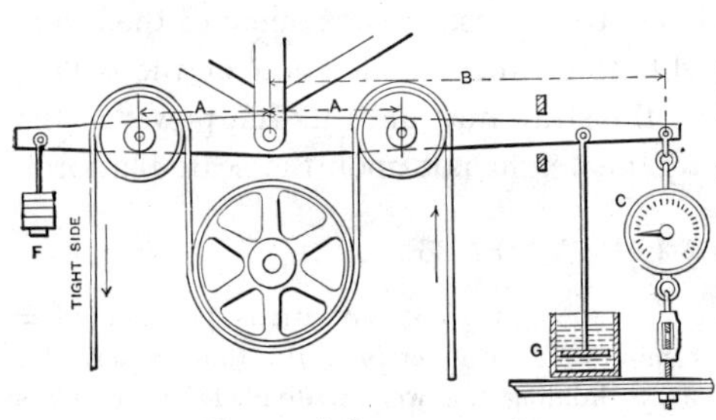

142. "TATHAM'S" DYNAMOMETER, for a vertical belt. A balance frame lever, pulleys, and dash-pot.

The work of the belt is:

$$\frac{W \times \frac{B}{A} \times S}{33{,}000} = \text{H. P.}$$

W = weight on scale.
B = length of lever.
A = fulcrum to pulleys which should be equal.
S = speed of belt in feet per minute.

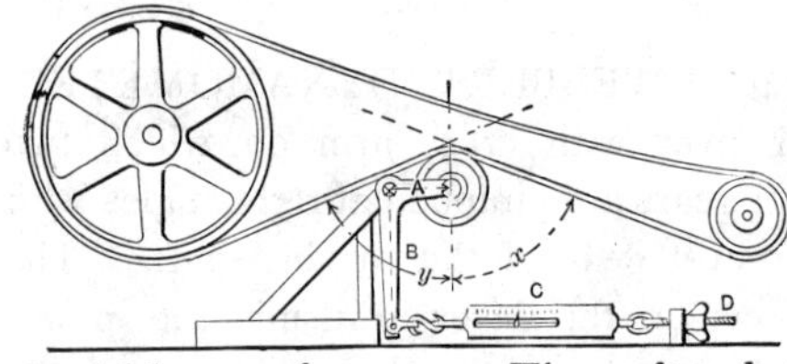

143. BELL-CRANK DYNAMOMETER.—Applied to the power side of a high-speed belt for driving electric generators. The angles of the belt over the bell-crank pulley should be equal, $y = x$. Then after deducting the weight to balance the pulley and belt when not running from the weight when running, the formula will be:

$$\left.\frac{W \times \frac{B}{A}}{2 \text{ cosine } \times}\right\} \times \text{ speed of belt in feet per}$$

minute = foot-pounds, which divided by 33,000 = H. P. B, long arm; A, short arm of lever.

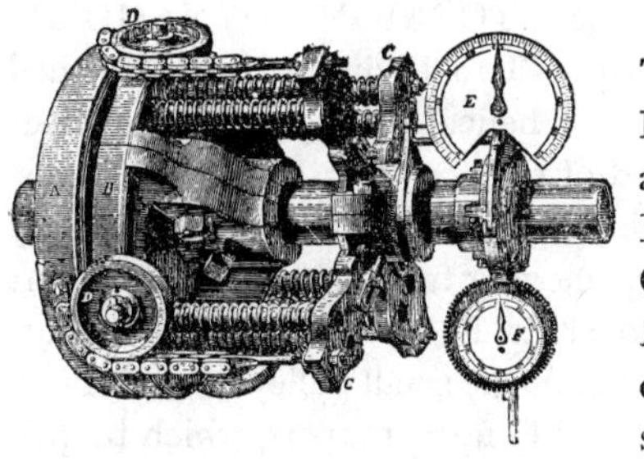

144. "NEER'S" ROTARY TRANSMITTING DYNAMOMETER.—A shaft is disconnected at a coupling and the discs A and B are clamped one to each shaft. Chains are attached to the disc A and roll around pulleys on the disc B, and are attached to the spider C. The chain tension is resisted by the helical springs and is recorded on the dial E. The dial F indicates revolutions.

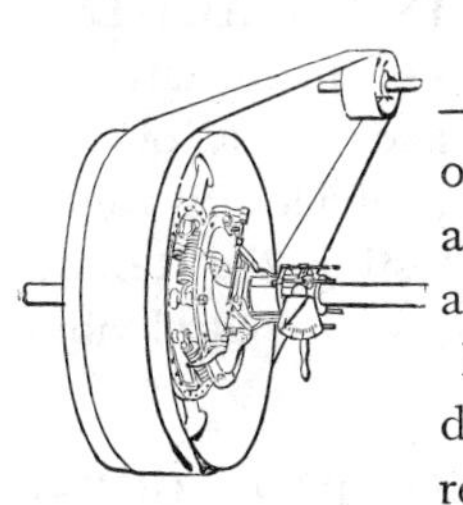

145. "VAN WINKLE'S" POWER METER.—A set of helical springs attached to two discs, one of which is made fast to the pulley, unkeyed and loose on the shaft; the other disc and hub are clamped to the shaft. A set of levers on a rock haft transmits the strain on the springs to an index and dial indicating the horse-power per 100 revolutions of the shaft.

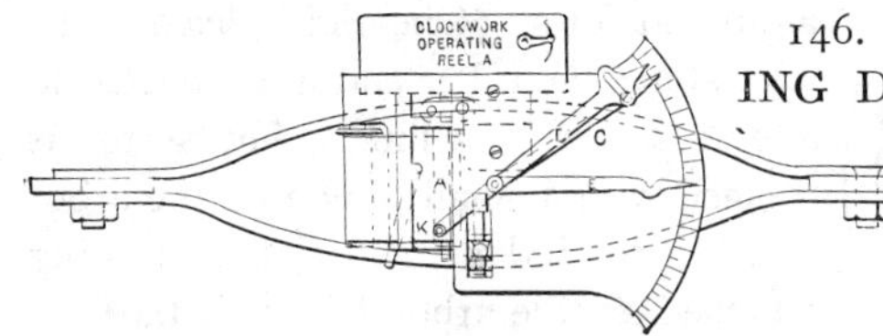

146. TRACTION RECORDING DYNAMOMETER.—The draft-pull compresses the elliptic-shaped springs, moving the index hand D, which carries a pencil at its opposite end K. A paper ribbon is drawn under the pencil and wound on a drum, driven by clockwork, making a continual record, to be measured by a suitable scale for the average work.

147. FRICTION MACHINE, for testing the friction of wheels at various speeds and loads. The adjustable circular balance holds the wheels or vehicle in place. The pounds tension on the scale multiplied by the peripheral velocity in feet per minute gives the foot-pound draft or friction.

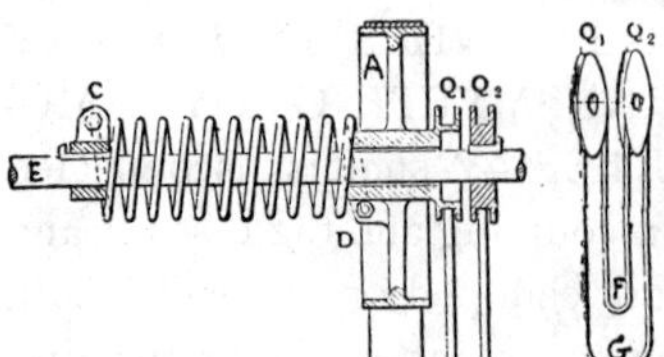

148. TORSION DYNAMOMETER.—To a driving shaft E is attached at C a helical spring. To the other end of the spring is attached a transmission pulley A and a small pulley Q_1, moving freely on the shaft E. At Q_2 another small pulley is fixed to shaft E. The tension of transmission displaces the relative position of the small pulleys and through an endless belt draws the loops F and G farther apart, which by pulleys and index not shown indicates the power transmitted.

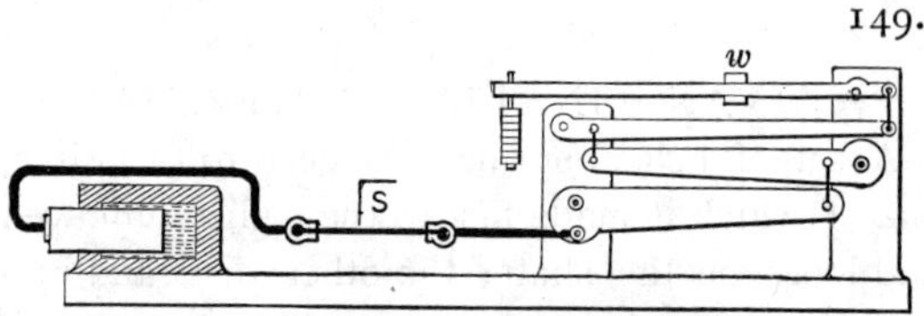

149. TENSILE TESTING MACHINE.—A hydrostatic ram and system of compound levers, used in testing the tensile strength of metals. S, article to be tested; *w*, stops to control vibration of levers; W, weight.

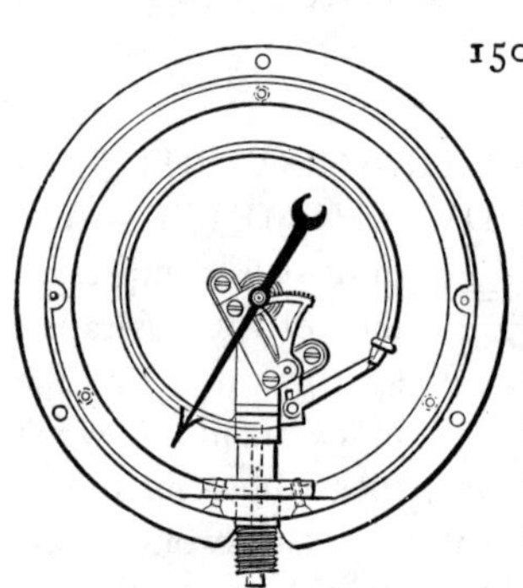

150. BOURDON PRESSURE GAUGE.—A flattened spring metal tube is bent to a circular form. One end is fixed to the inlet stud; the other end is connected to a lever sector by a link. The sector is meshed with a small pinion on the arbor carrying the index hand. A hair spring attached to the arbor keeps all the pivoted joints drawn in one direction for accuracy of pressure indication.

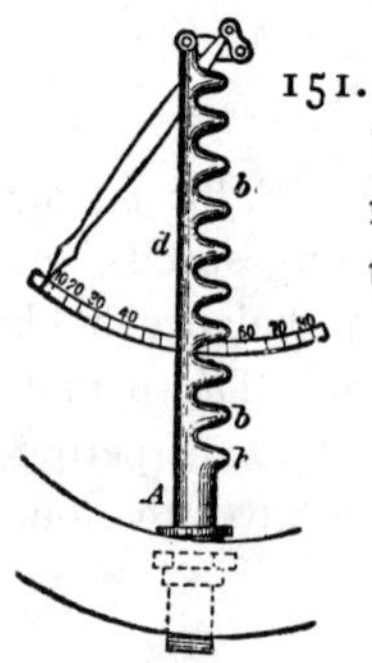

151. CORRUGATED TUBE-PRESSURE GAUGE.—The pressure within the tube expands it on the corrugated side and through the link connections with the index hand moves the hand.

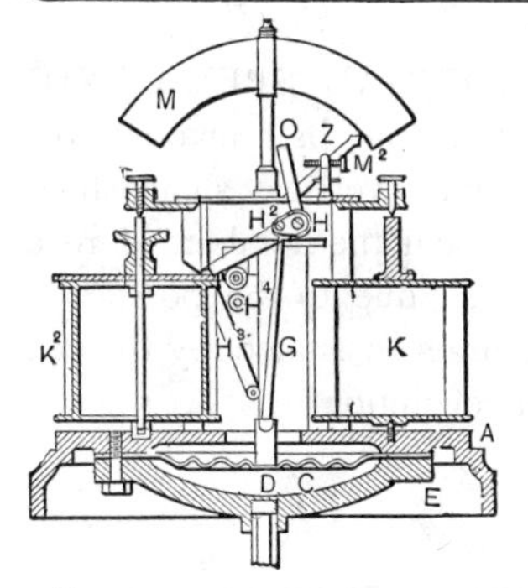

152. RECORDING PRESSURE GAUGE, "Edson" model.

D, corrugated diaphragm bearing the pressure; G, connecting rod from diaphragm to crank-pin, on the shaft on which the index hand is fixed, as also the arm and pencil bar, H^2, H^3, in front of the record sheet; K, K, winding barrels for the record sheet driven by a clock movement; M, index dial.

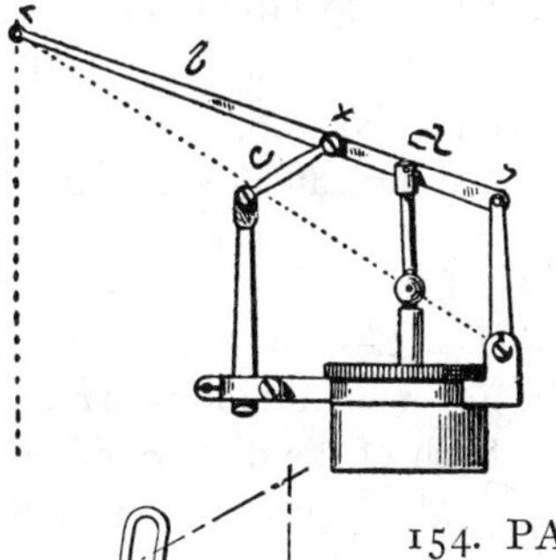

153. PARALLEL MOTION OF THE INDICATOR.

Proportions: $c : d :: d : b$ – nearly.

154. PARALLEL MOTION FOR AN INDICATOR.—The curved slot is made proportional to the length of the two arms of the pencil lever.

155. "AMSLER" PLANIMETER.—E is the fixed point; F the tracer. The disc has a sharp edge and a cylindrical section divided and read from a vernier scale. A worm screw and index wheel indicate the number of revolutions of the rolling disc.

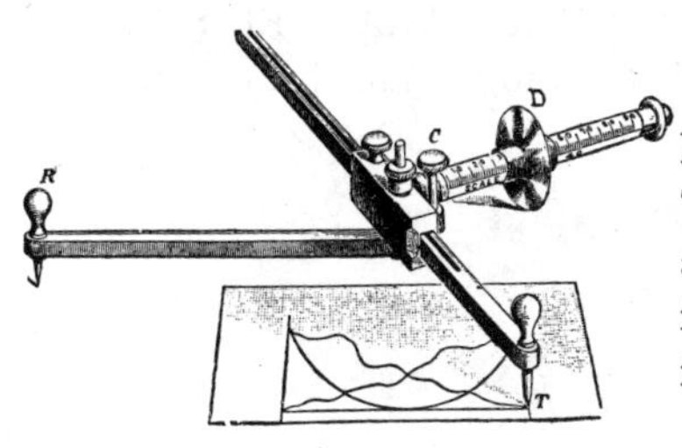

156. "LIPPINCOTT" PLANIMETER.—R is the fixed point; T the tracer; c is a smooth round arm on which a scale is laid off from the axis; D is a disc with a free motion on the scaled arm.

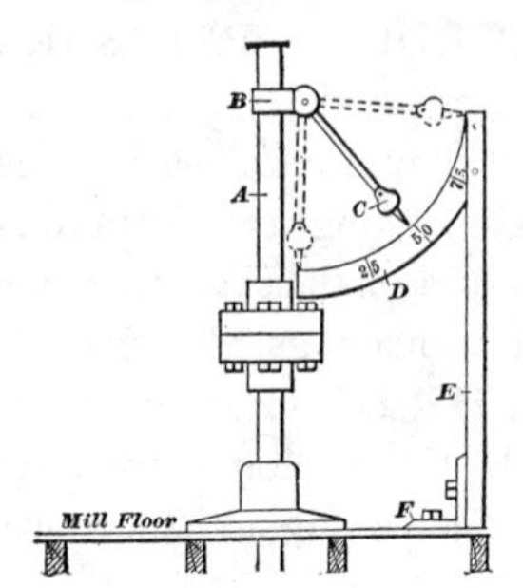

157. CENTRIFUGAL SPEED INDICATOR.—An arm and ball pivoted to a clamp on a revolving vertical shaft shows on a curved index bar the number of revolutions per minute, due to the position of the ball and pointer, assumed by the centrifugal force of revolution.

158. SPEED INDICATOR.—An application of the screw gear. The screw dial counts to 100, right or left. The second dial indicates the number of hundreds.

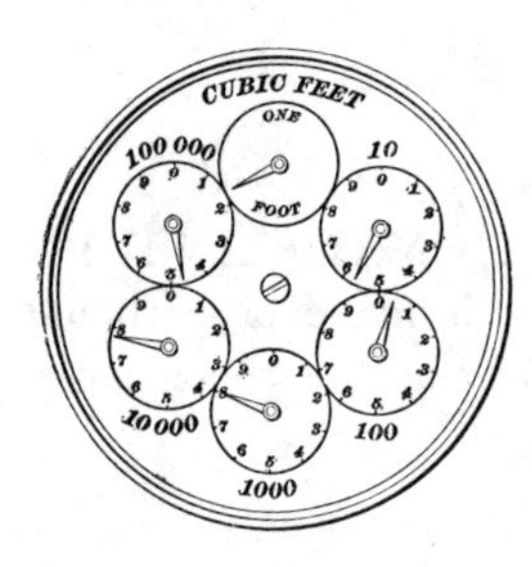

159. METER DIAL—how to read it. A revolution of the upper hand is a measure of one cubic foot. Each of the dials represents a multiple of ten. The figures following the motion of the index hands are to be noted, and if reading to the right must be put in inverse order, and if read to the left must be put in serial order. Thus the dial here represented reads 47,805 cubic feet.

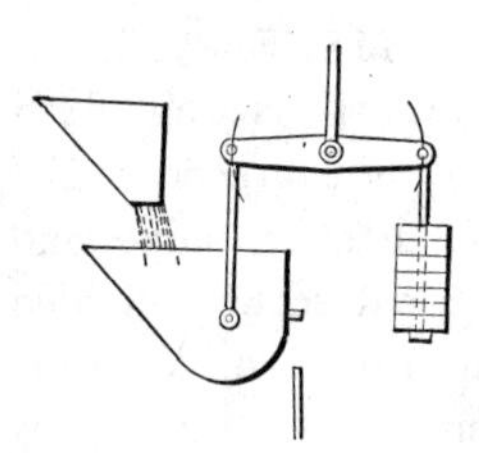

160. AUTOMATIC TIPPING SCALE, for measuring grain or water.

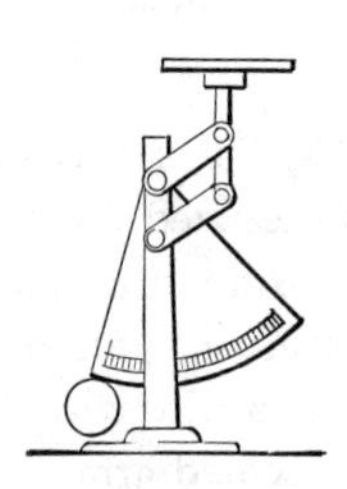

161. DOUBLE LINK BALANCED SCALE.—The upper link is fixed to the radial index plate.

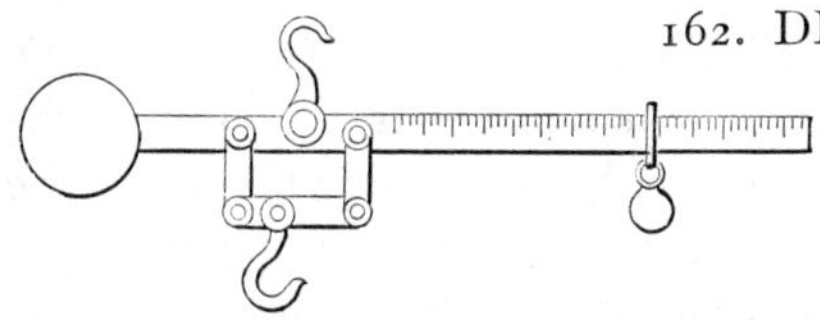

162. DIFFERENTIAL WEIGHING BEAM.—The link connection to the lower hook allows the V-bearings to be brought much nearer together than on a single bar.

163. ENGINE COUNTER.—A series of counter gears as in the following figures, may be placed overlapping, as here shown; each spindle mounted with a number dial and all covered by a perforated plate, showing the top figures of each dial. The spring pawl checks the first wheel in the train, to hold the number in place while the lever pawl takes its back motion.

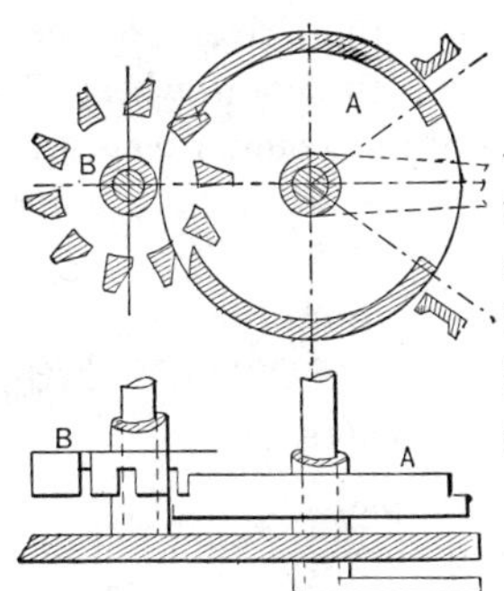

164. OPERATION OF A COUNTER.—The wheel B, with its ten pin teeth, is thrown one tooth at each vibration of the arm of the sector rim A. The wheel B also has a sector rim fixed to and revolving with it that throws the next pin-tooth wheel one tooth at each revolution, and so on.

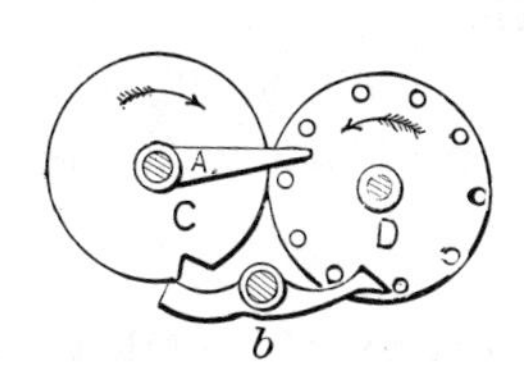

165. INTERMITTENT ROTARY MOTION, for counters and meters. The tappet A, revolving with the wheel C, carries the wheel D one pin notch per revolution. The pawl *b* is released by the notch in the wheel C while the tappet is in contact with the pin.

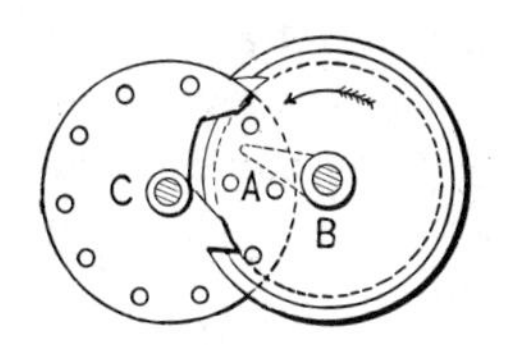

166. INTERMITTENT ROTARY MOTION, for counters and meters. B, driving wheel, the rim of which has an entering and exit notch for pins in the driven wheel and locks the wheel C at each revolution of wheel B

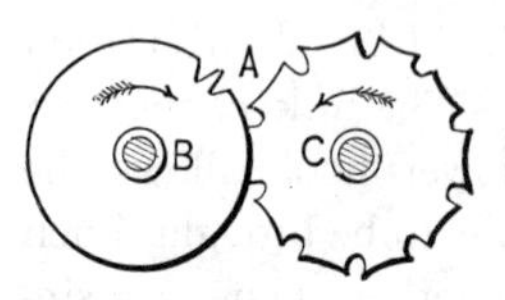

167. INTERMITTENT ROTARY MOTION, for counters and meters. A, the driving tooth in the wheel B; C is stopped by the concave sections that fit the periphery of the wheel B. The tooth A projects beyond the peripheral radius of wheel B, and the notches relieve the inverted curves of wheel C, allowing it to turn one notch at each revolution of wheel B.

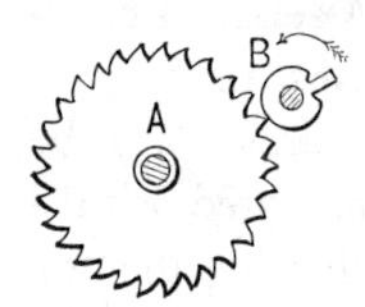

168. INTERMITTENT ROTARY MOTION, for counters and meters. In this form the largest number of revolutions of the single tooth pinion B, for one revolution of wheel A, may be obtained.

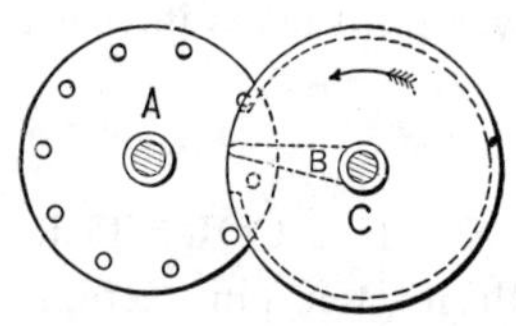

169. INTERMITTENT ROTARY MOTION, for counters and meters. Wheel C and its arm tooth B is the driver. A rim, shown by the dotted circle on wheel C, catches a pin tooth of the counter wheel A at each revolution. The opening in the rim allows the pin to enter and leave the inside of the rim.

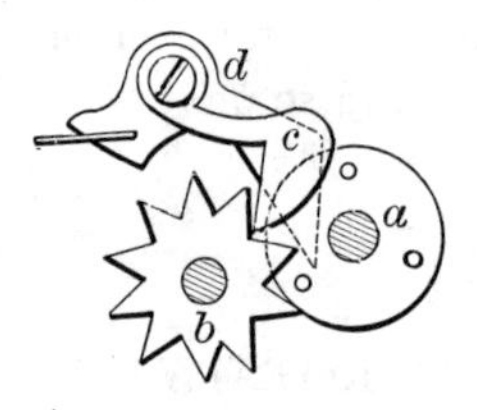

170. INTERMITTENT ROTARY MOTION, for counters and meters.

a, driving pin plate.
b, star wheel counter.
c, pawl.
d, spring latch.

The latch *d* is on the back of the pin plate and holds the star teeth, after rotation, by a light spring. *c* is a pawl that locks the teeth; pawl is lifted by pins in pin wheel.

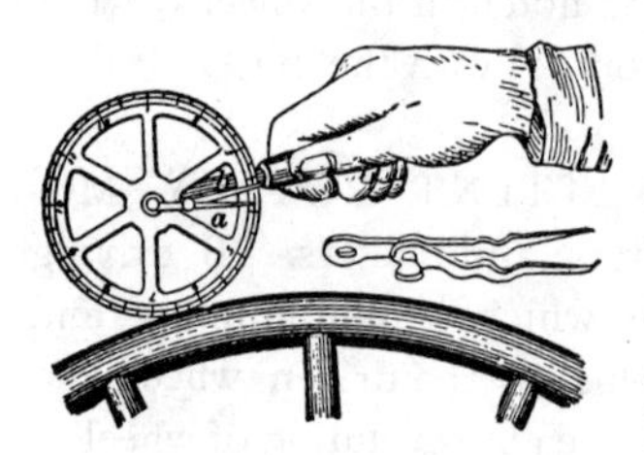

171. TIRE MEASURE COUNTER.—A wheel running freely in the forks of a handle, is made of a size that will roll exactly two feet to a revolution, and graduated in feet and inches. The supplementary index is set to allow for lap in welding the tire.

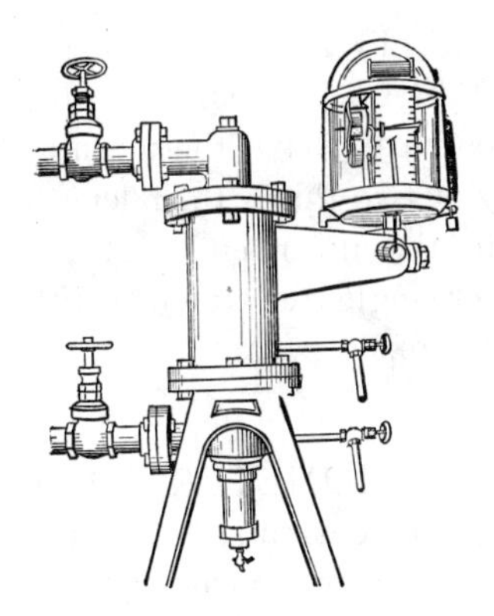

171*a*. THE ST. JOHN'S STEAM METER.—An automatic recording meter of the amount of steam passing through it for all purposes. The lifting of a conical valve by differential pressure operates a marking index through the lever on a small transfer shaft through the projecting arm from the cylinder. See detailed figure.

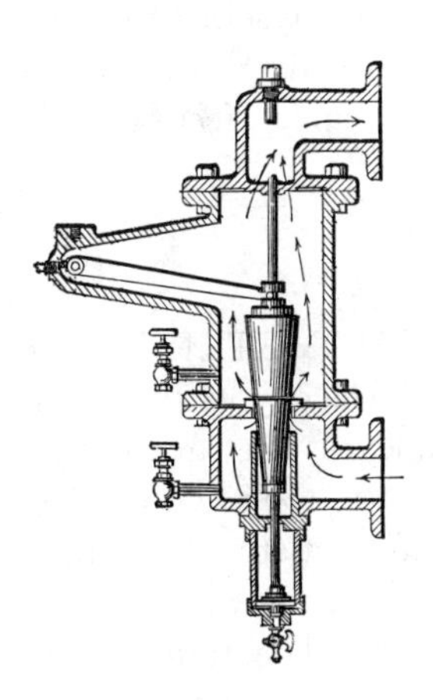

171*b*. DETAILS OF THE ST. JOHN'S STEAM METER.—The lifting of the conical valve by differential pressure allows the required quantity of steam to pass through the annular area, which is the measure under the initial pressure. The valve lift is recorded on a strip of paper moved by a clock; the mean of record curves being the measure for the time. The small chamber at the bottom is the dash pot filled with water and keeps the valve from chattering.

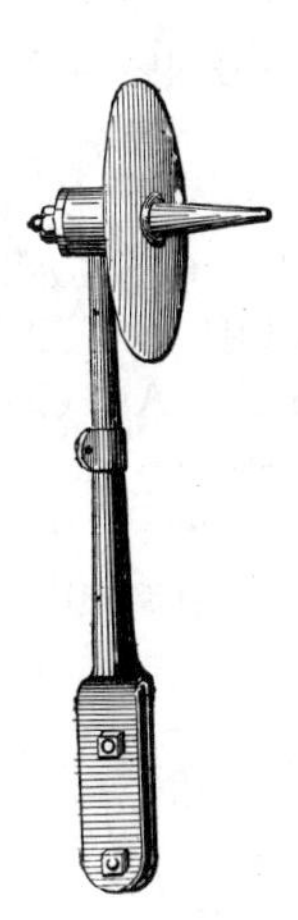

171*c*. BELT SHIPPER.—A taper pin with a flange at the large end and attached to a pole. This handy device enables the throwing of a belt off or on with safety.

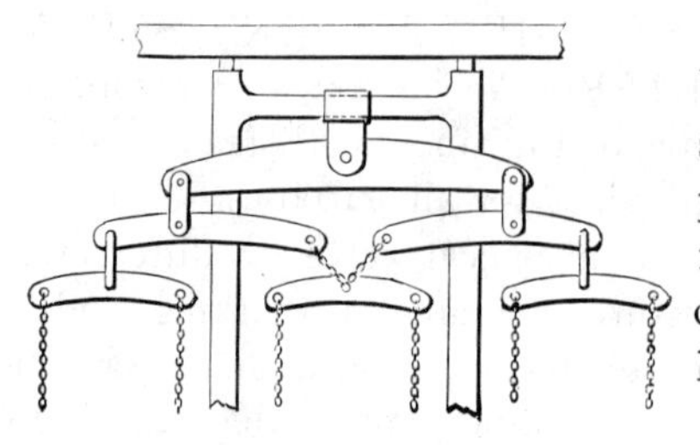

171*d*. THREE HORSE WHIFFLETREES. — The second pair have their center pins at two-thirds their length from the inner end and the center single tree attached with loose links.

171*e*. ANEMOMETER for measuring air currents. A small windmill but a few inches in diameter geared to a series of dials which by known air velocities are graduated in cubic feet of air passed per minute.

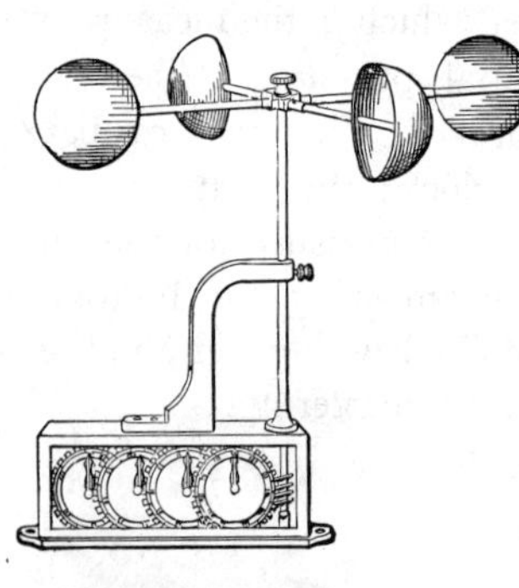

171*f*. ANEMOMETER for measuring the velocity of the wind. The dial indexes are geared by tenths, as 1, 10, 100, 1,000 miles, which by differentiating the time gives the velocity of the wind in miles per hour. The ratio of the wind velocity to the center of the cup velocity is usually about 3 to 1.

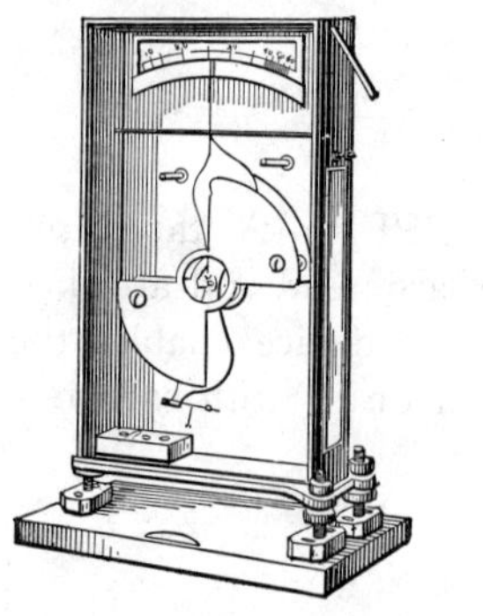

171*g*. ELECTROSTATIC VOLTMETER FOR MEASURING PRESSURES.—An electrostatic voltmeter is shown herewith. In this form the meter is constructed to measure pressures up to 20,000 volts.

Section IV.

STEAM POWER.

Boilers and Adjuncts, Engines, Valves and Valve Gear, Parallel Motion Gear, Governors and Engine Devices. Rotary Engines, Oscillating Engines.

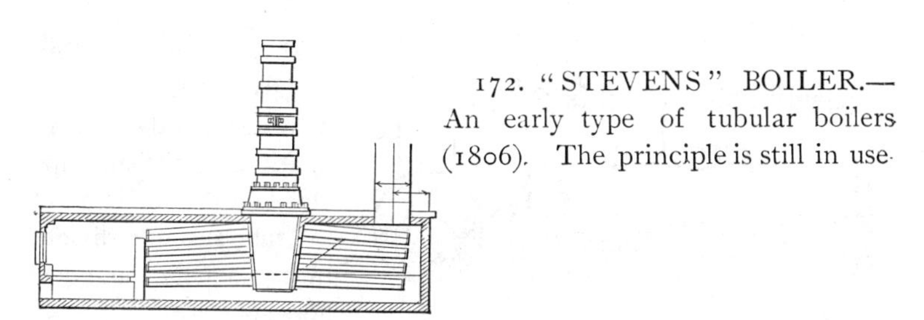

172. "STEVENS" BOILER.—An early type of tubular boilers (1806). The principle is still in use.

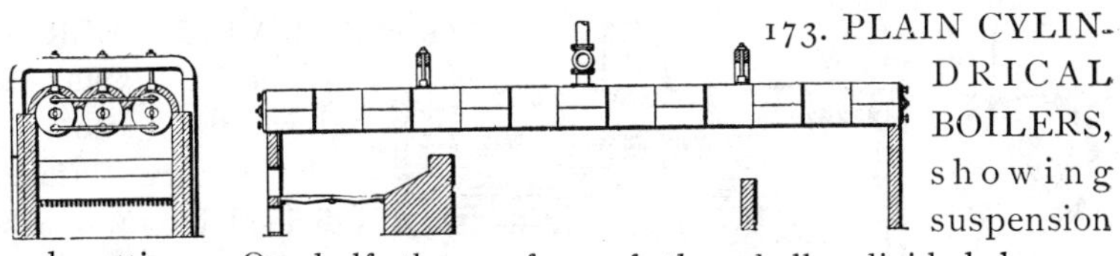

173. PLAIN CYLINDRICAL BOILERS, showing suspension and setting. One-half the surface of the shells, divided by 10, equals boiler horse-power.

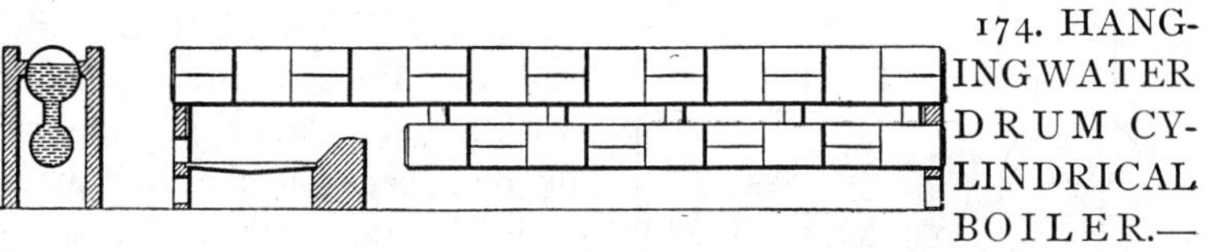

174. HANGING WATER DRUM CYLINDRICAL BOILER.—The drum, hanging from the main boiler by necks, gives a large increase of heating surface. One-half of shell and all of drum surface, divided by 12, equals boiler horse-power.

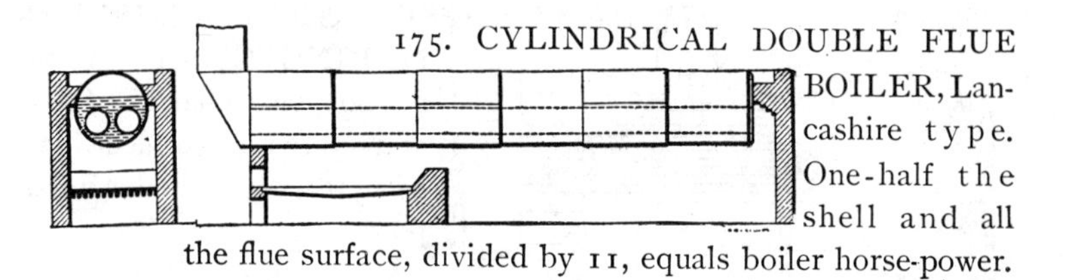

175. CYLINDRICAL DOUBLE FLUE BOILER, Lancashire type. One-half the shell and all the flue surface, divided by 11, equals boiler horse-power.

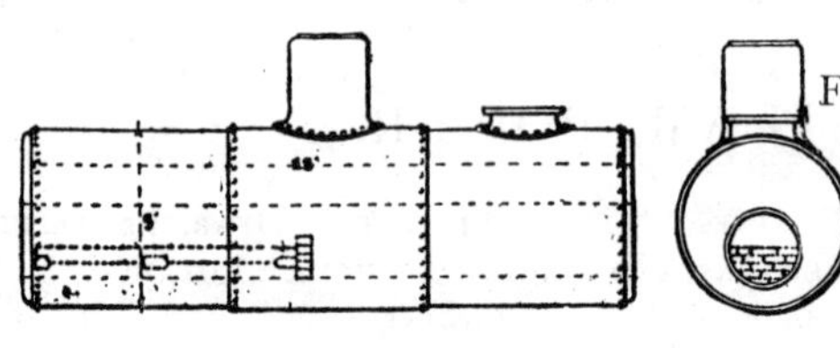

176. INTERNALLY FIRED FLUE BOILER.—The flue and half the shell surface, if exposed to heat, divided by 14, equals horse-power.

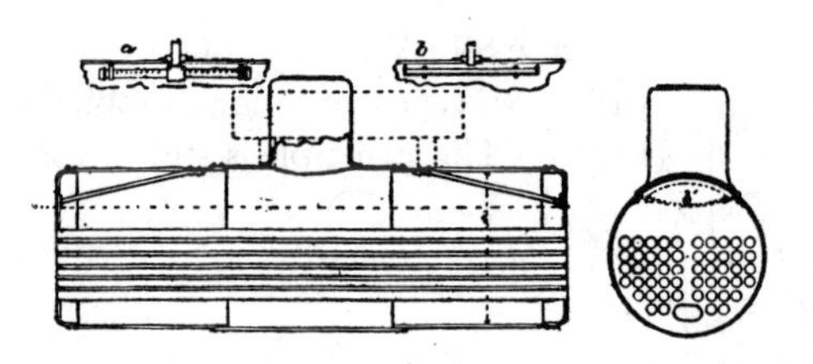

177. HORIZONTAL TUBULAR BOILER, with steam and dry steam pipe. *a*, Dry steam pipe. One-half the shell and all the tube surface, divided by 14, equals the boiler horse-power.

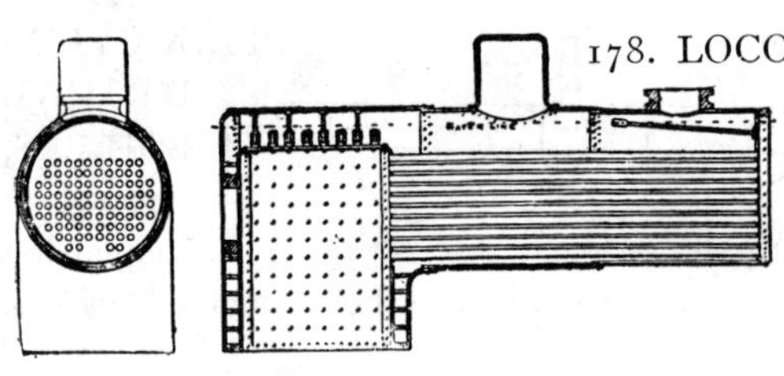

178. LOCOMOTIVE BOILER.—All the fire-box surface above the grate and all the tube surface, divided by 14, equals the boiler horse-power.

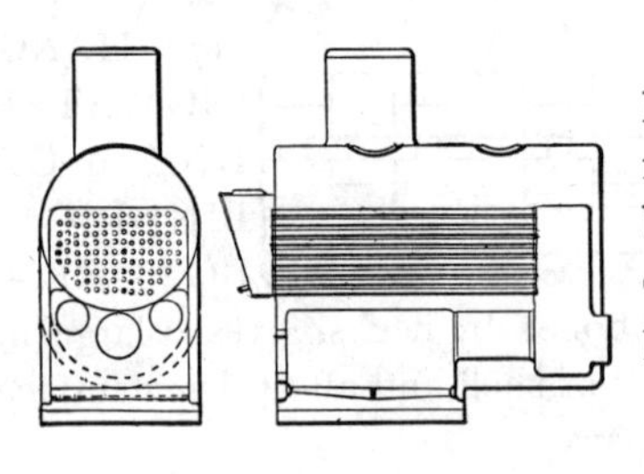

179. MARINE BOILER, with locomotive fire-box, three flues and return tubes. The area of the fire-box, flues, back chamber, and tubes, divided by 12, equals boiler horse-power.

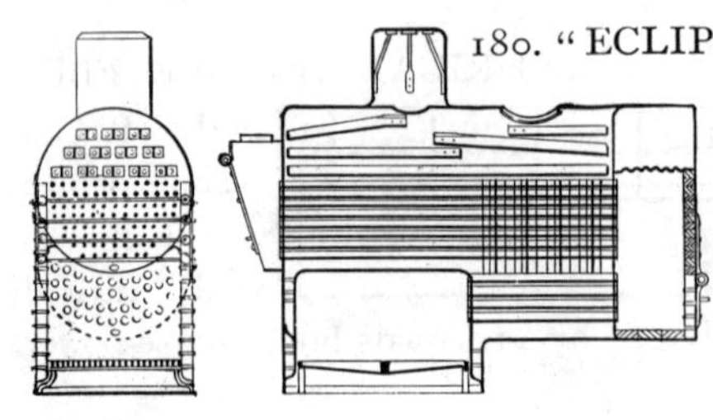

180. "ECLIPSE" RETURN TUBULAR MARINE BOILER—All the fire-box, back chamber, direct and return tube surface, divided by 12, equals boiler horse-power.

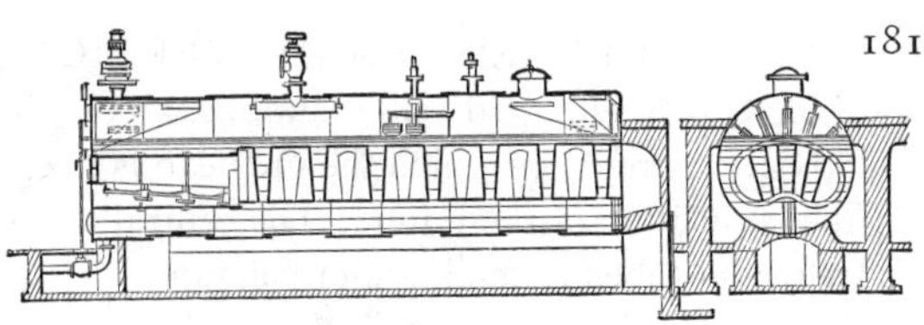

181. "GALLOWAY" BOILER.—An internally fired oval flue, with small conical tubes set diagonally across the flue.

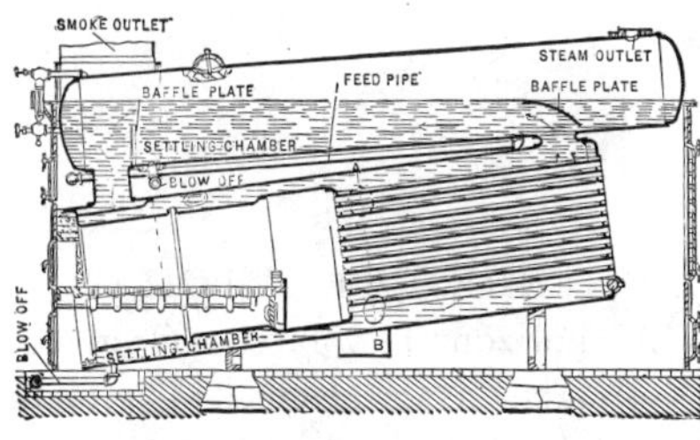

182. INTERNAL FIRED CYLINDRICAL TUBULAR BOILER.—Lower shell slightly inclined to facilitate circulation. Fire surface of tubes, fire-box, and all of shell exposed to heat, divided by 12, equals boiler horse-power.

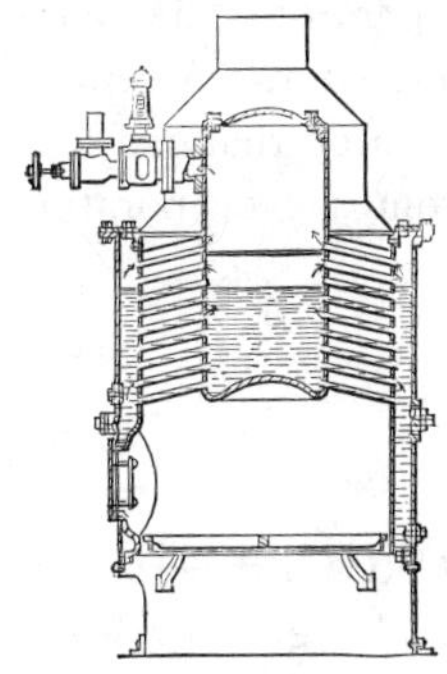

183. "DION" VEHICLE BOILER.—A central water and steam drum enclosed within an annular drum, and connected by short inclined tubes. A light and quick-firing boiler for a motor carriage.

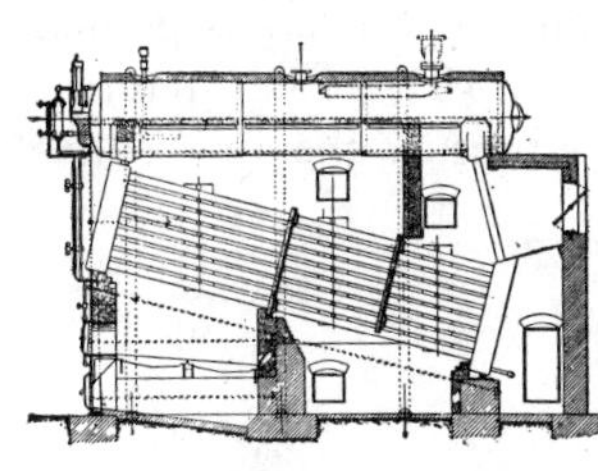

184. "BABCOCK & WILCOX" WATER TUBE BOILER.—Inclined straight tubes expanded in vertical steel headers, connected to a steam drum above. Partitions repass the flame through the tube spaces.

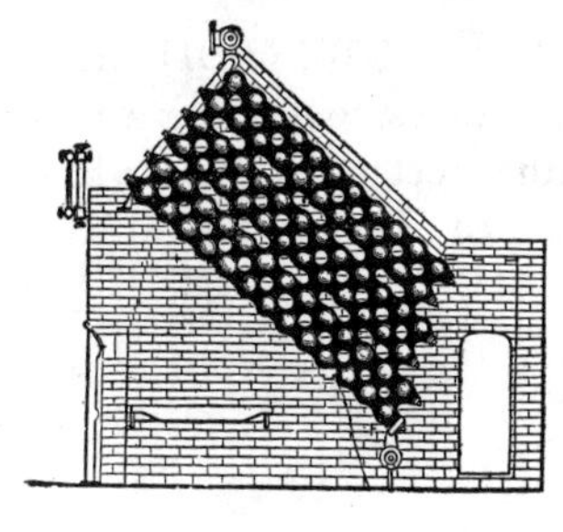

185. "HARRISON" BOILER.—A series of cast-iron globes with ground joints, held together by through bolts.

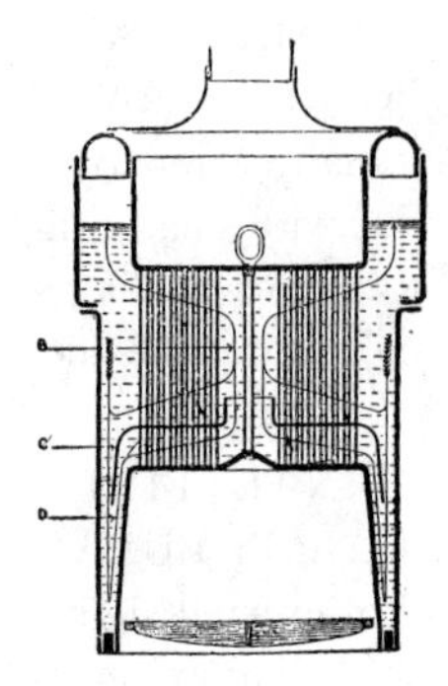

186. SUBMERGED HEAD VERTICAL BOILER, with enlarged water surface, and a circulating diaphragm by which the fire head is swept by the circulation of the water. The central space is free from tubes to facilitate circulation.

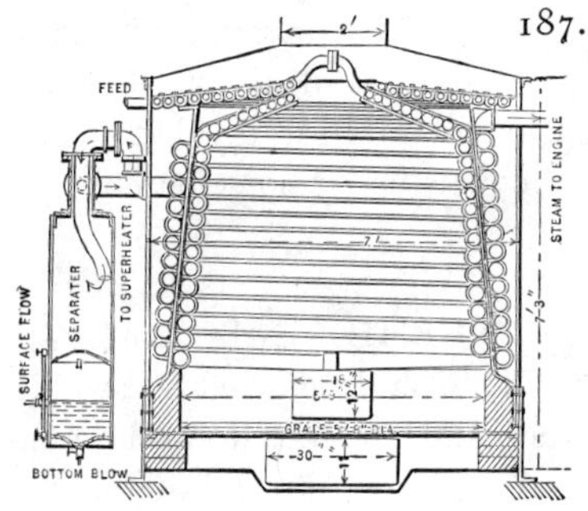

187. "HERRESHOFF" BOILER.—A horizontal volute coil at the top acts as a heater. The inner coil is the evaporator; the outer coil is the superheater. A separator entraps the water that may be carried over from the evaporating coil.

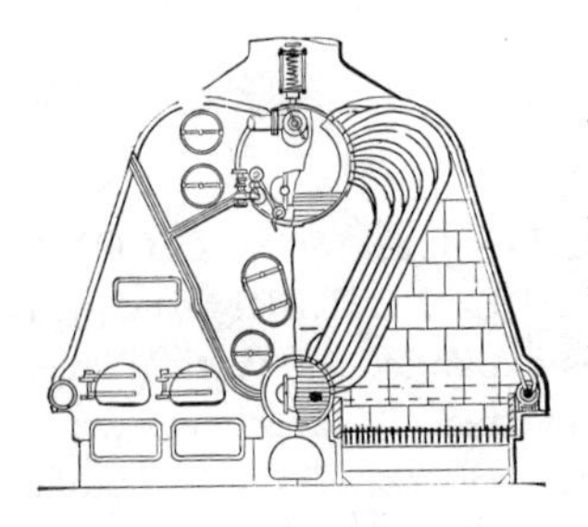

188. "THORNYCROFT" BOILER.—A large steel drum above and a water drum below, connected with a large number of bent tubes. The water return is through a large tube at the rear end of the boiler.

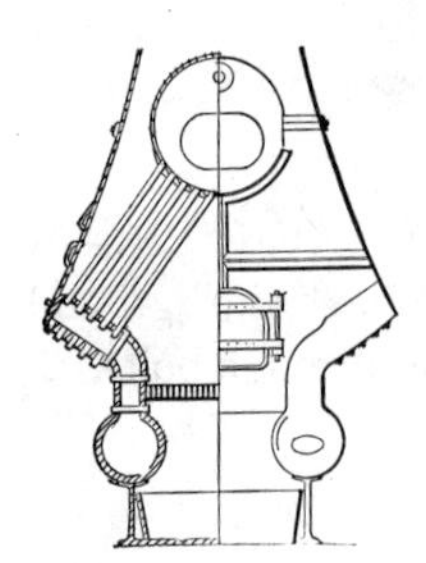

189. "SEE" WATER TUBE BOILER.—A series of short straight tubes connecting a horizontal steam drum with a rectangular water base on each side of the furnace. Return tube at back of boiler.

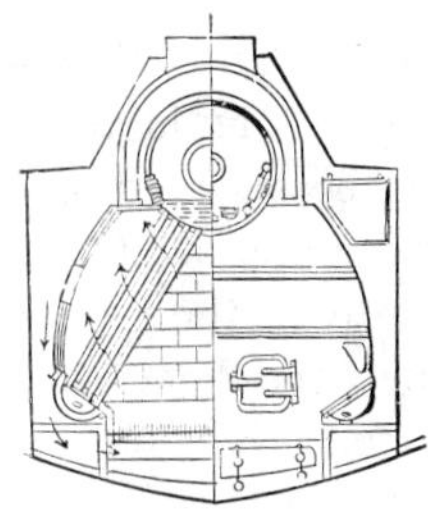

190. "YARROW" WATER TUBE BOILER.—Inclined sections of straight tubes from water-headers each side of the fire grate to a large steam drum above. Iron casing lined with fire tile. This design is for a marine boiler.

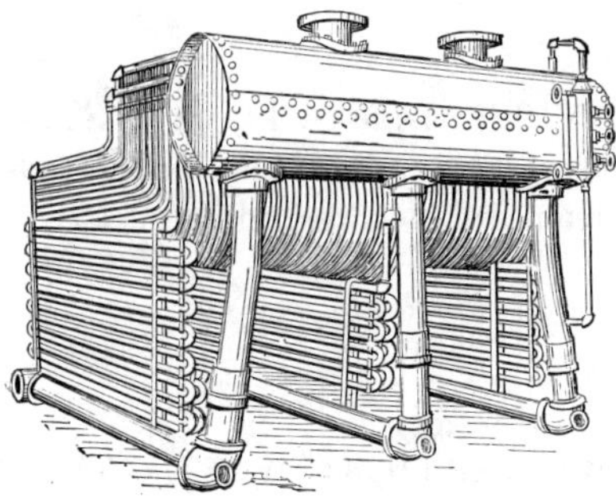

191. "BOYER'S" WATER TUBE BOILER.—Furnace walls are coils of pipe. Coils over the fire are connected from circulating pipes to steam drum.

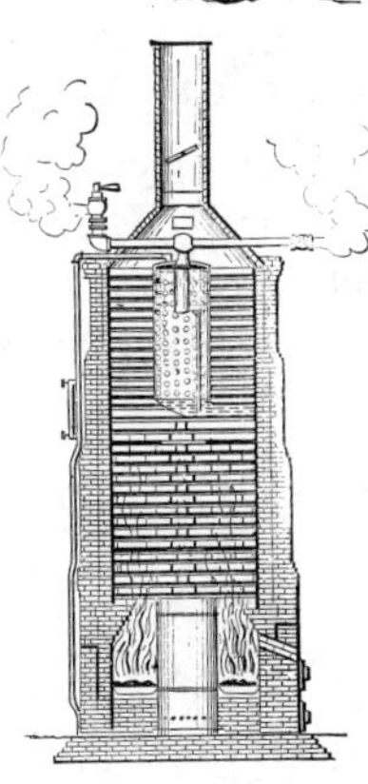

192. "HAZELTON" BOILER.—A central vertical drum in which tubes, with closed ends, are screwed radially. The grate is beneath the radial tubes and around the base of the drum.

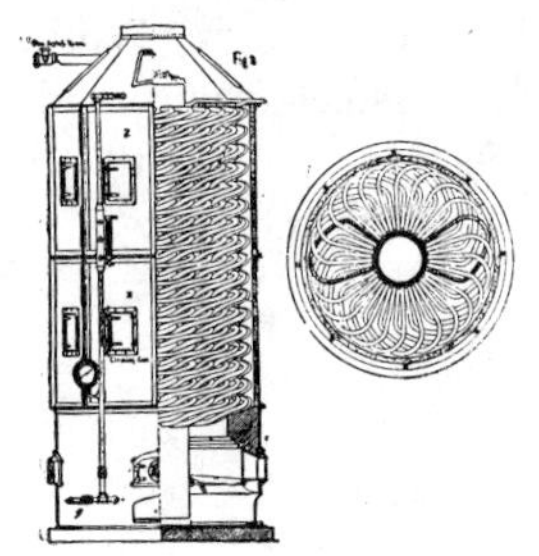

193. "CLIMAX" BOILER.—A central vertical water and steam drum, with bent tubes expanded in it, and inclined to facilitate circulation.

194. Section showing bent tubes.

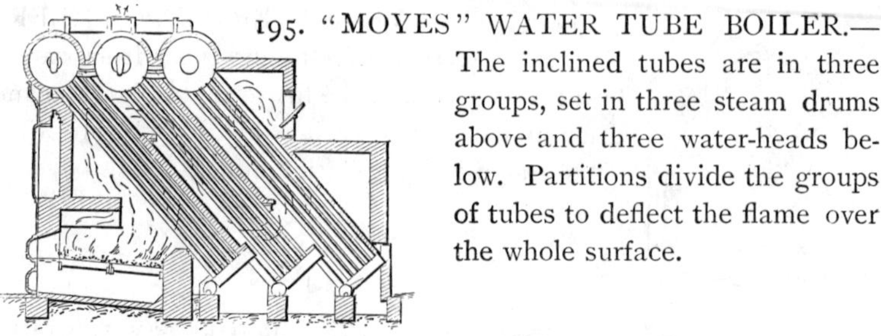

195. "MOYES" WATER TUBE BOILER.—The inclined tubes are in three groups, set in three steam drums above and three water-heads below. Partitions divide the groups of tubes to deflect the flame over the whole surface.

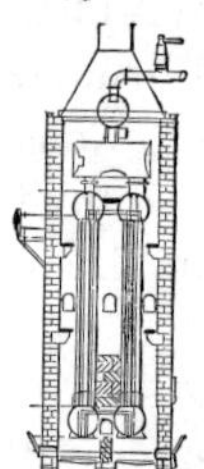

196. "WHEELER" VERTICAL TUBE BOILER.—Two sections of straight vertical tubes, with drum-heads top and bottom, and a steam drum connected by necks.

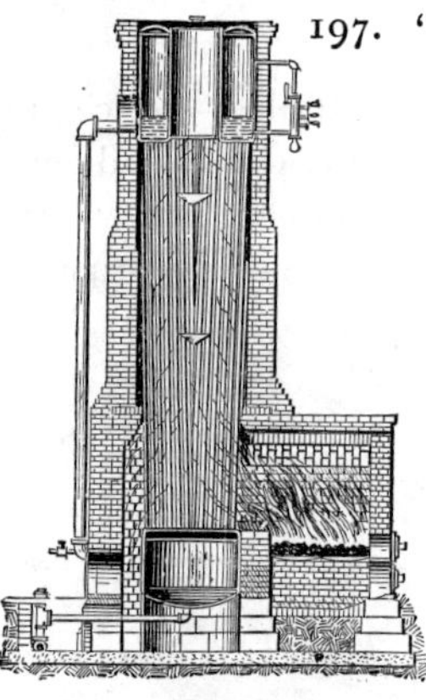

197. "CAHALL" VERTICAL WATER TUBE BOILER.—A water drum at the bottom forms the lower head for the tubes. An annular drum at the top forms the upper head, through which the smoke passes. The furnace and combustion chamber are outside, as is also the circulating pipe, as shown in the cut.

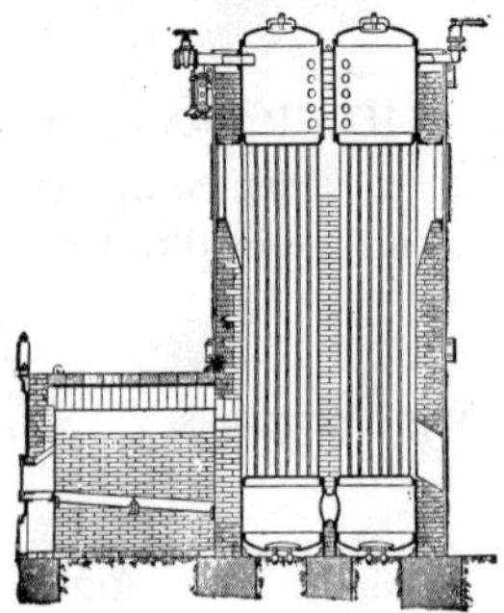

198. VERTICAL WATER TUBE BOILER (Philadelphia Engineering Works). Straight tubes between steel drums and a wall between the sections, so that the fire sweeps the length of all the tubes.

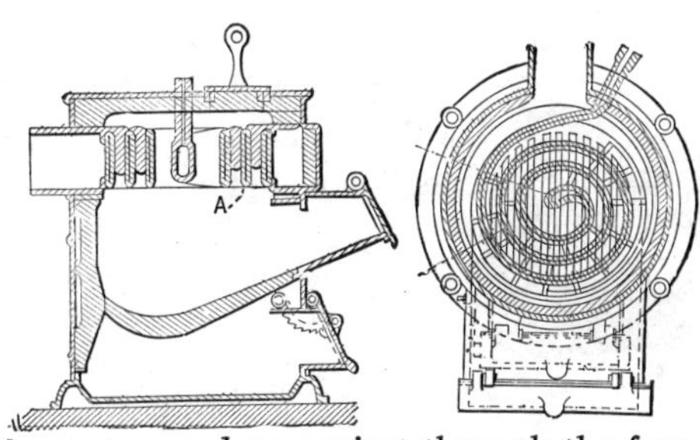

199-200. BOILER OF THE "SERPOLLET" TRICYCLE.—The steam generating surface is made of iron pipe, flattened and corrugated, then coiled into a volute form with the inner end turned up, and the outer end to project through the furnace shell. The cuts show a vertical section and horizontal plan.

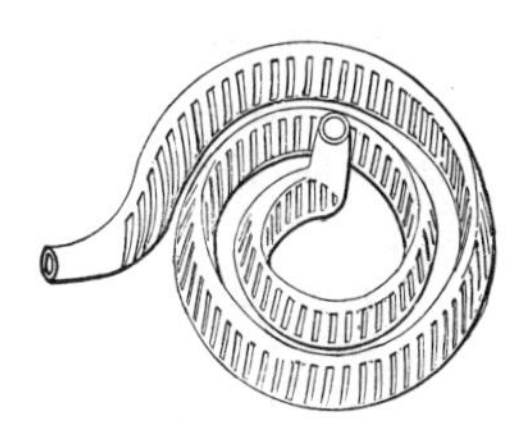

201. "SERPOLLET'S" STEAM GENERATOR, showing the corrugated flattened tube coiled into a volute. The width of the internal space is less than one-eighth of an inch.

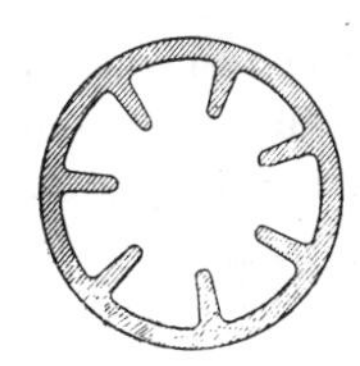

202. "SERVES" BOILER TUBE.—The projecting ribs enlarge the area of the fire surface of the tube.

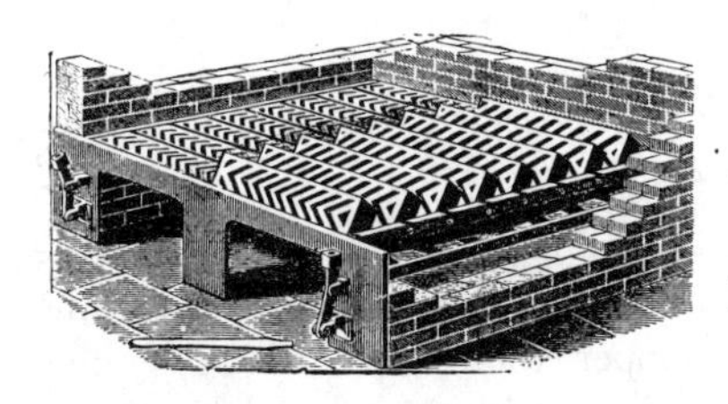

203. SHAKING AND TIPPING FURNACE GRATE, "Tupper" model. Each section rocks on trunnions by a hand lever and connecting bar.

204. SHAKING GRATE for a boiler furnace.— The flanges are strung upon square bars to form each grate section, which are shaken or dumped by a key wrench at the front.

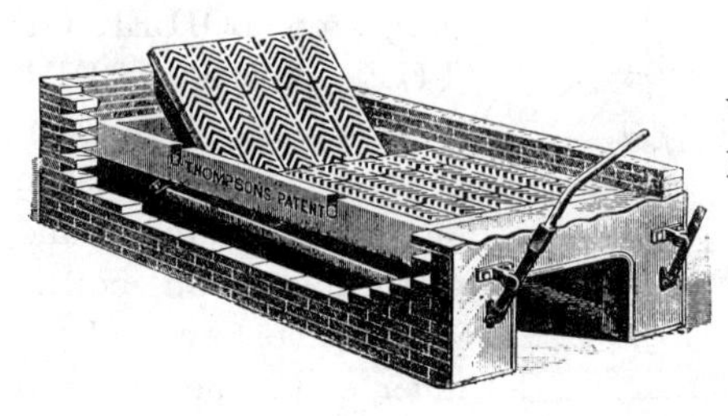

205. FURNACE GRATE, with dumping sections. "Tupper" model grate.

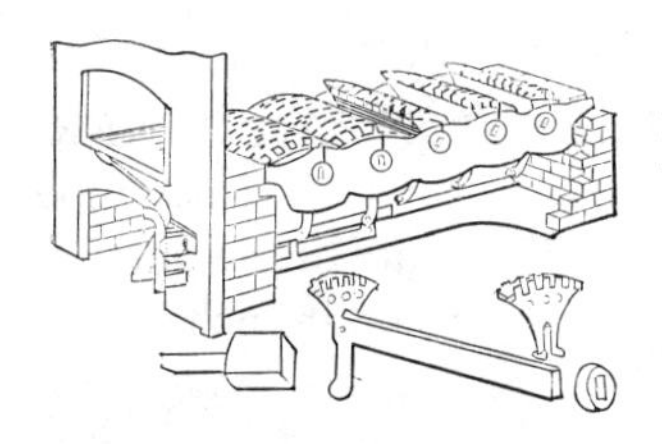

206. SHAKING GRATE for a boiler furnace. The sectors are astride cross bars, and are rocked by a lever and connecting rod to each tier of sectors.

207. SHAKING AND TIPPING FURNACE GRATE.—The front and back sections can be shaken separately by the double connections and levers.

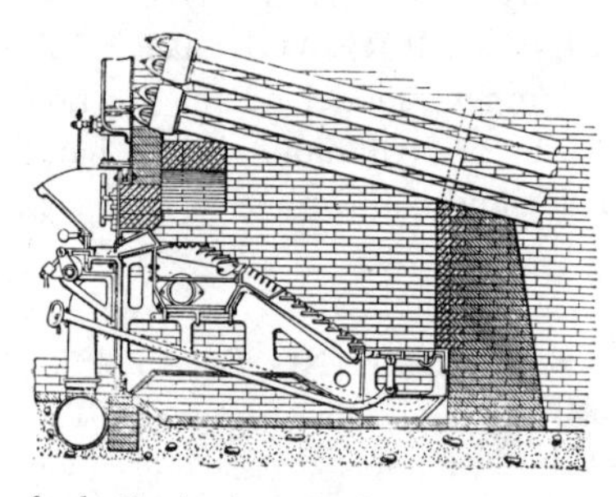

208. "COLUMBIA" STOKER, for soft coal. The coal is filled into the hopper on the outside of the furnace, and from the bottom of the hopper there is carried a chute which inclines upward into the furnace. A pusher pushes the coal upward along this chute and discharges underneath the burning fuel, displacing the latter and causing it to bulge upward and then slide down the inclined grates. Air is delivered under pressure from the air pipe, and, passing through the openings in the blast grates in the front portion of the furnace, mixes with the gases distilled from the coal before they pass through the burning fuel above.

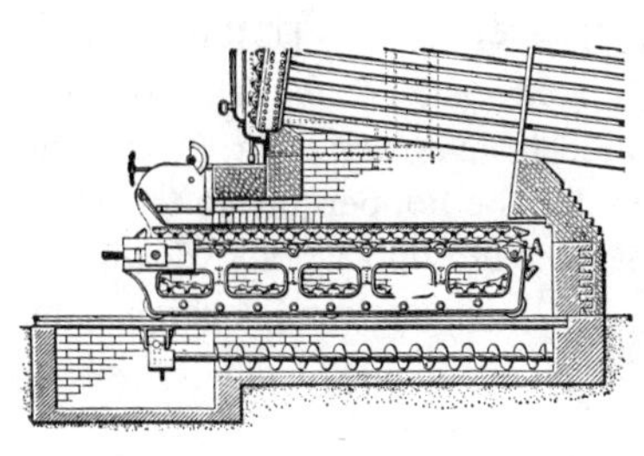

209. "PLAYFORD" MECHANICAL STOKER, for soft coal. A link grate moved by a sprocket shaft carries the coal, fed by a hopper, forward under the boiler, returning over a drum shaft at the bridge wall. A screw conveyer brings the ash and clinker forward to the pit.

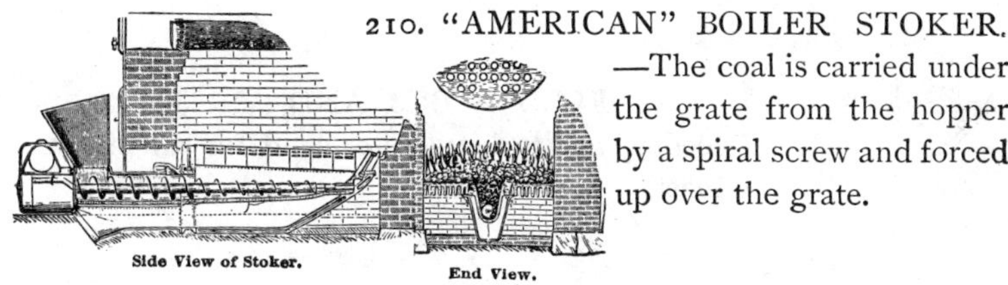

Side View of Stoker. End View.

210. "AMERICAN" BOILER STOKER. —The coal is carried under the grate from the hopper by a spiral screw and forced up over the grate.

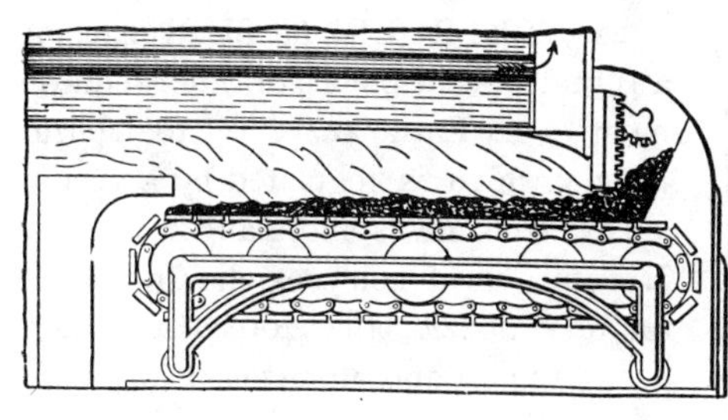

211. MECHANICAL STOKER for a boiler furnace, "Playford" model. The coal is carried into the furnace from a hopper by a travelling grate. A gate with rack and gear, operated by a lever, regulates the depth of the coal-feed.

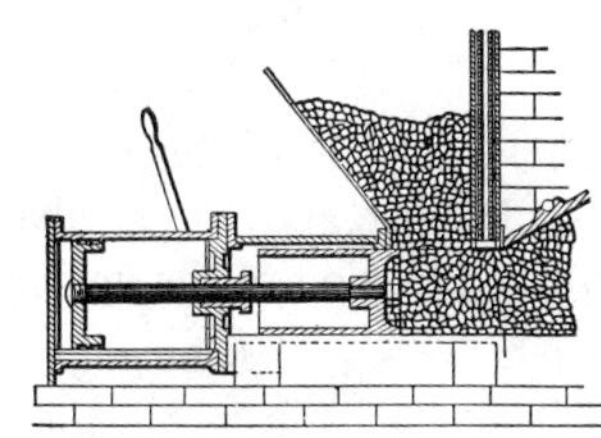

212. MECHANICAL STOKER for a furnace, "Jones" model, underfeed to the grate. A plunger, which may be operated directly by a steam piston, pushes a charge of coal, falling from the hopper, on to the fore plate of the grate, where it is coked, the smoke and gases being drawn into the hot fire and burned.

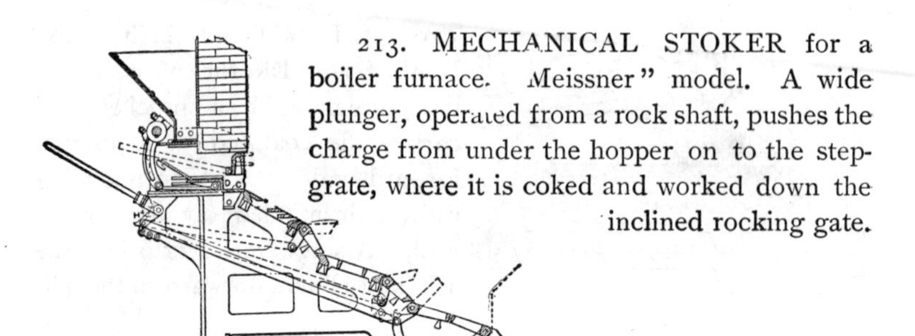

213. MECHANICAL STOKER for a boiler furnace. "Meissner" model. A wide plunger, operated from a rock shaft, pushes the charge from under the hopper on to the step-grate, where it is coked and worked down the inclined rocking gate.

214. FEED WORM AND AIR BLAST, for feeding fuel to furnaces or sand for an air sand blast.

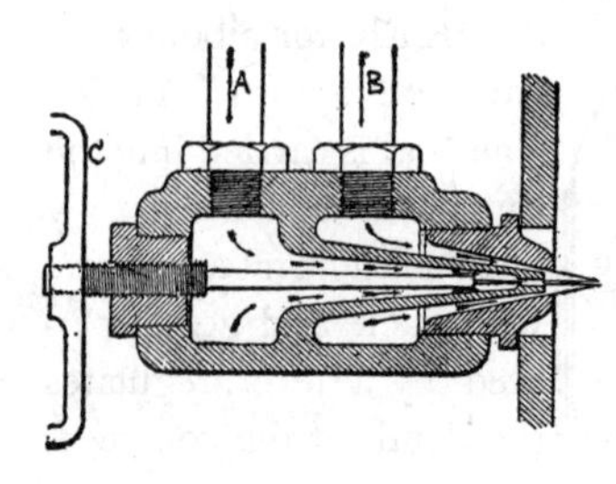

215. PETROLEUM BURNER, for a furnace, for a boiler, or other requirements. A, Entrance of oil to central nozzle, which is regulated by a needle valve with screw spindle and wheel, C; B, entrance of compressed air to the annular nozzle, the force of which draws the oil and atomizes it for quick combustion.

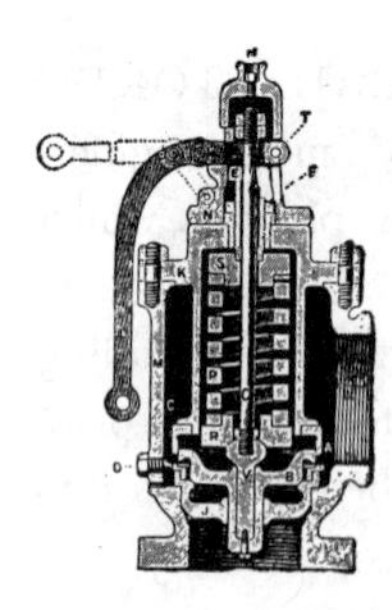

216. POP SAFETY VALVE.—The "Lunkenheimer," an enlarged lip disc above the valve disc, equalizes the increased tension of the spring when the valve opens.

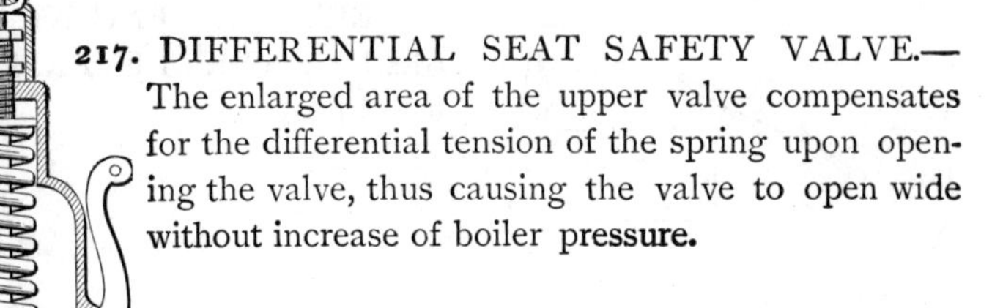

217. DIFFERENTIAL SEAT SAFETY VALVE.—The enlarged area of the upper valve compensates for the differential tension of the spring upon opening the valve, thus causing the valve to open wide without increase of boiler **pressure.**

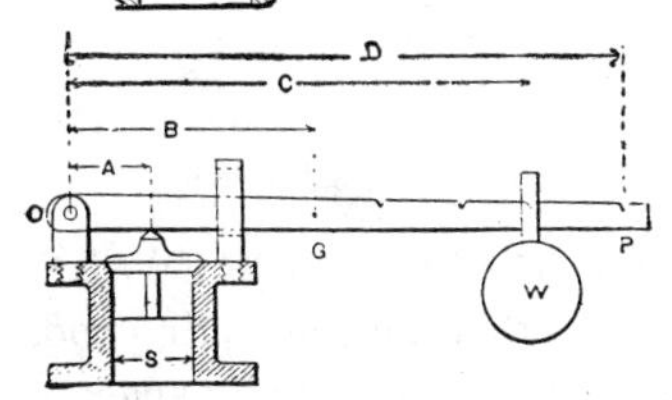

218. SAFETY VALVE.—Lever is of the third order. A, Short lever; B, centre of gravity of lever from fulcrum; C, distance of weight from fulcrum; S, diameter of valve; P, pressure per square inch; G, weight of the lever at its centre of gravity; W, weight of ball; V, weight of valve and spindle.

$$W = \frac{S^2 \times .7854 \times P \times A - (G \times B) - (V \times A)}{C}$$

$$C = \frac{S^2 \times .7854 \times P \times A - (G \times B) - (V \times A)}{W}$$

219. ORIGINAL FORM of the Æolipile or Hero's Steam Engine, 130 B.C. A reaction power, suitable for operation by the use of any gaseous or fluid pressure. The original type of several **modern motors.**

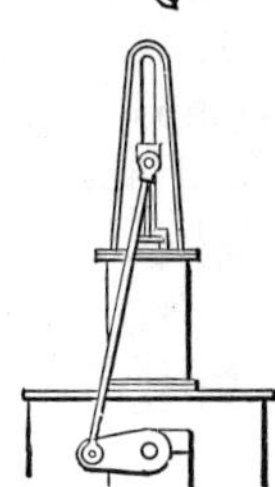

220. STEEPLE ENGINE, with **cross-head and slides.**

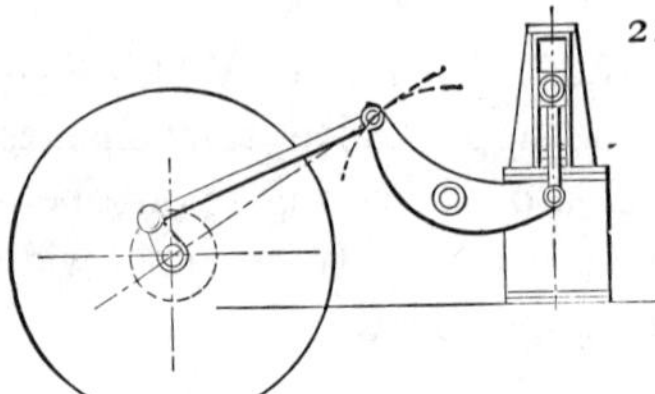

221. VERTICAL ENGINE, WITH BELL-CRANK LEVER, for stern-wheel boat.

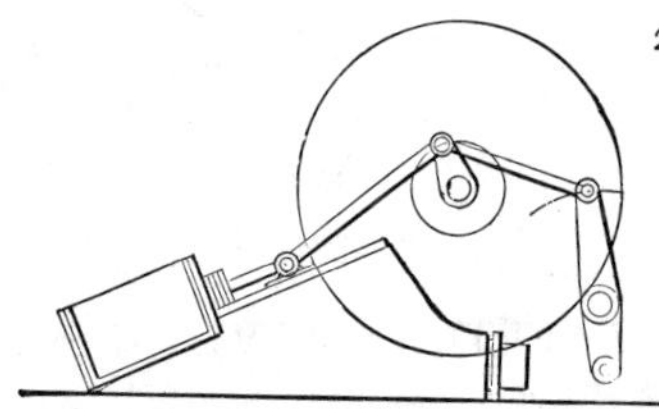

222. INCLINED PADDLE-WHEEL ENGINE, with upright crank-connected beam for driving air pump.

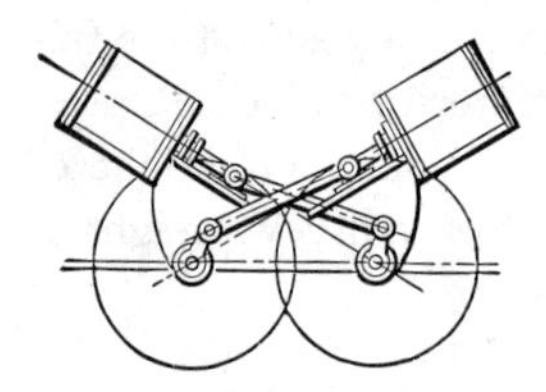

223. DIAGONAL TWIN-SCREW ENGINE, arranged so that the connecting rods cross each other, thus economizing space.

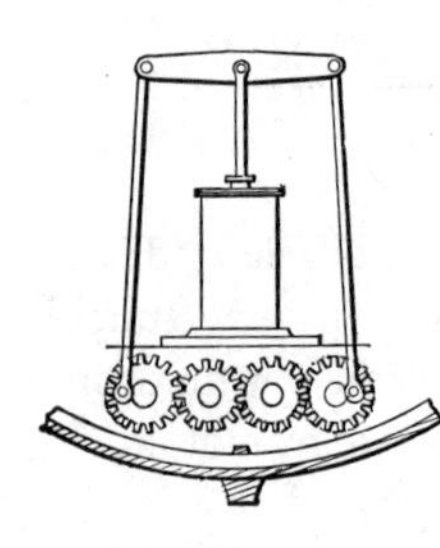

224. TWIN-SCREW VERTICAL CYLINDER ENGINE.—The outer gears are on the screw shafts; the inner gears are idlers to keep the beam even.

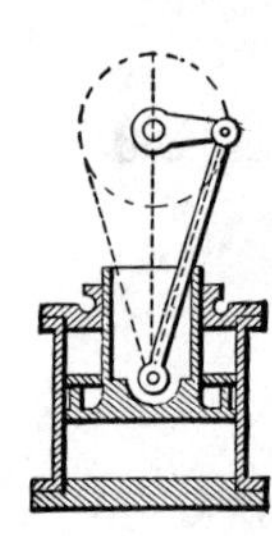

225. TRUNK ENGINE.—Does away with the slides and cross-head. It is also used for compounding by using the initial pressure at the trunk end and expanding beneath the piston.

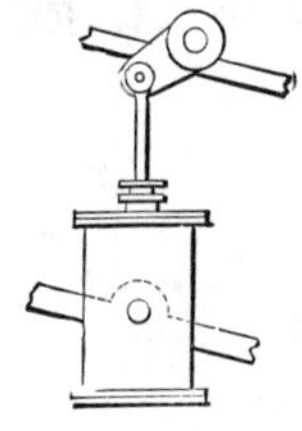

226. OSCILLATING ENGINE, with trunnions on middle of cylinder.

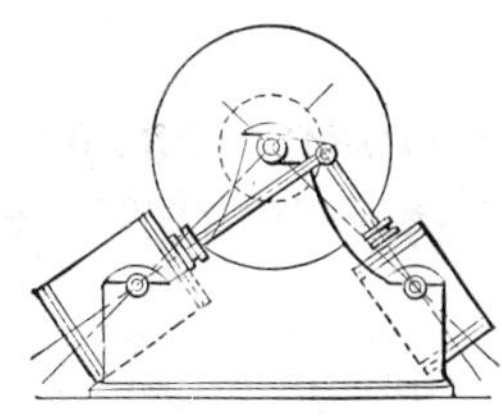

227. COMPOUND OSCILLATING ENGINE.—Cylinders at right angles.

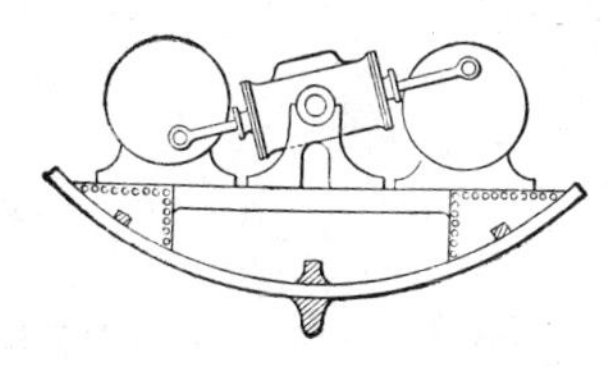

228. TWIN-SCREW OSCILLATING ENGINE.—A through piston rod connects directly to crank-pins on the shaft face plates. Suitable for small boats.

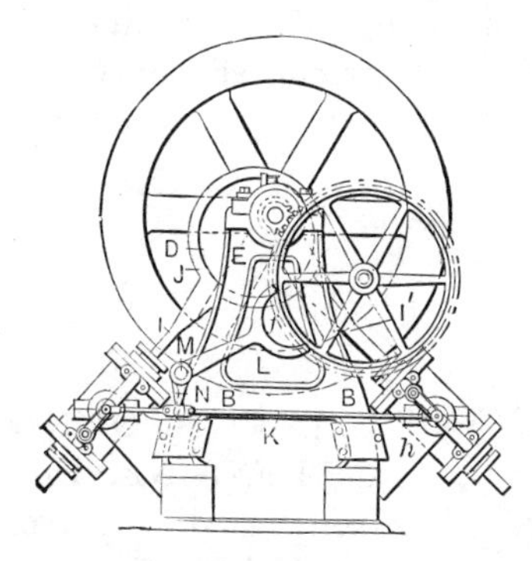

229. OSCILLATING HOISTING ENGINE.—The piston rods are attached to an eccentric strap; one fixed, the other pivoted. A lever operated by the same eccentric strap, through a short connecting rod, operates the valve gear of each cylinder alternately.

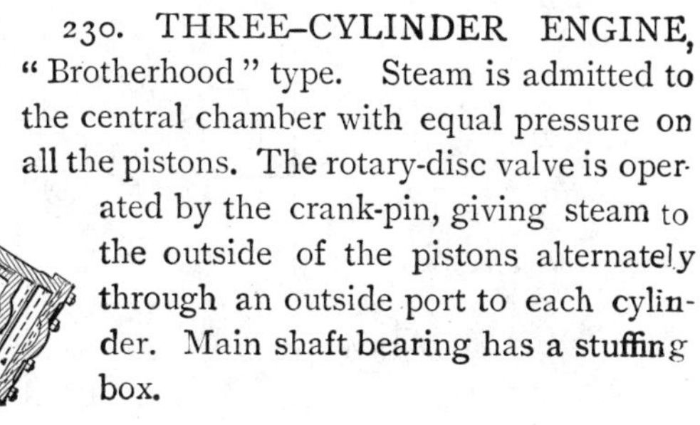

230. THREE-CYLINDER ENGINE, "Brotherhood" type. Steam is admitted to the central chamber with equal pressure on all the pistons. The rotary-disc valve is operated by the crank-pin, giving steam to the outside of the pistons alternately through an outside port to each cylinder. Main shaft bearing has a stuffing box.

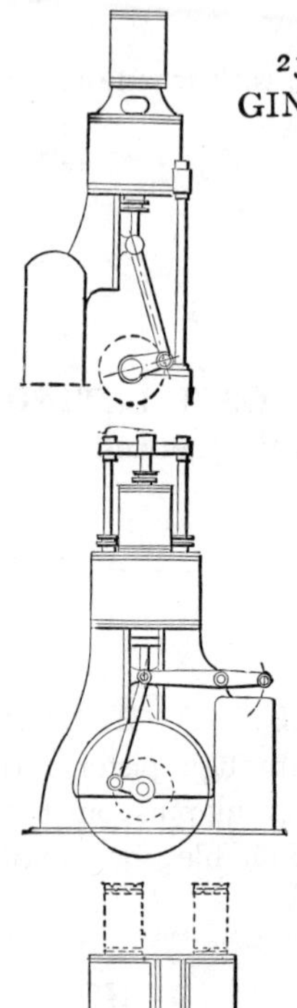

231. TANDEM COMPOUND VERTICAL ENGINE, with continuous piston rod.

232. TANDEM COMPOUND VERTICAL ENGINE, with cross-head and two piston rods for low-pressure piston

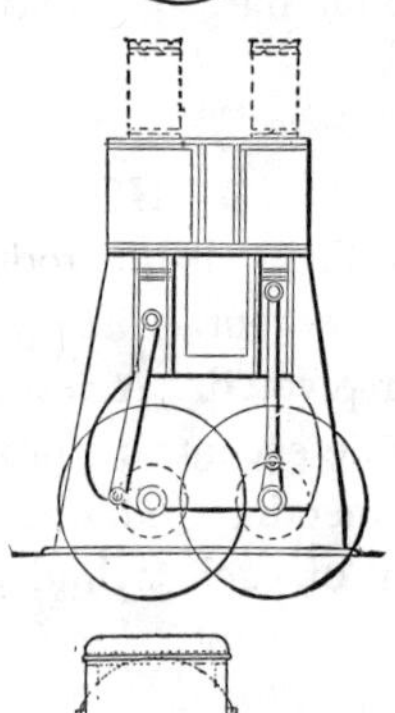

233. COMPOUND ENGINES for twin screws. There may be one or two pair of compound cylinders. The dotted lines represent cylinders of the tandem model.

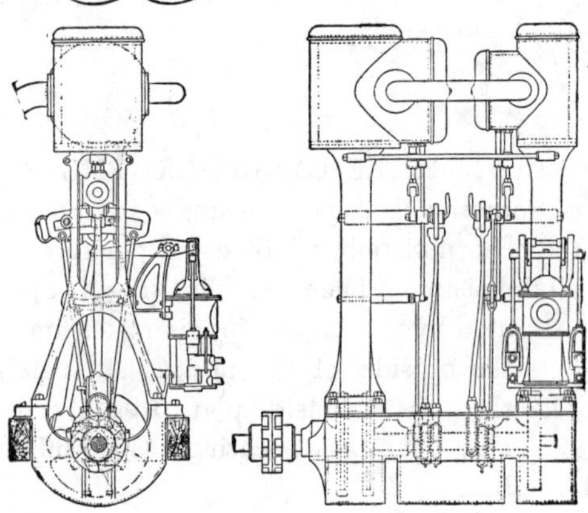

234. COMPOUND YACHT ENGINE, "Herreshoff" model. Direct receiver pipe. End and longitudinal elevation.

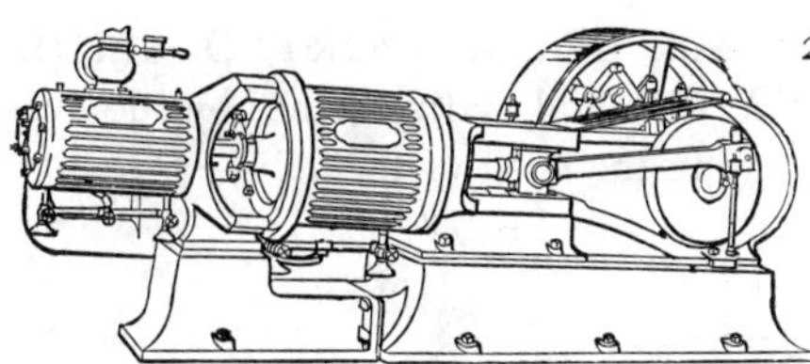

235. HIGH-SPEED TANDEM COMPOUND ENGINE, "Harrisburg" model.

236. TANDEM COMPOUND ENGINE, "Phoenix Iron Works" model. A direct pipe connection between the high and low pressure cylinder.

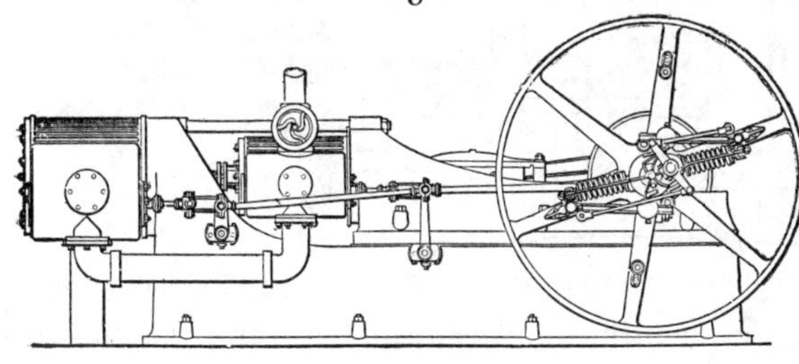

234. MODERN HIGH-SPEED ENGINE, with pulley governor, "Atlas" model.

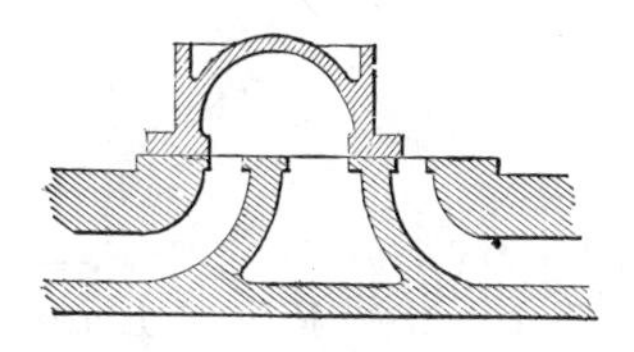

238. SINGLE D SLIDE VALVE, with lap. The length of the valve over the length from outside to outside of steam ports is double the lap.

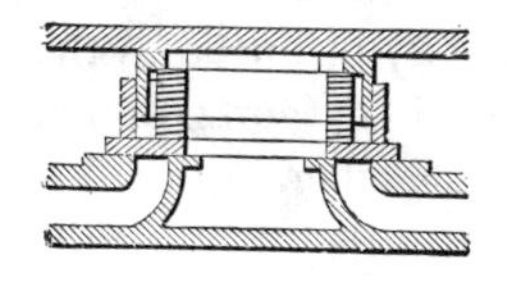

239. BALANCED SLIDE VALVE.—A ring in a recess of the valve rides against the steam chest cover, held by a spring.

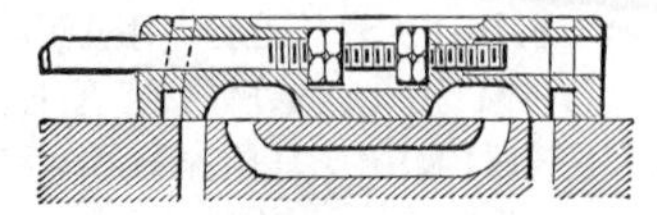

240. DOUBLE-PORTED SLIDE VALVE and adjustment by double nuts in the back of the valve.

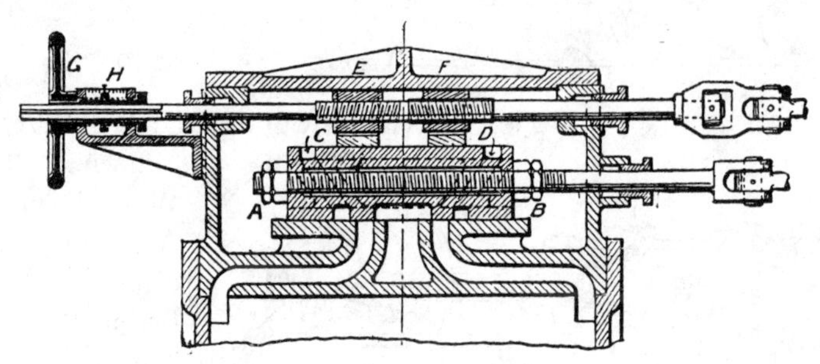

241. "MEYER" CUT-OFF VALVE.—C, D, Slide valve with perforated ports. The supplementary or cut-off valves are adjusted to the required distances, to meet the required cut-off, by a right and left screw, which has an index H, and wheel G, for turning the screw for cut-off adjustment on the outside of the steam chest.

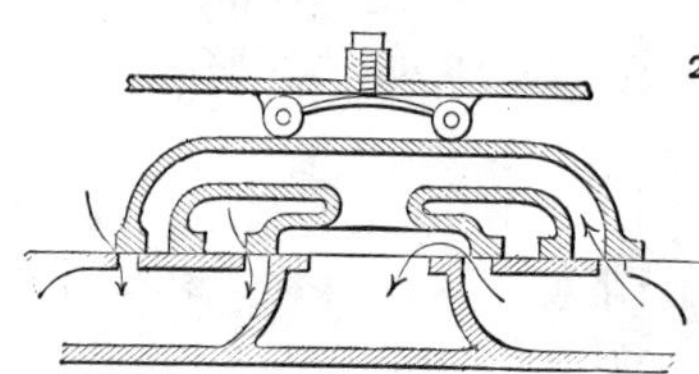

242. SINGLE D, SLIDE VALVE, with double steam and exhaust ports. Central steam ports open into steam chest at the side of the valve.

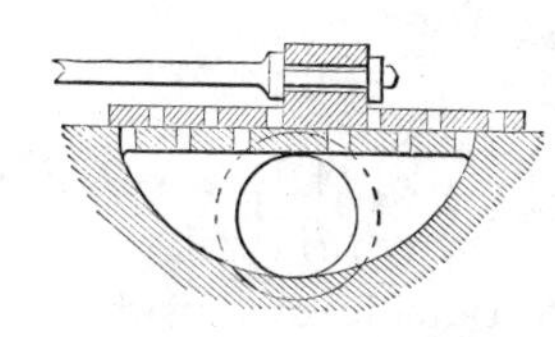

243. GRIDIRON SLIDE VALVE, for large port area with small motion of the valve.

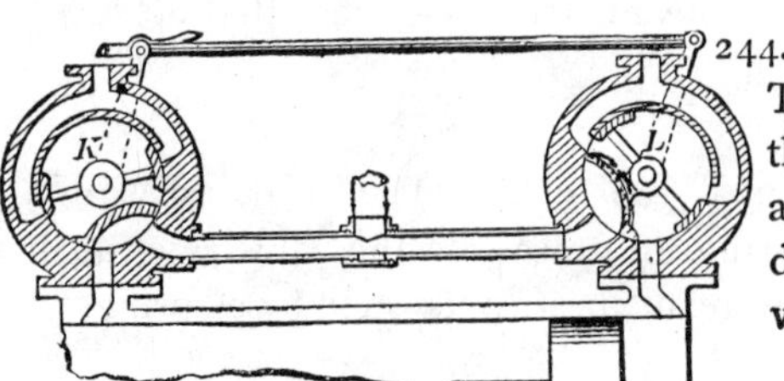

244. ROTARY VALVES.—The valves K and L are three-winged cylinders, and are nearly balanced by the double inlet ports of the valve chamber.

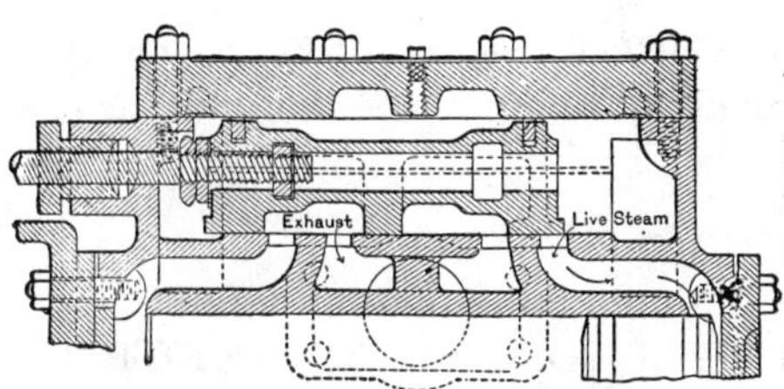

245. STEAM ENGINE VALVE CHEST.—Double ported exhaust; shortens the steam passages. "Erie City Iron Works" model.

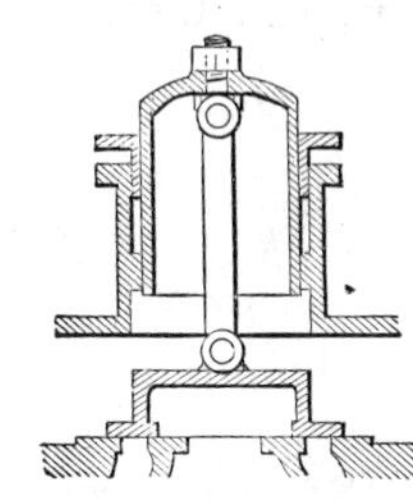

246. BALANCED SLIDE VALVE.—A bell-shaped piston, riding in a packed gland in the steam chest cover, is connected to the top of the valve by a link.

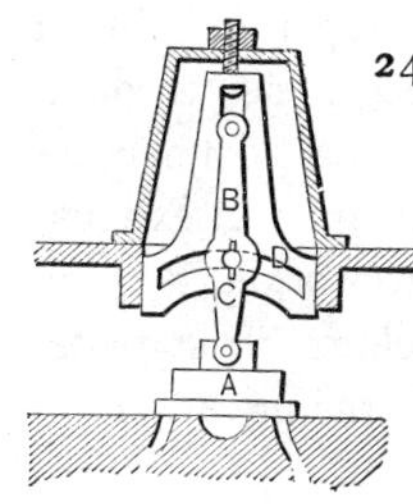

247. BALANCED SLIDE VALVE, "Buchanan & Richter's" patent. The arm B carries a roller in the curved slot of the supporting piece D. The pressure is relieved by the nut and screw in the cover.

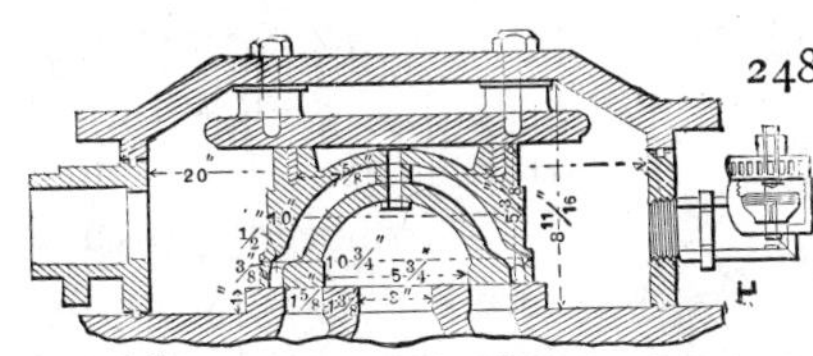

248. "RICHARDSON-ALLEN" BALANCED SLIDE VALVE.—The valve slides under an adjustable plate fixed to the steam chest cover, and is balanced by a recess in the back of the valve that is open to the exhaust port.

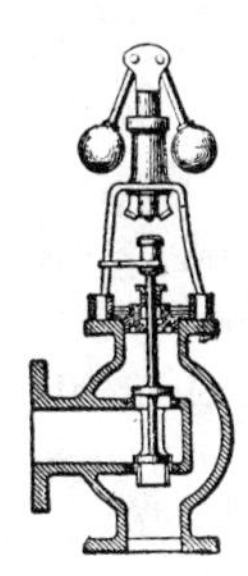
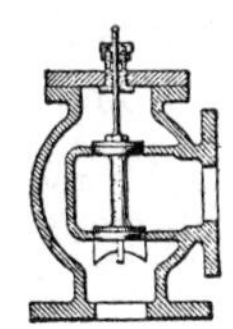

249. BALANCED THROTTLE VALVE, with direct governor connection.

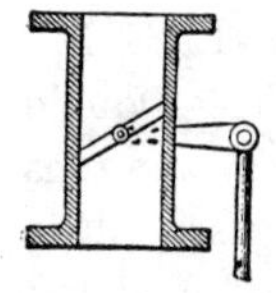

250. WING THROTTLE VALVE, or Butterfly Throttle, operated by direct connection with a governor.

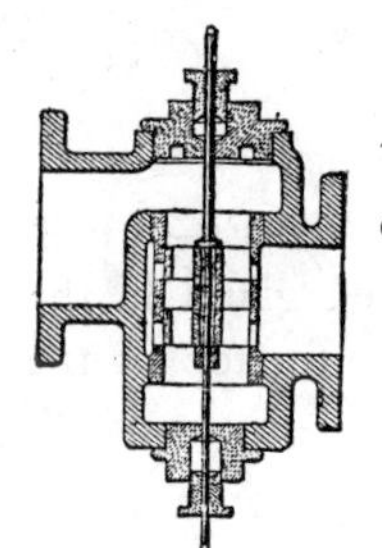

251. MULTIPLE PORT PISTON THROTTLE VALVE.—A perfectly balanced valve with through connecting rod.

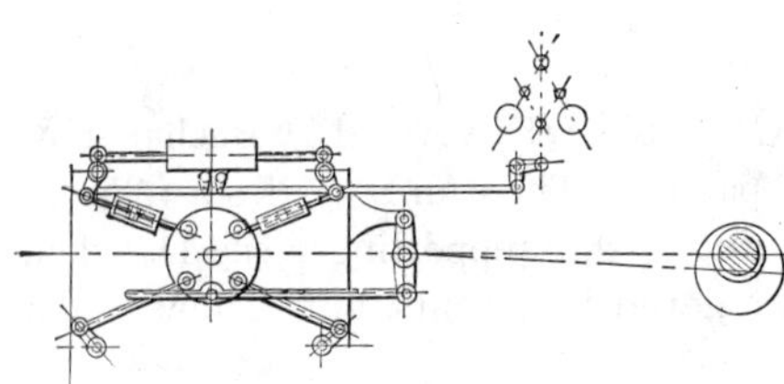

252. "CORLISS" VALVE GEAR.—Operated by a single eccentric through a lever and connecting rods. Steam and exhaust valves are worked by pins on a rocking wrist plate. The trips on the steam-valve gears are controlled by the governor.

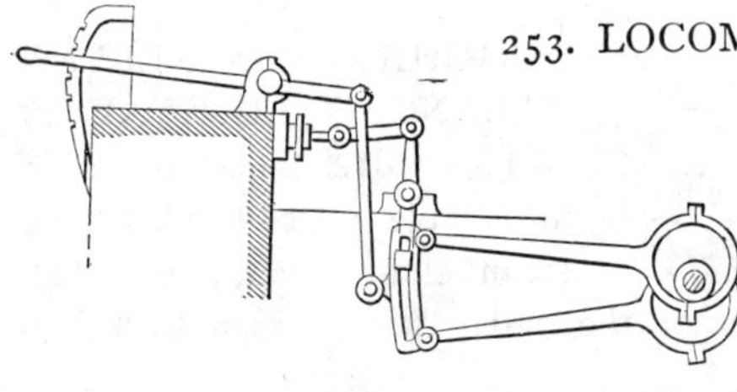

253. LOCOMOTIVE LINK-MOTION VALVE GEAR.—In this arrangement the slotted link is moved up and down over the wrist pin block by the lever and connecting rod; the lever, locking in the toothed sector, allowing for a close connection to the valve stem by a lever and short connecting rod.

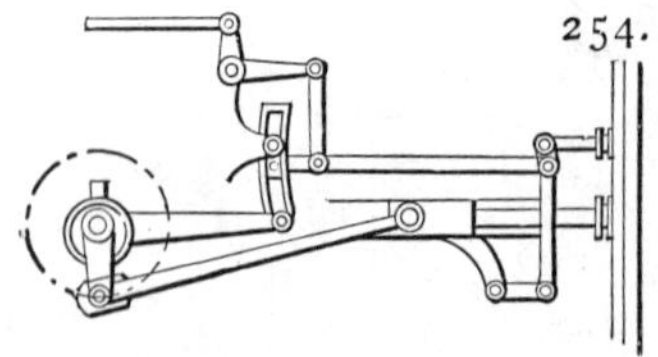

254. WALSCHAERT'S VALVE GEAR.—The slotted link is hung at its centre on a fixed pin. The valve-rod block is raised or lowered by the bell-crank lever. Lead is made by the cross-head link and lever.

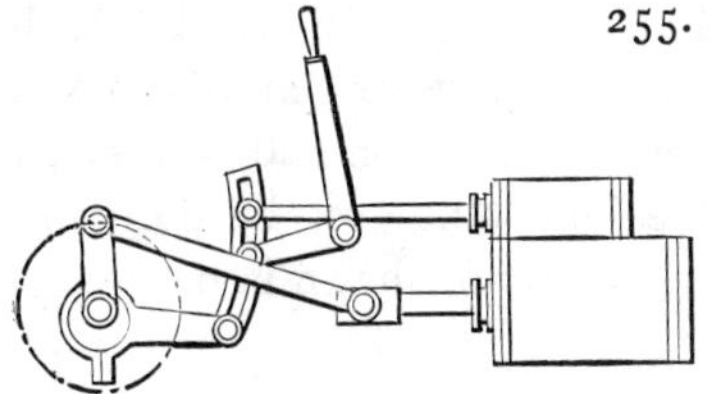

255. REVERSING LINK MOTION.—The slotted link is pivoted to the end of the eccentric rod and is moved up and down by the bell-crank lever. The block carrying the valve rod is stationary in the slot.

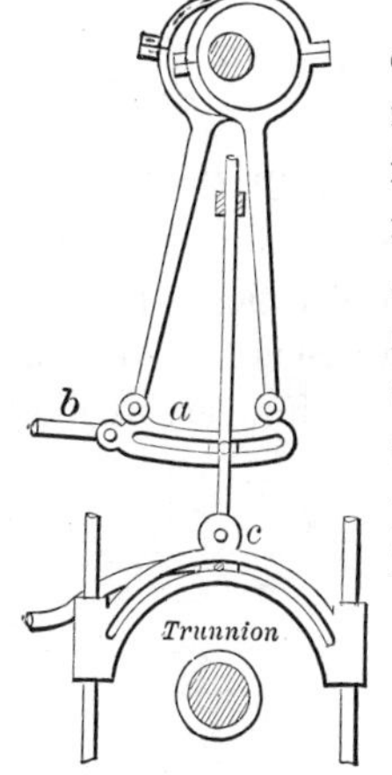

256. VALVE GEAR of an oscillating marine engine. The slotted link *a*, receives a rocking motion from the eccentrics and rods, and is thrown from its centre either way for forward or back motion of the engine by the lever connecting rod *b*. A block and pin attached to the valve rod freely traverse the link slot. The circular slotted frame *c* is concentric with the cylinder trunnions and the valve rod by a sliding block and pin to accommodate the oscillation of the cylinders.

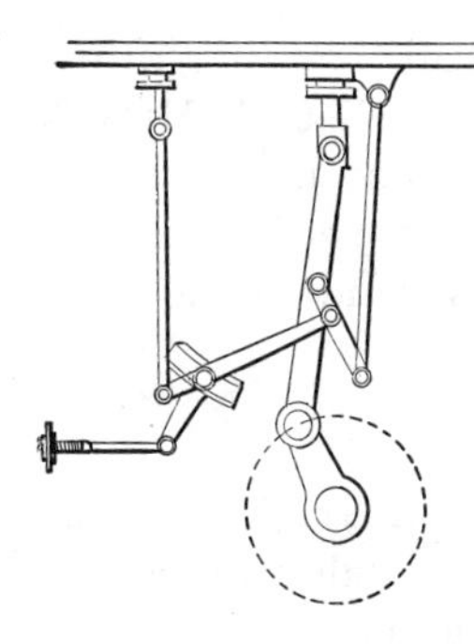

257. "JOY'S" VALVE GEAR for a vertical engine. Operated from a pin in the connecting rod. Reversal is made by changing the position of the slotted link

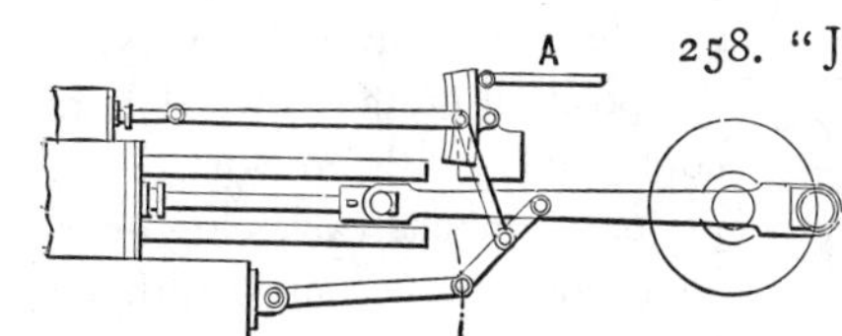

258. "JOY'S" VALVE GEAR for a horizontal engine. Adjustment is made by the angular position of the slotted link. Valve motion by crank rod and links.

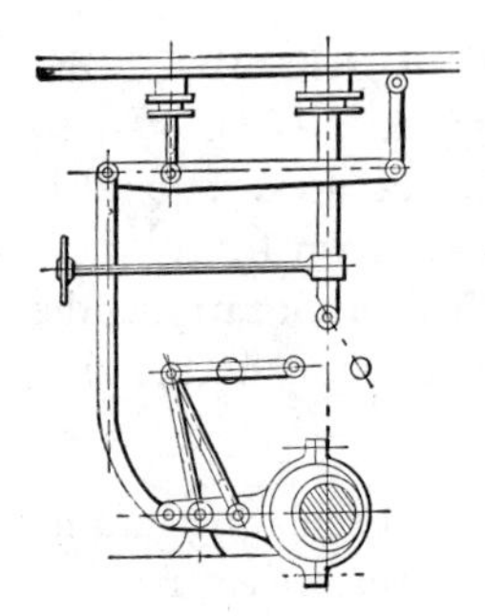

259. "BREMME" VALVE GEAR with single eccentric. The eccentric arm is rocked by the double link connection and is reversed by throwing the link joint over by the hand screw and sector arm, not shown in cut.

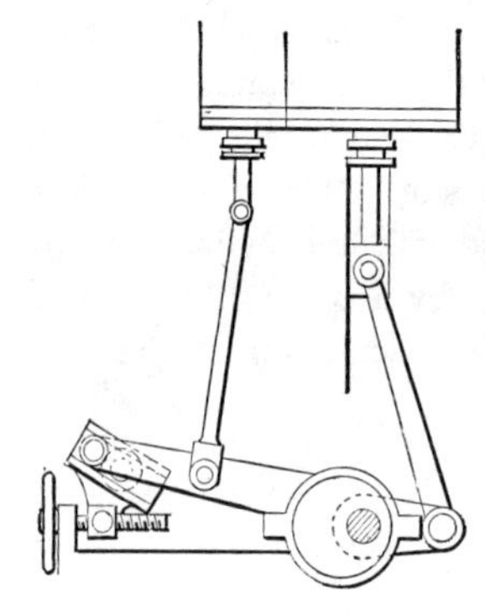

260. SINGLE ECCENTRIC VALVE GEAR, with variable travel, adjustable by a hand-wheel. The eccentric drives a block in a slotted link, which is rocked on a central pivot by the screw for varying the throw of the valve.

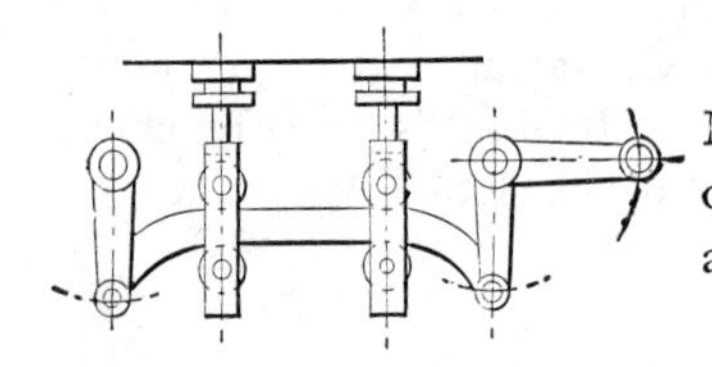

261. CAM-BAR VALVE MOVEMENT.— The horizontal movement of the cam bar by the bell-crank lever alternately moves the two valves.

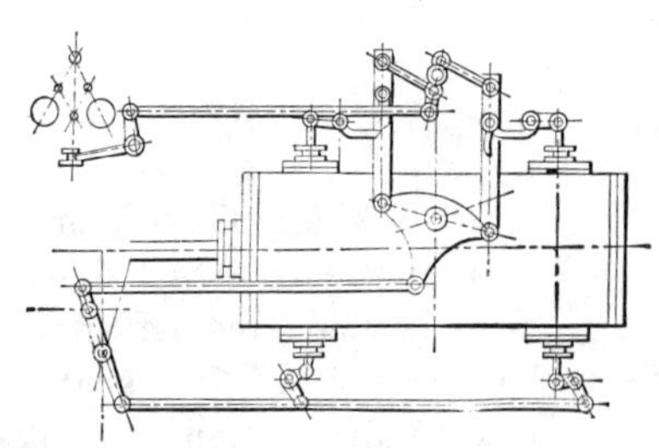

262. VALVE GEAR of a Cornish engine, with trip poppet valves for steam. The governor releases the valves by varying the position of the vertical bars connected to the rocking wrist plate. Exhaust valves are operated from the eccentric through the lever that operates the steam valves.

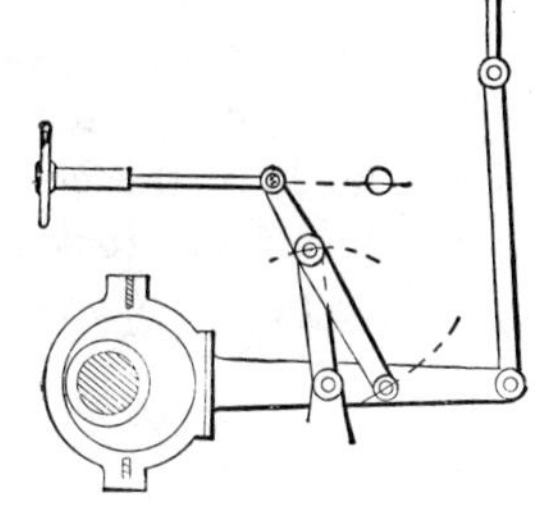

263. VARIABLE EXPANSION GEAR, with one eccentric. The movement of the fulcrum of the eccentric bar lever by the screw changes the throw of the valve.

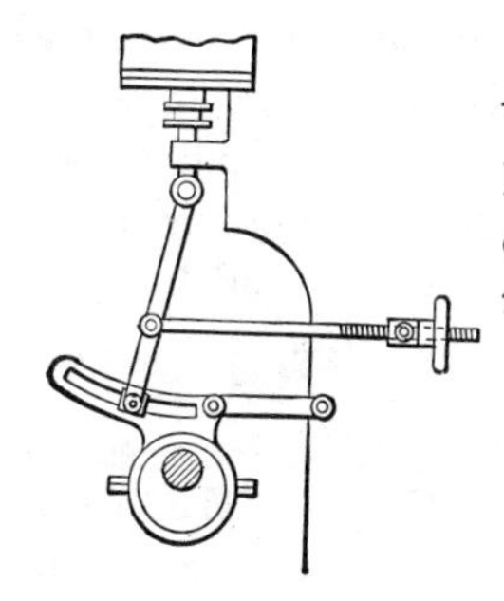

264. SINGLE ECCENTRIC VARIABLE VALVE THROW.—"Fink" link gear for a D valve. The link block is moved in the curved slot of the link for variation of valve throw, adjustable by the hand-wheel.

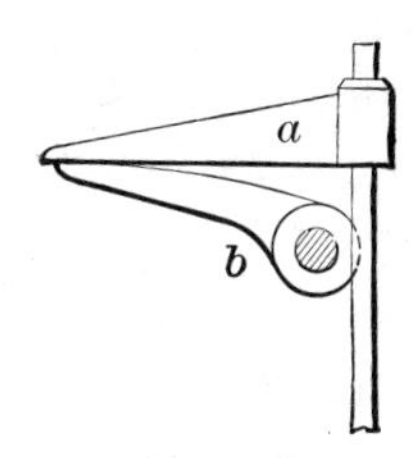

265. "ALLEN" VALVE LIFT OR TOE.—*a*, The valve lifter and rod to which the valves are attached; *b*, the toe on the rock shaft, operated from a cam on the engine shaft.

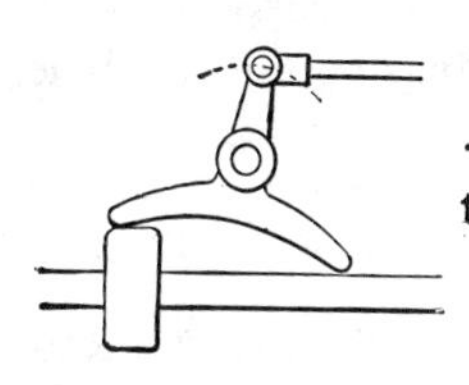

266. TAPPET LEVER VALVE MOTION. —Used on pumps, rock drills, and percussion tools.

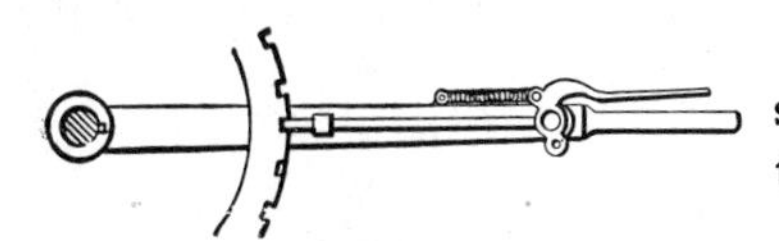

267. STARTING LEVER, with spring to hold the bolt in the sector notches.

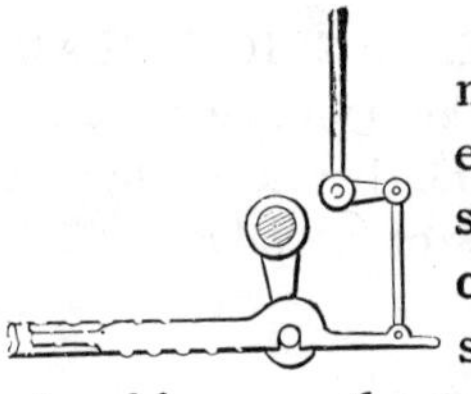

268. SIMPLE UNHOOKING DEVICE, much in use on the engines of side-wheel steamers. The turning down of the handle of the short bell-crank lever lifts the hook in the eccentric rod off from the wrist pin of the rock-shaft crank,—when the engine can be worked by a hand lever on the rock shaft.

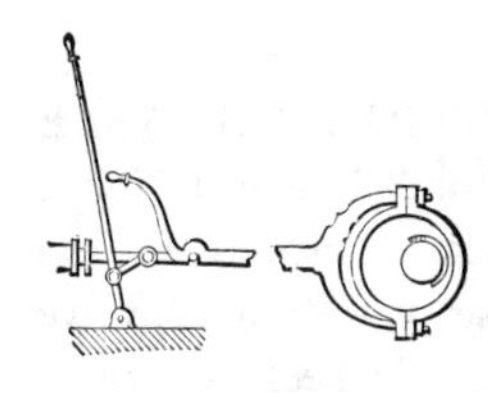

269. SIMPLE REVERSING GEAR for steam engines. On raising the eccentric rod the valve spindle is released from the hook, when the engine can be reversed by the hand lever; the eccentric then runs back by friction a half turn, it being loose on the shaft, and the key shoulder cut away to allow the eccentric to turn half over.

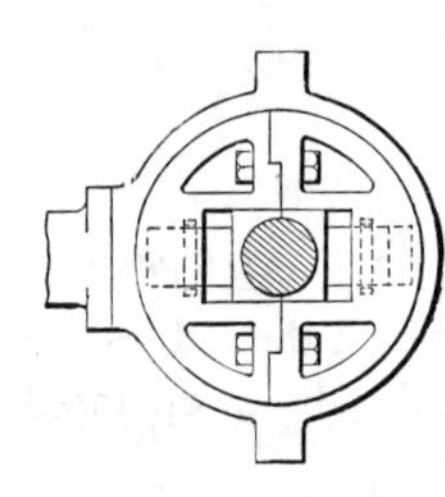

270. "JOY'S" HYDRAULIC SHIFTING ECCENTRIC.—The centre block is keyed to the shaft; pistons on each side of the block work in cylinders in the eccentric. Oil is pumped to one or the other piston through holes in the crank shaft and piston, for reversal of the engine.

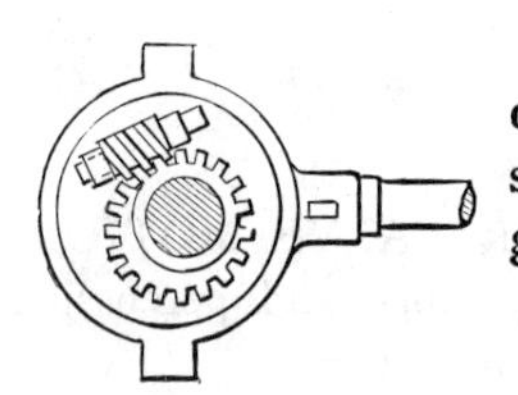

271. SHIFTING ECCENTRIC.—The eccentric is movable on worm gear and its sleeve, which is keyed to the shaft. The tangent worm is pivoted in lugs on the eccentric.

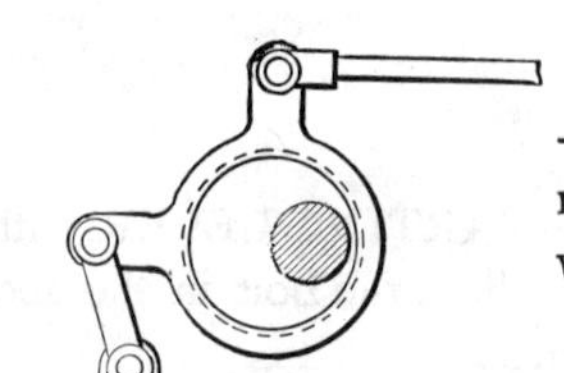

272. VALVE MOTION ECCENTRIC.—The rocker connecting link increases the motion of the valve rod and travel of the valve.

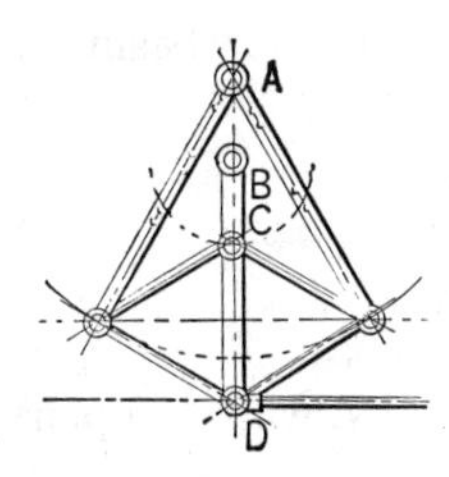

273. " PEAUCELLIER'S " PARALLEL MOTION.—A, B and B, C are of equal distances, when the connecting rod will move in a straight line. When B is connected with the outer joint of the link quadrangle the inner joint C will have a straight-line motion.

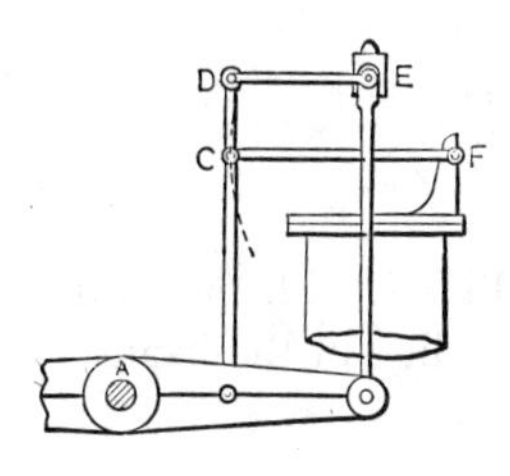

274. PARALLEL MOTION, used on side-lever marine engines.

E, cross-head.
C, F, radius bar.
D, E, parallel bar.

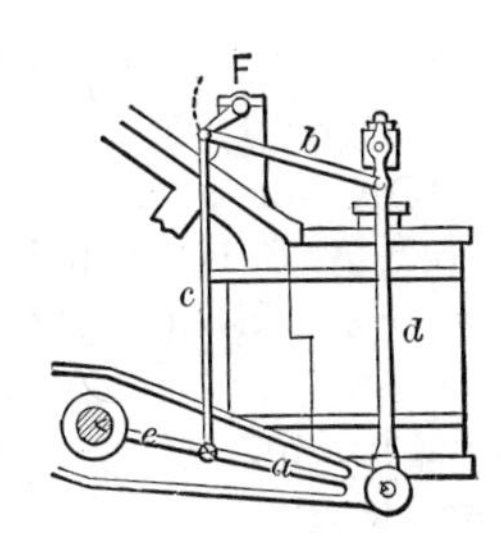

275. PARALLEL MOTION, for a side lever marine engine.

a and *b* are of equal length.
c and *d* are of equal length.

Radius of rocker-shaft crank $F = \frac{b}{e}$

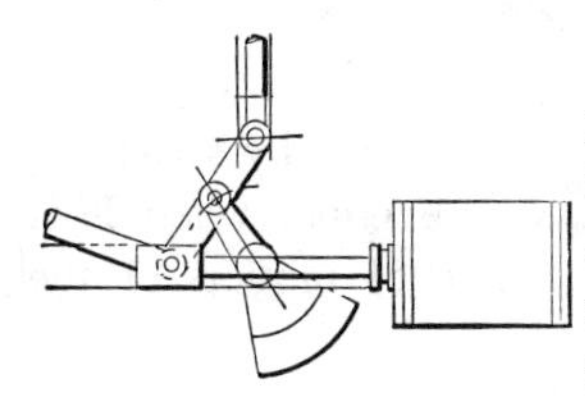

276. PARALLEL MOTION and compensation weight for steam engines, " Forney's " patent. The link from the cross-head traverses the slot at right angles to the engine centre, and is pivoted at its centre to the swinging link and weight.

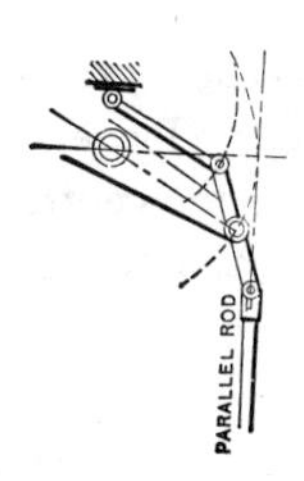

277. PARALLEL MOTION.— Length of radius bar equal to beam radius. Link radii are equal. Distance of radius bar pivot above beam centre is equal to link radius.

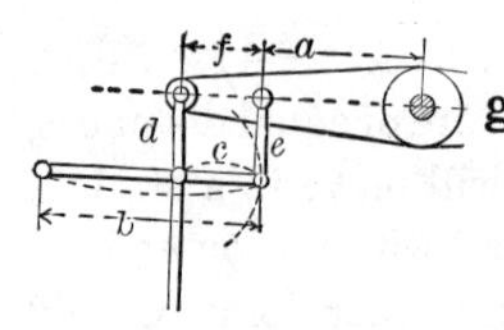

278. PARALLEL MOTION for beam engines, in which
a and *b* are of equal length.
c and *f* are of equal length.
d and *e* are of equal length.

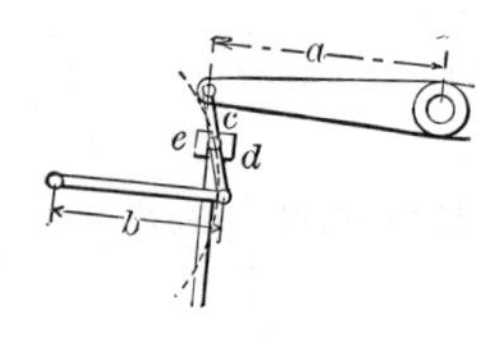

279. PARALLEL MOTION, with two pairs of connecting bars.
a and *b* are of equal length.
c and *d* are of equal length.
e, cross-head.

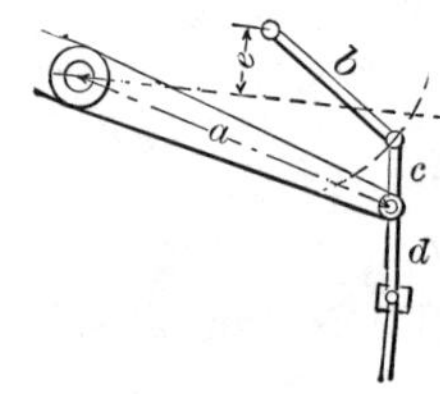

280. PARALLEL MOTION, with the radius bar pivoted above the centre line of the beam.
c and *d* are of equal length.
$e = c$ or d.
$b =$ half a.

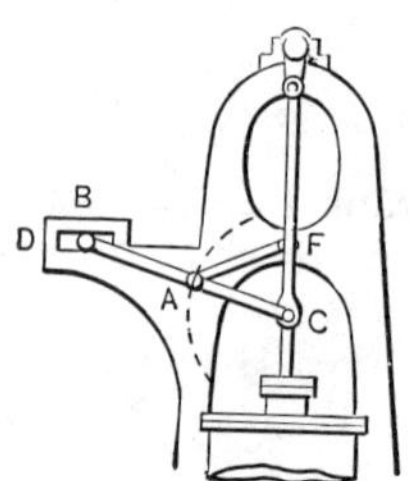

281. PARALLEL MOTION for a direct-acting engine. The radius bar, A, F, is pivoted to the frame on the centre line and at right angles to the slot, B.
A, C and A, F are of equal length.
A, B and A, C are of equal length.

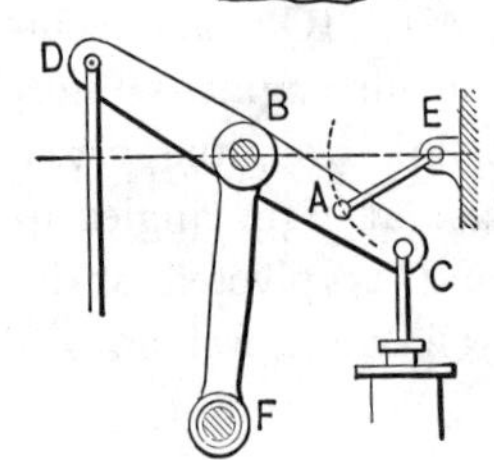

282. PARALLEL MOTION by a rocking beam. A, E and A, C are equal when E is pivoted in the centre line of motion of the piston rod.

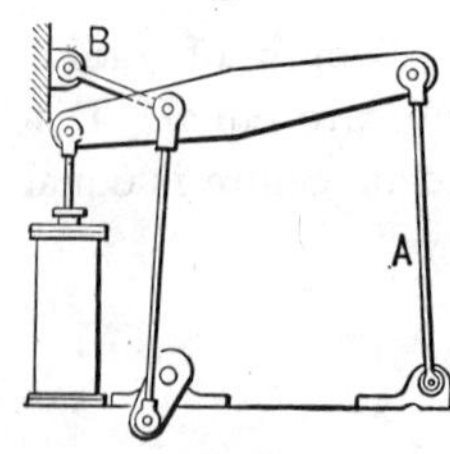

283. PARALLEL MOTION.—The "grasshopper" movement of one of the early locomotives. B, the radius bar, pivoted in the centre line of motion of the piston rod; A, the rocker rod.

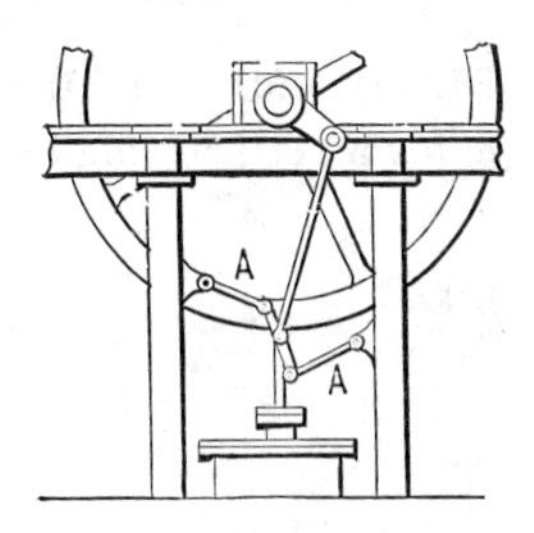

284. PARALLEL MOTION for a vertical engine. A, A, radius bars pivoted to engine frame opposite to the middle of stroke.

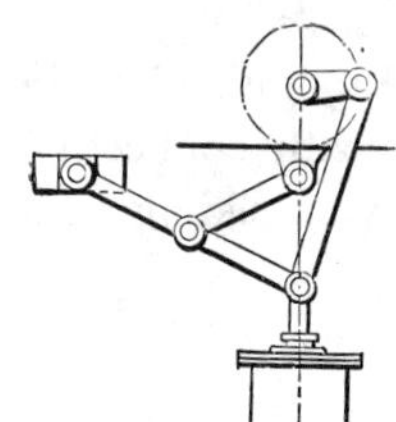

285. PARALLEL MOTION for an engine. The radius bars are of equal lengths from the centre line of the engine and sliding pivot of the long bar. Both fixed and sliding pivots at right angles with the centre line when at half stroke.

286. PARALLEL MOTION of a piston rod by direct connection with a spur gear rotating upon the wrist pin of the crank. The crank-pin gear meshes in a fixed internal toothed gear of double its diameter. One of the curiosities of old-time engineering.

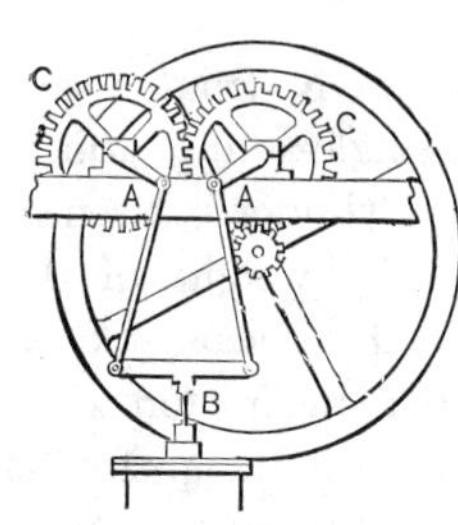

287. "CARTWRIGHT'S" PARALLEL MOTION for steam engines by geared wheels. A free cross-head on piston rod and connected to two cranks on shafts with equal spur gears from which power is transmitted through a third spur wheel. *Very old (1787).*

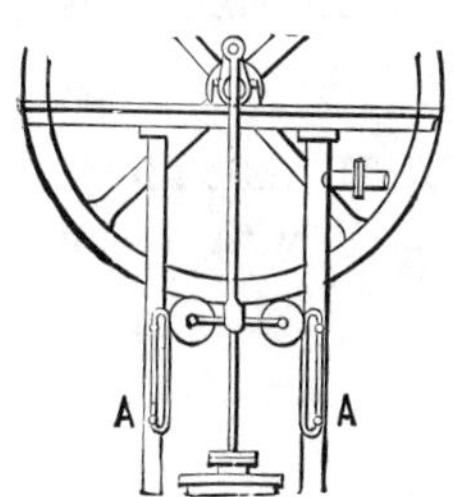

288. PARALLEL MOTION by a cross-head and rollers running against guide-bars. *Old.*

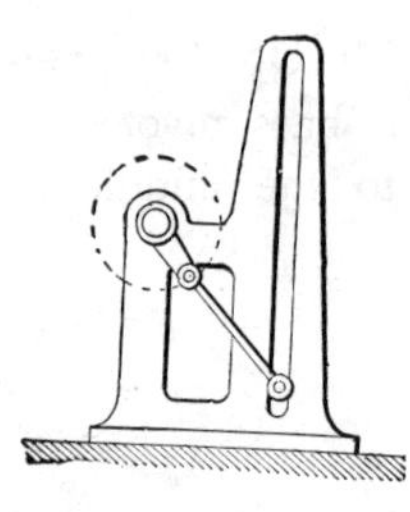

289. CROSS-HEAD SLIDE athwart the shaft. An obsolete design for a vertical engine in a side-wheel steamer.

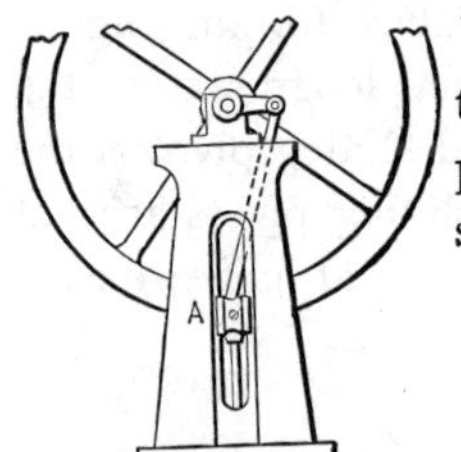

290. PARALLEL MOTION by guide bars in the frame of a vertical engine, with connecting piston rod and crank. Cross-head sliding in a slot in the frame. *Old.*

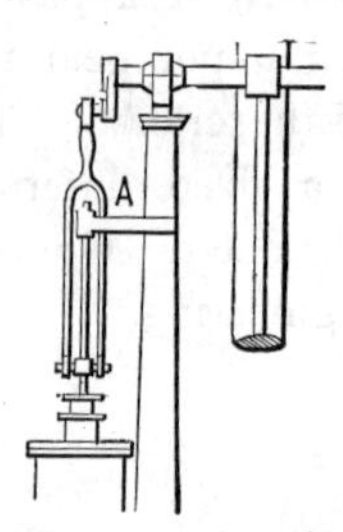

291. PARALLEL MOTION to piston rod and cross-head by prolonging the piston rod through a fixed guide and connecting to the crank with a forked rod. A very old device and much in use now on pumps.

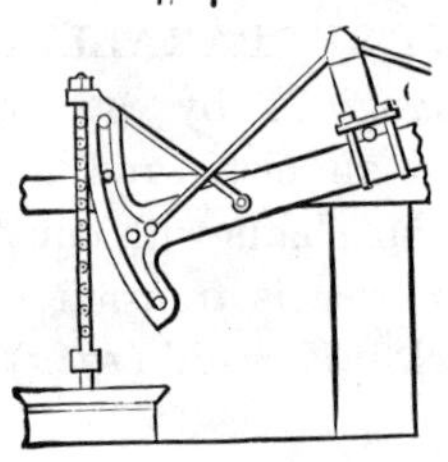

292. PARALLEL MOTION from a sector beam. Used on old, single-acting, atmospheric pumping engines. Cylinder is open at top. Piston is lifted by the weight of the pump rods on the other end of beam. Low-pressure steam follows the rising piston when a jet of water condenses it, and the piston is drawn down by atmospheric pressure.

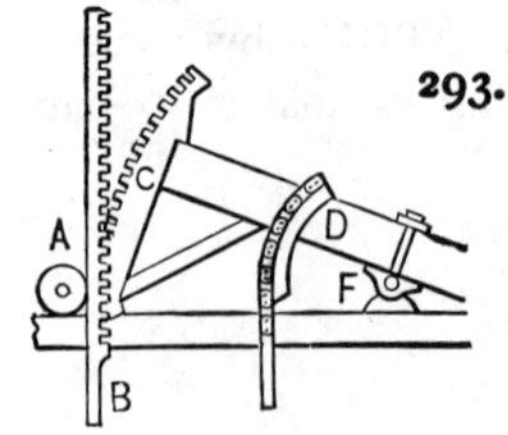

293. RACK GEAR PARALLEL MOTION.—An old pumping device used with a single-acting beam engine.

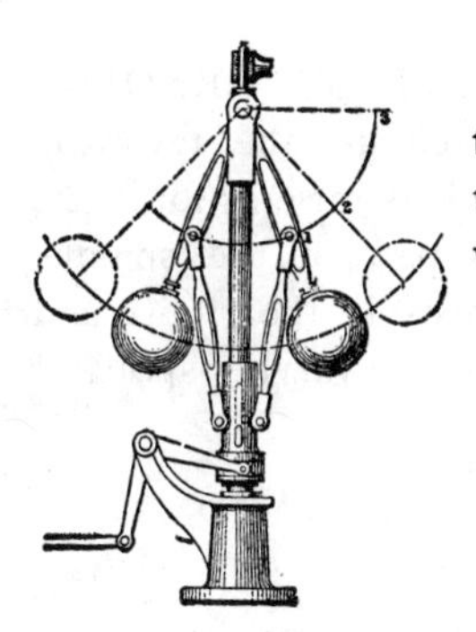

294. "WATT" GOVERNOR.—The centrifugal action of the balls lifts the sleeve and, through the bell crank, operates the throttle valve.

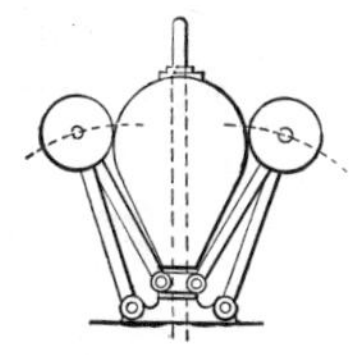

295. COMPENSATING GOVERNOR, "Dawson" patent (English). Intended to be isochronous in its movement. The central weight is connected directly with the throttle-valve stem.

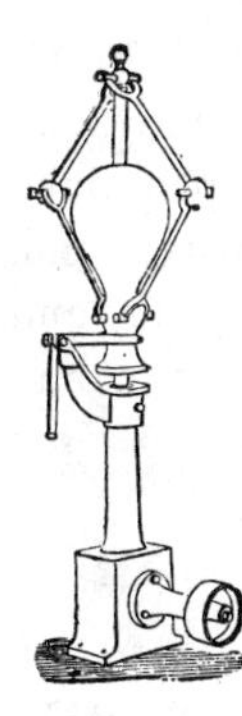

296. GRAVITY CENTRIFUGAL GOVERNOR.—The weight on the central rod is lifted by the centrifugal action of the light balls and moves the lever that controls the valve gear. A high-speed governor.

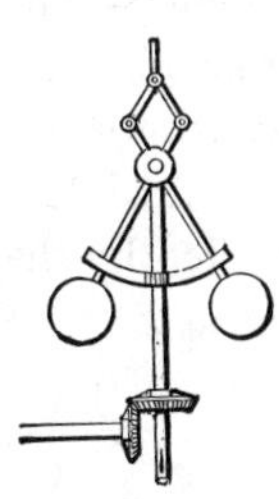

297. ENGINE GOVERNOR, in which the arms cross each other and are extended above in a link movement. The arms are guided in a slotted sector.

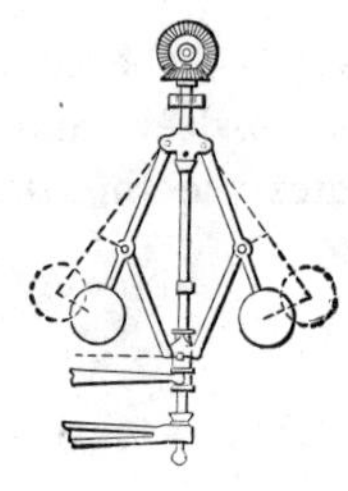

298. CENTRIFUGAL BALL GOVERNOR.—The balls, with arms pivoted to the revolving spindle, through their connections raise or lower the grooved sleeve on the lower part of the spindle. The yoke of the valve lever rests in the groove and thus controls the valve gear by the varying speed of the governor.

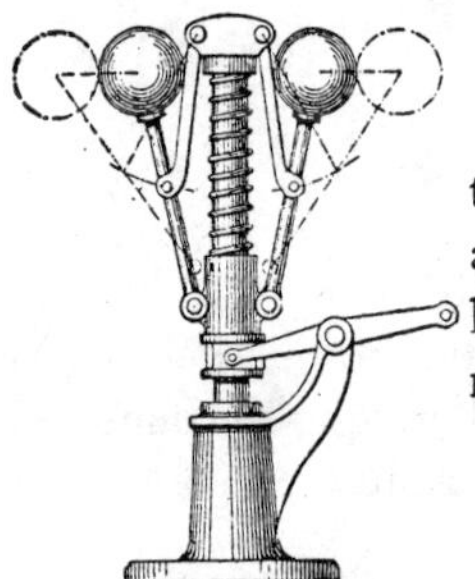

299. INVERTED GOVERNOR.—The centrifugal force of the balls is resisted by a spring around the spindle. The extension of the balls lifts the lever spool through the toggle-joint movement.

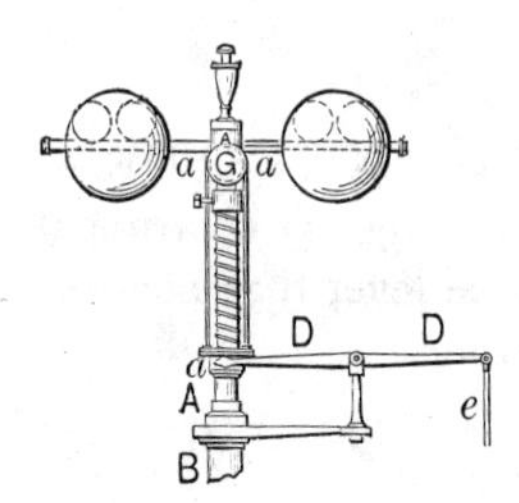

300. DIRECT-ACTING CENTRIFUGAL GOVERNOR.—The balls traverse the radial arms *a*, *a*, on friction rollers and are restrained by steel ribbons that pass over a pair of pulleys at G, and are attached to the spring and grooved collar that operates the lever and throttle valve.

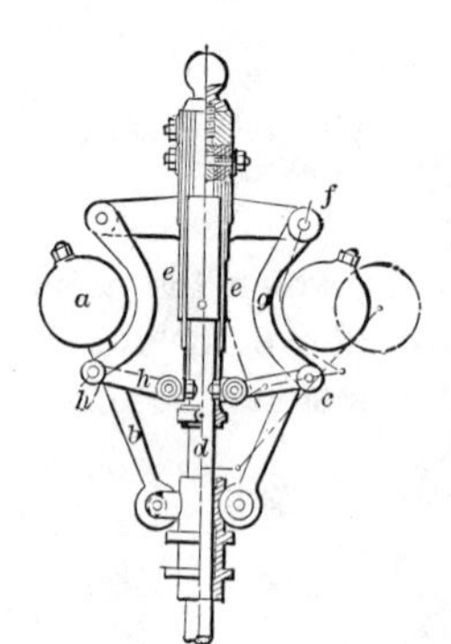

301. SPRING BALANCED CENTRIFUGAL GOVERNOR, "Proell" patent.—The balls are attached to the inverted arms *b*, *b*, and raise the collar sleeve by their outward throw. The movement is restrained by the vertical leaf springs and links. The lift is controlled by the curved links hung from the cross bar at *f*.

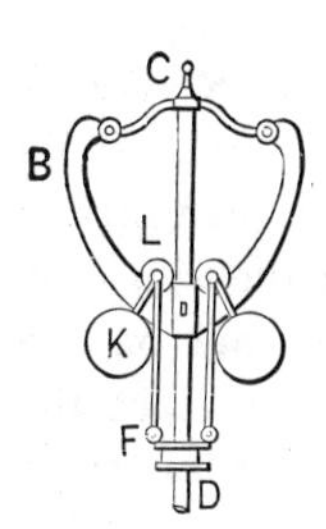

302. PARABOLIC GOVERNOR.—One of the many curious devices for governing steam engines. The parabolic form of the guide arms is intended to equalize the motion of the grooved slide by modifying the effect of centrifugal force in the position of the balls. Also called an isochronous governor, producing equal valve movement for equal change in the speed of the engine.

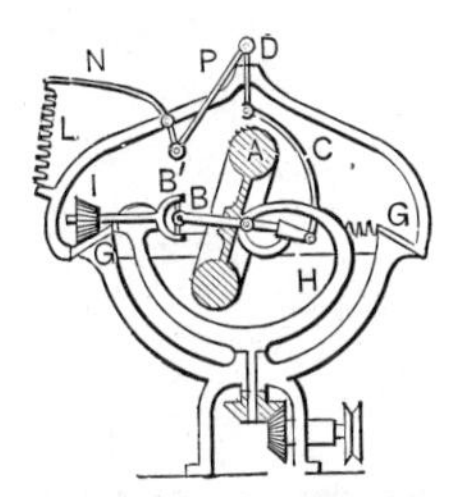

303. "ANDERSON'S" GYROSCOPE GOVERNOR for steam engines. A, The gyroscope wheel; B, its spindle connected to its driving shaft by the universal joint B', and revolved at high velocity by the pinion I rolling around the fixed bevel gear G. H, a frame holding the gyroscope wheel and its flexible shaft and revolving it on the vertical axis by the bevel gear and band from the engine shaft. The outer end of the spindle B is held in a jointed arm of the frame H, to allow of the retaining action of the spring L, through the bell crank N, connecting rod P, and rod and bow D, C, pivoted with a free vertical movement in the fixed frame. A swivel at D allows the rod and bow to turn freely with the wheel and frame H. By the rapid rotation of the wheel on its own axis and its counter rotation on the vertical axis of the carrying frame H, its own axis has a strong tendency toward a vertical position, which is balanced by the spring L, causing the rod D to take a vertical motion, corresponding to variation in speed, and transmitting it to the valve gear.

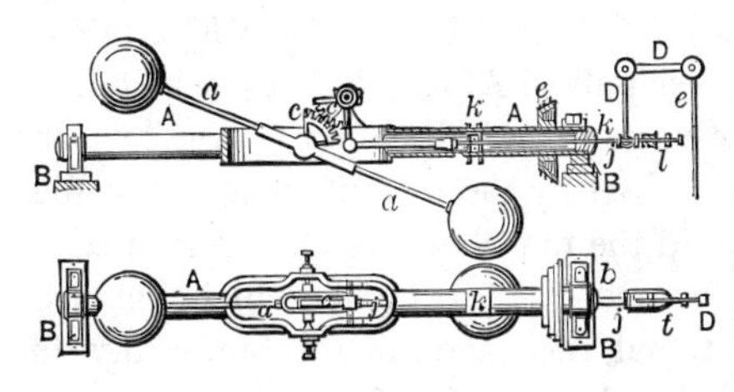

304. HORIZONTAL CENTRIFUGAL GOVERNOR, "Bourdon" model. The balls are balanced on a rigid arm pivoted to the horizontal spindle. A sector *c* on the ball arm meshes with a sector pivoted on the hollow spindle of the governor, which operates a lever and push rod to the throttle. As the balls move only by centrifugal force of revolution, they are wholly controlled by a helical spring in the hollow spindle.

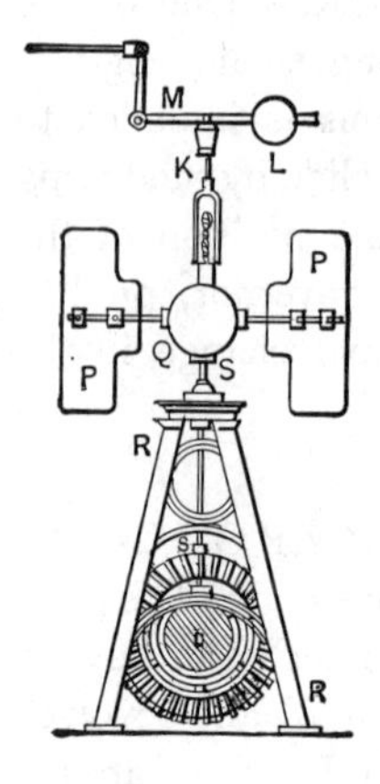

305. VANE OR WING GOVERNOR.—The resistance of the vanes P, P to the air by their variable speed from the engine gear, lifts or depresses the ball Q, connected with the wings, by means of a quick-pitch thread and nut on the revolving spindle, causing a movement of the weighted bell-crank lever M, L, and by its action controls the throttle valve.

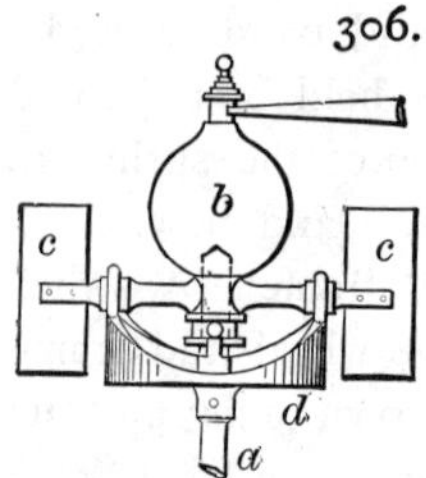

306. GOVERNOR FOR A STEAM ENGINE (old).—A revolving spindle, *a*, carries with it a pair of cylindrical inclined planes, *d*. The ball *b*, frame and wings *c*, slide freely upon an extension of the spindle. The varying air resistance given to the wings *c*, *c* by the revolution of the spindle lifts the ball; the friction rollers on the cross-arm moving up and down the incline as the speed varies, moving the valve lever or an internal valve spindle.

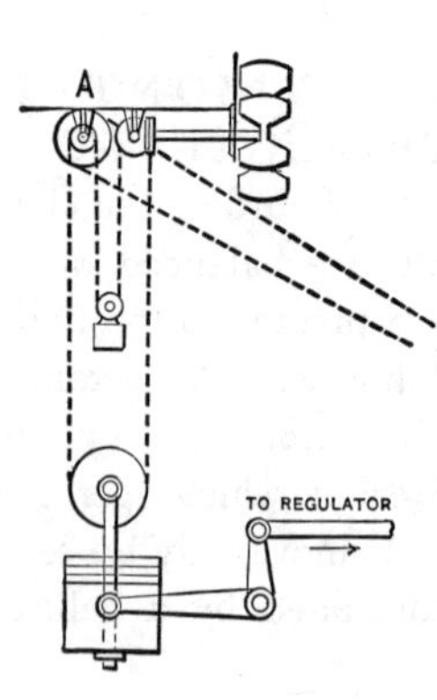

307. DIFFERENTIAL GOVERNOR.—The larger pulley, A, is driven by a belt from the motive power, winding up the larger weight which is offset by the revolution of the smaller pulley and the fan wheel, which is regulated by the difference in the weights which balances the frictional resistance of the fan. Any difference in the speed of the motive power raises or lowers the large weight, moving the bell crank.

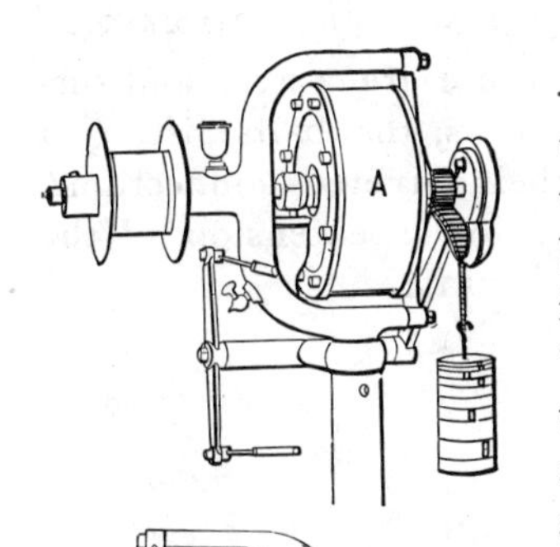

308. "HUNTOON" GOVERNOR. —A ribbed cylinder, A, is partly filled with oil. A paddle wheel, B, is revolved by the pulley and shaft which by fluid friction moves the ribbed cylinder and pinion in the same direction. The pinion meshes in the toothed sector, which is counterbalanced by an adjustable weight. The sector rock shaft operates the steam throttle valve through its arm and connecting rods.

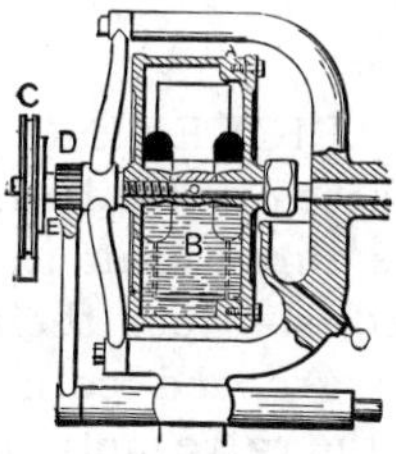

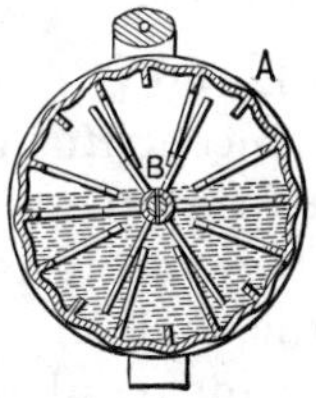

309. Vertical Section.

310. Cross Vertical Section. Showing ribs and paddle wheel.

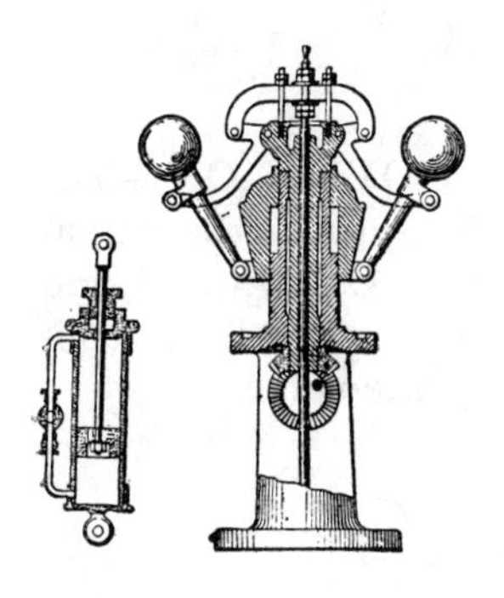

311. "PROELL" GOVERNOR.—In addition to the weight lifted by the centrifugal balls, an air dash pot is used in the line of the central rod connected at the top by a yoke pivoted to the bell-crank arms. The dash pot with bye-pass is shown at the left.

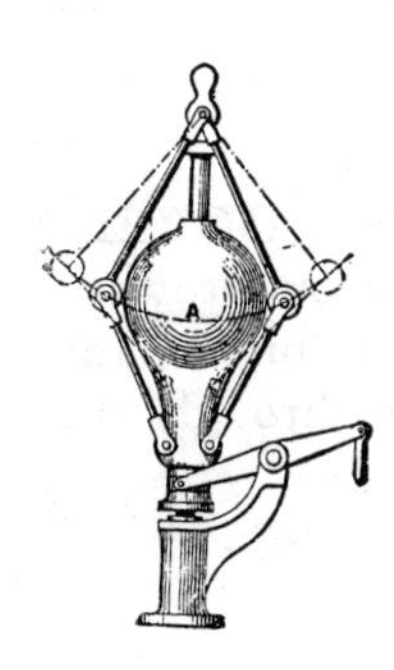

312. "PORTER" GOVERNOR.—The centrifugal balls lift a central weight, A, by the toggle-arm connection. A high-speed governor.

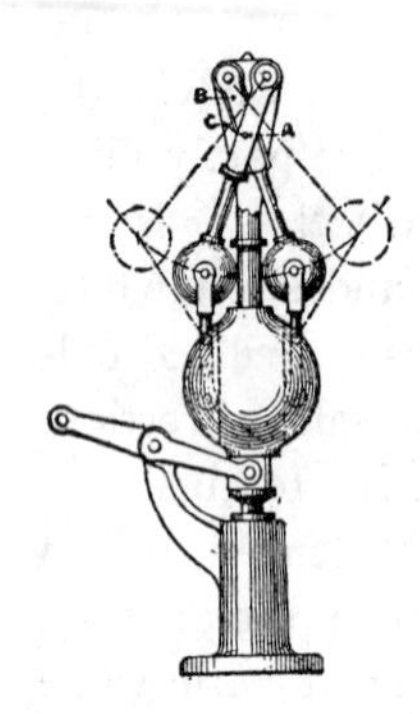

313. "RICHARDSON" GOVERNOR.—The arms in this governor are crossed and suspended from two points, the balls lifting a central weight by their pivoted connections. The groove on the lower extension of the weight operates the throttle.

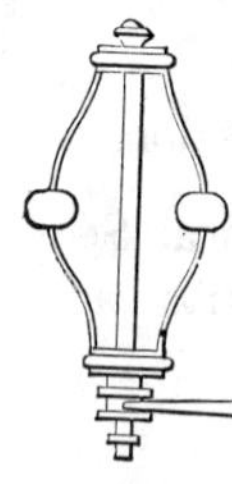

314. PRINCIPLE OF THE "PICKERING" GOVERNOR.—The centrifugal force of the balls revolving with the central spindle throws out the springs to which they are attached, shortens their length on the spindle, and lifts the grooved collar that carries the lever for regulating the valve motion.

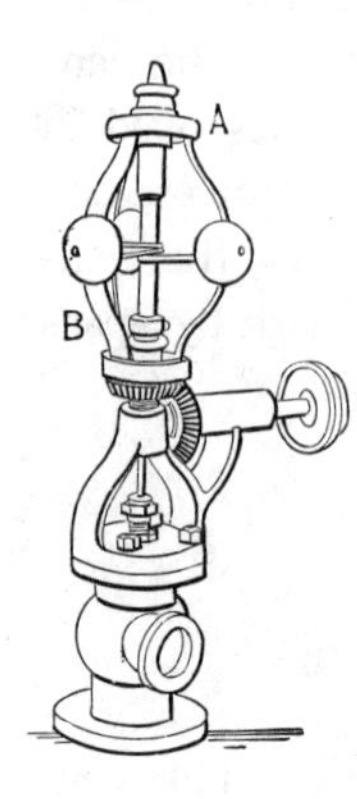

315. "PICKERING" GOVERNOR.—The revolving balls are held by springs, the extension of which draws the cap, A, downward and with it the central valve rod, with direct connection to the balanced throttle valve.

316. PULLEY OR FLYWHEEL GOVERNOR, "Sweet's."—The eccentric moves toward the centre by the centrifugal action of the weight restrained by the spring through the connecting link.

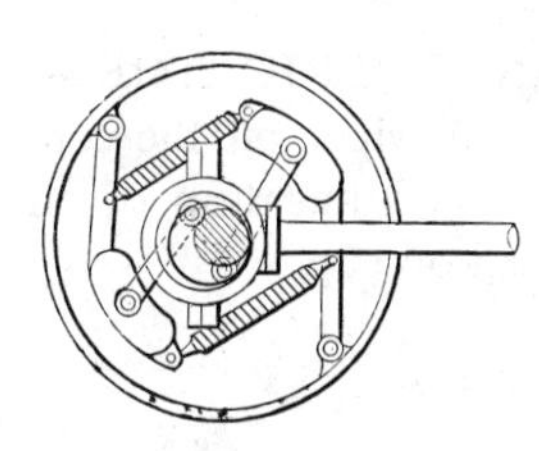

317. CRANK-SHAFT GOVERNOR.—The centrifugal action of the weights, balanced by the springs, shifts the position of the inner eccentric to vary the throw of the main eccentric.

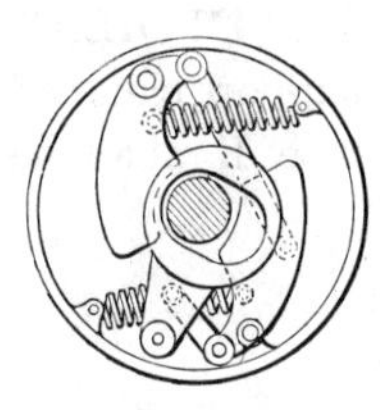

318. CRANK-SHAFT GOVERNOR.—The centrifugal action of two hinged weights, balanced by springs, varies the eccentric by moving it toward the centre by excess of speed. Eccentric is hinged to an arm of the pulley or fly wheel.

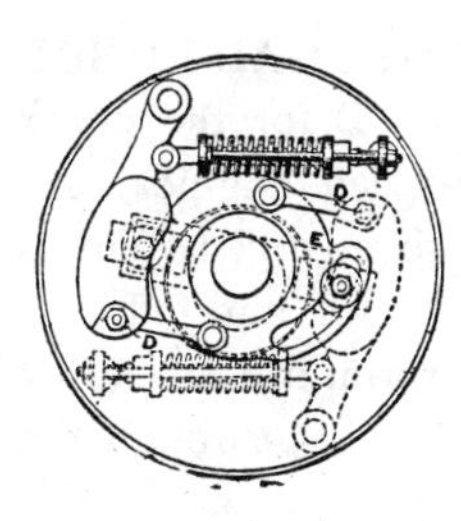

319. FLY-WHEEL OR PULLEY GOVERNOR.—The centrifugal force of two pivoted weights connected to a spiral-slotted face plate, in which a wrist pin on the arm of the eccentric sets it forward or back; controlled by the adjustment of the holding springs.

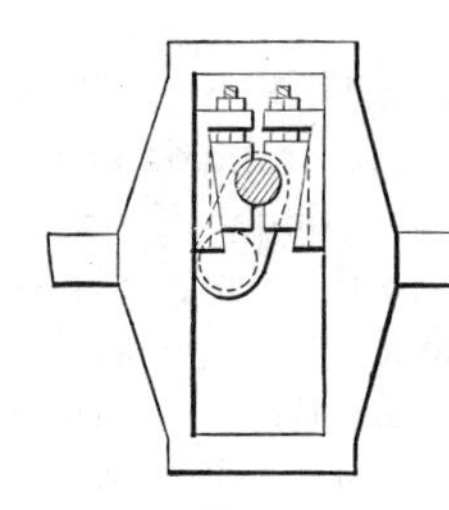

320. SLOTTED CROSS-HEAD, with "Clayton's" adjustable wrist-pin box. Two taper half-boxes and sliding taper gibs, with heads carrying screws for adjusting the boxes to both slide cross-head and wrist pin.

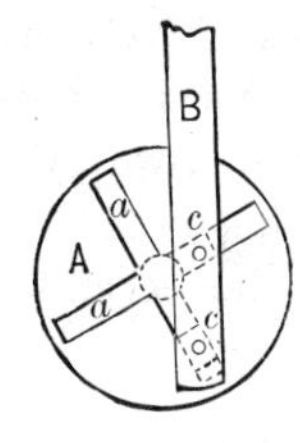

321. TRAMMEL CRANK.—The pins *c*, *c* on the rod B traverse the two right-angled slots in the revolving face plate, producing a reciprocating motion of the rod B.

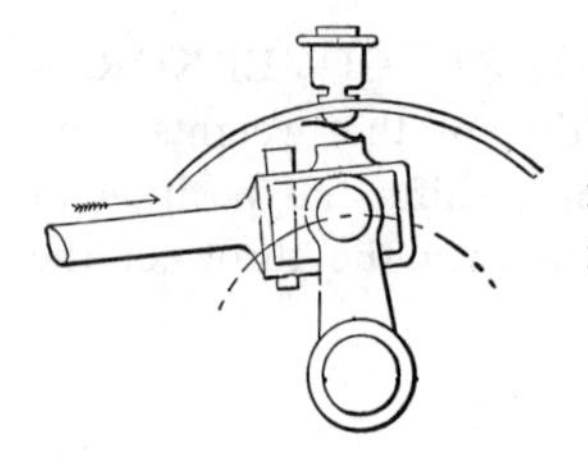

322. CRANK-PIN LUBRICATOR.—The oil cup is fixed. A wiper on the connecting rod end takes off the drop of oil from the capillary feed oil cup.

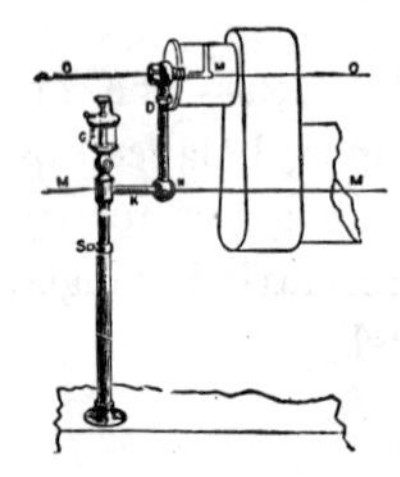

323. CENTRIFUGAL CRANK-PIN OILER made adjustable by the sliding support clamped at S, so that the revolving feed pipe K shall be aligned with the axis of the shaft.

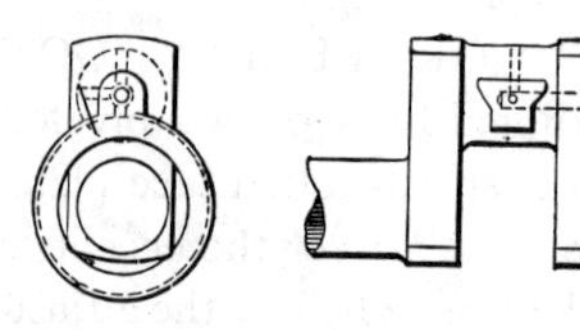

324. CENTRIFUGAL LUBRICATING DEVICE for the crank pin of a high-speed engine. An annular cup with an open front is fastened to the crank and fed by a drip spout at A. The oil is thrown to the outer rim of the cup by the centrifugal force of revolution and to the oil holes through the crank pin.

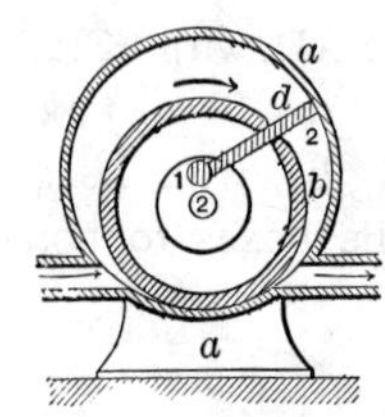

325. "COCHRANE" ROTARY ENGINE.—A wing piston rotating around the central axis of an outer shell or cylinder. A hollow cylinder of smaller diameter is pivoted eccentric to the wing axis to keep one side in contact with the shell. The steam pressure revolves the wing and shaft with a force due to the varying area of the wing outside of the inner cylinder.

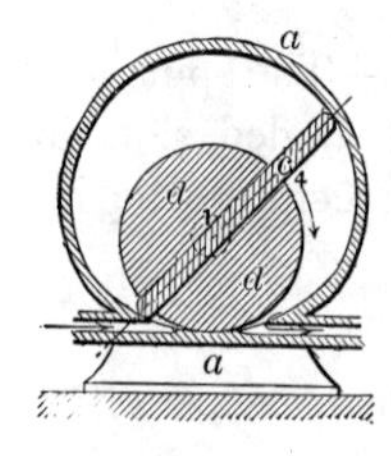

326. "FRANCHOT" ROTARY ENGINE.—A slotted concentric cylinder carries a continuous solid wing across and in contact with the interior surface of an ovoid shell, shaped for exact diameter in all directions on the eccentric axis of revo ution.

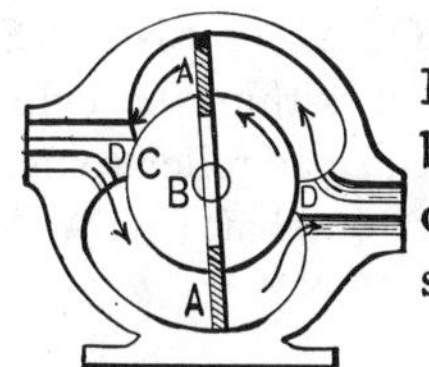

327. DOUBLE SLIDE PISTON ROTARY ENGINE.—In this engine the shaft and piston barrel are concentric, while the walls of the steam chambers are ovoid. A difficult form of construction.

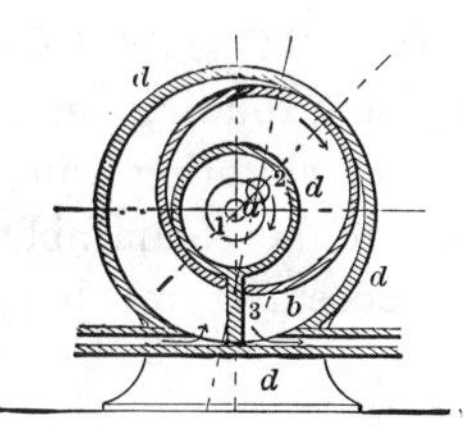

328. "LAMB" ROTARY ENGINE.—An annular cylinder with a fixed partition between the inlet and outlet. The piston is a hollow cylinder with a longitudinal slot, which slides up and down the partition, the outside of the cylinder wiping the inner surface of the shell. The centre of the traversing cylinder is pivoted to a crank pin, which carries it around a common centre shaft.

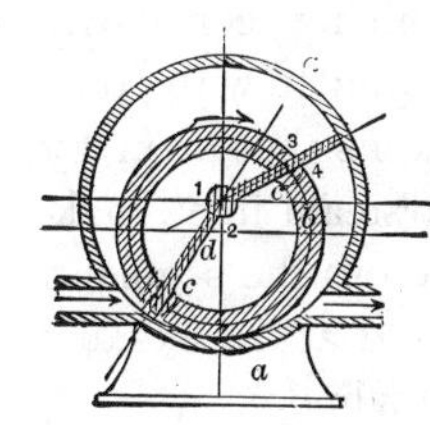

329. "COCHRAN" ROTARY ENGINE.—The wing pistons *d*, *d* are packed in the eccentric inner cylinder by a slotted rocking cylinder and revolve concentric with the outer cylinder or shell. The inner cylinder is pivoted eccentric to the shell, making a tight joint at the bottom.

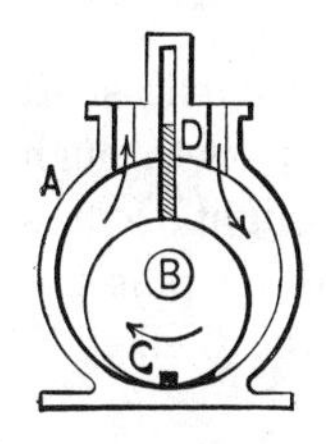

330. ROTARY ENGINE.—B, shaft; C, eccentric rotating piston; D, follower slide. The eccentric cylindrical piston operates the slide by its revolution.

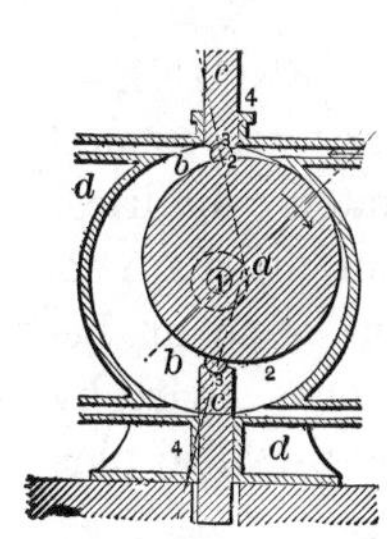

331. "NAPIER" ROTARY ENGINE.—An eccentric mounted cylinder on a shaft concentric with the shell. There are two sliding wings in slots in the shell, held to their bearings by springs or cam wheels on the shaft outside with connecting bars. There are two pair of ports.

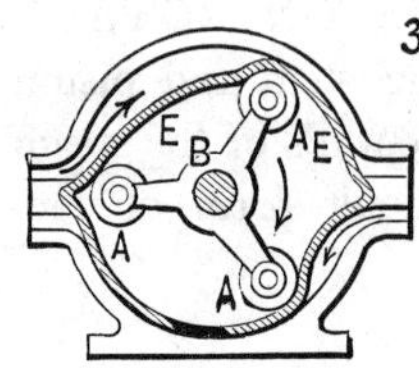

332. ROLLER PISTON ROTARY ENGINE. A rubber lining loosely placed within the cylinder is rolled over by the three-armed roller spider. E, E, rubber lining; B, spider on shaft; A, A, A, rollers.

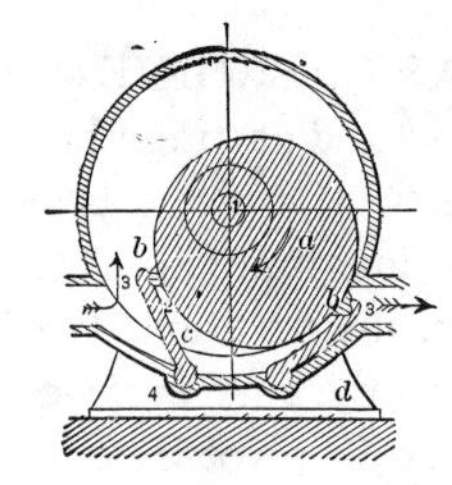

333. "COCHRANE" ROTARY ENGINE. —An eccentric cylindrical piston rotating on an axis central to the shell. The vibrating wings pivoted in the outer shell form the steam abutment by closing against the eccentric revolving cylinder.

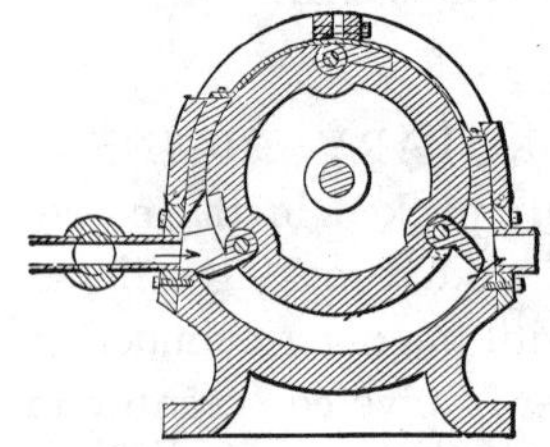

334. "BOARDMAN" ROTARY ENGINE.—A cylinder revolving concentric with an outer segmental cylinder, with pockets containing swing pistons that open by centrifugal action at the steam inlet, making a steam abutment across the segment. The swing pistons are closed at the exhaust port by contact with the small segment of the outer cylinder.

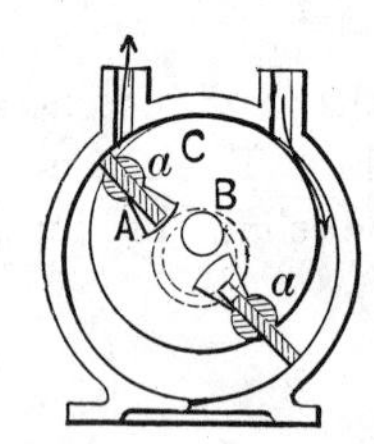

335. ROTARY ENGINE, with concentric shaft and wing barrel. The two wing slides pass through cylindrical rockers to give the slides a slight oscillating motion; slides are kept extended by pins traversing a circular slot concentric with the shell.

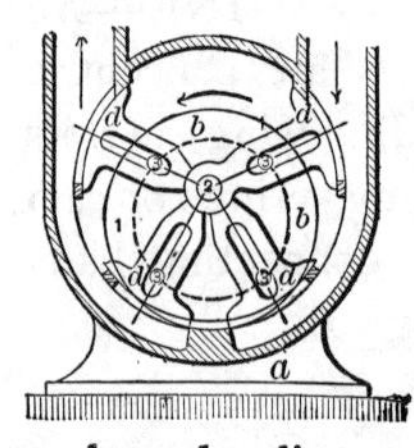

336. "SMITH" ROTARY ENGINE.—Four arms with cylindrical sectors are rotated around an axis central to a perforated cylindrical shell. The driven shaft and head discs are eccentric to the shell. The pressure of steam between the wings tends to push them apart, by which the differential leverage on the disc pins revolves the disc and shaft.

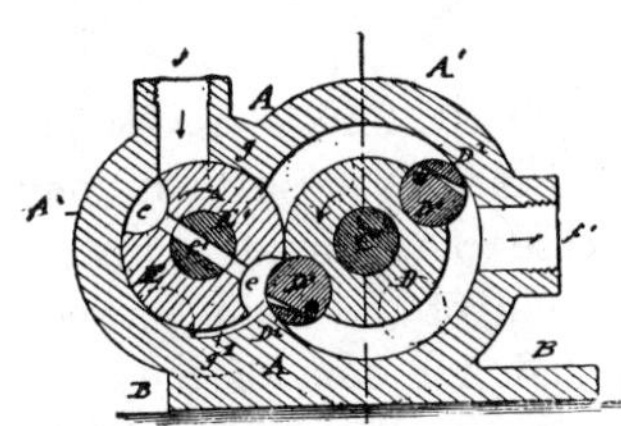

337. "BERRENBERG" ROTARY ENGINE.—Two intersecting cylindrical shells. The steam cylinder D has two cylindrical pistons, D′, D′, on opposite sides, that mesh in corresponding cavities in the cylindrical steam valve, both rotating in unison by equal external gearing. The steam port passes through the rotary valve E at the proper moment for the impulse. The supplementary sectors D^2 are hinged to the pistons D^1 to make a more perfect contact with the outer cylindrical shell.

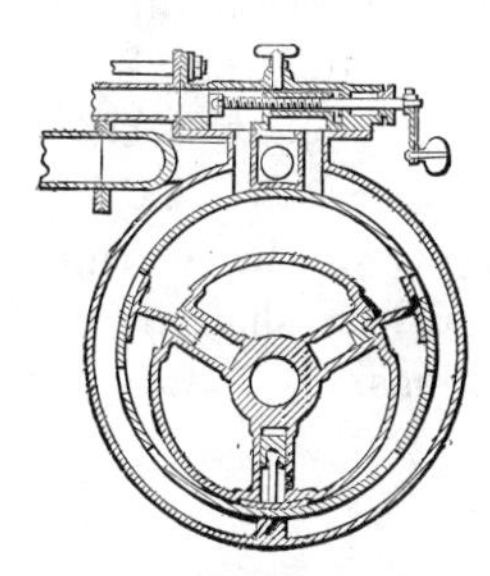

338. "FLETCHER'S" ROTARY CONDENSING ENGINE.—A hollow drum on a shaft eccentric to a double shell. Three slots carry slides and socketed arms as abutment wings, which are kept in contact with the cylindrical shell by a ring not shown. Steam ports on inner shell at the left side. Exhaust ports on the inner shell at the right.

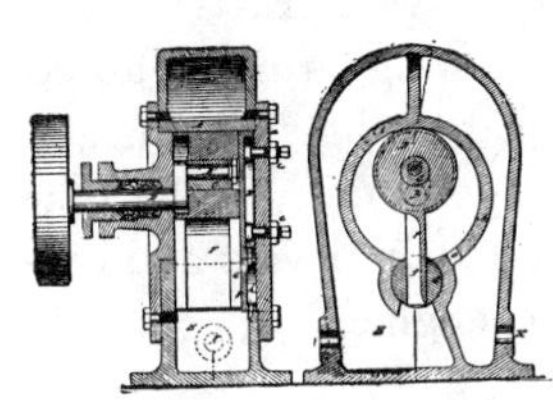

339. "BARTRUM & POWELL" ROTARY ENGINE.—A double shell divided for steam and exhaust. The inner shell cylindrical with a shaft and crank concentric. The crank pin carries a smaller winged cylinder, the wing sliding through a rocking joint. The end packing is made adjustable by a plate set up with screws. The crank pin has an eccentric sleeve which, by a slight rotation, compensates the wear of the rubbing surfaces.

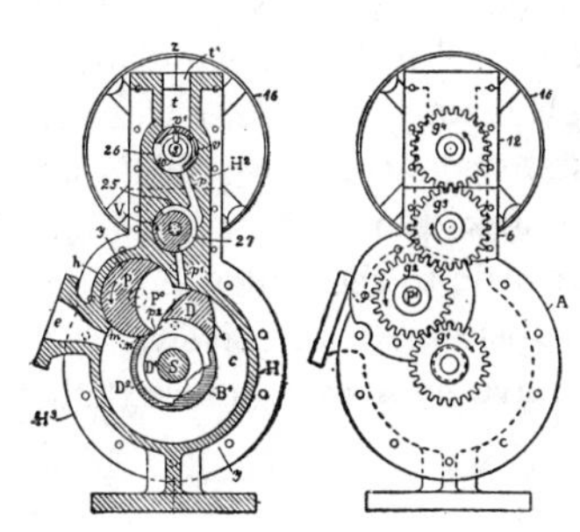

340. "RITTER" ROTARY ENGINE.— A revolving cylinder concentric with the shell, carrying an abutting lip or extension fitting the outer case. A revolving lunette controlled by gear on main shaft allows the lip to pass; a continuous gear train operates the valve.

341. Exterior with valve gears.

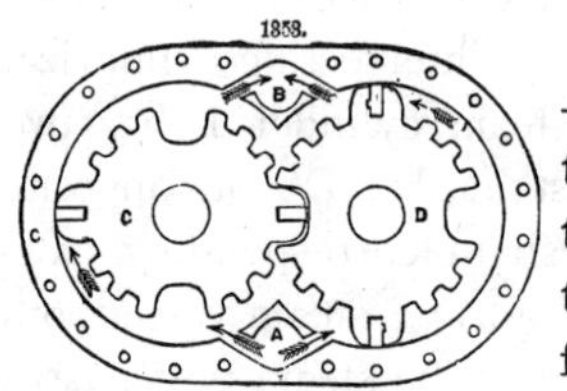

342. "HOLLY" ROTARY ENGINE. — The two geared pistons mesh their long teeth into the recesses of the opposite piston, thus making the sum of the radii between the centres less than the sum of the radii from each centre to its cylinder wall. Pressure rotates the gear in the direction of the longest leverage.

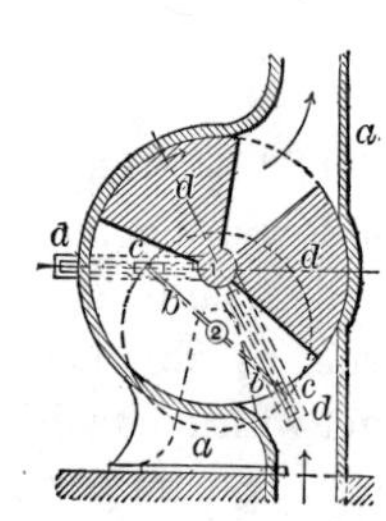

343. "STOCKER" ROTARY ENGINE.— The sector pistons are each connected through central concentric shafts to slotted cranks in which a sliding box and link connect to a crank on a shaft eccentric to the sector shaft. A differential motion of the sectors is produced while rotating which rotates the driven shaft by the outside slotted crank connections.

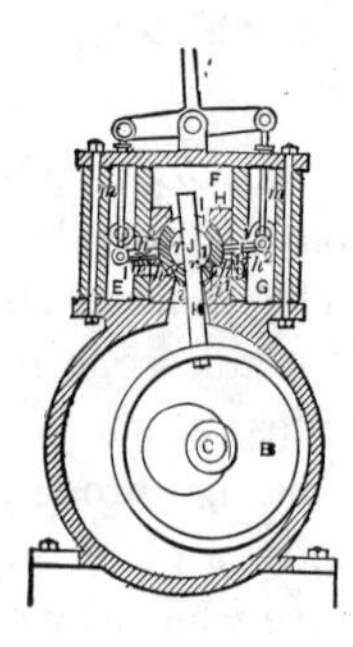

344. "FORRESTER" ROTARY ENGINE. — A cylindrical block and guard wing swing on an eccentric on the shaft. The guard wing slides in and operates the ports of a two-port rotary valve, the outer shell of which is operated by levers and connecting rods for reversing the engine.

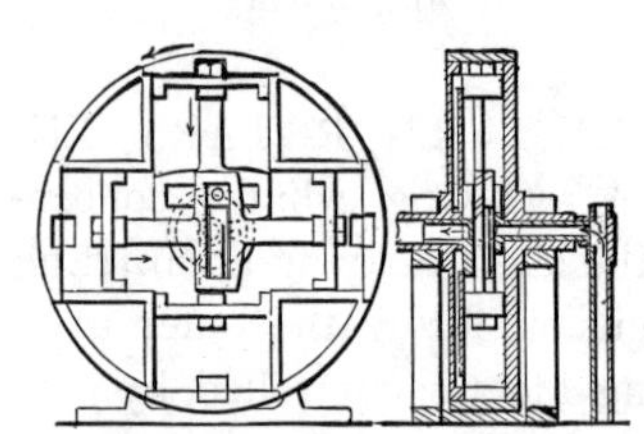

345. "KIPP" ROTARY PISTON ENGINE. — A broad pulley enclosing four single-acting cylinders with opposite pistons connected by yoked rods. A fixed crank pin and slide block placed eccentric to the pulley axis gives the propelling force by displacing the pistons successively. The steam follows through ports in a disc valve with inlet and exhaust through the hollow shaft.

346. Section.

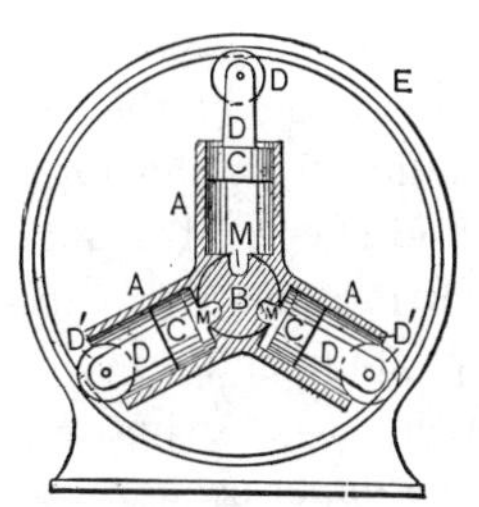

347. "RUTH'S" ROTARY ENGINE.—A revolving cylinder engine. Three cylinders, A, A, A, radiate from a shaft set eccentric to an outer circle or ring on which the piston connected sheaves revolve. The pistons take steam through the ports M, M, M, just past the shortest eccentric radius, and drives out the piston during a half revolution, when the exhaust is opened and the piston is pushed back by the eccentric ring.

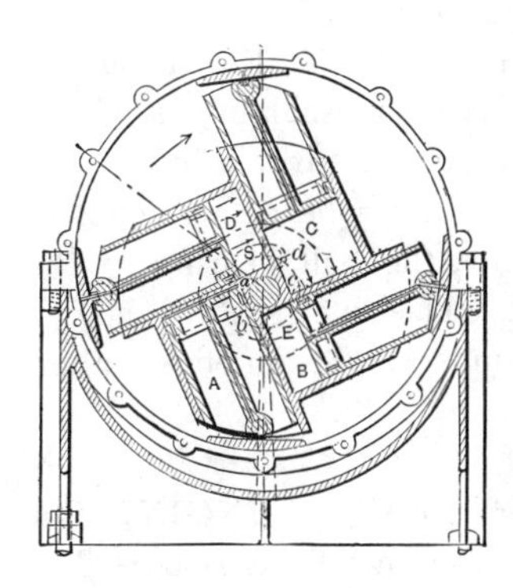

348. "ALMOND" ENGINE.—Four single-acting cylinders set tangent to a shaft which is central to an outer shell. The pistons have jointed segmental plates at their outer end that press against the outer shell and cause the cylinders and shaft to revolve by the eccentric direction of their pressure. Disc ports for steam and exhaust.

349. ROTATING CYLINDER ENGINE.—The cylinder rotates on trunnions with a through piston rod terminating with rollers running in an oval ring. Steam and exhaust ports in the trunnion. Pressure of the piston-rod rollers on the oval ring revolves the cylinder and fly-wheel on its runnion.

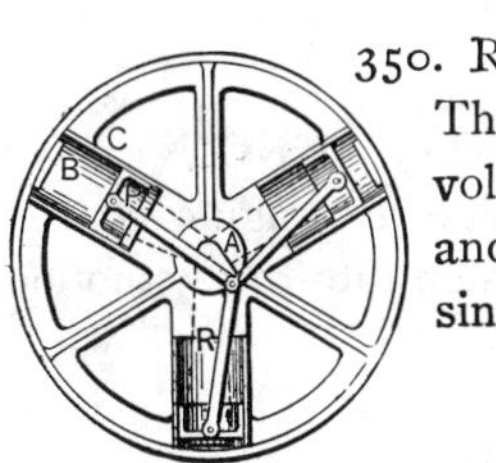

350. ROTARY MULTI-CYLINDER ENGINE.—Three or more cylinders are attached to and revolve with the fly-wheel. The crank is stationary and eccentric to the fly-wheel. Each cylinder is single-acting. Valves are on a central disc at A.

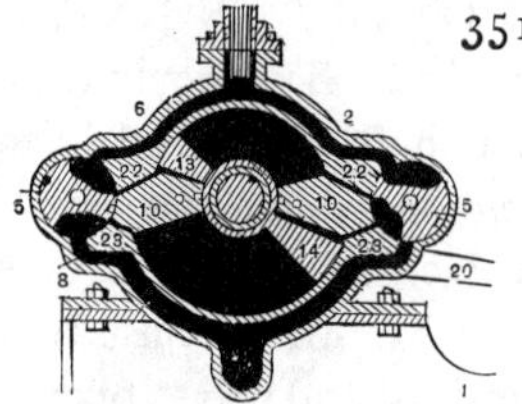

351. "BATES" COMPOUND VIBRATING ENGINE.—The upper section of the cylinder has a shorter radius than the lower section for the compound effect. The shaft and wings are concentric and vibrate between two stationary abutments, 10, 10. Opposite each abutment is a cylindrical valve, which by its motion admits the steam to the upper section, and transfers its exhaust to the lower section, and also the final exhaust.

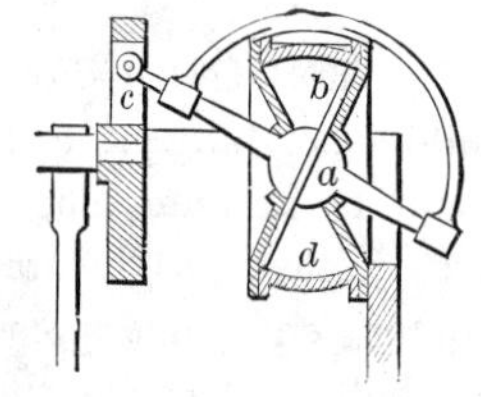

352. "DAVIE'S" DISC ENGINE.—A disc, *b*, is fixed to an oscillating shaft, *a*, which swings in a circuit pivoted in the disc crank, *c*. The cylinder heads are cones in the apex of which the ball bearing of the shaft oscillates. The outer shell of the cylinder, *d*, is spherical over which the disc moves. Steam enters alternately on either side of piston.

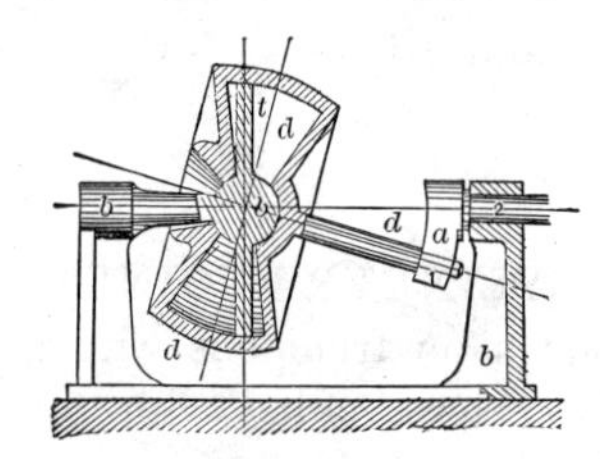

353. "REULEAUX" ENGINE OR PUMP.—A disc on a fixed shaft. The cylinder swings on a central spherical bearing, carrying an arm pivoted in a crank.

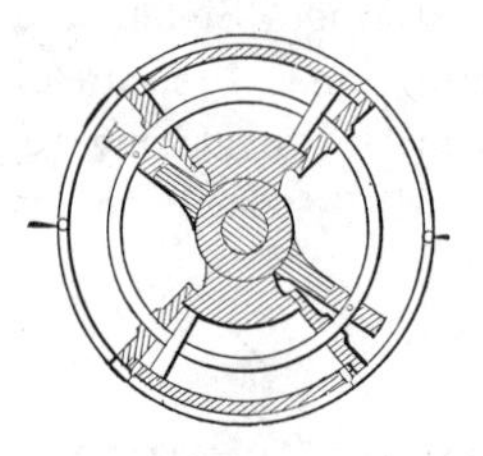

354. "LINK" VIBRATORY ENGINE.—A pair of curved cylinders with an annular piston rod to which is attached the arms from the central shaft. The reciprocal motion of the piston rocks the central shaft, the motion of which is made continuous by a link and crank, not shown.

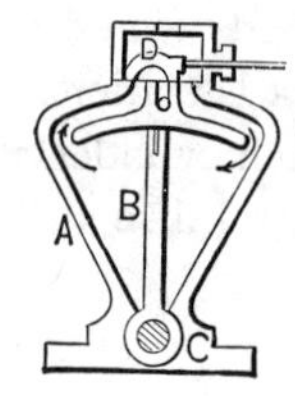

355. OSCILLATING PISTON ENGINE.—A crank and connecting rod outside the engine convert the oscillating motion of the piston into rotary motion.

356. VIBRATING PISTON ENGINE, "Parson's" model. Two sector pistons vibrating in a cylinder. One sector is fast on a central solid shaft, the other is fast on a concentric hollow shaft. At the other end of each shaft is a crank and link connection to a wrist pin at opposite positions on a face plate which is fast on a revolving shaft eccentric to the piston shafts. The exhaust port is in the circumference of the cylinder.

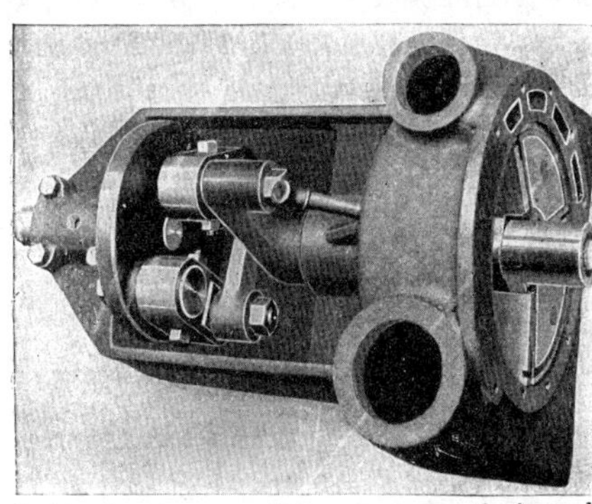

357. Shows the crank end of the vibrating shafts with the link connections. The steam port is in the cylinder head, which is the steam chest. During one-half of a fly-wheel revolution one of the sectors makes a large angular movement, while the other makes a relatively small angular movement, and during the second half, the two sectors reverse their relative movements—*i.e.*, the one going slow during the first half making the quick movement during the second half, and *vice versa*.

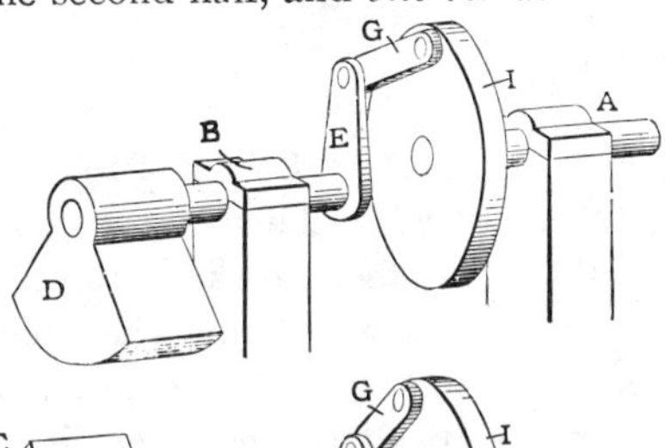

358. Shows the detail of one sector, piston, shaft, crank, and link connection with the eccentric revolving disc and shaft.

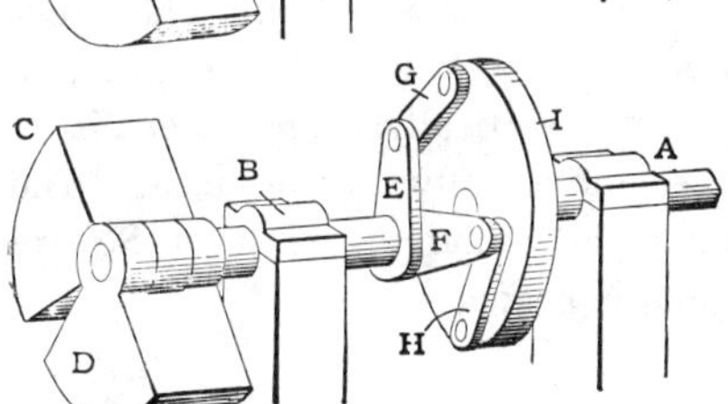

359. Shows both sector pistons, concentric shafts, cranks, and link connections to the opposite wrist pins on the revolving face plate of the constant velocity shaft.

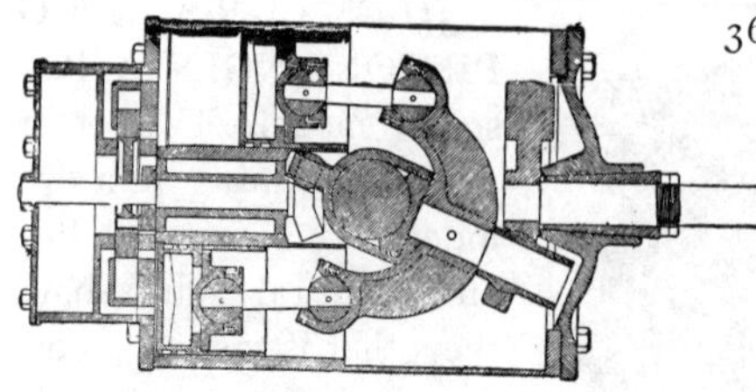

360. "KNICKERBOCKER" FOUR-PISTON ROTARY ENGINE.—A four-armed yoke is socketed on a centre common to the four pistons. Its spindle is a crank pin, and makes a conical circuit with the crank and shaft. The ends of the yoke are socketed to the pistons by connecting rods. The pistons take steam successively, making a continuous pressure on the circuit of the crank.

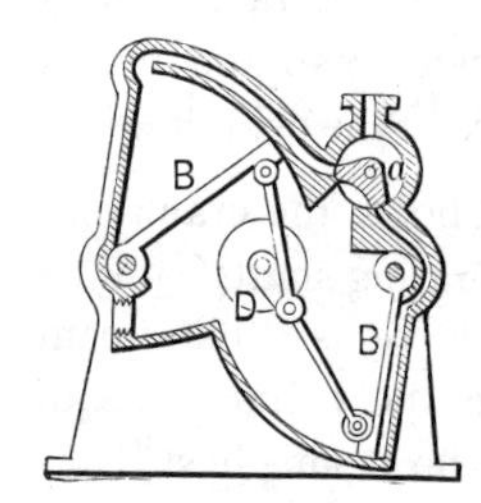

361. "ROOT'S" DOUBLE QUADRANT ENGINE.—In this design the two oscillating pistons are connected directly with the crank on the inside of the engine case, which is also the exhaust receiver. From the positions of the connecting rods at the end of the stroke of each piston the dead centre is eliminated.

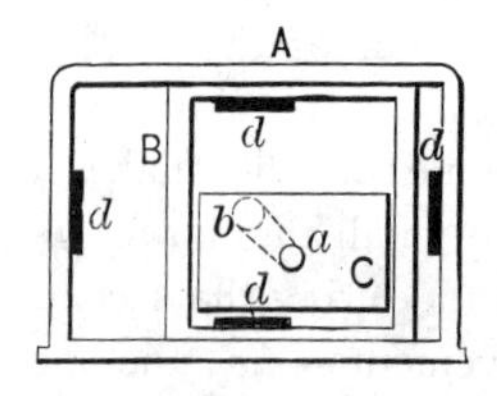

362. "ROOT'S" SQUARE PISTON ENGINE.—The oblong square box, A, is the cylinder proper. B, is a frame sliding freely in a horizontal direction by the force of the steam from the side ports, *d*, *d*. C is the inner rectangular piston, connected directly to the crank pin *a*, the shaft, *b*, being central to the range of the moving pistons. The piston, C, receives steam from the top and bottom ports, *d*, *d*, within the frame, B.

363. "DAKE" SQUARE PISTON ENGINE.—Two rectangular pistons, one within the other, working at right angles in the outer piston. The inner piston is connected to the crank pin, and moves vertically. The outer piston moves horizontally in the case. The principle is similar to the Root Square Engine, No. 362.

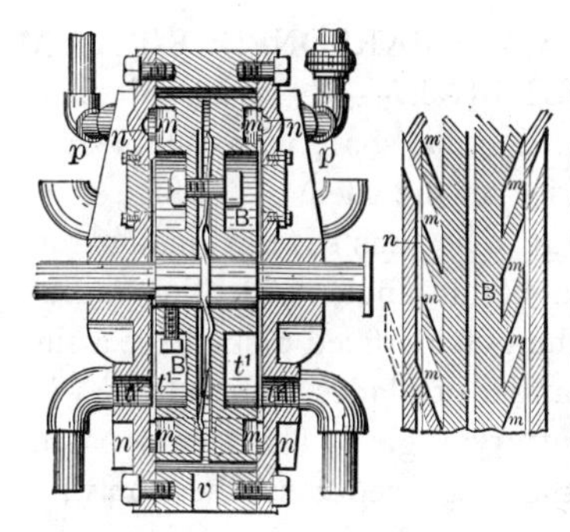

364. "WILKINSON'S" STEAM TURBINE.—Two rim-pocketed discs running against the disc surfaces of a shell with oblique steam ports. The discs are feathered on the shaft, and held against the faces of the shell by springs. A groove around the shell opposite the pockets allows the steam to pass around to the exhaust pipes.

365. Section showing steam pockets.

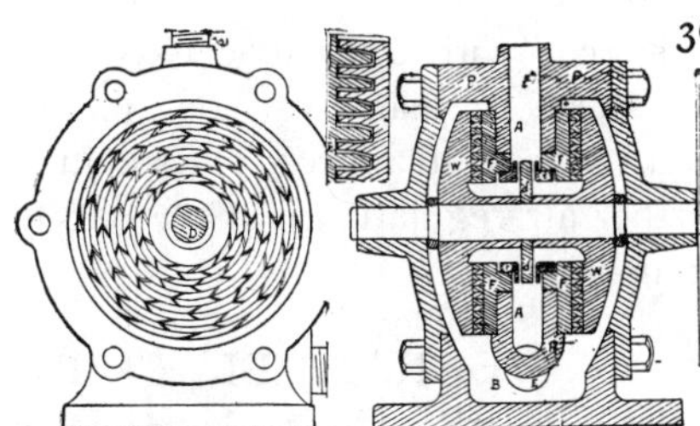

366. "DOW" STEAM TURBINE.—Two discs fixed to a shaft have on their face a series of circular grooves and tongues, meshed with a pair of fixed discs with grooves and tongues, as shown in small section 367. The tongues on the revolving discs are cut across at short distances in a slanting direction. The tongues on the stationary disc are cut in the opposite direction. The steam passes to the centre hub, and is forced through the openings across the tongues, giving motion to the discs and shaft.

368. Vertical section of engine.

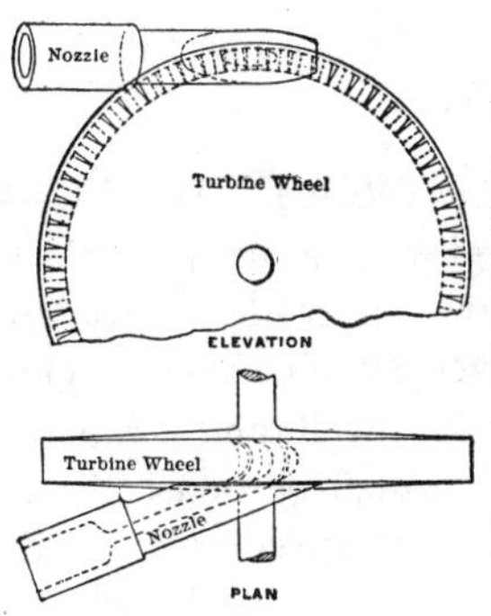

369. "DE LAVAL" STEAM TURBINE. —A jet or jets of steam impinge at a small angle upon the concave buckets at the periphery of a disc wheel, pass through the cavities between the buckets and exhaust at the other side. The buckets are lunette. The nozzle has an expanding orifice.

370. Plan showing nozzle at side of wheel.

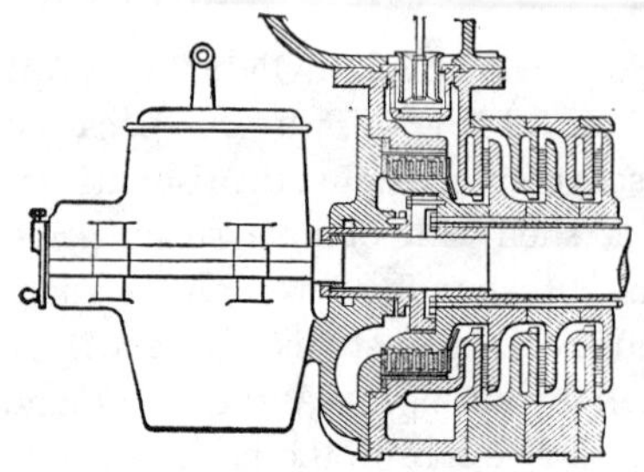

371. "PARSONS'" STEAM TURBINE.—A series of discs fixed on a shaft with intersecting discs on the shell. The face of the shaft discs has several small blades set at an angle with the radius. The outside fixed disks have a similar set of blades interlocking with the revolving blades and set at a contrary angle. The steam passes from the valve to the inner edge of the first fixed disc, then outward through the blades, and returns through the vacant space of the next pair and outward again.

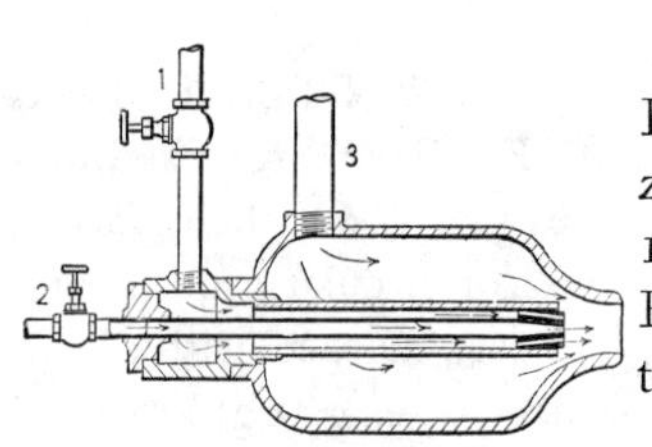

371*a*. CRUDE PETROLEUM BURNER with concentric fixed nozzles. One of many varieties in use. 1, oil feed; 2, steam feed; 3, air inlet. Further air regulaticn is made outside the nozzles.

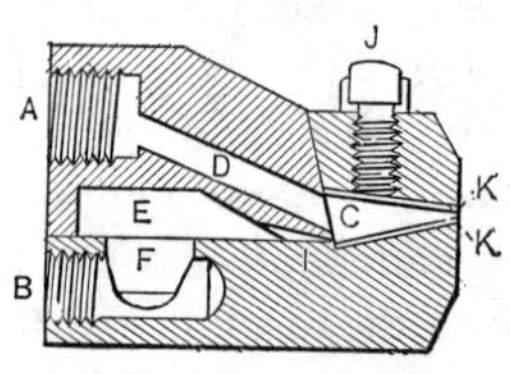

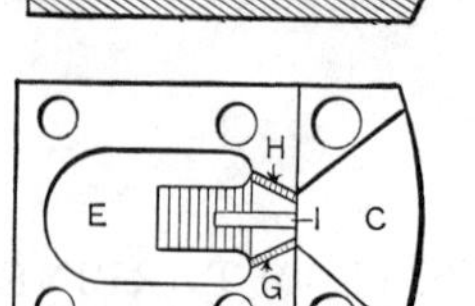

371*b*. THE HAMMEL CRUDE OIL BURNER. Gives a broad flame.

A, oil supply; B, steam supply; C, mixing chamber.

E, steam chamber connecting with steam ducts G, H, I.

K, K, steel plates which can be renewed when worked out.

J, set screw.

Gives a broad and powerful flame for boiler furnaces.

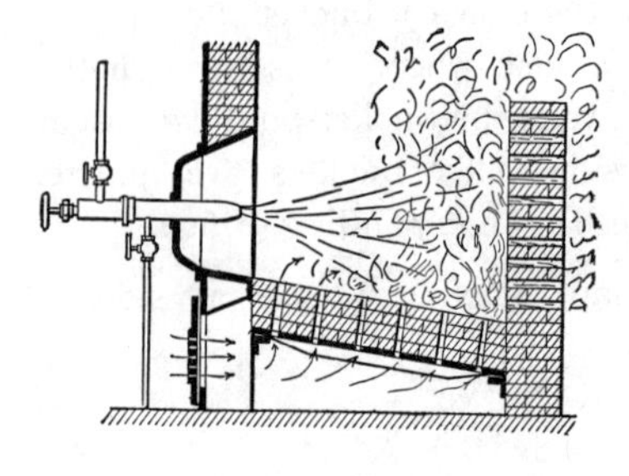

371*c*. PETROLEUM FURNACE.—For the most perfect combustion of crude petroleum the furnace should have a perforated back wall and grate of fire brick, which becomes highly heated and thus completes the combustion of the oil.

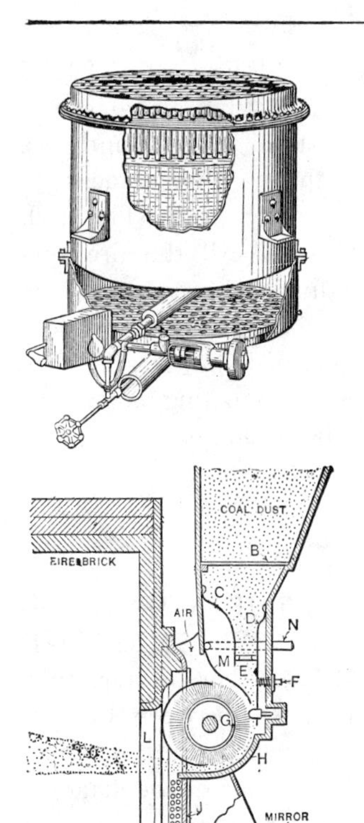

371*d*. AUTOMOBILE BOILER.—A baffle plate above the water line at the riveted joint prevents foaming or splashing of the water into the steam pipe by the vibration of the carriage. The burner is a tube perforated flat chamber with a vapcrizer, air mixing jet, regulator, and pilot light.

371*e*. FEEDING PULVERIZED OR DUST fuel to furnaces.

The coal dust is charged into the hopper, passes through a screen B, and regulated in its flow by the elastic plates C and D and the link at E. The screw F regulates the brush pad so that the brush throws both air and coal dust. J are the draft holes, K a screen with a mirror to view the fire. Brush makes 900 revolutions per minute.

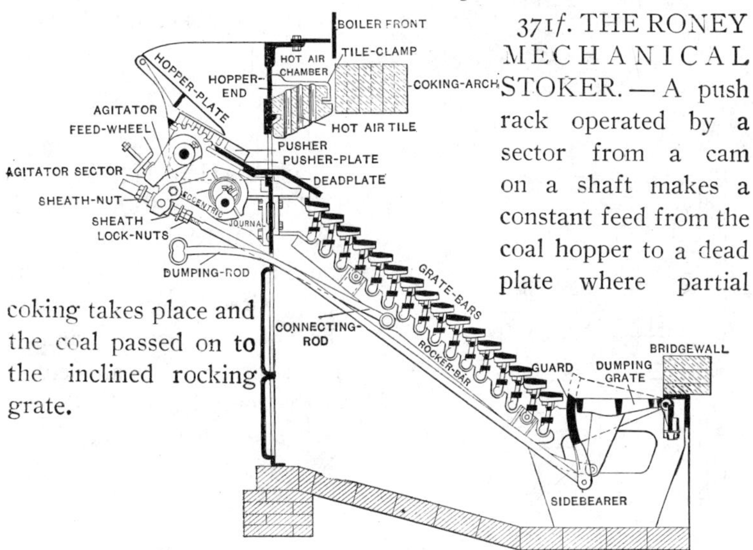

371*f*. THE RONEY MECHANICAL STOKER.—A push rack operated by a sector from a cam on a shaft makes a constant feed from the coal hopper to a dead plate where partial coking takes place and the coal passed on to the inclined rocking grate.

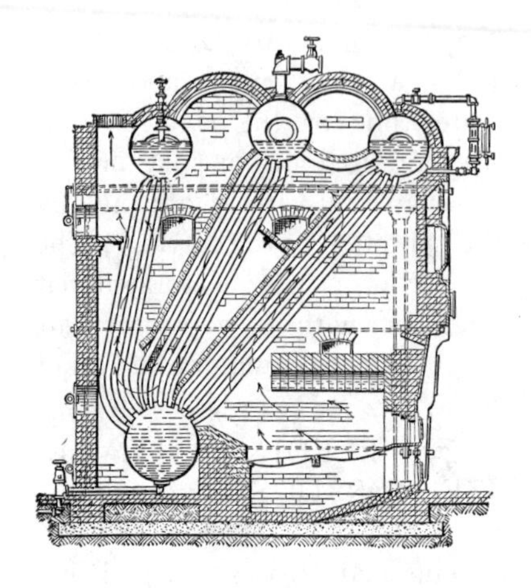

371*g*. THE STERLING BOILER.—The hot gases of combustion pass lengthwise through the three stacks of tubes guided by the fire brick partitions. All the fire surface divided by 12 equals the boiler horse power. Tubes are cleaned by steam blow pipes. Circulating pipes outside the setting.

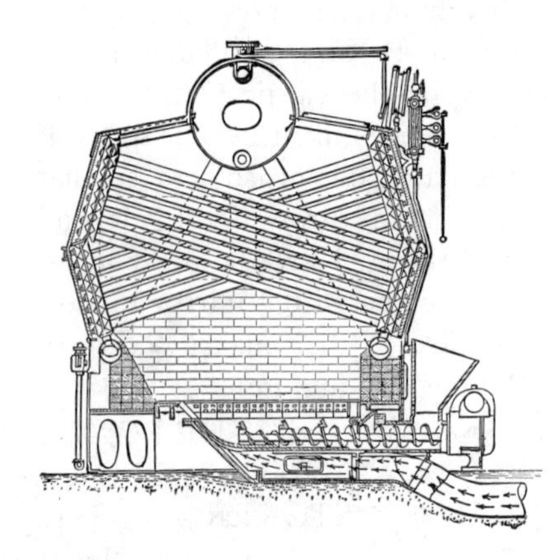

371*h*. THE WORTHINGTON WATER TUBE BOILER.—The water tube sections are between headers and cross each other in series; the lower ends of the diagonal sections are connected with a cross pipe for circulation from the steam drum. Has the American stoker attached.

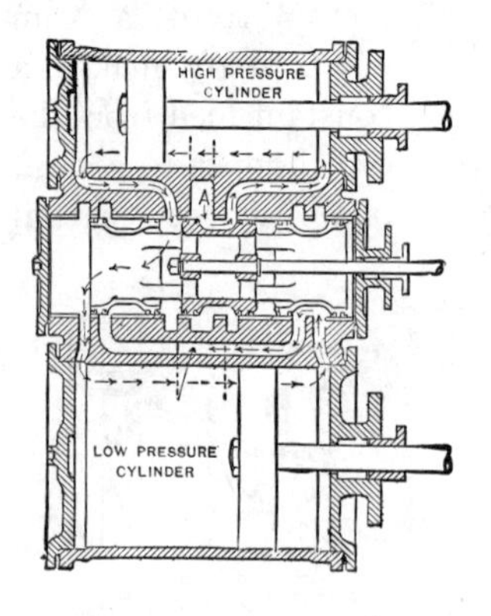

371*i*. VAUCLAIN'S COMPOUND LOCOMOTIVE CYLINDERS.—A single piston valve for both cylinders with direct steam passages through the valve chamber. High pressure steam enters at the central port A. Steam inlet and exhaust indicated by the arrows.

Section V.

STEAM APPLIANCES.

INJECTORS, STEAM PUMPS, CONDENSERS, SEPARATORS, TRAPS, AND VALVES.

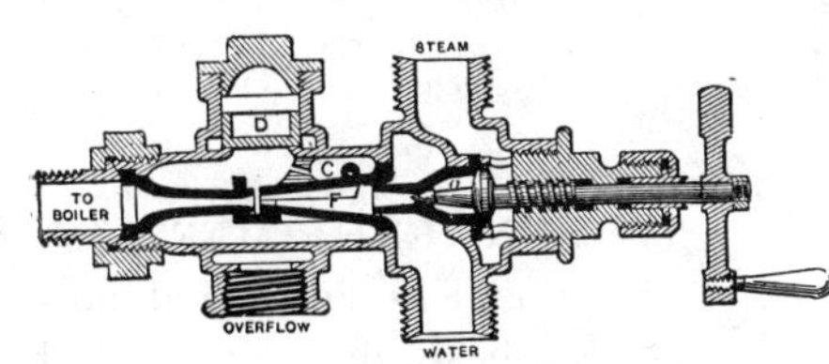

372. "PEERLESS" INJECTOR.—An exhaust steam injector. A hinged section of the combining tube allows a free flow of the exhaust until a water current is started, when the hinge closes and the overflow valve closes, as in other injectors.

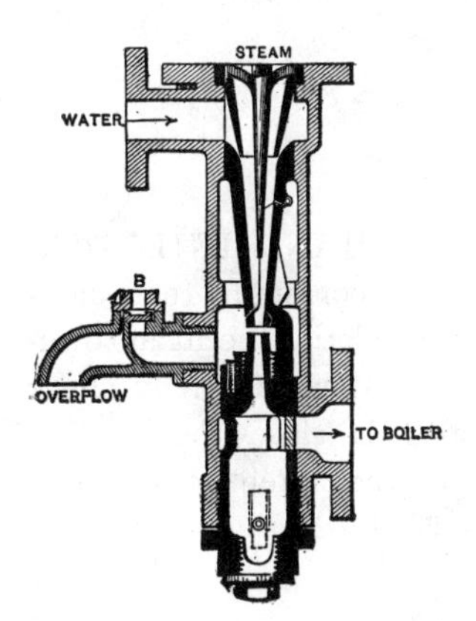

373. "SHAEFFER & BUDENBERG" INJECTOR.—An exhaust injector by which the exhaust steam establishes a feed jet to the boiler. A hinged section in the combining tube allows a free flow of steam to draw the water; the hinged section then closes and the injector operates the same as others for feeding a boiler.

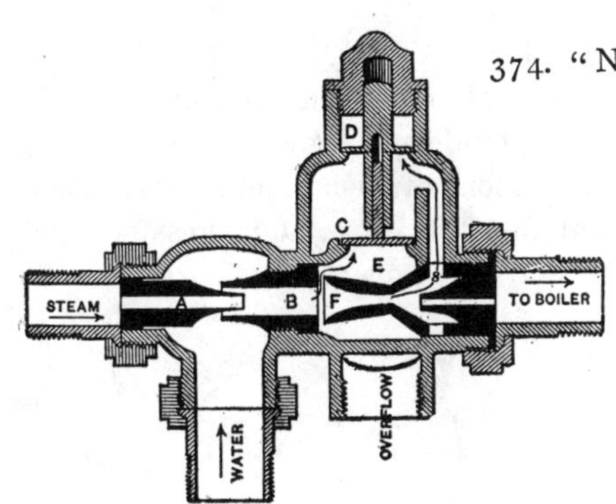

374. "NATIONAL" AUTOMATIC INJECTOR, has four fixed tubes. The two check valves, C, D, open and close successively as the lift is started and the current established.

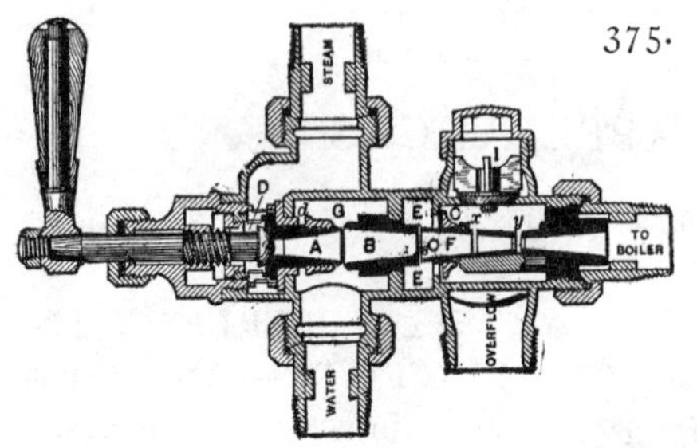

375. "METROPOLITAN" INJECTOR.—The steam is turned on by a screw spindle valve. It has three fixed nozzle tubes, A, B, F. A disc relief-check valve, C, and a wing check, I.

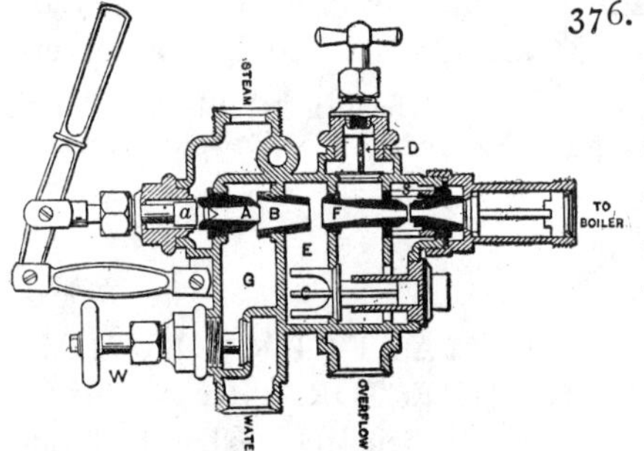

376. "LUNKENHEIMER" INJECTOR.—Four fixed nozzle tubes with a lever-moved valve, *a*; W, water-regulating valve; D, stop check to overflow; C, automatic check; W, water valve.

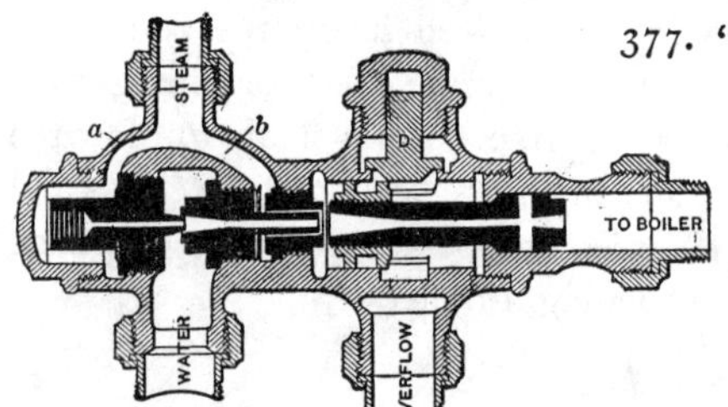

377. "EBERMAN" INJECTOR.—The combining tube slides for regulating the lift and overflow. A single gravity check valve, D, closes the overflow when the current to the boiler is established.

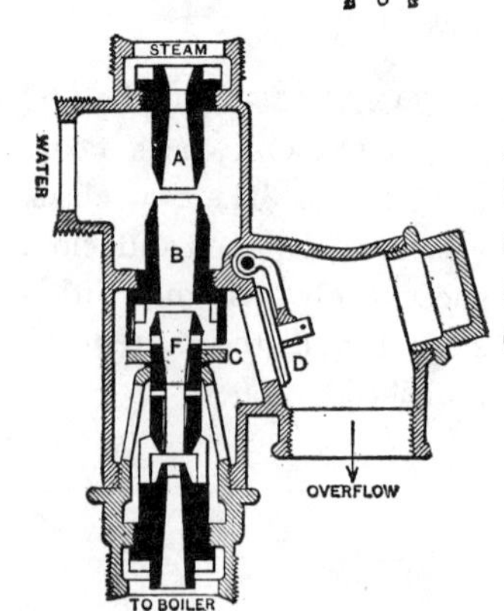

378. "NATHAN" INJECTOR.—A vertical model with four fixed nozzle tubes, tandem. A disc valve, C, closes at the moment the current is established, and the flap valve, D, makes the final closure of the overflow.

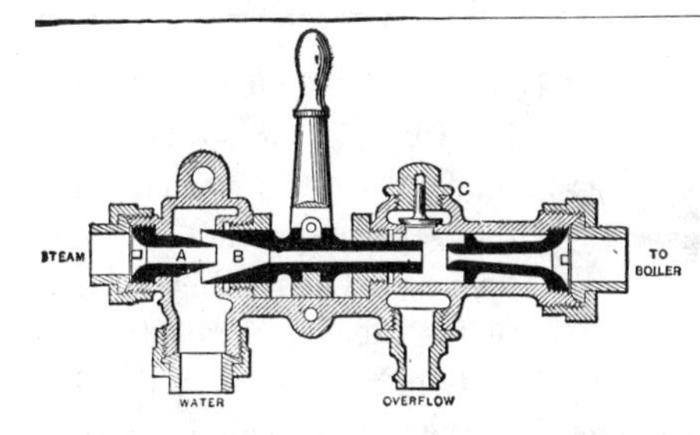

379. "LITTLE GIANT" INJECTOR.—This model has two fixed tubes. The central or combining tube is movable for adjustment. A single automatic check valve regulates the overflow.

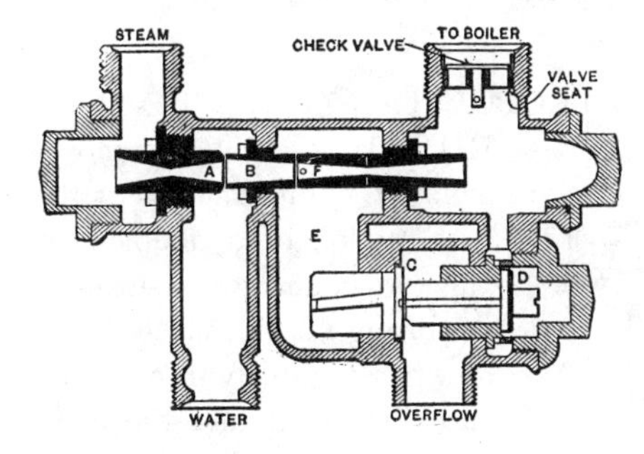

380. "PENBERTHY" SPECIAL INJECTOR.—Has three fixed nozzle tubes. The opening of a detached valve gives steam pressure in the chamber E, and opens both overflow check valves. When the current is established check valve C closes, followed by check valve D.

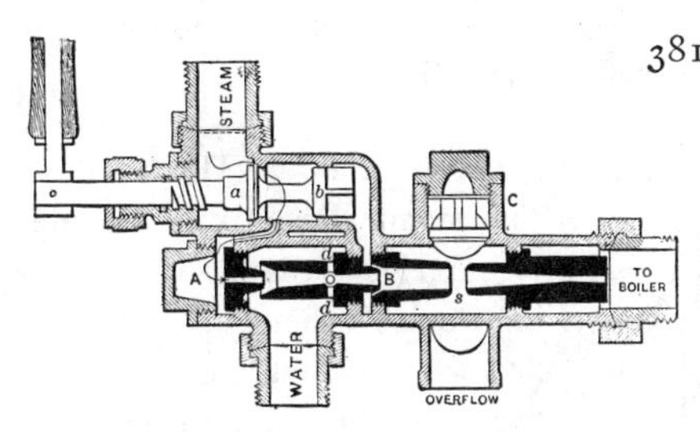

381. "PARK" INJECTOR.—A double tube in tandem, in which the handle has two movements to operate the lift and force nozzles. A self-lifting check valve governs the overflow.

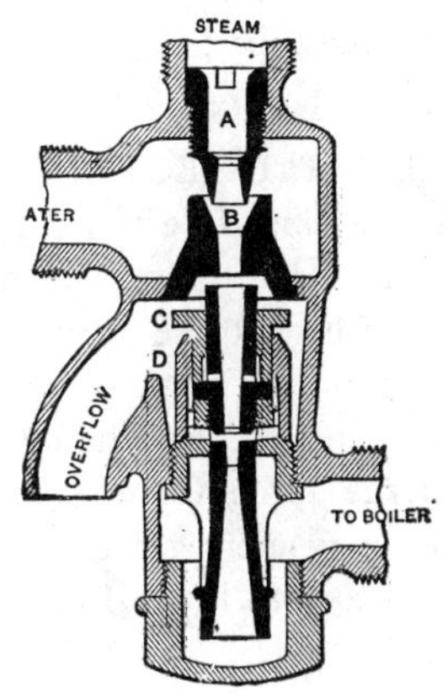

382. SELLERS'" RESTARTING INJECTOR.—In this model all the tubes are fixed. Two concentric check valves, C, D, guided by the combining tube, are operated by the pressure in the combining tube at the moment that the water reaches it, closing the overflow.

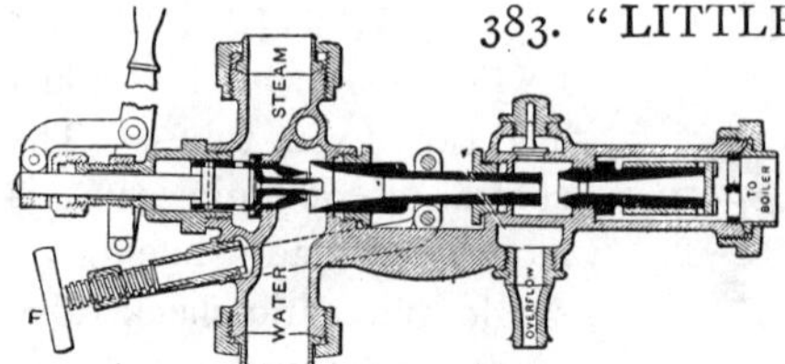

383. "LITTLE GIANT" LOCOMOTIVE INJECTOR.—In this model the lift is started when the separate steam valve is opened. The forcing or combining tube is movable for regulation by a screw and yoke, F. A movement of the handle opens the injection nozzle, and closes the lift nozzle ports.

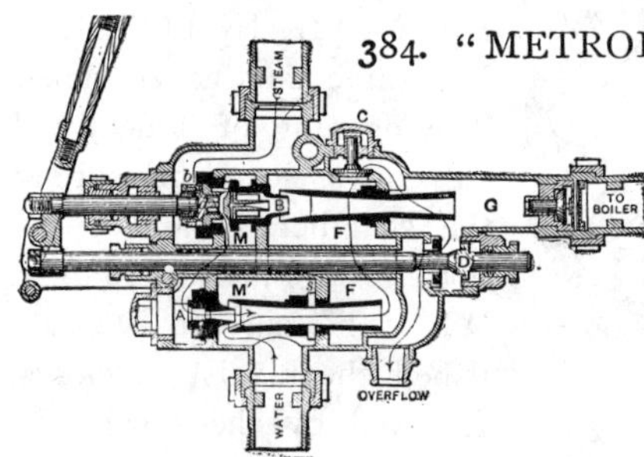

384. "METROPOLITAN" DOUBLE-TUBE INJECTOR.—The first movement of the handle opens the first section of a double-beat valve at *b*, and gives steam to the lifting nozzle A; the overflow passing freely through the check valve C, and the open valve at D. A further movement of the handle opens the second section of the double-beat steam valve B, and closes the overflow valve D.

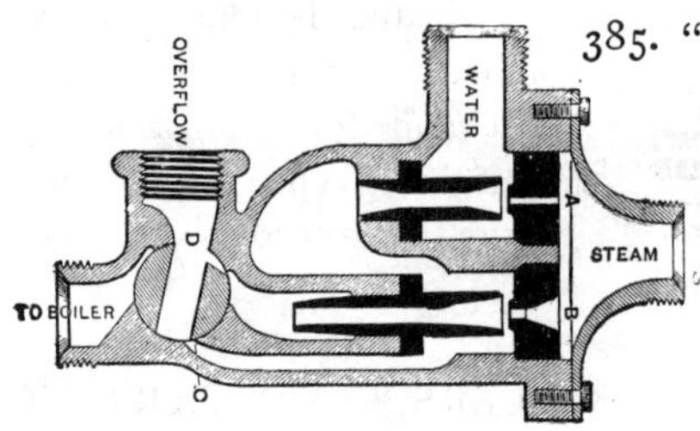

385. "BROWNLEY" INJECTOR.—The steam flows to the double-jet nozzles without any regulating device other than the overflow cock, which by this peculiar construction relieves both lift and force tubes.

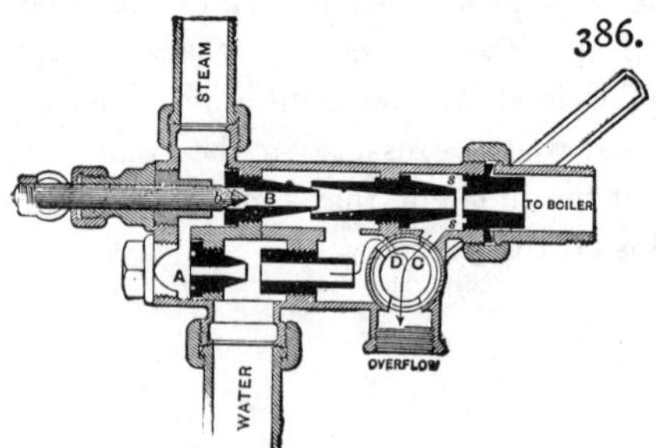

386. "LEADER" INJECTOR.—A double-tube injector. A separate valve gives steam to the lifting nozzle A, with the overflow cock open. The first movement of the handle opens the force valve *b*; a further movement closes the overflow to both lift and force tubes.

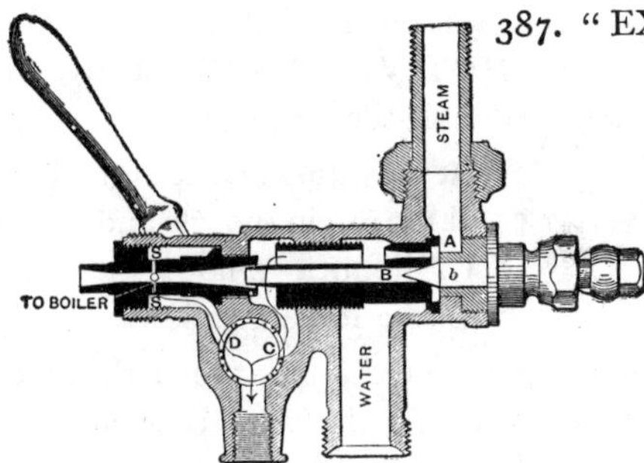

387. "EXCELSIOR" INJECTOR.—A separate valve gives steam to the lifting nozzle A, the overflow cock D C being open. The first movement of the handle opens the conical valve *b*; a further movement closes the overflow cock D C to both the lifting and force overflow S.

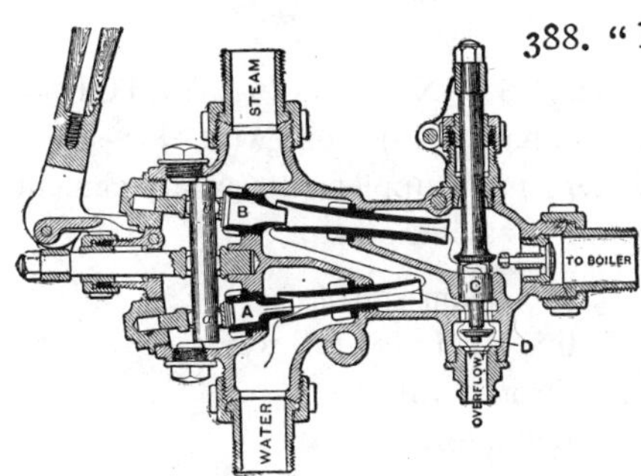

388. "KORTING" INJECTOR.—A double-tube automatic movement by which the difference in area of the valve discs at A and B allows the balance lever to open the lifting nozzle first and, by a further movement of the handle, opens the force nozzle B. The overflow is self-adjusting for both nozzles.

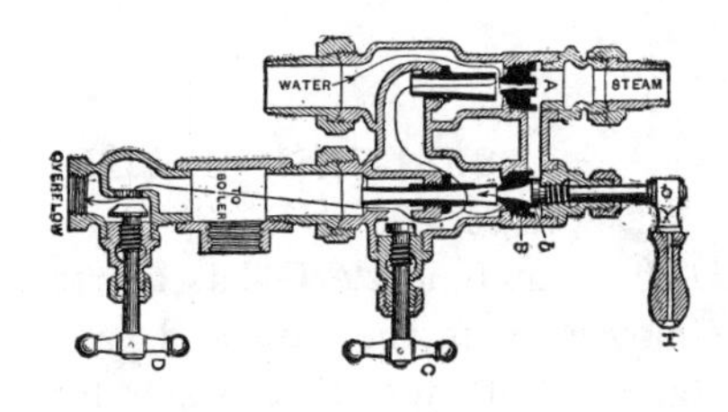

389. "HANCOCK" INSPIRATOR.—A double-tube injector. The tube A lifts the water and starts the circulation through the overflow, when the steam nozzle B is opened and valves C and D are closed.

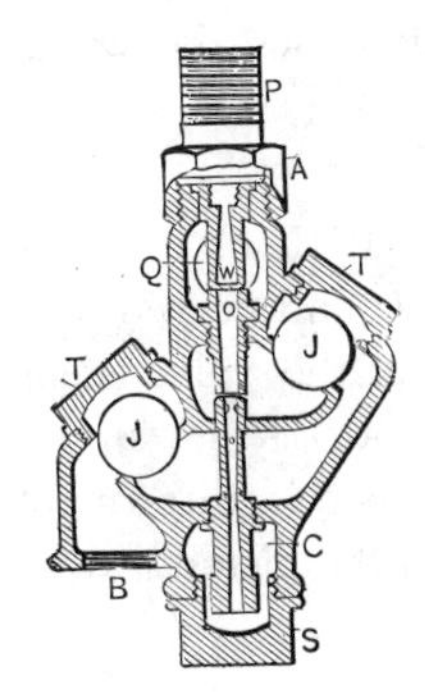

390. BALL-VALVE INJECTOR, automatic in action.

J, J, ball valves.
P, steam inlet.
W, inverted nozzle.
Q, suction inlet.
B, overflow.
C, side outlet to boiler.
S, cap.

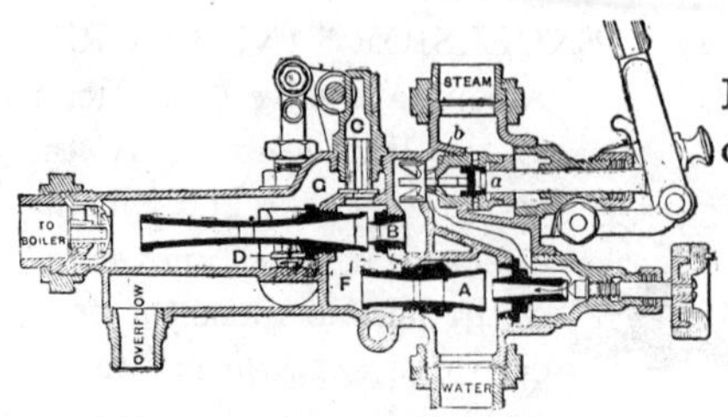

391. "HANCOCK" LOCOMOTIVE INSPIRATOR, a double-tube injector.

A, the lifting nozzle and tube.
B, the forcing nozzle and tube.
C, the lift overflow.
D, the force overflow.

Two movements of the handle are required for starting; the first opens the starting valve *a* and overflow D, with valve H open. A further pull of the handle opens the force valve *b*, and the pressure closes the overflow valve D.

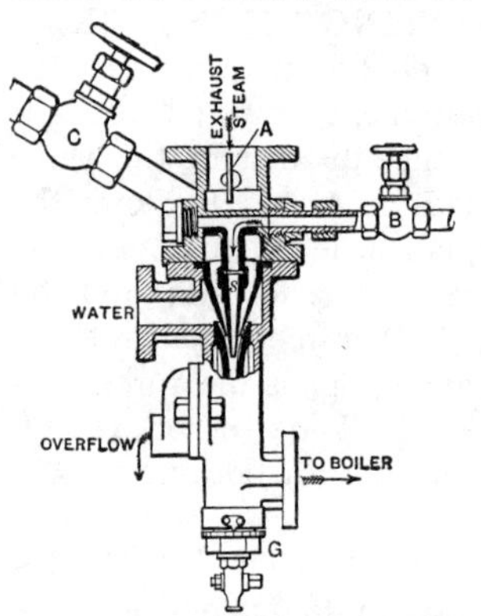

392. "STANDARD" INJECTOR.—An exhaust injector with live-steam starter and supplementary attachment for a live-steam injector.

B, live-steam starter.
C, live steam for full work.
A, throttle valve.
G, regulator.

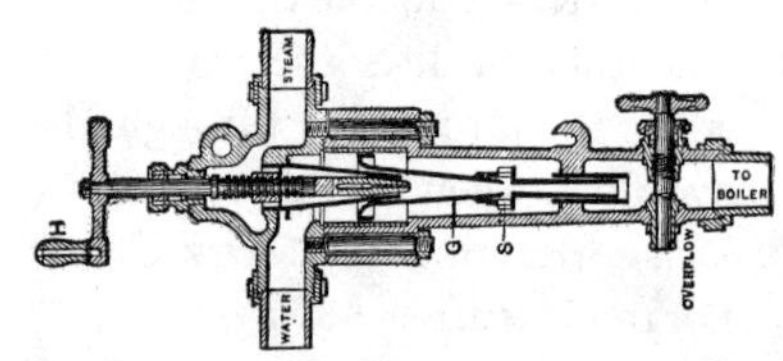

393. "SELLERS'" SELF-ADJUSTING INJECTOR.—The water nozzle G has a free movement in the case and cage at S. With too much water for the steam, the nozzle is pushed back and partially closes the water area. Self-adjusting.

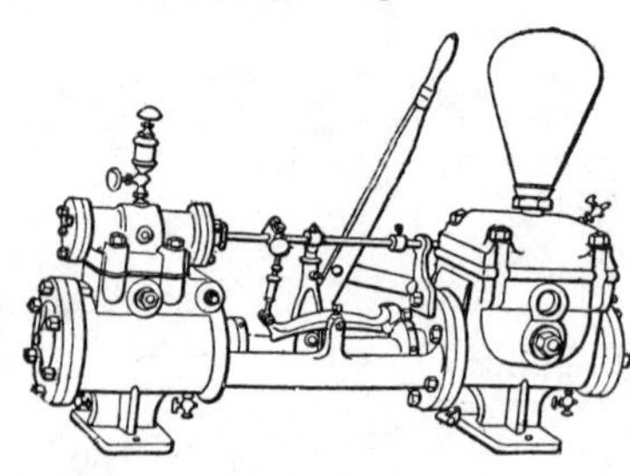

394. STEAM PUMP, with rotating piston valve and curved tappet. An arm on the valve stem is linked to the end of the curved tappet. The tappet is thrown by a roller clamp on the piston rod.

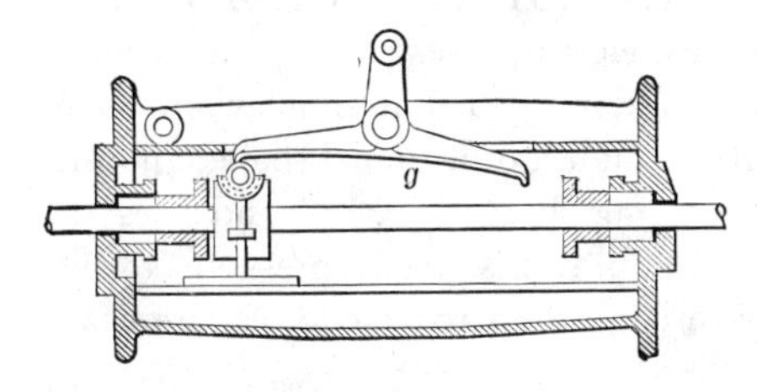

395. "MISCH'S" VALVE TAPPET, for a steam pump. A three-armed lever rocked by a roller travelling with the piston rod.

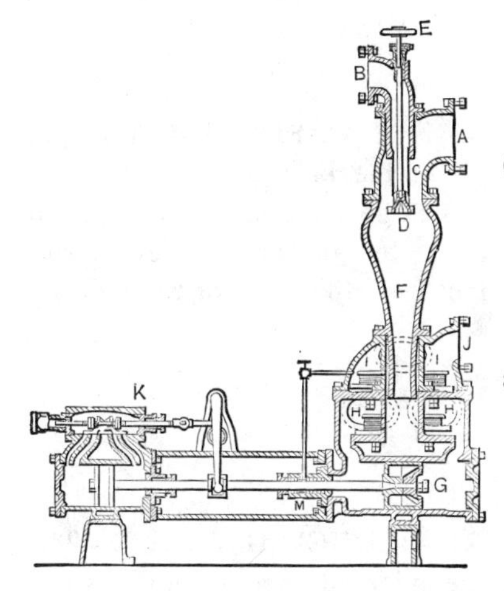

396. INDEPENDENT JET CONDENSER PUMP.

A, exhaust inlet from engine.

B, water inlet.

C, water nozzle.

D, spray valve regulated by screw spindle and wheel E.

F, spray chamber.

J, water discharge from pump.

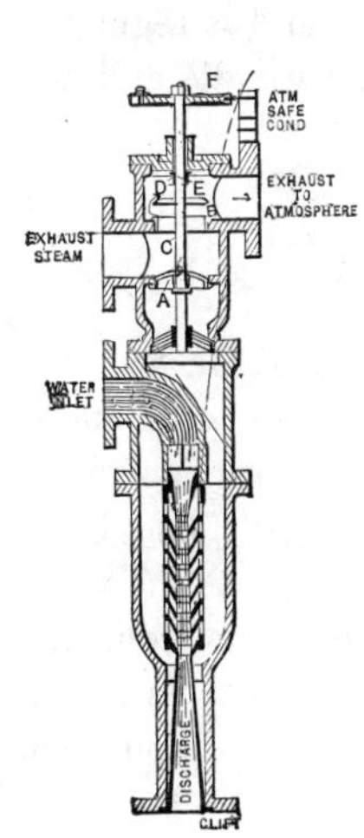

397. EJECTOR CONDENSER, with automatic three-way valve. By the operation of two valve discs on a single stem the exhaust steam is passed to the atmosphere, or is condensed by the multiple nozzle water jet. "Korting" model.

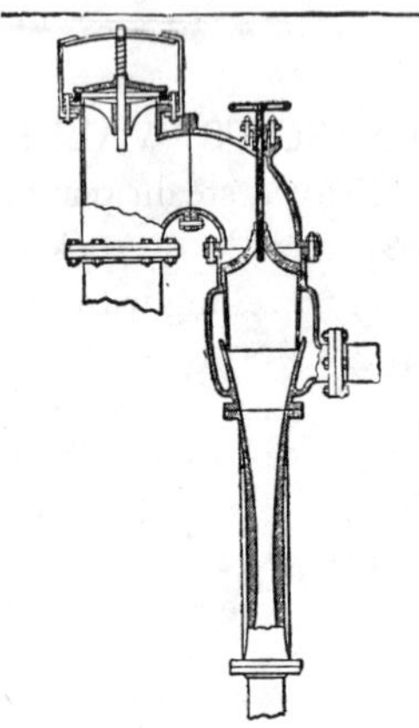

398. EXHAUST JET CONDENSER.—The exhaust steam passes through a cylindrical nozzle and meets a thin annular stream of water at the mouth of a funnel-shaped nozzle. The converging sheet of water condenses the steam, and prevents back pressure by its velocity through the narrow end of the nozzle.

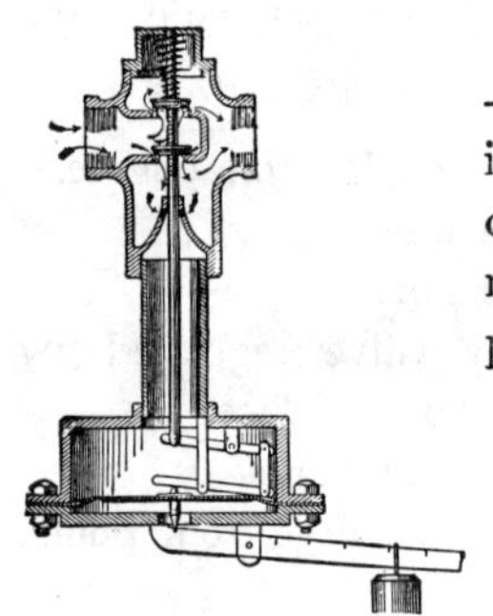

399. BALANCED REDUCING VALVE.—The spindle of the balanced throttle discs is attached to a large diaphragm by levers, and counterbalanced by an outside lever with movable weight for adjustment of the reduced pressure.

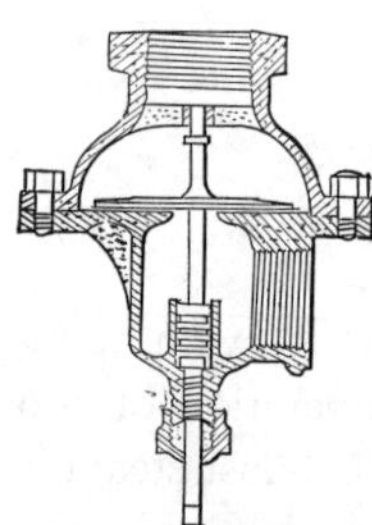

400. PRESSURE REDUCING VALVE.—The back pressure on the enlarged area of the disc valve regulates the flow of steam or air, and is regulated by the weight at the bottom of the spindle and the adjusting screw.

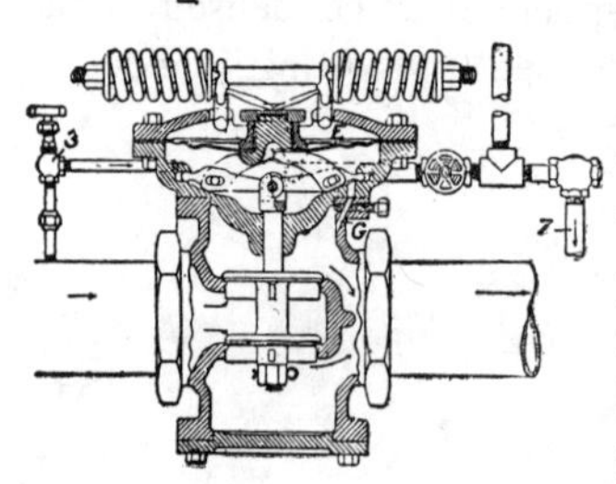

401. "FOSTER" PRESSURE REDUCING VALVE.—The balanced valve is opened by a diaphragm against the pressure of springs. The high-pressure connection, 3, starts the valve into position. The passage from the low-pressure side at G admits steam from low-pressure side to the diaphragm, which is connected to the valve spindle by toggle joints.

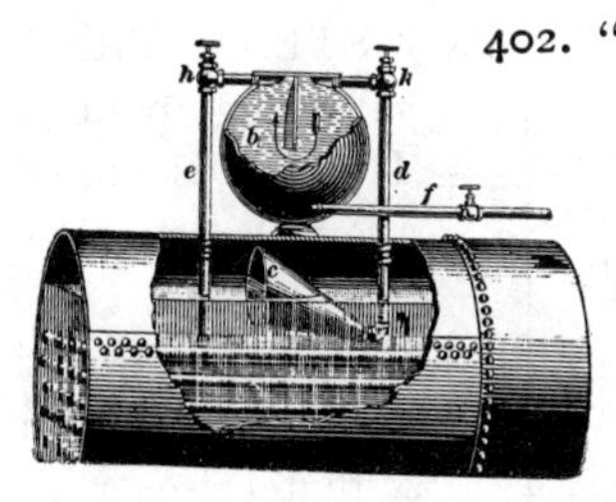

402. "HOTCHKISS" BOILER CLEANER, for removing the surface scum from steam boilers. The circulation through the settling globe is produced by the difference in temperature in the rising pipe, *d*, and the return pipe, *e*. The large area in the globe allows the dirt to settle, to be blown off through the pipe, *f*.

403. FEED-WATER HEATER and surface condenser. Exhaust steam enters at the top, and is condensed on the outside of the tubes. The feed water is circulated through the tubes.

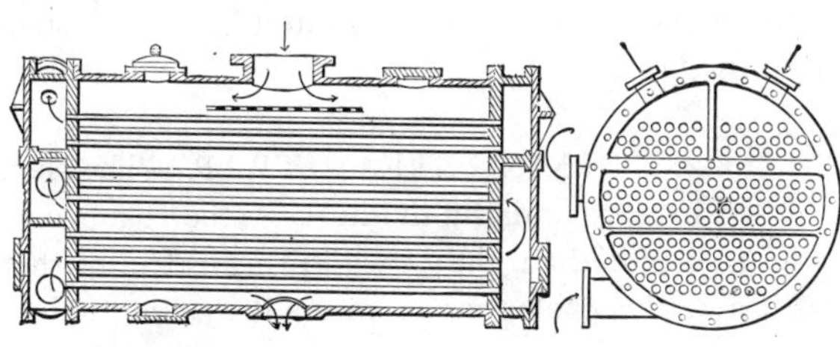

403 a. Cross section.

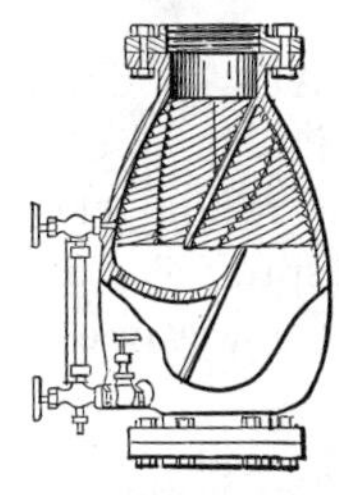

404. STEAM SEPARATOR.—The entrained water in the steam is lodged upon the rough walls, and drips to the strainer and into the pocket, and is drawn off through the valve. The glass gauge indicates the height of water in the pocket.

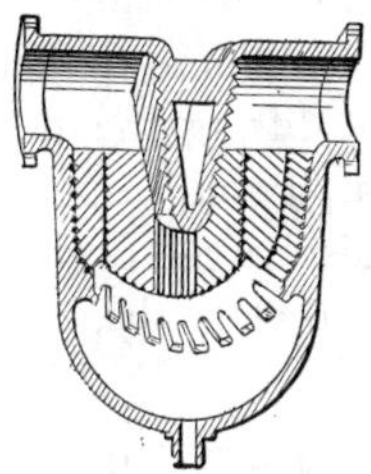

405. STEAM SEPARATOR, in line for horizontal pipes. The corrugated surface catches the water of condensation, which falls through the grating to the recess below. "Austin" model.

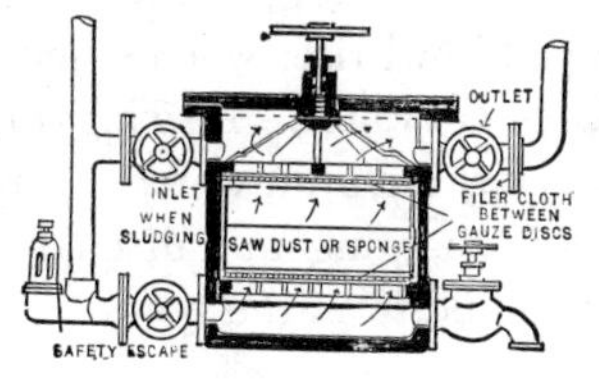

406. FILTER FOR BOILER, feed water. An upward flow. Water enters from the left and flows through felt held between wire gauze and perforated plates. The space may be filled with sponge or coarse sawdust.

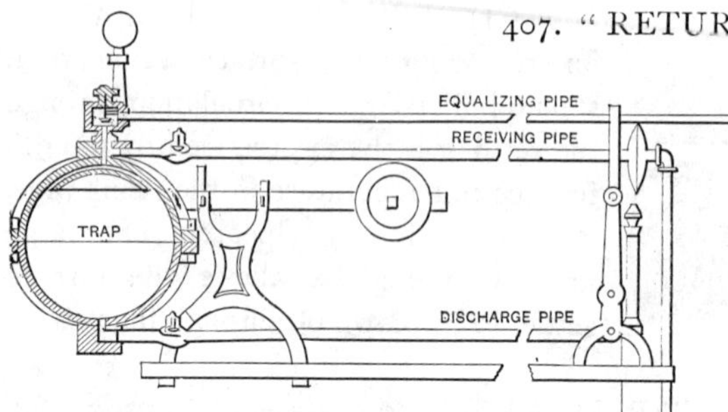

407. "RETURN STEAM TRAP, "Blessing" pattern. The trap is placed above the water line of the boiler. The globe is balanced on a weighted lever so that it rises when empty and falls when filled with water. The movement of the globe up and down trips valves that alternately charge the globe with the water from a heating system and discharges it into the boiler.

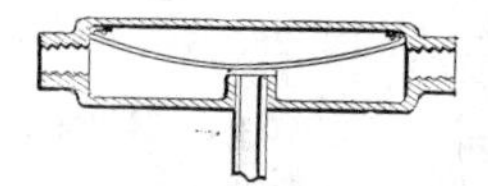

408. SPRING STEAM TRAP.—The shell of iron expands by the heat of the steam at a less rate than the brass spring valve, so that the hot steam closes it and the cooler water opens it by contraction.

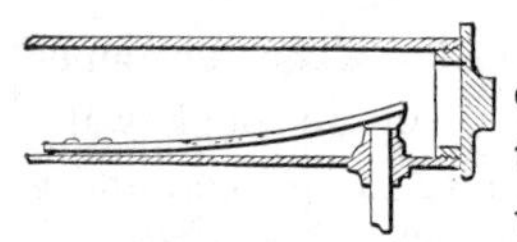

409. SPRING STEAM TRAP.—A differential expansion of the spring itself causes it to open with the water temperature and close with steam temperature. The spring is made of two strips of metal, the upper one of brass and the lower one of steel, riveted together.

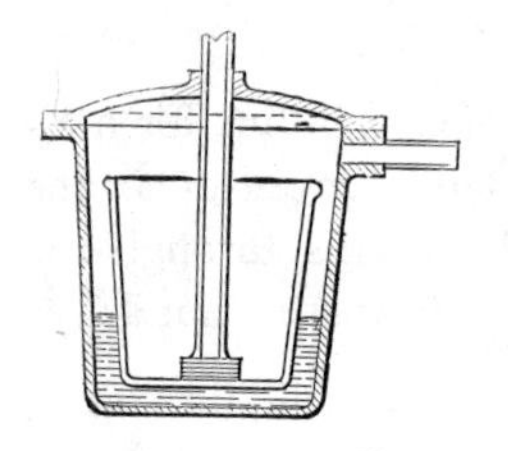

410. STEAM TRAP.—The water condensed in a heating system flows into the trap case and closes the valve by lifting the float. By the overflow into the float, it sinks, opening the valve, and the water is discharged from the float, allowing it to rise and to close the valve.

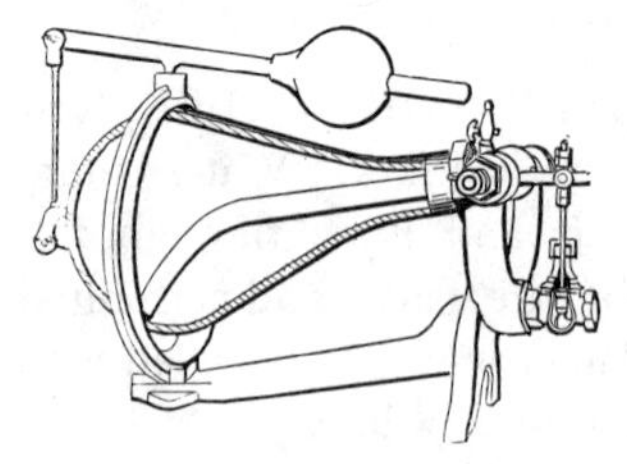

411. "BUNDY" STEAM TRAP.—The pear-shaped bowl rises when empty, and falls when full of water. It swings on trunnions carrying an arm, which operates a valve for charging and discharging the water to and from the bowl.

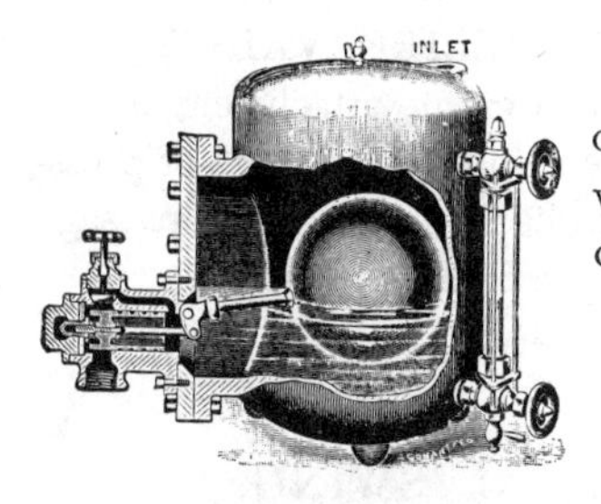

412. STEAM TRAP WITH VALVE, operated by a float. The ingress of water lifts the float and opens the discharge valve. "Curtis" model.

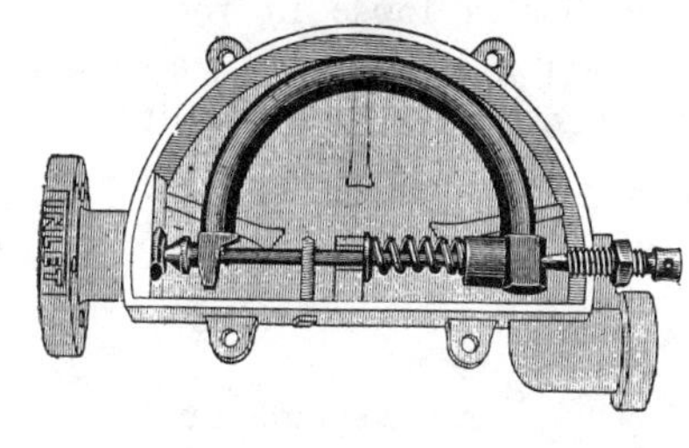

413. "HEINTZ" STEAM TRAP.—The differential expansion of two metals in the semi-circular arc opens or closes the inlet valve. Adjustment is made by the set-screw.

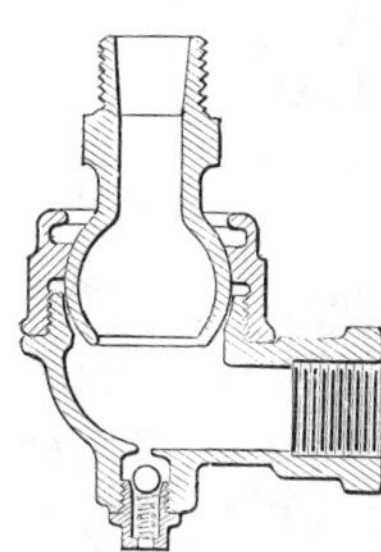

414. "MORAN'S" FLEXIBLE STEAM JOINT and automatic relief valve. A ground globular pipe fitting held in a spherical union joint.

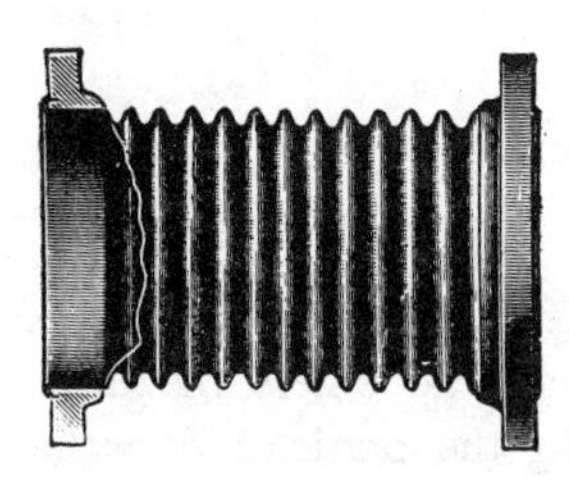

415. CORRUGATED EXPANSION COUPLING, "Wainwright's" model. A hard brass tube, corrugated, gives the tube a longitudinal elasticity to take up the expansion of steam pipes.

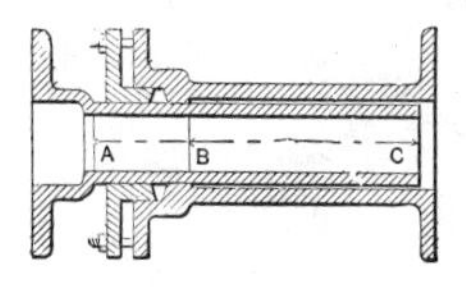

416. FLANGED EXPANSION JOINT.—Used in pipe lines to take up the change in length due to difference in temperature.

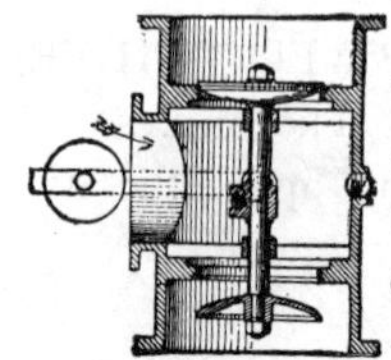

417. AUTOMATIC RELIEF VALVE.—The valve is kept closed by a crank attachment to the spindle and weighted lever outside. Excess of pressure raises the stem and discs, throttling the passage of steam and relieving the back pressure.

418. HORIZONTAL SWING CHECK VALVE.—The disc is loose in the swing frame and may be reground tight by a socket wrench passed through the plug opening.

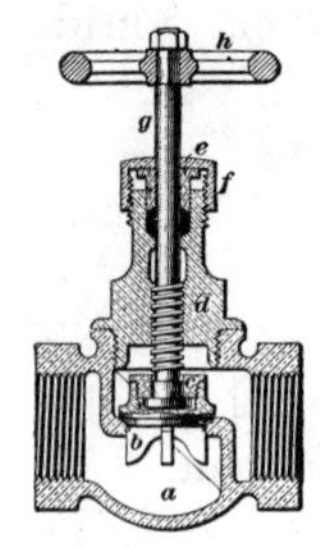

419. GLOBE VALVE.

a, the body.	*c*, the spindle nut.
d, the bonnet.	*e*, gland.
g, the spindle.	*f*, gland nut.
b, the winged disc.	*h*, wheel.

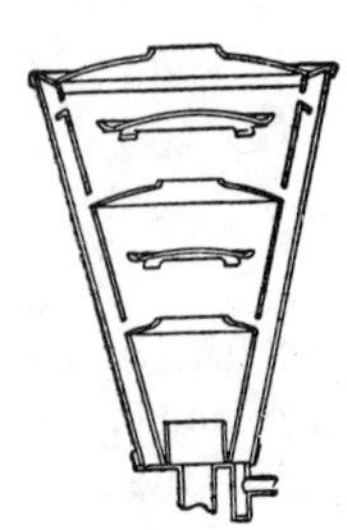

420. EXHAUST STEAM HEAD. — The exhaust steam is deflected by perforated discs and cap plates, which separate the water to drip between the inner and outer shell.

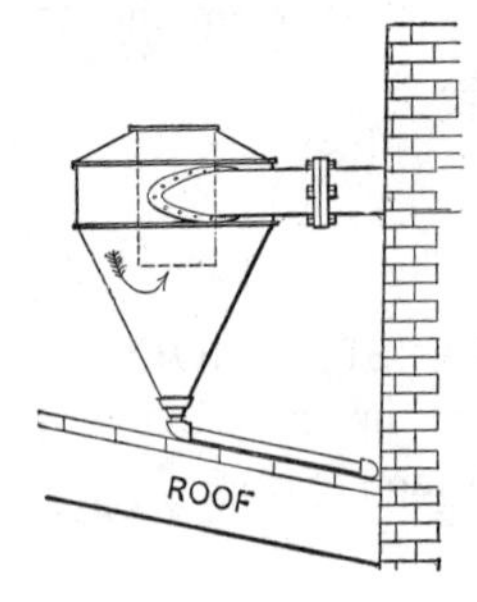

421. CENTRIFUGAL EXHAUST HEAD. —The exhaust steam head enters the drum tangentially, throwing the particles of water against the outer surface to drip to the bottom.

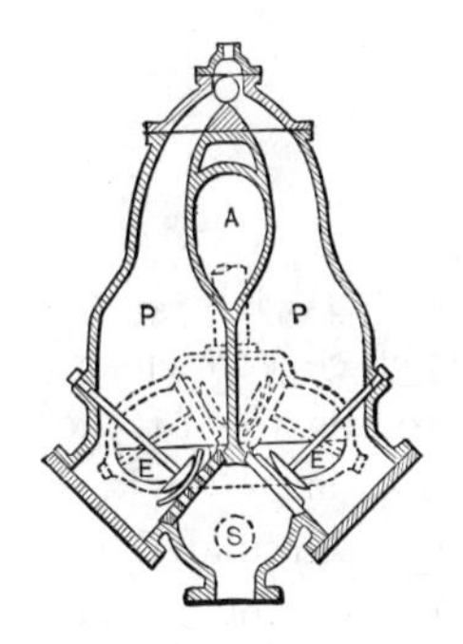

421*a*. THE PULSOMETER STEAM PUMP.—Water is forced from each chamber alternately by the steam pressure, while the opposite chamber is filled by the vacuum caused by the condensation of the steam in contact with the wet surface. The ball valve is very lightly balanced and is thrown over by the alternating vacuum and steam pressure.

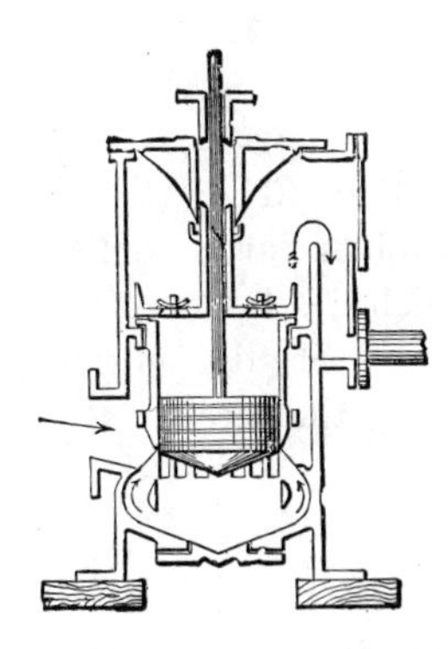

421*b*. THE EDWARDS AIR PUMP.—Has no suction valves. The ports in the cylinder are opened by passing the piston to the bottom of the cylinder. The water and air enter above the piston and is discharged above. The discharge valves are sealed by water held back by the dam. The piston rod is sealed by a water filled cup.

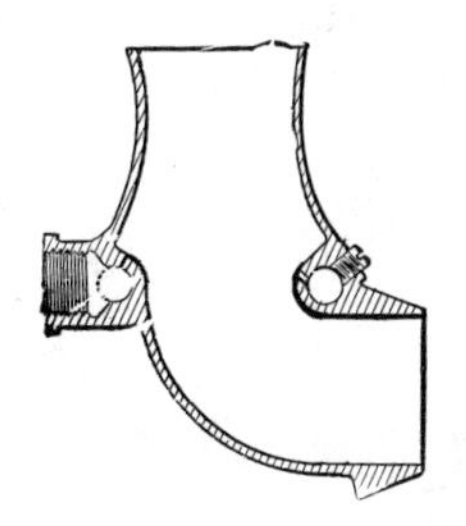

421*c*. STEAM SOOT SUCKER for cleaning boiler tubes by drawing the soot and ashes from the tubes by an annular steam jet.

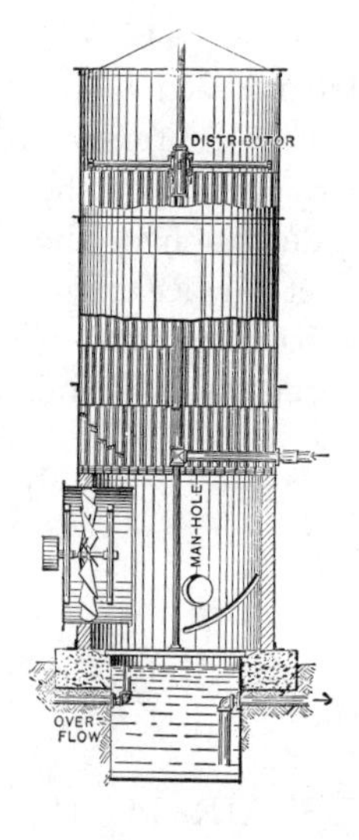

421*d*. AIR COOLING TOWER.—For cooling the water of a surface condenser. The hot water is forced to the top of the tower and distributed over a large surface of tile through which air is circulated by the large fan at the bottom of the tower. The water much cooled drips to the tank below from which it is pumped for use again.

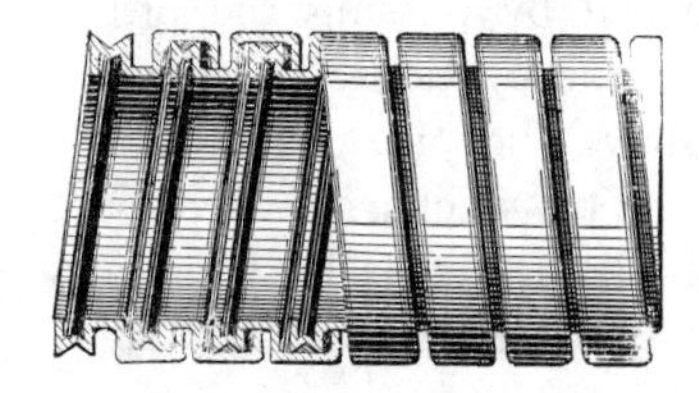

421*e*. FLEXIBLE METALLIC HOSE.—The joints are packed with rubber, which lies between the overlapping edges of the corrugated tape forming the screw.

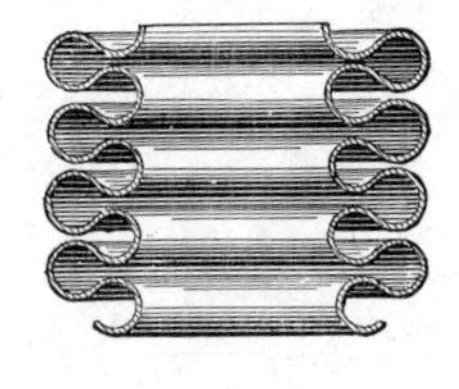

421*f*. FLEXIBLE METALLIC TUBING.—The corrugations are deep indented rings spun or pressed from a plain tube. It may be also made spiral.

Section VI.

MOTIVE POWER.

Gas and Gasoline Engines, Valve Gear and Appliances, Connecting Rods and Heads.

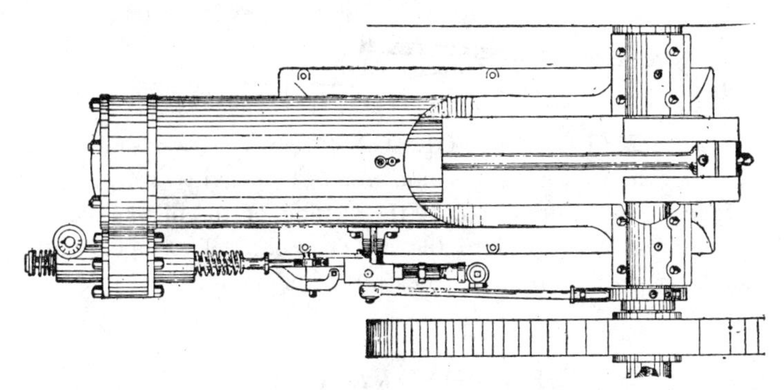

422. GASOLINE ENGINE, "Olds" model. Plan showing location of valve chest and valve gear, operated from an eccentric with an alternating sector gear for an impulse at every other revolution.

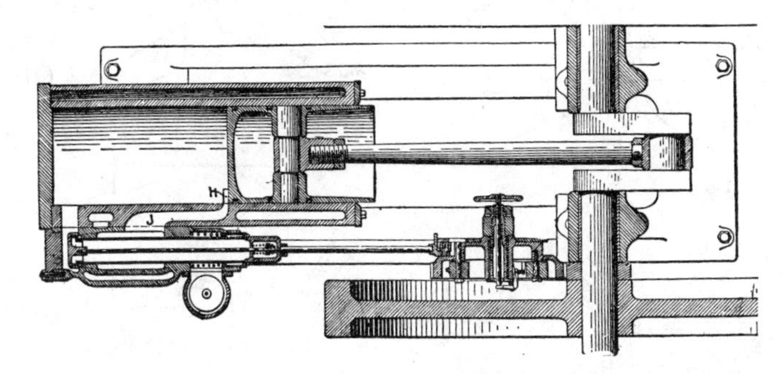

423. SECTIONAL PLAN OF A GASOLINE ENGINE.—Four-cycle type, with exhaust port opened by the piston at the end of the stroke, and continued exhaust through an annular valve around the inlet valve. The charge is heated and vaporized in the valve chamber by the exhaust. "Olin" model.

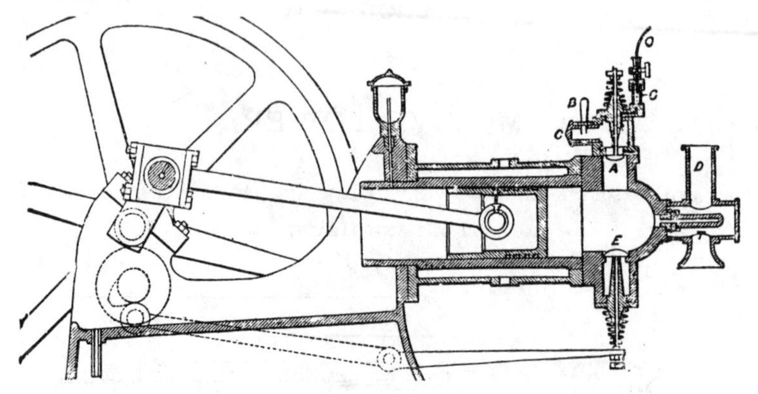

424. SIMPLE GAS OR GASOLINE ENGINE.—A, inlet valve; E, exhaust valve; gasoline enters by gravity at G, regulated by a faucet. Air enters at B by the suction of the piston, atomizing the gasoline as it drops into the air chamber. The tube igniter is heated by a gasoline burner beneath the bell mouth.

425. GASOLINE ENGINE VALVE GEAR.—The centrifugal action of the weights on the reducing gear operates a bell crank that directs the exhaust push rod on or off the cam. "Olin" model.

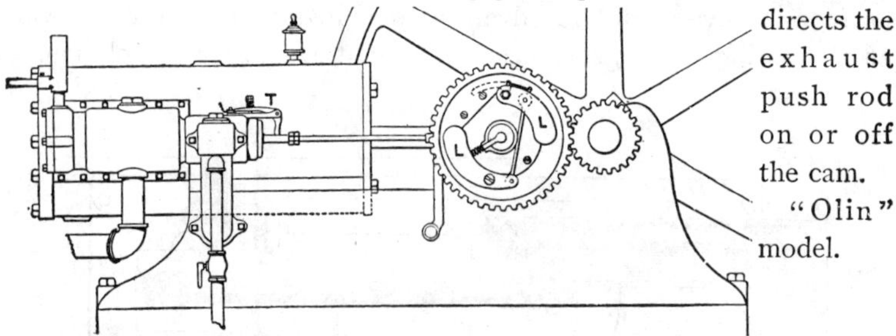

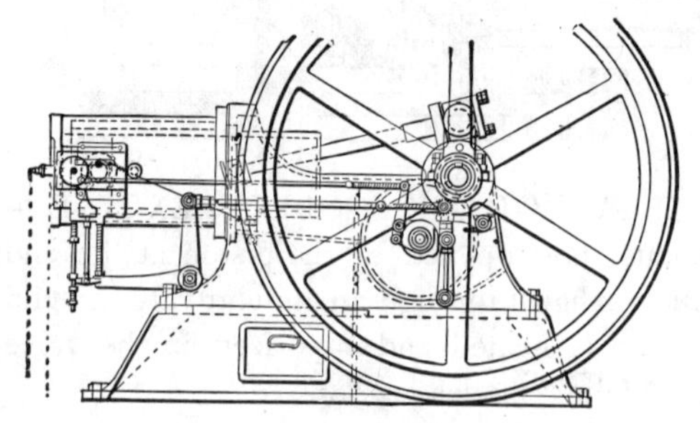

426. GAS ENGINE, "Union" model. A four-cycle motor with half-reducing gear; push-rod lever and two push rods for governing charge and exhaust.

427. GASOLINE CARRIAGE MOTOR.—Four cycle or compression type. Ribs on cylinder for air cooling. H is the carburetter with wire-gauge atomizer; O, gasoline feed-pipe. Warm air is drawn into carburetter from the pipe over the Bunsen burner, G, by the suction of the piston; it is then saturated with gasoline vapor, and returned by a separate pipe to the inlet valve, C.

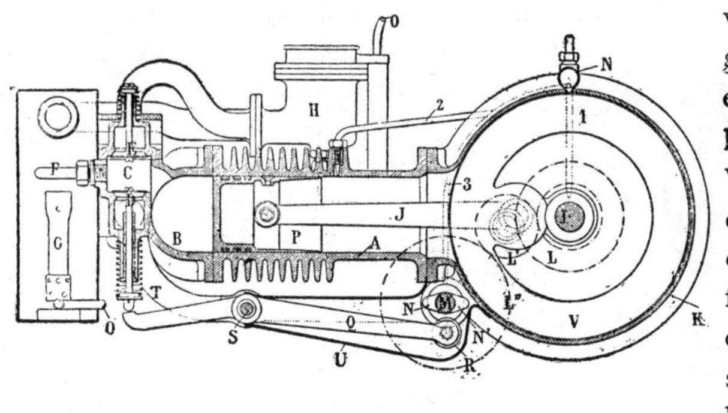

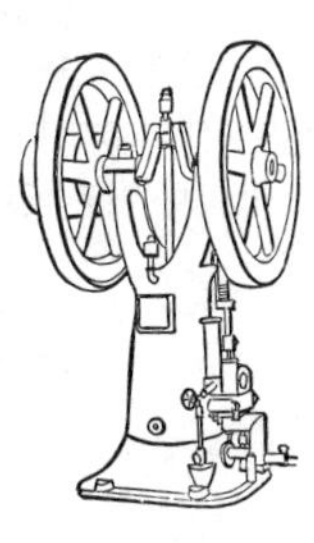

428. VERTICAL GASOLINE ENGINE, "Webster" pattern. The cylinder and water jacket form part of the framework of the engine. A four-cycle type.

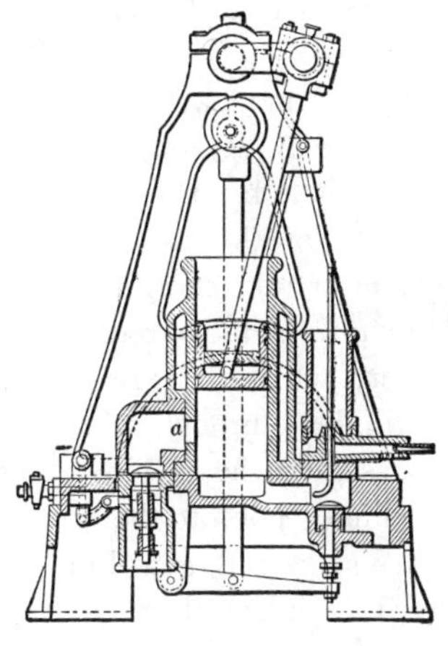

429. VERTICAL GAS ENGINE, "Root" model. Four-cycle compression, with double explosion. *a* is a secondary chamber and port, closed about half-stroke, shutting off part of the charge during compression, which is exploded during the impulse stroke of the piston.

430. VERTICAL KEROSENE OIL ENGINE, "Daimler" model. The oil is vaporized by the heat of the exhaust, and forced into the cylinder, with the proper proportion of air for explosive combustion, by the downward stroke of the piston and compression in the crank chamber. The upward stroke charges the crank chamber with air and vapor.

431. "DIESEL" MOTOR.—A, cylinder; *p*, air pump; *y*, air-pump lever; T, air receiver. Air is compressed by the pump to 450 lbs. per square inch, and stored in the receiver. Oil is fed by a small pump to the inlet-valve chamber, where it is atomized by entering the cylinder with the compressed air. Explosion every other revolution.

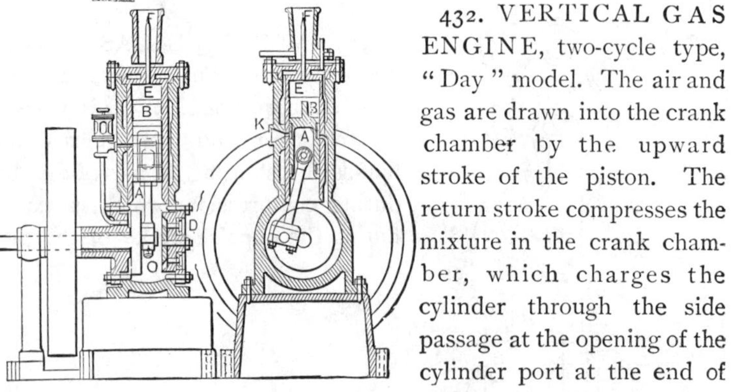

432. VERTICAL GAS ENGINE, two-cycle type, "Day" model. The air and gas are drawn into the crank chamber by the upward stroke of the piston. The return stroke compresses the mixture in the crank chamber, which charges the cylinder through the side passage at the opening of the cylinder port at the end of the down stroke of the piston. E, clearance space; B, guard on piston; A, crank chamber; F, tube igniter; D, O, inlet valves.

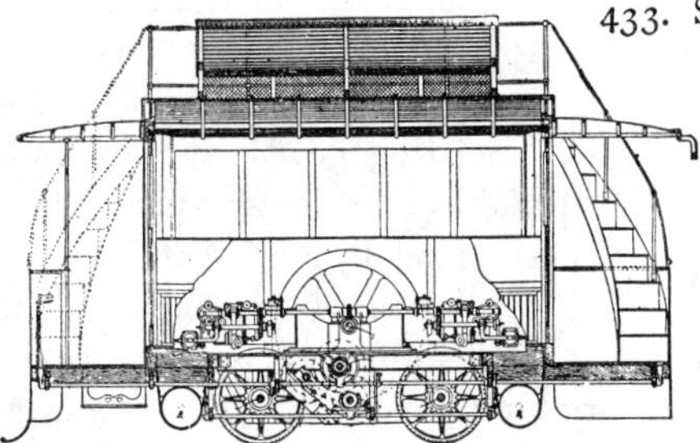

433. STREET RAILWAY GAS MOTOR PASSENGER CAR, German model. The motor consists of two cylinders on opposite sides of the crank shaft, placed under the seats. The fly-wheel is behind the seats. The power is transmitted to the axles through gears, sprockets, and chains, with friction regulation. Motor runs continually. Compressed gas is stored in cylinders under the car floor.

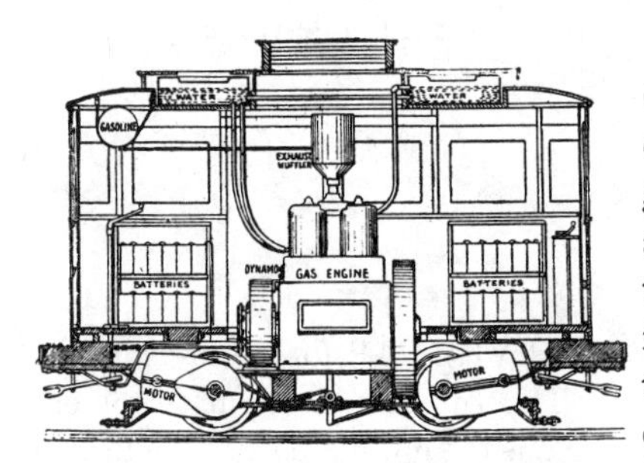

434. GASOLINE MOTOR CAR.—The gasoline motor runs constantly, operating an electric generator which charges the storage batteries, that in turn supply the current as required for the intermittent or variable work of the electric motors geared to the car axles.

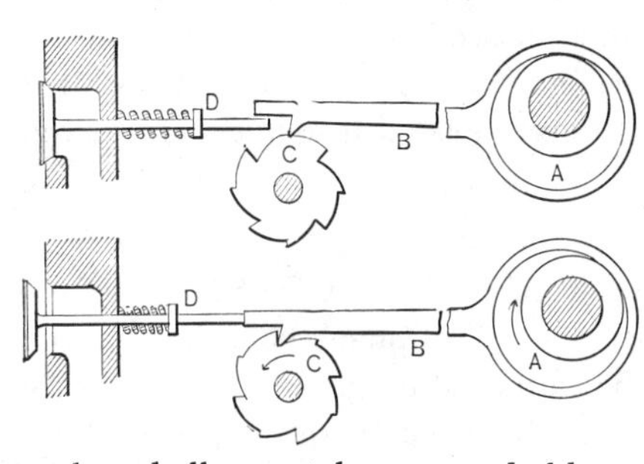

435. VALVE GEAR for a gas engine.—A simple device for opening the exhaust valve of a four-cycle motor. The eccentric gives the push rod a forward stroke at each revolution of the shaft. The ratchet wheel C has a friction resistance, with every other tooth a shallow notch, so as to hold up the lip of the push rod at every second revolution of the shaft and make a miss-hit on the valve rod. At the next revolution the lip falls into a deep notch and the push rod opens the exhaust valve.

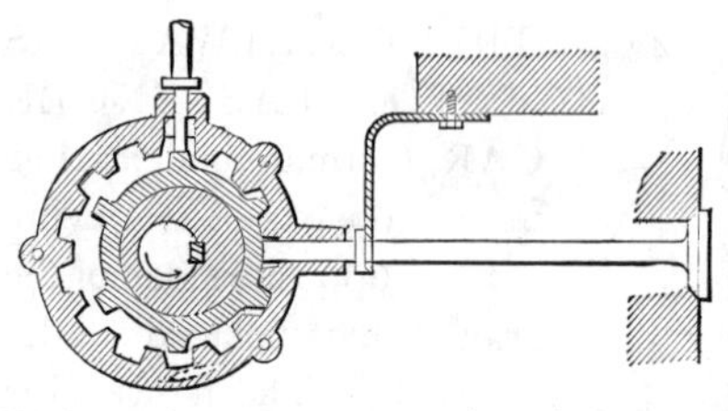

436. VALVE GEAR, for a four-cycle gas engine. The cam is fixed to the engine shaft. The inner ring gear is swept around within the outer fixed gear, skipping by one tooth at each revolution of the engine shaft. This makes a contact of a ring-gear tooth with the exhaust-valve rod at every other revolution, necessary for the operation of a four-cycle motor.

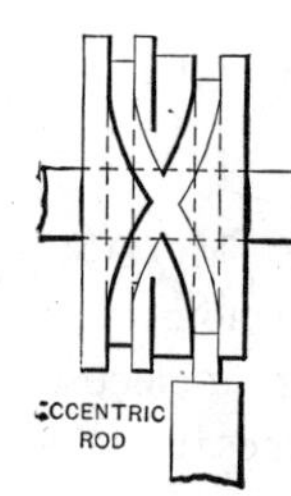

437. DOUBLE-GROOVED ECCENTRIC, for two lengths of rod thrown alternately by traversing the push rod in the cross grooves, also for single-valve rod throw for four-cycle gas engine.

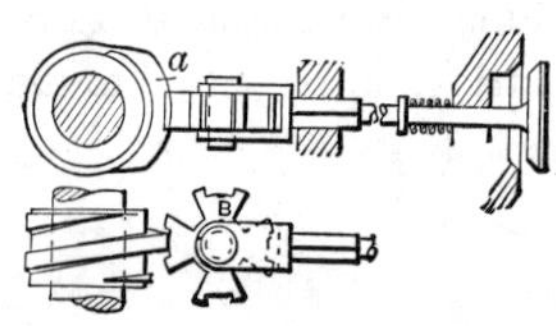

438. VALVE GEAR for a four-cycle gas engine. The two-thread worm on the engine shaft has the middle part of the thread extended to form a cam. The four-part gear, B, revolves by the action of the worm, and at every other revolution the cam section of the worm runs into the recess of the revolving gear, and the valve rod is not operated, thus opening the exhaust valve at every second revolution as required.

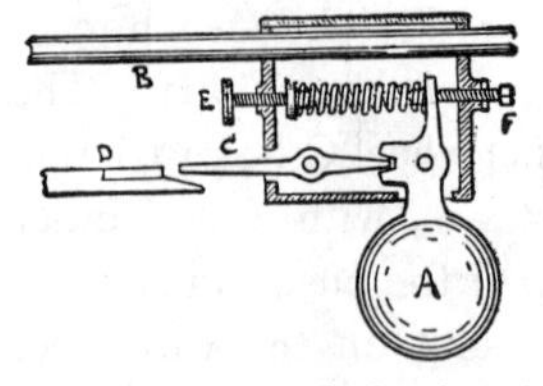

439. PLUMB-BOB GOVERNOR for a gas engine. The plumb-bob, A, is pivoted in a box attached to the exhaust valve push rod. The back motion of the push rod produces a forward motion of the bob, acting like a pendulum, and a downward motion of the pick blade, C, bringing it in contact with the valve spindle, D. The spring-end screws, E and F, are for the adjustment of the motion of A.

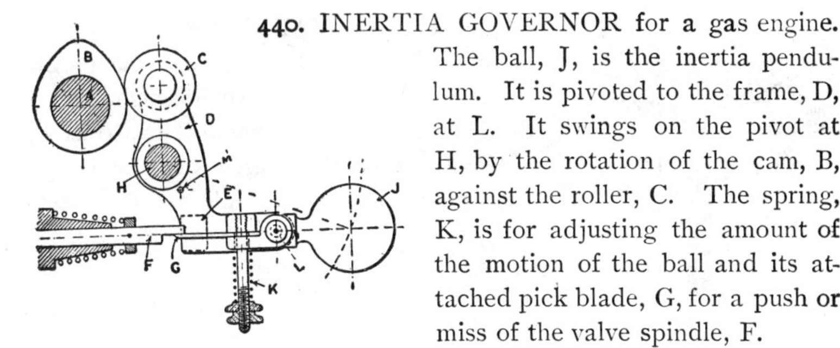

440. INERTIA GOVERNOR for a gas engine. The ball, J, is the inertia pendulum. It is pivoted to the frame, D, at L. It swings on the pivot at H, by the rotation of the cam, B, against the roller, C. The spring, K, is for adjusting the amount of the motion of the ball and its attached pick blade, G, for a push or miss of the valve spindle, F.

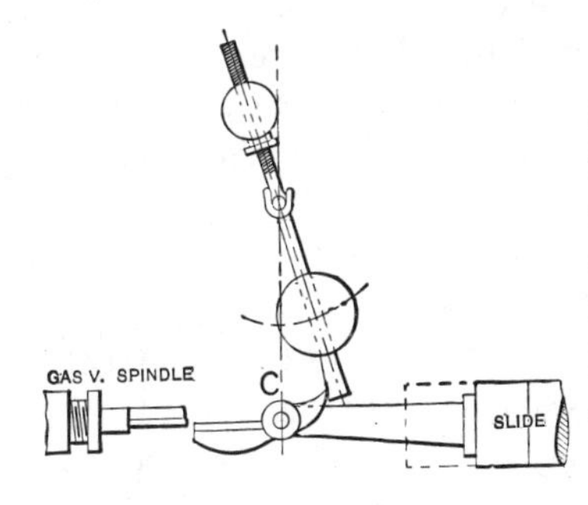

441. PENDULUM GOVERNOR for a gas engine. The pendulum is adjusted by the distance of the small compensating ball to vibrate synchronously with the push rod at the required speed of the engine. Increased speed releases the clip, and a miss charge is made.

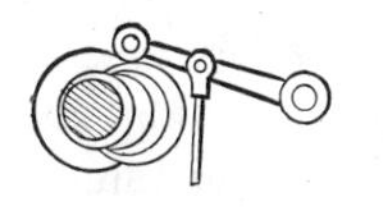

442. DIFFERENTIAL CAM THROW, by the transverse motion of a rolling disc on a lever or by direct thrust. Much used on the valve gear of gas engines. The rolling disc is traversed by the governor from one cam to another.

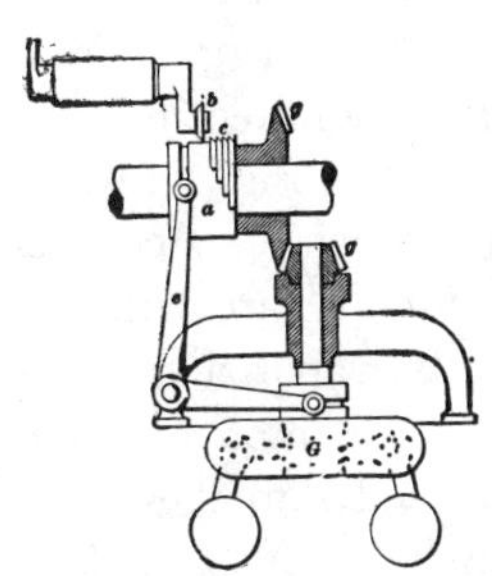

443. GOVERNOR AND VARIABLE CAM for a gas engine. The centrifugal movement of the governor balls slides the sleeve on the governor shaft, and also the variable cam sleeve, *a*, on the driving shaft, by the bell-crank lever, *e*. The disc roller, *b*, on an arm of a rock shaft, rolls upon one or the other cams at *c*, thus varying the movement of the inlet valve, which is connected to another arm of the rock shaft.

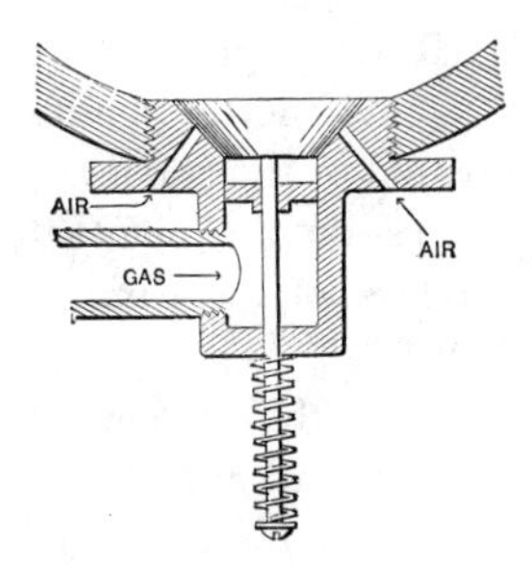

444. INLET VALVE for gas engine. A valve disc slightly held in contact with the seat by the spring. Air holes should be drilled close together around the valve seat, so that combined air area shall be larger than the area of the gas inlet.

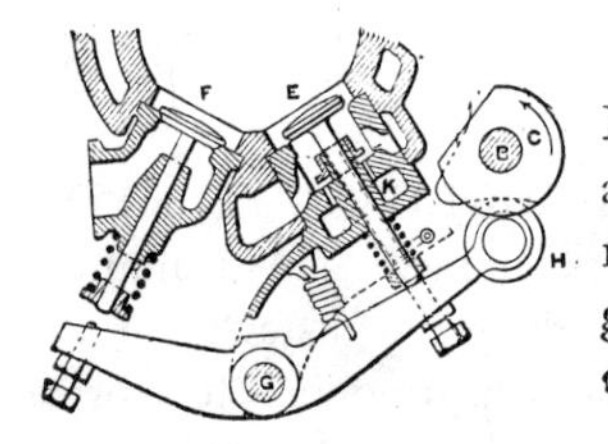

445. GAS ENGINE VALVE GEAR.—E, Inlet valve; F, exhaust valve. Valves are operated by a bent lever, with sliding roller H and double cam C, which by a groove rides the roller alternately on to the cams.

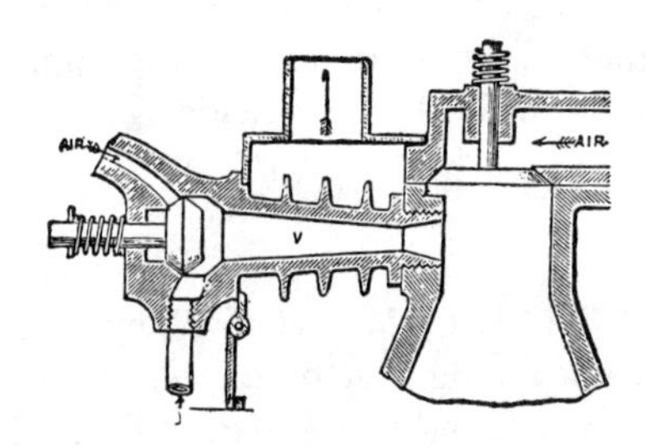

446. GASOLINE VAPORIZER. —The inlet nozzle, V, is ribbed on the outside and is enclosed in a chamber through which the exhaust passes. Gasoline and air are drawn into the nozzle regulated by the small valve, and additional air for the explosive mixture is drawn by the piston through the large valve. "Capitaine" motor.

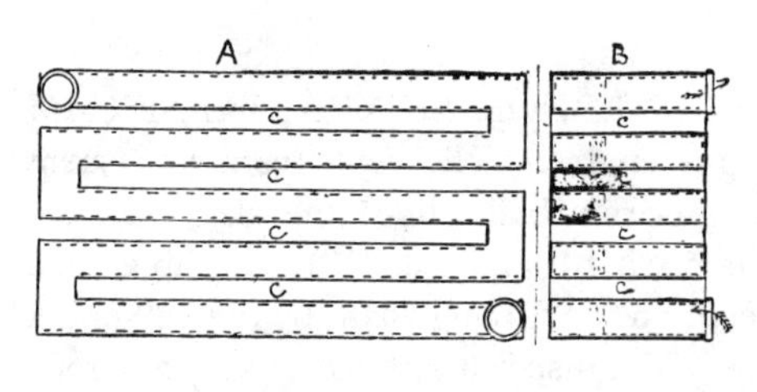

447. CARBURETTER for making air gas from gasoline; non-freezing. A, plan—a zig-zag series of chambers with spaces between for air circulation to keep its vaporizing walls warm; B, a vertical section; *c, c, c*, open spaces. Canton or other flannel wrapped over wire gauze frames is pushed into the longitudinal spaces before the ends are soldered; may be made of tinplate.

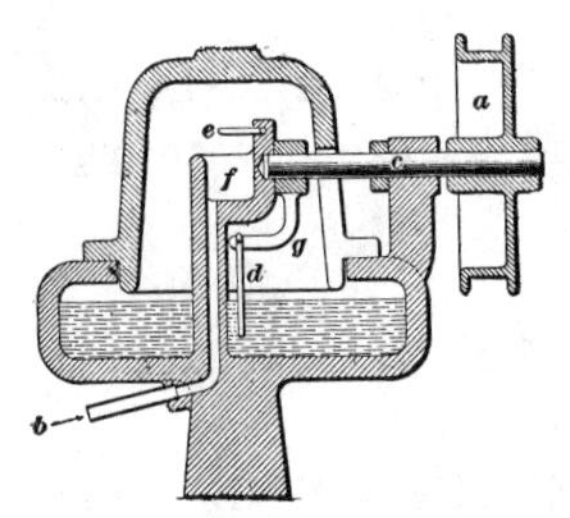

448. AUTOMATIC OILER.—Much in use on explosive motors. Shaft *c*, and crank *g*, with the dip wire *d*, are revolved by a belt dropping the oil on the wiper *e*, into the small tank *f*, from which it flows to the cylinder.

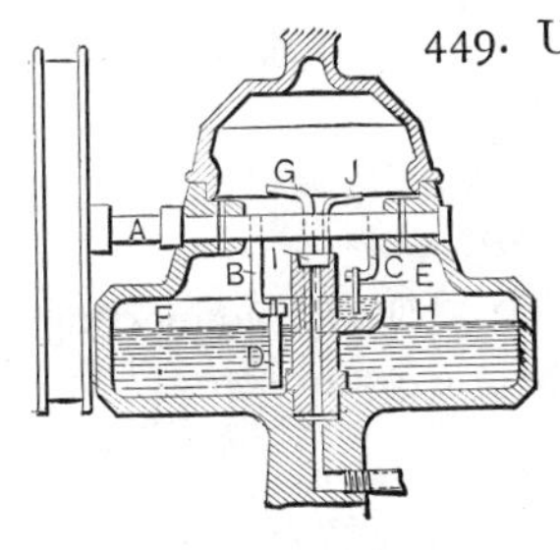

449. UNIFORM AUTOMATIC OILER.—Used on gas engines. The shaft, driven by a belt from the valve-gear shaft of the engine, carries two hooks and dip wires, one of which raises the oil from the variable level below to a constant level oil reservoir, from which the second hook and dip wire feed the wiper that leads the oil to the cylinder.

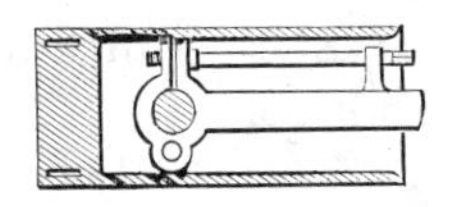

450. CRANK-ROD HEAD ADJUSTMENT for trunk pistons. A jointed brass tightened by a long-armed screw.

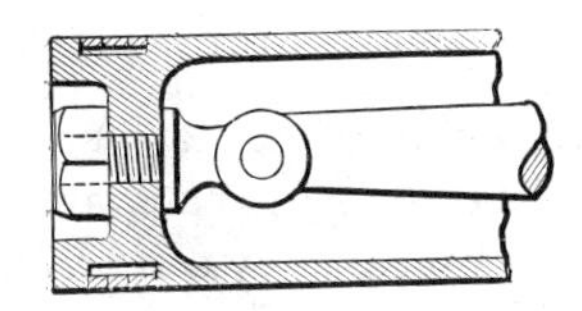

451. TRUNK PISTON ROD connection for a gas engine.

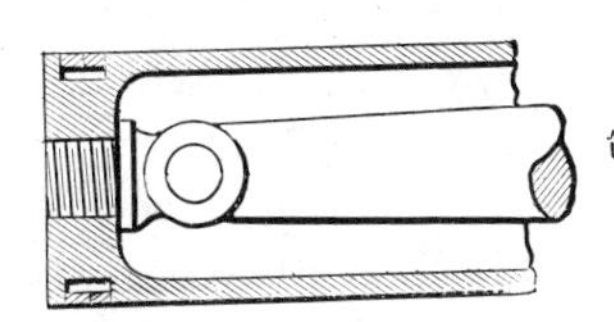

452. TRUNK PISTON ROD connection for a gas engine.

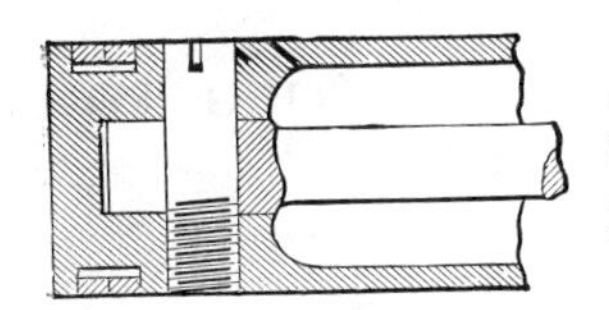

453. TRUNK PISTON ROD connection for a gas engine. Most reliable form. Head of screw pin should be keyed.

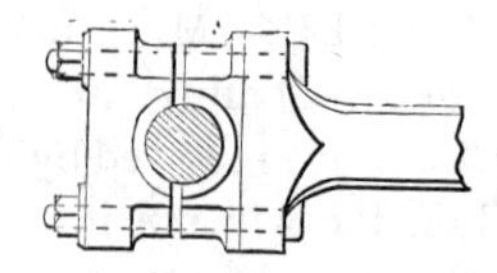

454. CONNECTING ROD HEAD, with full split brasses, held by cap and through bolts.

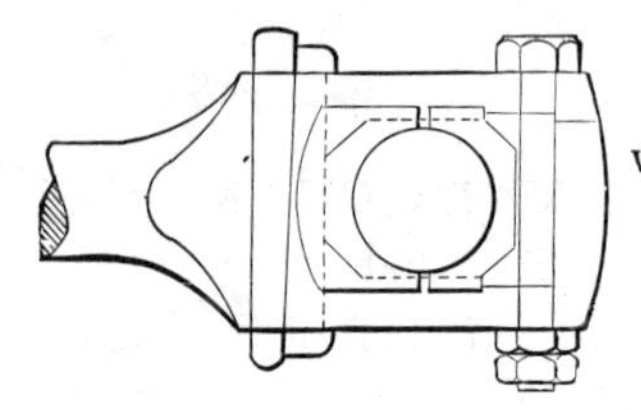

455. CONNECTING ROD END with set-in end block.

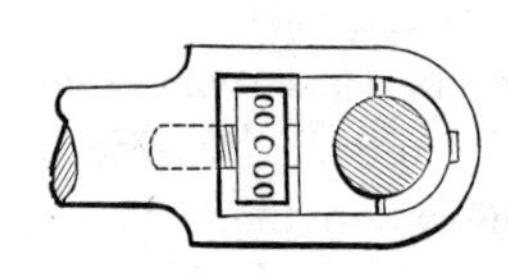

456. SOLID STRAP END, for connecting rod. Brasses set up by a capstan screw.

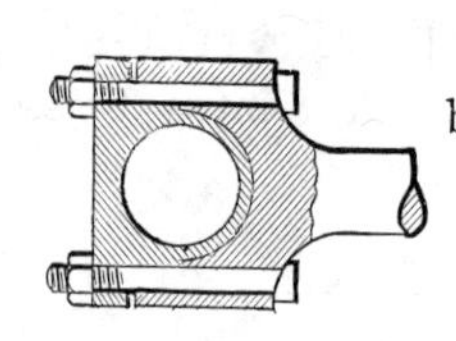

457. CONNECTING ROD END, with half brass and brass cap. Through bolts.

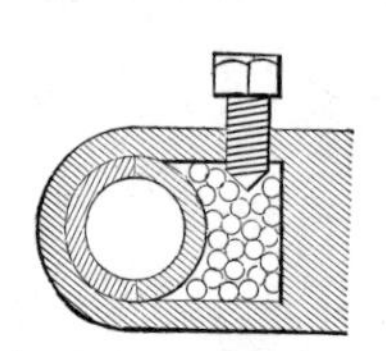

458. STEEL BALL ADJUSTMENT for connecting rod brasses. A number of steel balls are enclosed in a chamber and compressed by a screw.

459. SOLID END CONNECTING ROD.—Brasses slip in sidewise, and are locked in by the key.

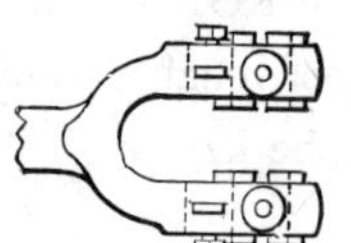

460. FORKED END CONNECTING ROD, with keys and set screws.

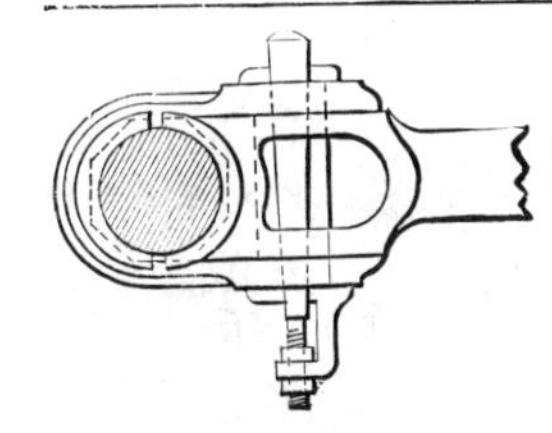

461. CONNECTING ROD END with locknut key.

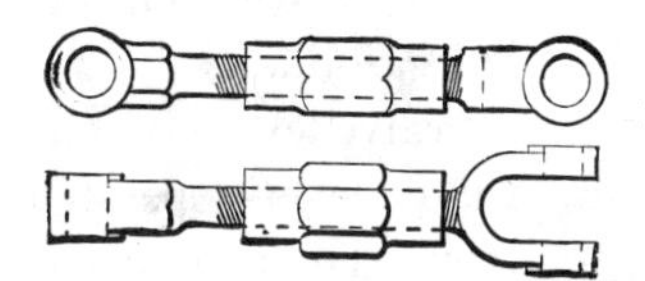

462. ADJUSTABLE LINK with right and left screw coupling.

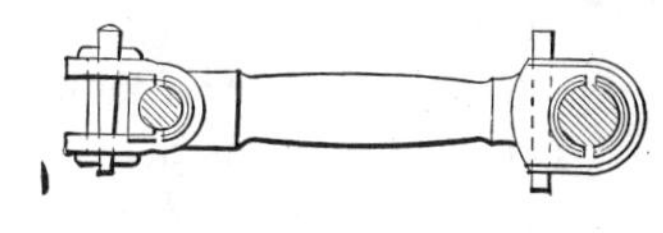

463. LINK OR CONNECTING ROD, with adjustable brasses. Keys inside and outside of pins.

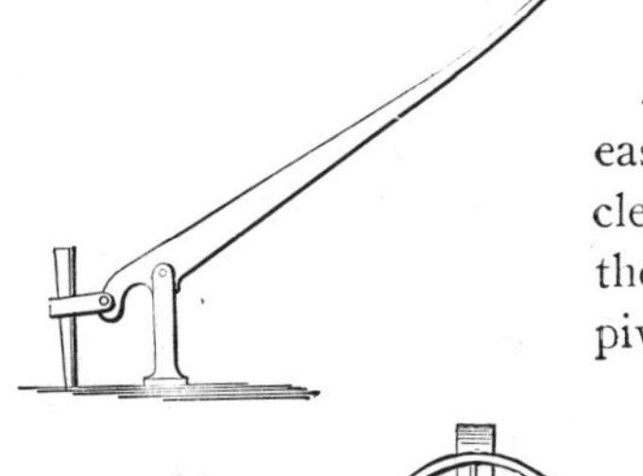

463*a*. STAKE PULLER. — An easy way to pull stakes and posts. A clevis to pinch the stake or post against the end of the lever with the lever pivoted to the foot post.

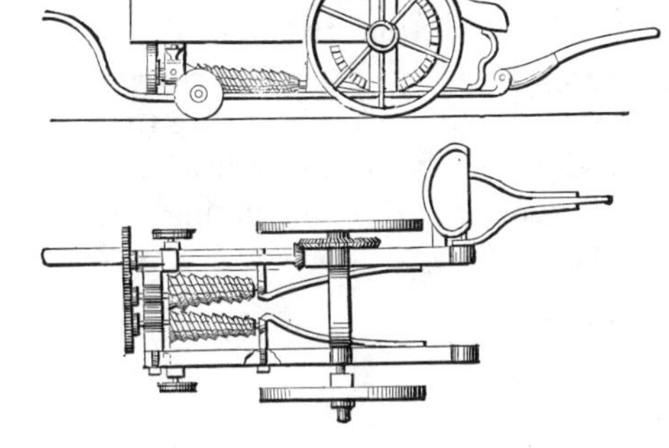

463*b*. STALK PULLER. — The conical spiked drums catch the stalks and throw them off at one side. The cones are driven by gearing and shaft from the large wheel. Will pull cotton, hemp and other stalks that are planted in rows.

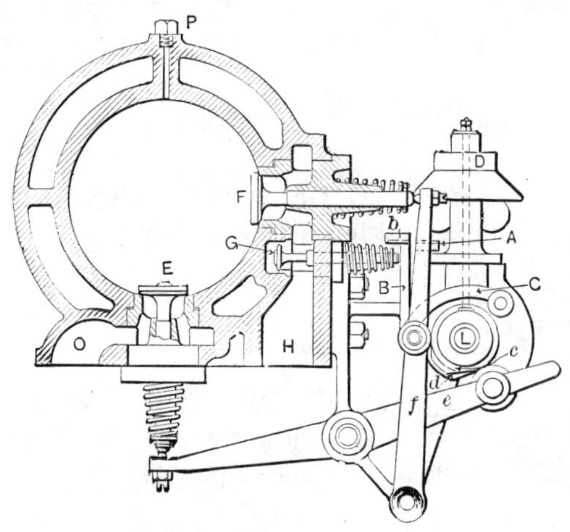

463*c*. VALVE GEAR FOR EXPLOSIVE MOTORS.—H, air inlet; F, air valve; G, gas or gasoline valve; *f*, air valve lever; B, gas valve lever operated from the cam at C; O, exhaust; E, exhaust valve; *e*, exhaust valve lever, operated by cams at *c*.

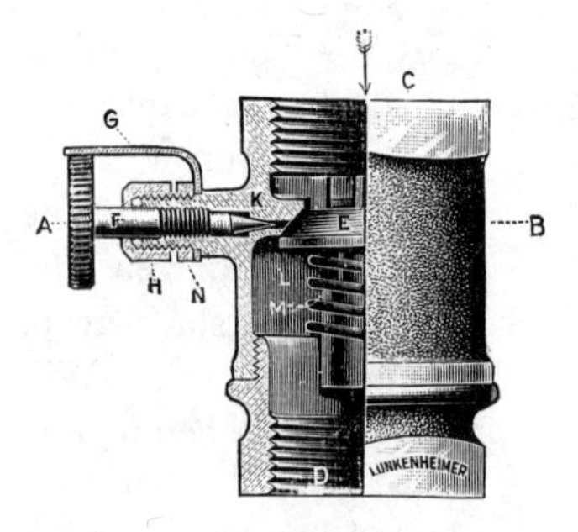

Section on A-B.

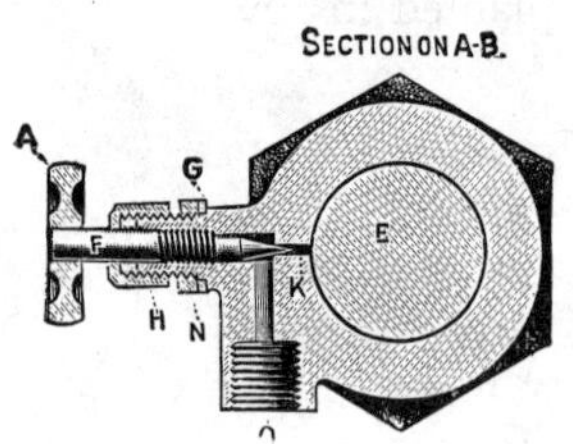

463*d*. GASOLINE ATOMIZER.—By injection through the valve seat K which has a grooved passage around it to distribute the gasoline evenly to the indraft of the piston A, the regulating needle valve.

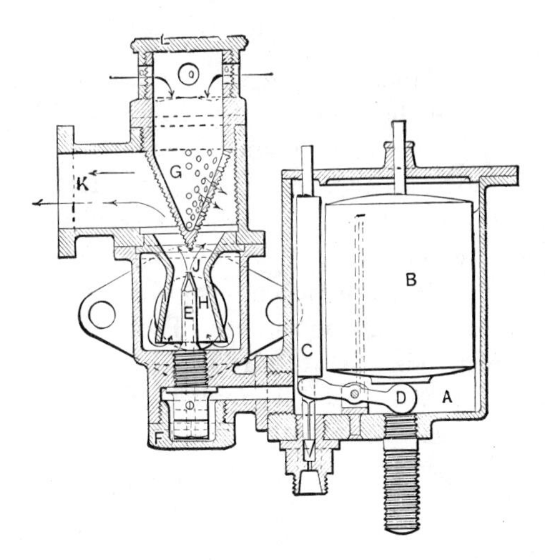

463*e*. GASOLINE ATOMIZER, of the constant feed type.—A, receiving tank; B, float; C, counter weight and valve; E, jet nozzle; H, air inlet; G, perforated cone with air regulating cap L.

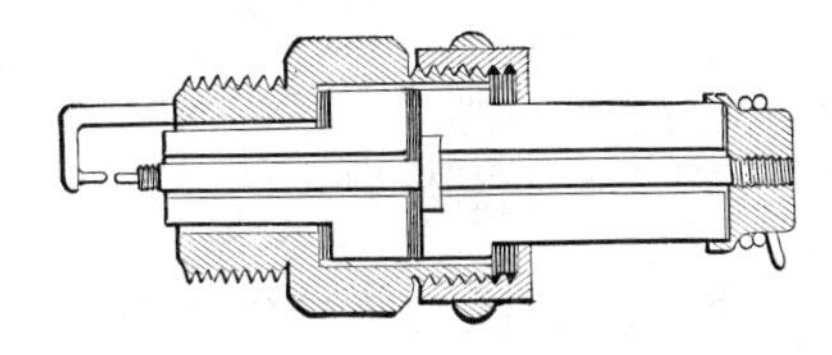

463*f*. ELECTRIC IGNITION PLUG, for a gas or gasoline motor. Electrodes of platinum; copper spindle with collar; insulation porcelain or lava with mica disk between.

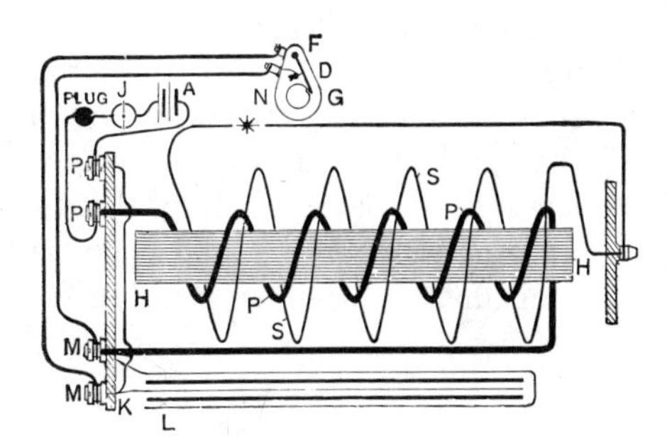

463*g*. JUMP SPARK COIL for gas and gasoline engines.—H, H, iron wire core; P, primary coil; S, secondary coil; L, condenser; D, spark breaker; A, battery; J, switch; P, M, binding posts.

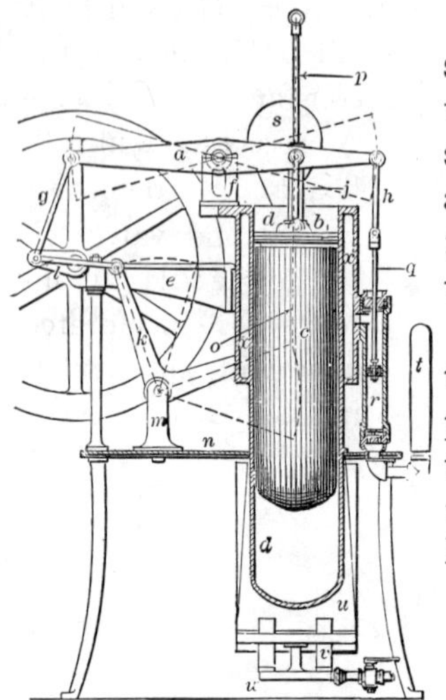

463*h*. CALORIC ENGINE, Ericsson Model.—*d, d,* the cylinder in which the transfer piston moves with space between it and the cylinder to allow the air to be quickly transferred from the hot end to the cool end and vice versa.

b. Impulse piston attached directly to the walking beam. The transfer piston is operated by a yoke connection with the bell crank lever *k,* and rod *p*. *x,* water jacket, *r,* pump, *u,* Bunsen burner.

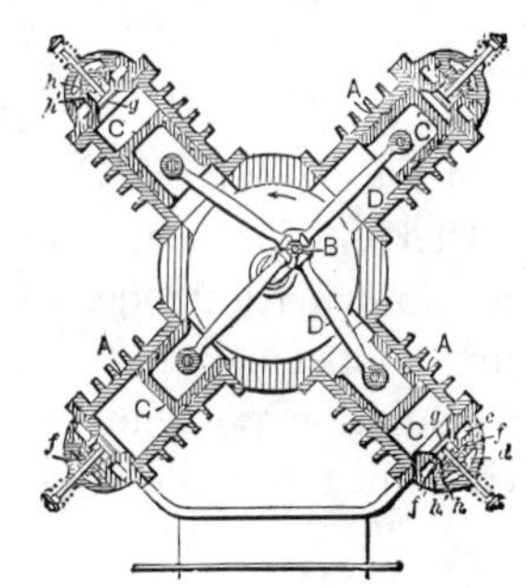

463*i*. FOUR CYLINDER GASOLINE MOTOR.—Four cycle, air cooled type. The successive impulses in the four cylinders require only a very light fly wheel to regulate the motion. A French design.

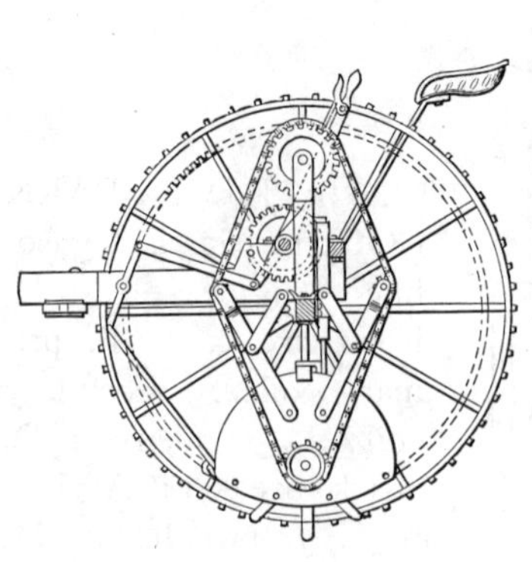

463*j*. HARROW AND CLOD CRUSHER. — Two portions of this machine co-operate and reduce the soil to a pulverized condition, ready to receive the seed. The machine is mounted on a pair of high wheels having serrations or teeth projecting from the tread surface, and the inner portion of the rim carries a continuous gear adapted to rotate the drag, through the small gear wheel and chain arrangement on either side of the machine.

Section VII.

HYDRAULIC POWER AND DEVICES.

WATER WHEELS, TURBINES, GOVERNORS, IMPACT WHEELS, PUMPS, ROTARY PUMPS, SIPHONS, WATER LIFTS, EJECTORS, WATER RAMS, METERS, INDICATORS, PRESSURE REGULATORS, VALVES, PIPE JOINTS, FILTERS, ETC.

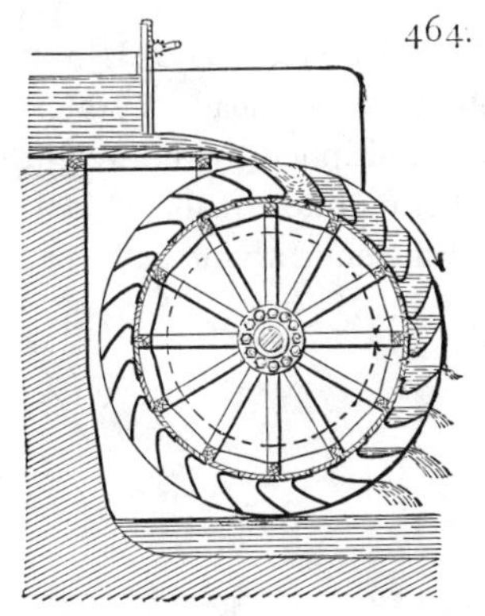

464. OVERSHOT WATER WHEEL, with steel buckets. With the gate chute impinging upon the buckets an efficiency of from seventy to seventy-five per cent. may be obtained.

$$\frac{h \times w}{33{,}000} \times .70 = \text{horse-power.}$$

h, Total height of water-fall from race; **w**, weight of water falling per minute.

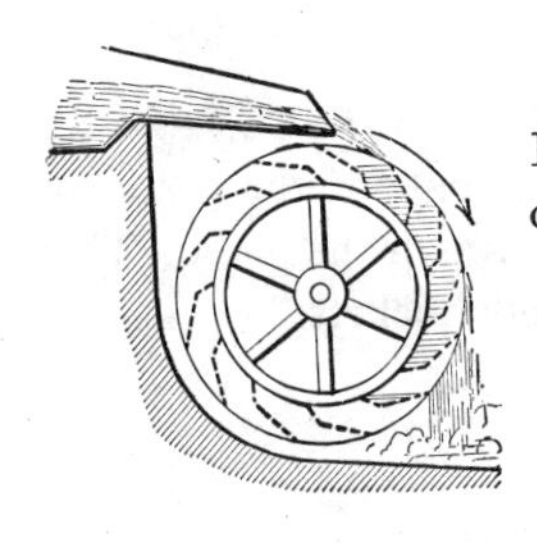

465. OVERSHOT WATER WHEEL.—Power equals about sixty per cent. of the value of the water-fall flowing over the wheel.

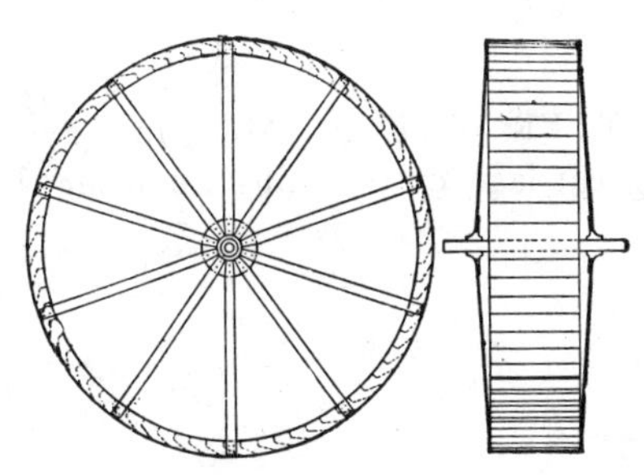

466. IRON OVERSHOT WHEEL.—The frame and buckets are made of iron or steel. The lightest wheel of its kind. "Leffel" model.

467. Front view.

468. UNDERSHOT WATER WHEEL.—Power equals about forty per cent. of the value of the water-fall flowing under the gate.

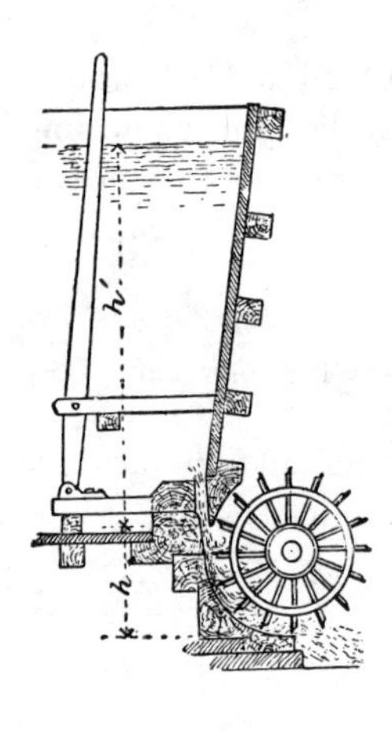

469. SAW-MILL WATER WHEEL and flume. $h + h'$ represents the head of water. The total head in feet multiplied by the weight of water discharged per minute equals the foot-pounds of power. Efficiency about sixty per cent.

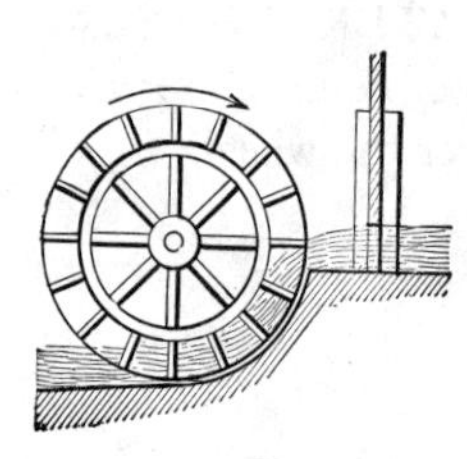

470. BREAST WATER WHEEL.—Power equals about forty per cent. of the value of the water-fall flowing through the gate. This form should have housed buckets.

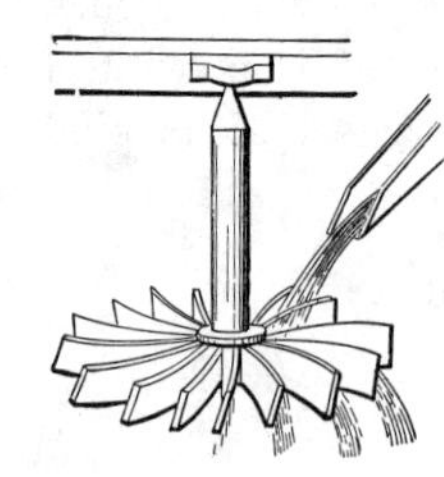

471. FLUTTER WHEEL.—Much in use to back the log carriage of saw-mills. Efficiency very low.

472. BARKER WHEEL.—A reaction water wheel. The reaction of the water escaping from the tangential orifices at the ends of the arms under the pressure of the water head in the hollow shaft gives impulse to the wheel. Very low efficiency.

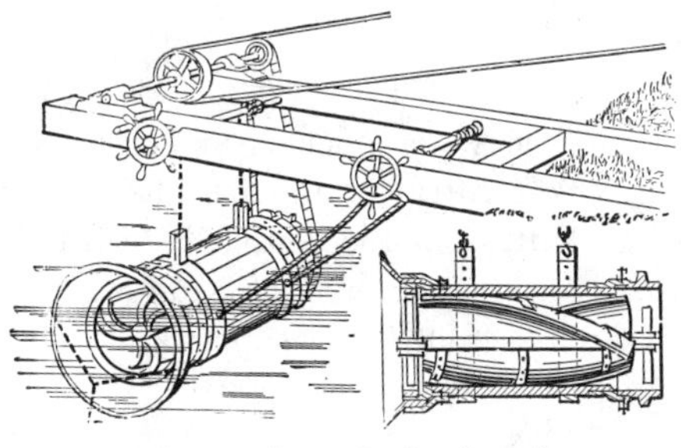

474. Section of wheel and case.

473. CURRENT MOTOR.—A propeller revolving within a case with expanding mouth to increase the force of the current. A sprocket-wheel on the rear end of the propeller shaft with chain transmission to shaft on suspension frame.

475. CURRENT WATER WHEEL.—The most efficient velocity of the wheel periphery is forty per cent. of the current velocity. The horse-power is :

$$\frac{\text{Area of immersion of blades}}{150} \times (V - S)^2$$

V = Velocity of the stream; S = velocity of periphery of wheel,—both in feet per second.

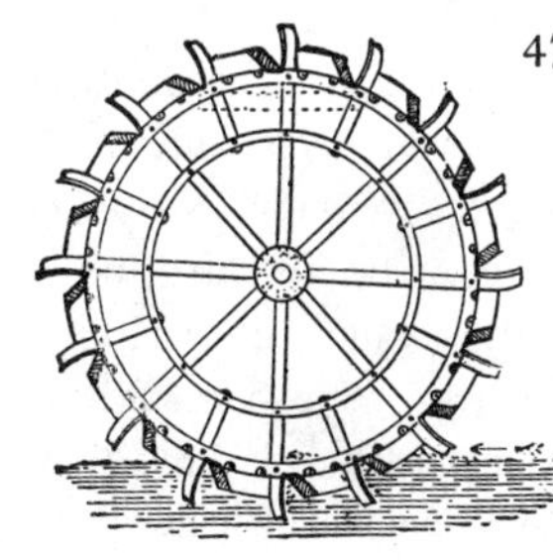

476. FIXED BUCKET WATER-RAISING CURRENT WHEEL.—Long rectangular buckets are attached across the rim of the wheel with side openings, indicated by the hatched spaces. At the top the water flows over the side of the wheel into a trough.

477. BUCKETED WATER-RAISING CURRENT WHEEL.—The buckets are pivoted to the outside rim of the wheel, and tilted into the trough at the top by a tail-piece on the bucket striking the trough.

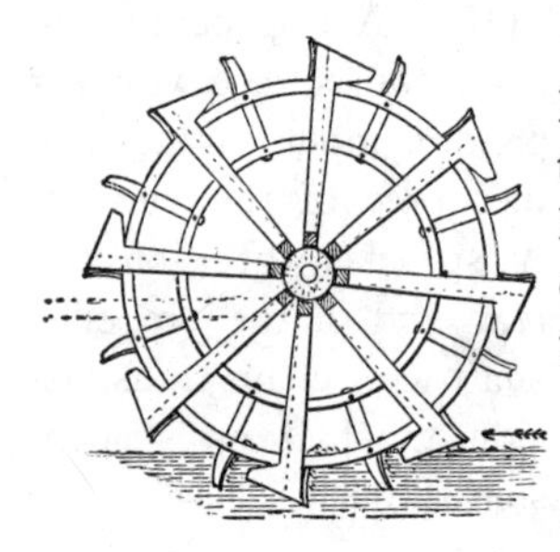

478. CURRENT WHEEL WATER LIFT.—The water buckets and arms are troughs that carry the water to the central hollow shaft, from the end of which it is discharged into a trough. Used for irriga tion and low-grade water supply.

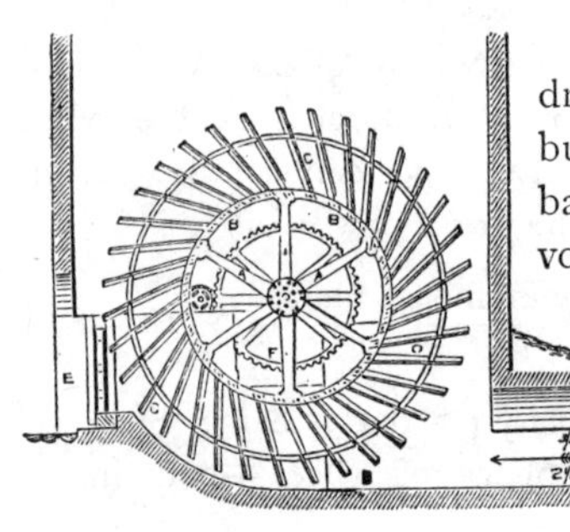

479. DRAINAGE WHEEL, used for draining fens and lowlands. Broad buckets on a power-driven wheel with a back or tangential slope, the wheel revolving in a current shield. Such wheels, at proper speed, will lift a large volume of water to a height of nearly half their diameter.

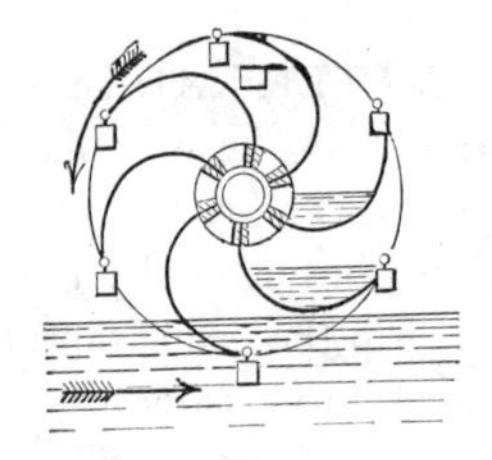

480. PERSIAN WHEEL.—A current-driven water lift; used in Eastern countries. A hollow shaft, with curved arms and floats, with buckets suspended at their periphery. The current carries the floats forward, filling the buckets and at the same time dipping water into the curved arms. The water follows the arms in their revolution and discharges through the hollow shaft, while the buckets are tipped at the top of the wheel into a trough.

481. ANCIENT WATER LIFT.—A series of earthen pots lashed to the periphery of a wheel revolving in a stream. The long pots are so inclined to the axial line of the shaft that they dip and fill while in the stream, and empty while passing the trough.

482. "ARCHIMEDIAN" SCREW WATER LIFT.—A water wheel on an inclined hollow shaft is driven by the current. A spirally wound pipe in or outside of the shaft conveys the water to an elevated trough.

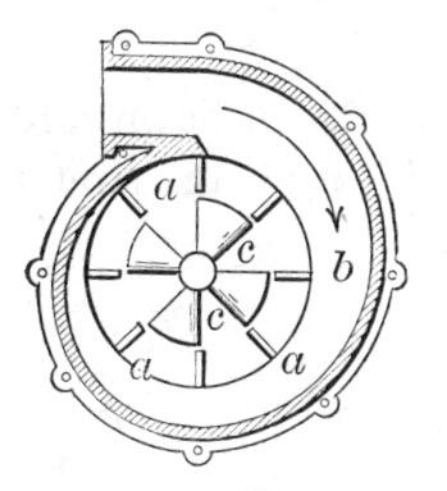

483. VOLUTE TURBINE.—The water, under pressure of its head, passes along the volute, striking the radial buckets *a*, *a*, *a*, flows inward and down through the central inclined buckets *c*, *c*. Efficiency about eighty per cent.

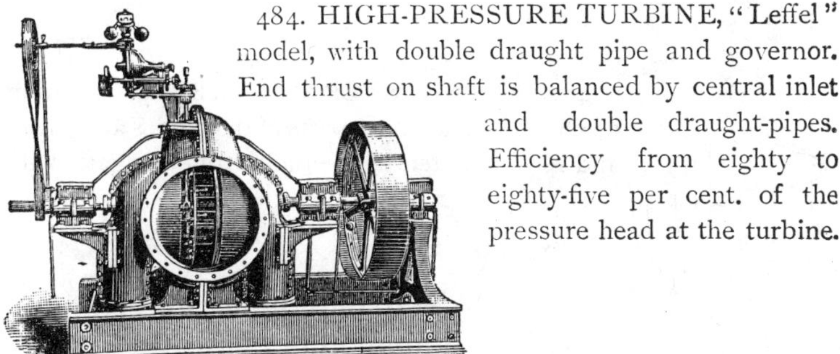

484. HIGH-PRESSURE TURBINE, "Leffel" model, with double draught pipe and governor. End thrust on shaft is balanced by central inlet and double draught-pipes. Efficiency from eighty to eighty-five per cent. of the pressure head at the turbine.

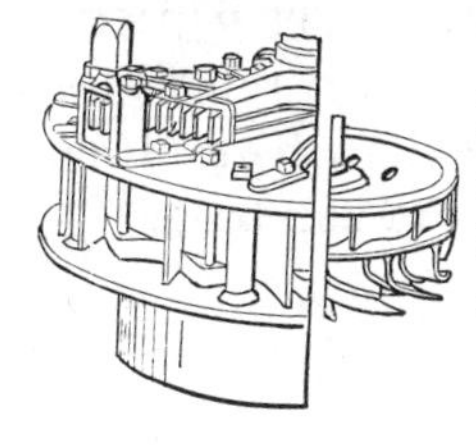

485. "LEFFEL" DOUBLE-RUNNER TURBINE.—The upper section of the running-wheel discharges inward and down the centre. The lower section has curved blades to discharge downward. One register gate for both sections.

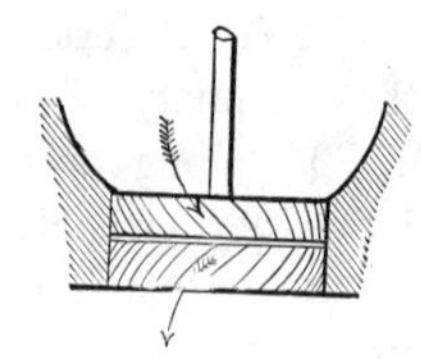

486. "JONVAL" TURBINE. — The upper inclined blades are fixed. The lower inverse blades form the wheel.

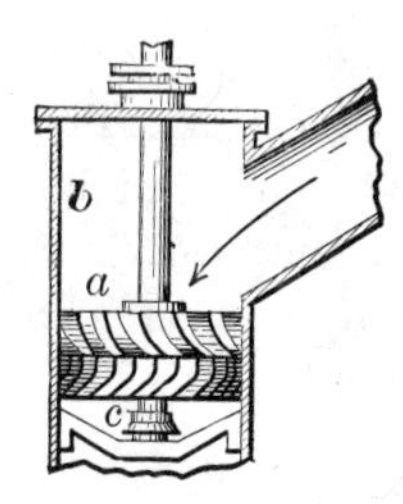

487. "JONVAL" TURBINE.—*b*, The case; *a*, the chute or directrix, fixed; *c*, the wheel buckets. The curved buckets are set slightly tangent and curve downward in parabolic or cycloidal form. Water discharges downward. Efficiency from eighty to eighty-five per cent.

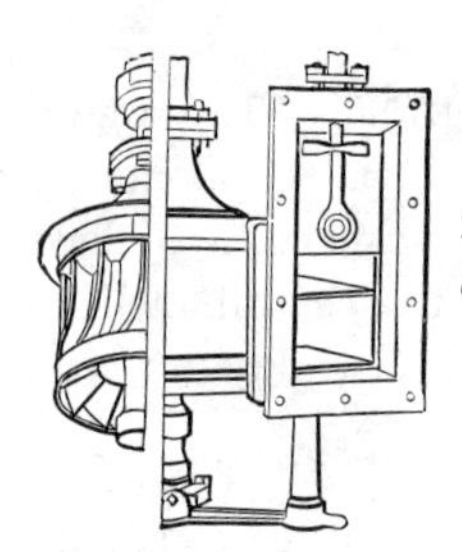

488. TURBINE AND GATE.—A downward flow from angular fixed guides in the water chamber.

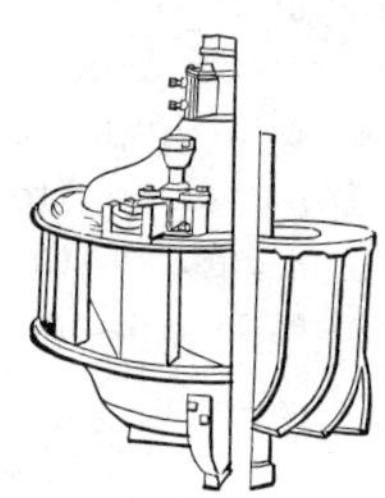

489. "LANCASTER" TURBINE, downward discharge. The upper parts of the blades are vertical. and receive water tangentially from the gate plates.

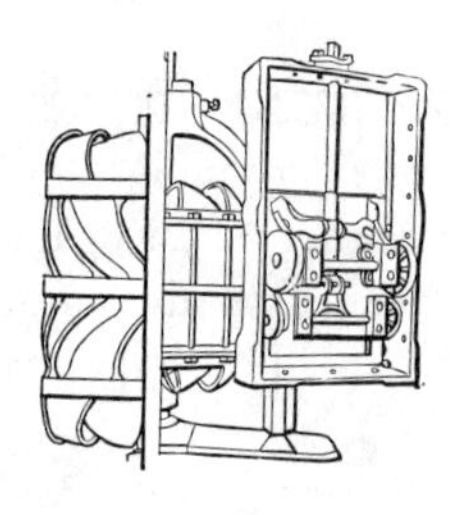

490. "MUNSON" DOUBLE TURBINE.—The water discharges both upward and downward through curved guide blades, to reverse curves in the top and bottom wheel blades.

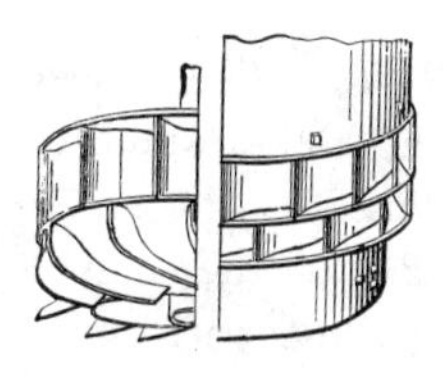

491. "CAMDEN" TURBINE, has two independent sets of buckets. The upper set is inward and central discharge, the lower set is curved backward, with tangential discharge.

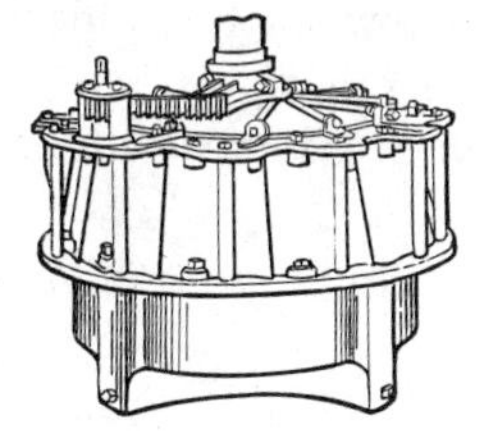

492. "MODEL" TURBINE.—The running-wheel has a downward discharge. The register gates are pivoted and operated by arms from a sector.

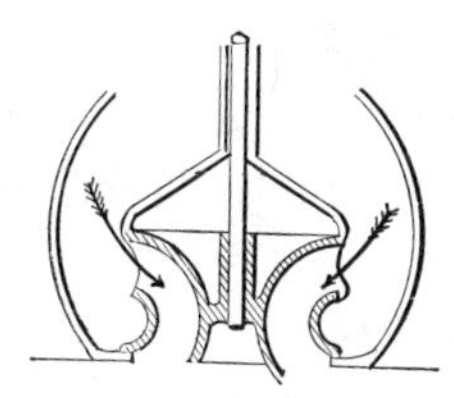

493. "SWAIN" TURBINE.—Inward and downward flow, with continuous curved blades.

494. "WARREN" CENTRAL-DISCHARGE TURBINE.—Plan: The wheel revolves on the inside of a fixed directrix. Water enters from outside, and discharges into and beneath the wheel. *a*, Directrix; *b*, wheel.

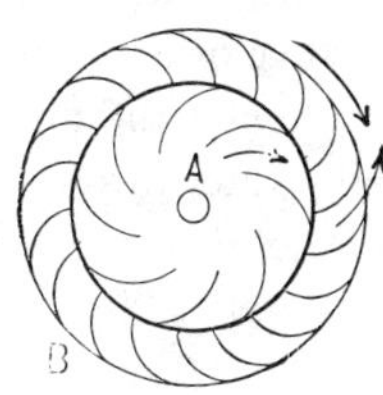

495. "FOURNERON" TURBINE.—The rim of outer buckets revolves around the inner directrix, the water moving outward. Efficiency, about eighty per cent.

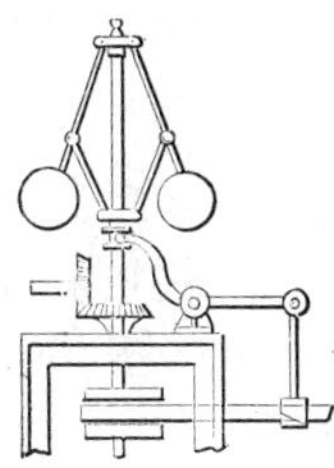

496. BELT WATER-WHEEL GOVERNOR.—The middle pulley on the governor spindle is loose, the outside pulleys are tight. The action of the governor balls operates a belt shipper which throws the belt upon the upper or lower tight pulley at abnormal speed. A corresponding set of tight and loose pulleys operate a pair of bevel gear that oper or close the gate.

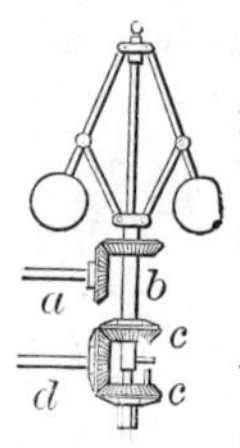

497. WATER WHEEL GOVERNOR.—The wheel motion drives the bevel gear at *a* and the hollow spindle, *b*, revolving the balls and connecting arms. The small central spindle has a vertical motion, due to the centrifugal force of the balls. The central spindle carries a pin which slides in a slot in the outer hollow spindle, which at abnormal speed catches one or the other pins in the loose bevel gears, *c*, *c*, which, acting on the bevel wheel and shaft, *d*, opens or closes the gate.

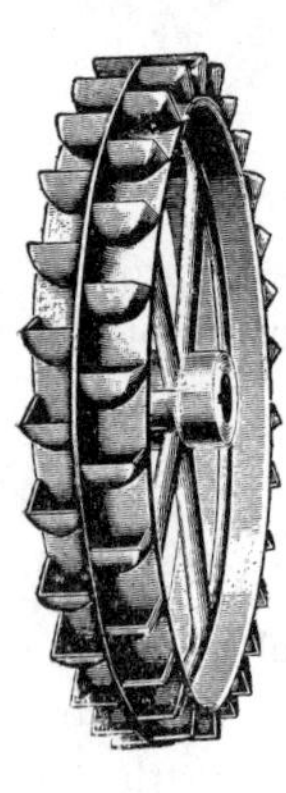

498. IMPACT WATER WHEEL, "Leffel" pattern. Step buckets. Efficiency, eighty-five per cent.

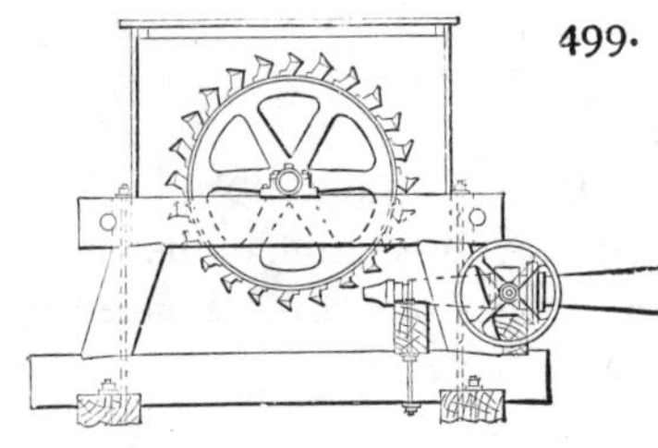

499. PELTON WATER WHEEL.—An impact wheel driven by the force of a high-pressure water jet. Efficiency, eighty-five per cent. of the product of the height and weight of flowing water through the jet, less the friction head.

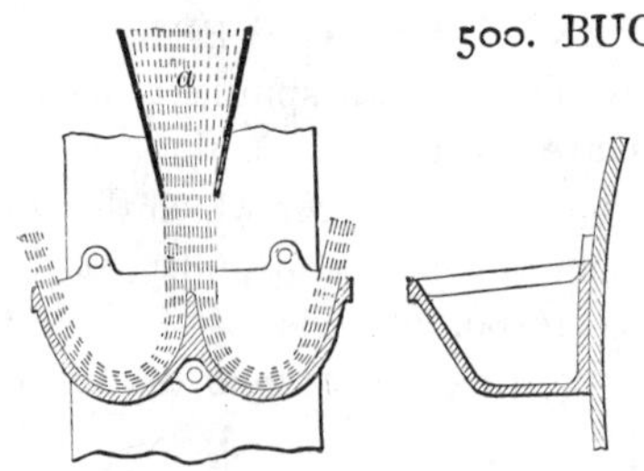

500. BUCKETS OF A PELTON WATER WHEEL.—Showing the method of separating the jet and returning the parts nearly in line with the impact jet, thus gaining about eighty-five per cent. of the total power of the jet.

501. Section of bucket.

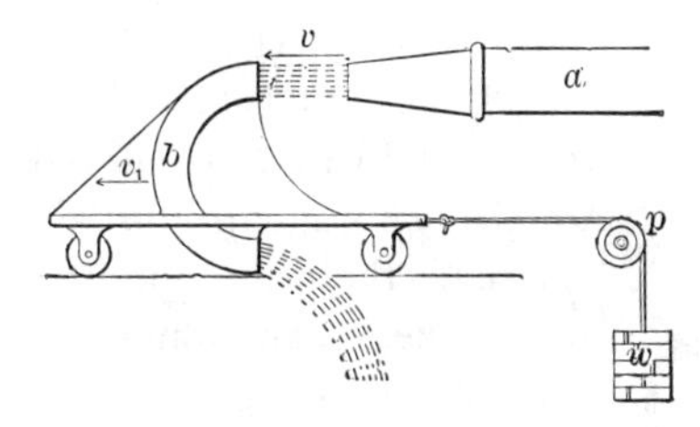

502. POWER OF WATER.—Apparatus for measuring the force of a water jet when discharged through a semicircular tube or trough. The total force is measured by the weight *w*.

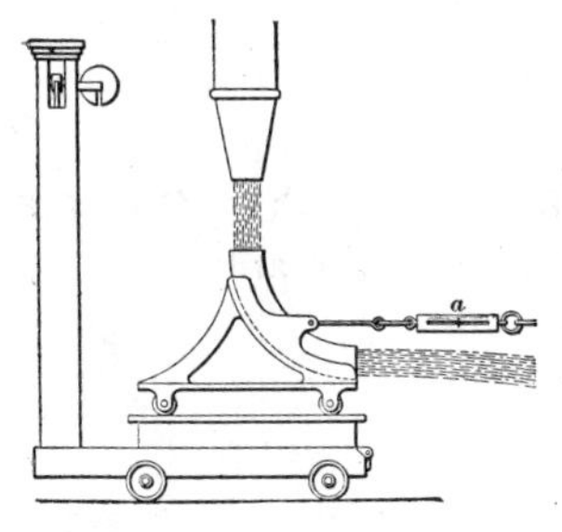

503. POWER OF WATER.—Apparatus for measuring the force of a water jet when turned to a right angle by a bent trough. *a*, A spring scale. The vertical force is weighed on the platform scale, the horizontal force by the spring scale.

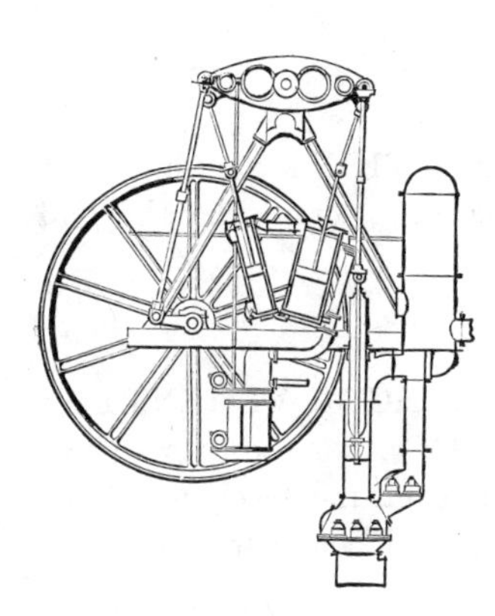

504. COMPOUND BEAM PUMPING ENGINE for water works. The high- and low-pressure cylinders are inclined, to make room for direct connection of the pump and crank rods.

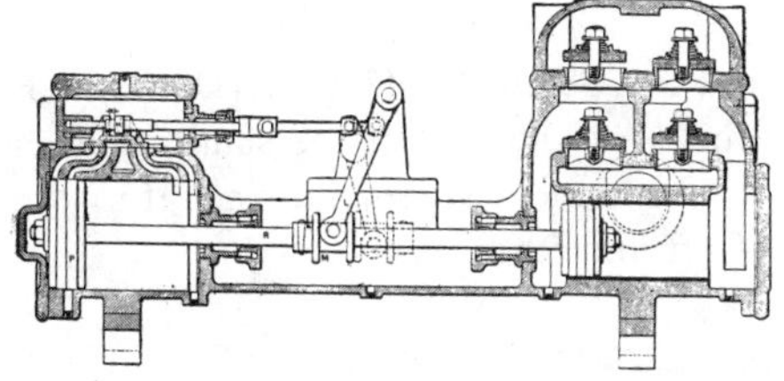

505. "DEAN" STEAM PUMP.—The valve gear of the Duplex pump. A lever and rock shaft, moved by a spool on the connecting rod, operates the valve of the opposite cylinder, for alternating the strokes of the pistons.

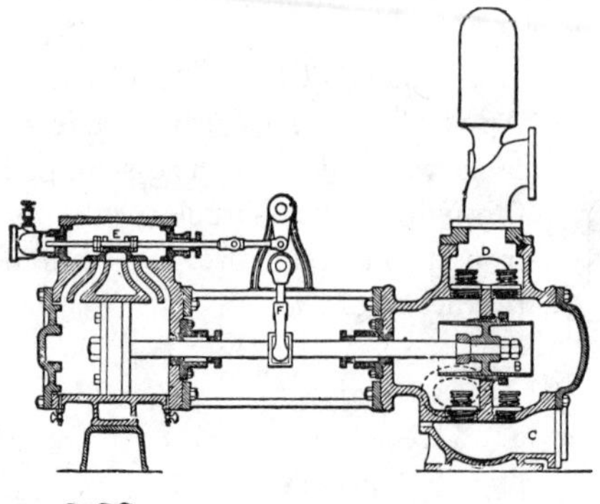

506. WORTHINGTON DUPLEX PUMP. — Two rock shafts with arms moved by the opposite piston rod alternate the valve motions and strokes. The water piston is of the plunger form.

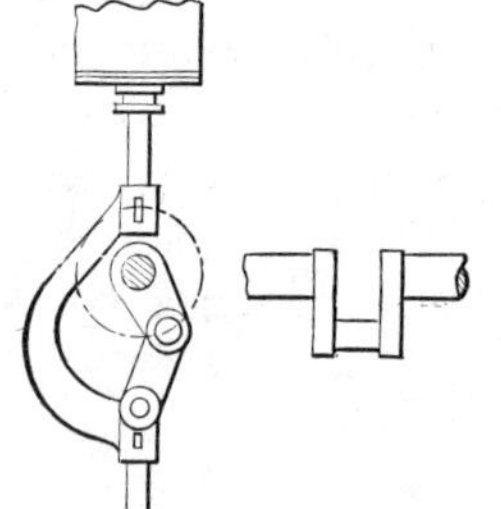

507. HALF-YOKE CONNECTION for pump piston rods with central crank.

508. The centre crank.

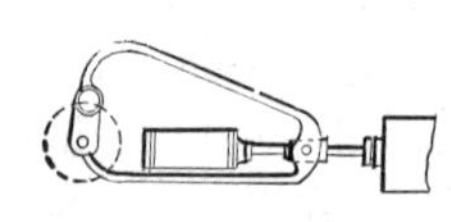

509. YOKE CONNECTION for a continuous piston rod and outside crank; crank shaft beyond the steam cylinder.

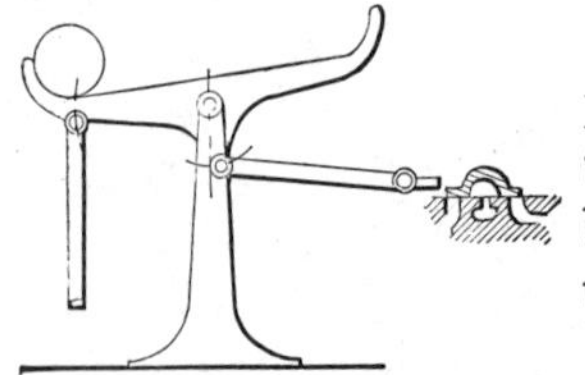

510. REVERSING MOVEMENT for a pump valve. The piston-rod trip carries the ball frame beyond the level, when the ball rolls across and completes the valve throw.

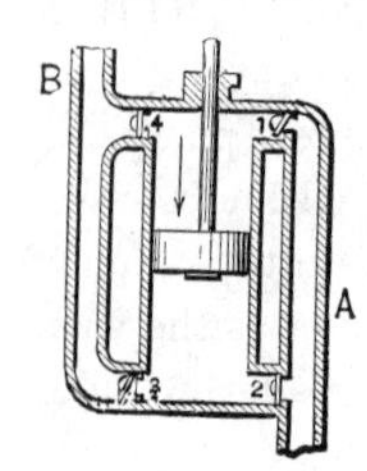

511. DOUBLE-ACTING LIFT AND FORCE PUMP.—In this form the work is the same for each stroke of the piston, and the pressure equal to the total height of lift and force.

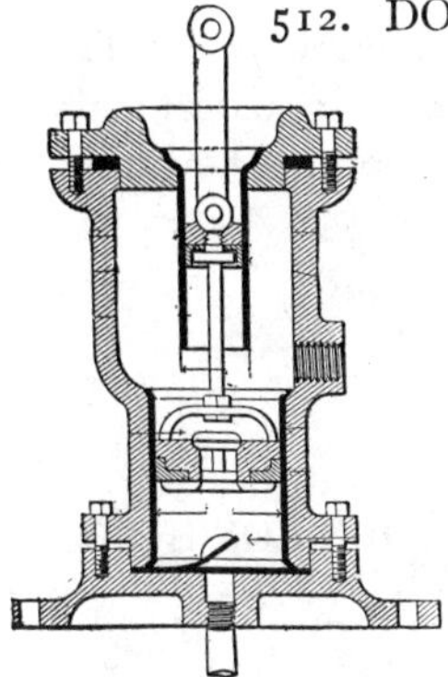

512. DOUBLE-ACTING DIFFERENTIAL PUMP. —The lower section is of the same construction as the ordinary lifting pump. The upper section has a solid piston connected by rod to the lower bucket piston, and moving in an open cylinder projecting down from the cover, thus making the upper part of the pump an air chamber.

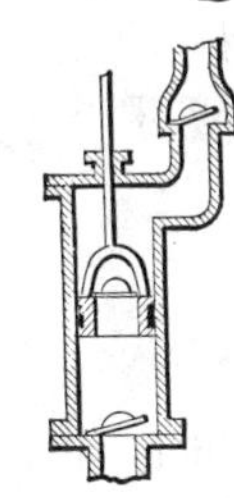

513. LIFT AND FORCE PUMP.—The limit of lift or suction is practically twenty-five feet. The force may be to any desired height, according to the strength of working parts and applied power. Total power is on the up-stroke of the piston.

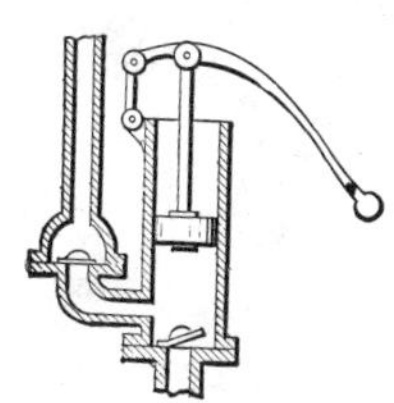

514. LIFT AND FORCE PUMP, with solid piston. In this form the power is divided; the up-stroke is equal to the lift or suction, and the down-stroke equal to the force required for any height.

515. TRAMP PUMPING DEVICE, sometimes called the Teeter pump. A self-evident illustration of an obsolete practice.

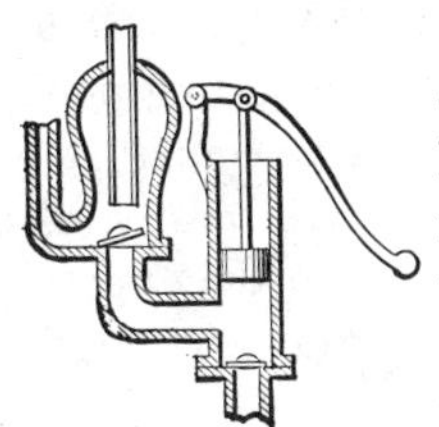

516. LIFT AND FORCE PUMP with air chamber. The air chamber is required for long lines of pipe to prevent reaction and water hammer. Water under pressure absorbs more air than at atmospheric pressure, often depriving the air chamber of its air cushion when recharging becomes necessary.

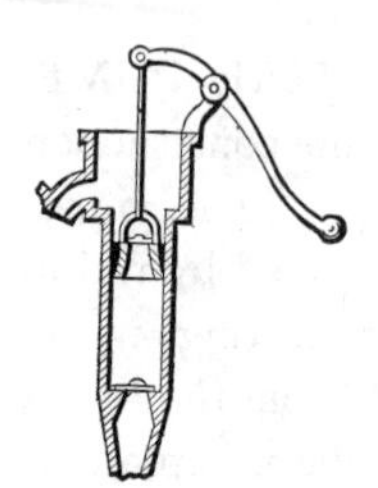

517. LIFT PUMP.—The limit of water lift in this pump is about thirty feet, but practically about twenty-five feet is its available working height.

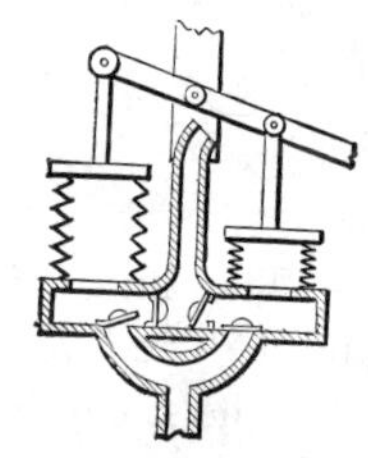

518. DOUBLE-LANTERN BELLOWS PUMP OR BLOWER.—A very ancient device for water and for a blower of air for forges. Will make a constant blast by using one side as a receiver, dispensing with the valves and connection on receiver side.

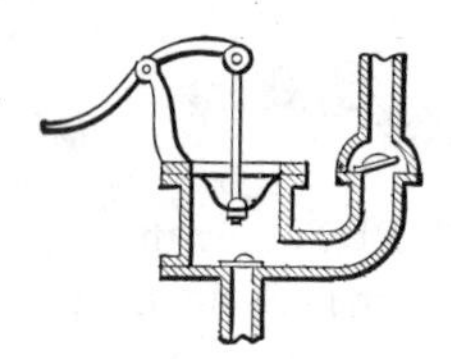

519. DIAPHRAGM PUMP, in which a flexible diaphragm is used instead of a piston.

520. "FAIRBURN" BAILING SCOOP, for low-lift drainage or irrigation. The tilting scoop may be connected to a walking beam or directly to a vertical engine.

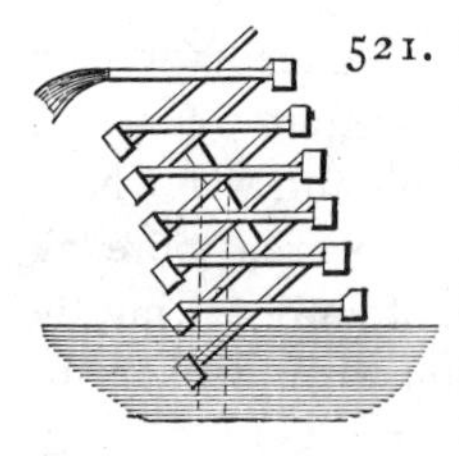

521. PENDULUM WATER LIFT.—A double series of scoops with flap valves and connecting pipes. The swinging of the pendulum frame alternately immerses the lower scoops, and at the next stroke raises the water by its transfer to the opposite scoop, when the next oscillation transfers to the next opposite scoop, and so on.

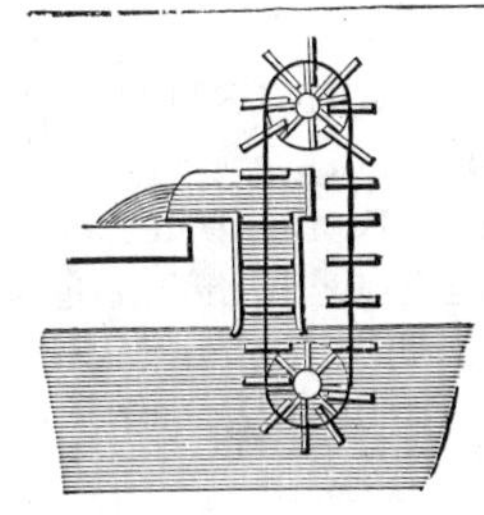

522. CHAIN PUMP.—An old device for raising water, now in use in many modifications.

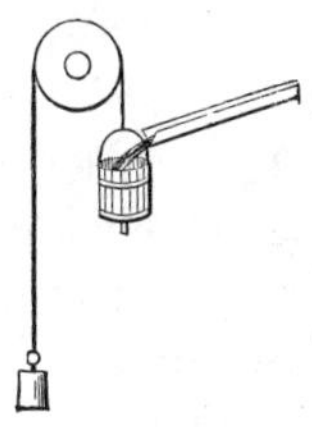

523. RECIPROCATING MOTION by the automatic action of a fall of water. A bucket with a valve in the bottom, which lifts and discharges the water by the contact of the valve spindle with a stop at the bottom of the bucket run; the weight lifting the bucket again to the spout. *Very old.*

524. WELL PULLEY AND BUCKETS.—Buckets are balanced empty.

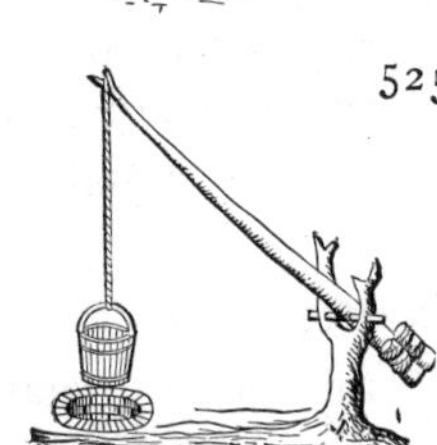

525. SWAPE, OR NEW ENGLAND SWEEP.—A very ancient as well as modern method of raising water from wells. The weighted end of the pole overbalances the bucket, so as to divide the labor of lifting the water.

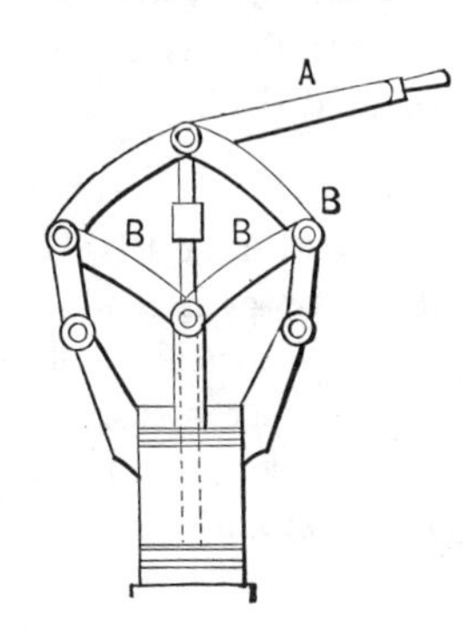

526. PARALLEL MOTION for double piston pump. A, The lever handle; links equal lengths.

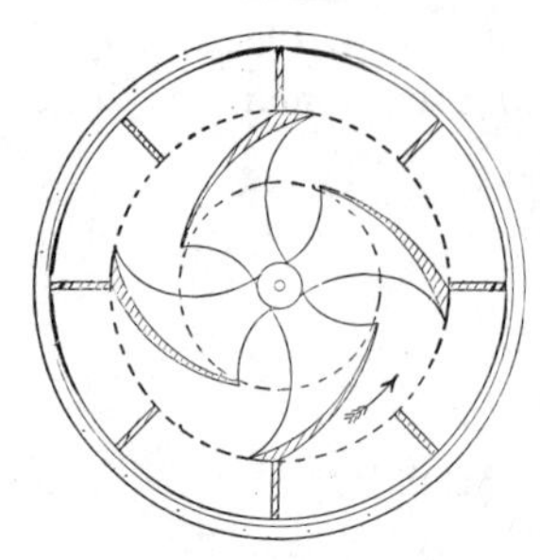

527. "GOLDING" CENTRIFUGAL PUMP.—Four volute blades are attached to the shaft by arms. To the outer case are attached radial blades with their edges nearly touching the revolving volute blades. Suction at centre; discharge at sides of outside shells.

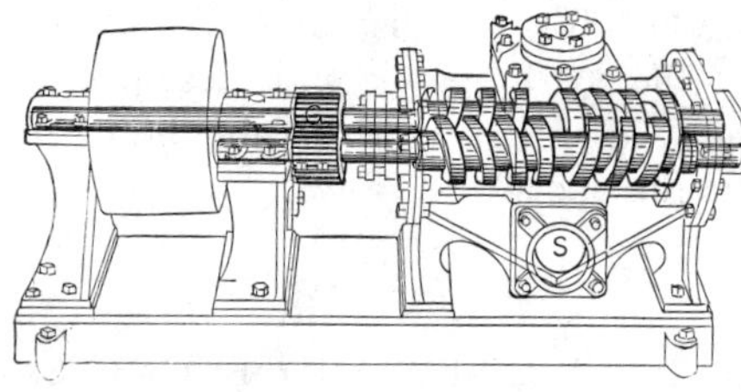

528. "QUIMBY" SCREW PUMP.—The screws revolve, meshed in each other, and are enclosed in a close-fitted case. Suction at each end from S, and discharge from the middle at D. End thrust is neutralized by the screws on each shaft being right- and left-handed.

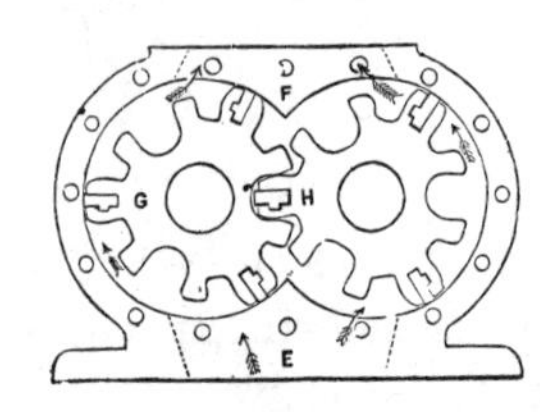

529. ROTARY PUMP, "Holley" system. Similar in design to the steam engine, No. 342, only each piston has three long teeth meshing into the recesses of the opposite gear piston. Used in combination with No. 342 in the Silsby fire engine.

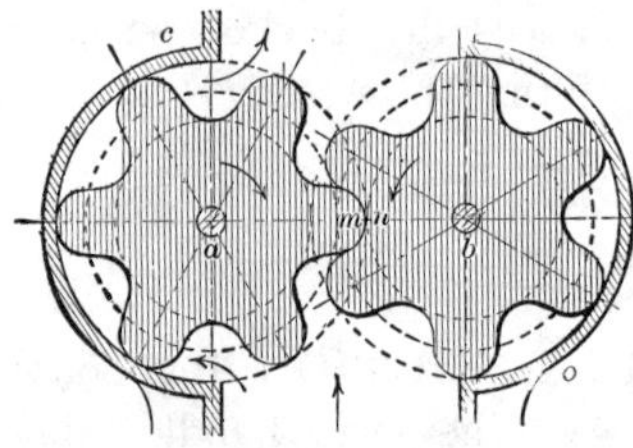

530. "PAPPENHEIM" ROTARY PUMP.—One of the earliest rotary devices for raising water. Two deep cog-wheels with their teeth meshed and rotating in a close-fitted shell.

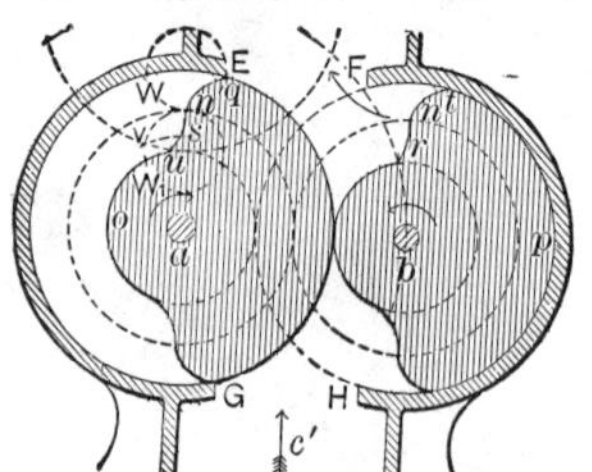

531. "REPSOLD" ROTARY PUMP.—Two differential sector cylinders revolving in contiguous cylindrical shells. The greater and smaller sector surfaces match and alternately close the area between the centres of revolution.

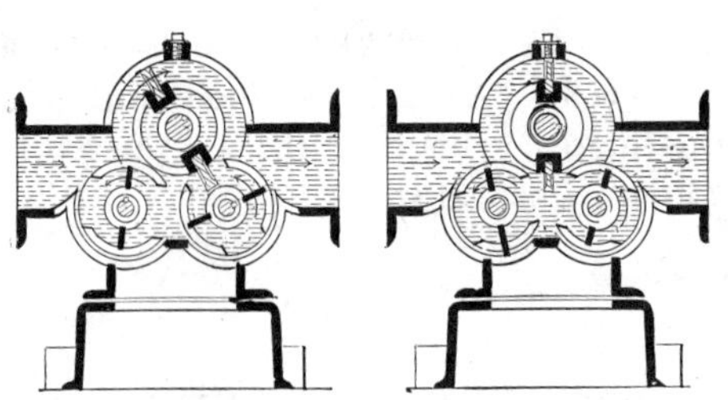

532–533. TRI-AXIAL ROTARY PUMP.—A late French invention. The upper cylinder receives the power and rotates the lower chambered cylinders through three spur gears. The wings of the upper or power cylinder are set fast and are the only rubbing surfaces. The cylindrical surfaces roll on each other with equal velocity. The extended surface of the lower cylinders furnishes a water packing that is practically tight.

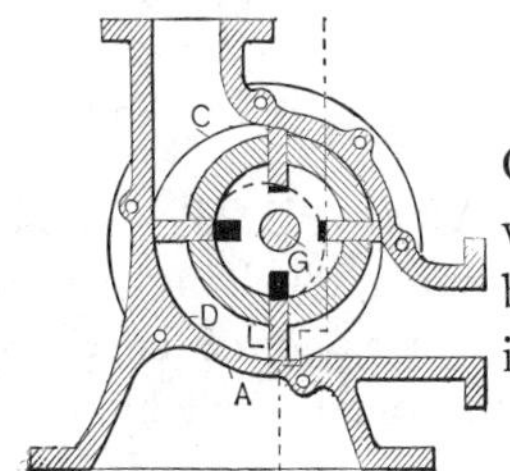

534. ROTARY PUMP OR MOTOR.—Can be run in either direction. The shell and wing drum are eccentric. The wings are guided by projections running in a concentric groove in each head.

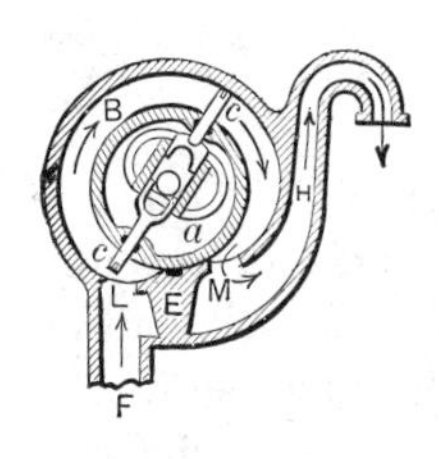

535. "CARY" ROTARY PUMP.—A rotating drum concentric with the outer fixed cylinder and a fixed heart-shaped cam groove in which the sliding wings are guided. A stop, E, closes the suction and force side of the chamber. The form of the outer cylinder wall is spiral.

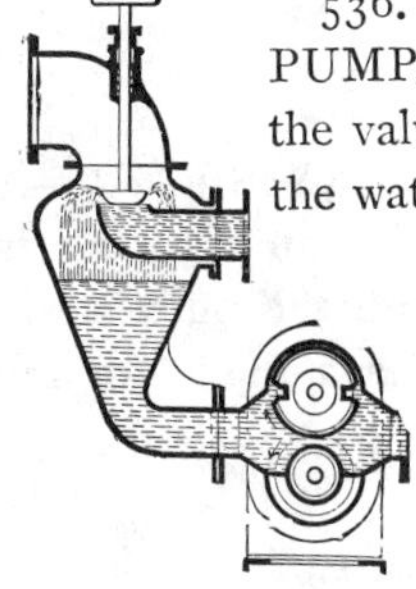

536. VACUUM JET CONDENSER AND ROTARY PUMP.—The jet in the vacuum chamber is regulated by the valve. The rotary pump, being entirely immersed in the water below, is water-packed.

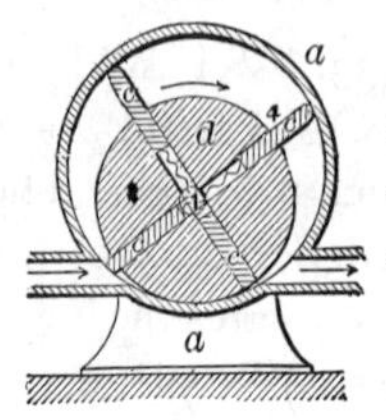

537. "RAMELLI" ROTARY PUMP.—One of the earliest (1588). A slotted cylinder with four wings eccentric to a cylindrical shell. The wings are pushed out by helical springs.

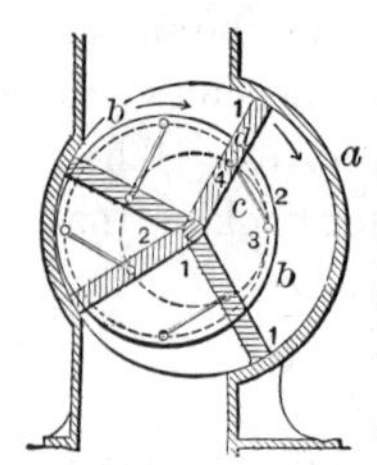

538. "HEPPEL" ROTARY PUMP.—Four wings are jointed concentric with the cylindrical shell. A disc and shaft are set eccentric to the cylindrical shell. The wings are linked to the eccentric disc as shown, so that the wings on the upward stroke move faster than the wings moving downward on the opposite side.

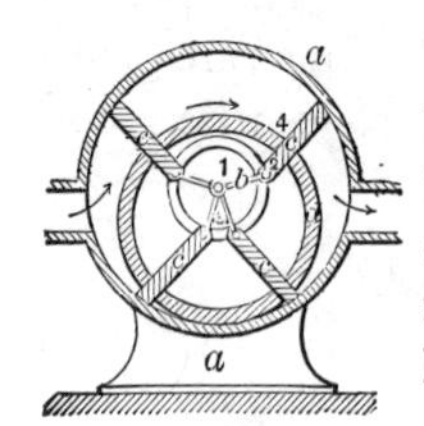

539. "EMERY" ROTARY PUMP.—Four wings driven by a hollow cylinder revolving eccentric to the outer shell. The inner ends of the wings are guided concentric with the outer shell by pins moving in a slot or groove in the shell heads, and kept in position by a toggle-joint connection.

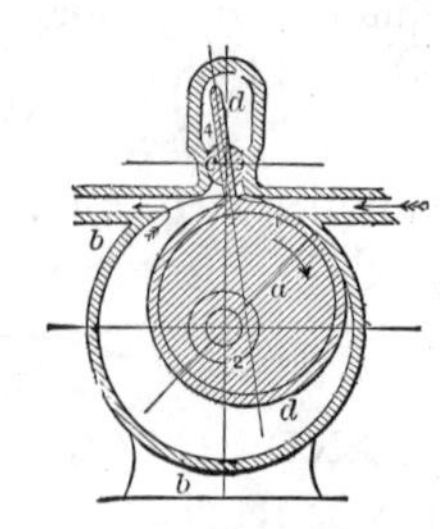

540. "KNOTT" ROTARY PUMP.—A hollow winged cylinder within which an eccentric revolves on an axis central with the shell, causing the winged cylinder to wipe the inner surface of the shell. The small slotted cylinder makes a packing for the wing.

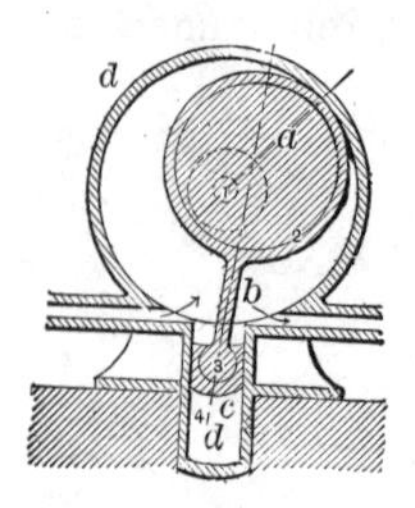

541. "PATTISON" ROTARY PUMP.—A hollow winged cylinder in which an eccentric is rotated on an axis central with the outer shell. The piston and socket serve as a guide for the wing.

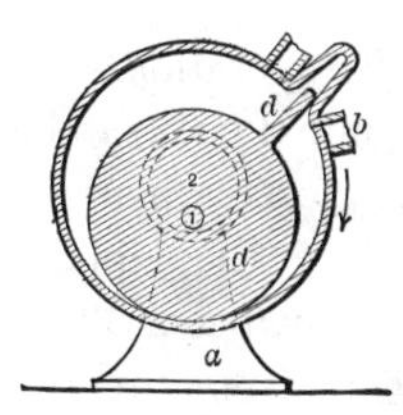

542. "COCHRANE" ROTARY PUMP.—A slide pocket in the outer shell receives the piston wing of the inner eccentric cylinder, which swings in contact with the shell on its centre, 2, carried around by a cam crank.

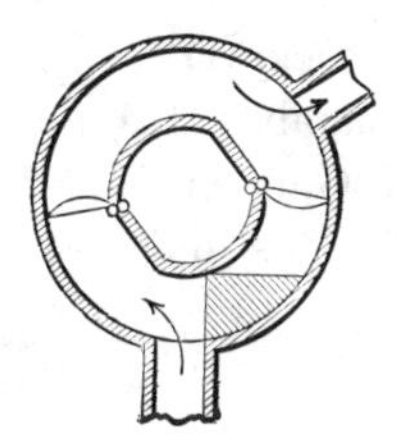

543. One of the early forms of Rotary Pumps. Only suitable for free flow to the pump. Will not lift. Obsolete.

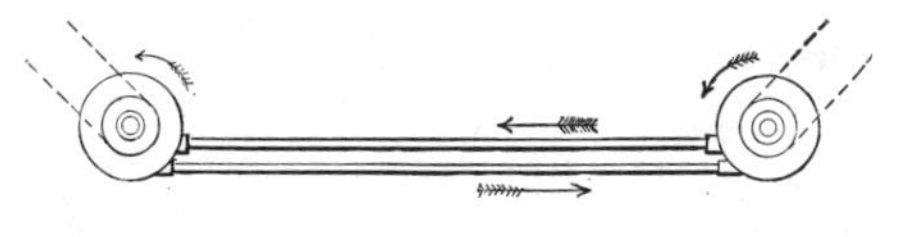

544. HYDRAULIC TRANSMISSION OF POWER.—A driving rotary pump connected by a flow and return pipe to a driven rotarv motor at any convenient distance. Has been applied to bicvcles.

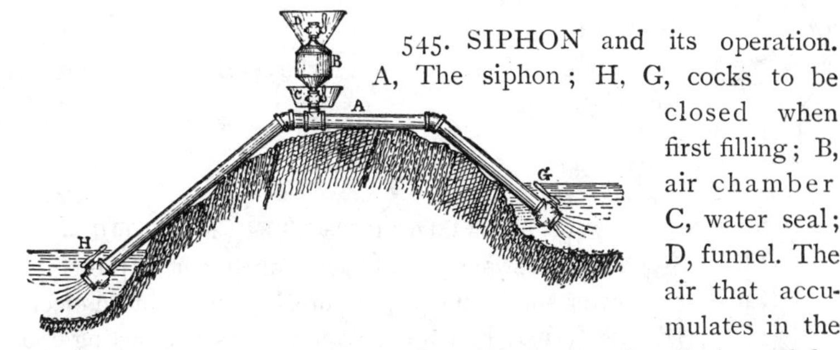

545. SIPHON and its operation. A, The siphon; H, G, cocks to be closed when first filling; B, air chamber C, water seal; D, funnel. The air that accumulates in the chamber, B, by the operation of the siphon, may be discharged by closing cock C, opening cock D, and filling the chamber with water. Close D and open C, when any air below C will rise into the chamber, and water will take its place without stopping the running of the siphon.

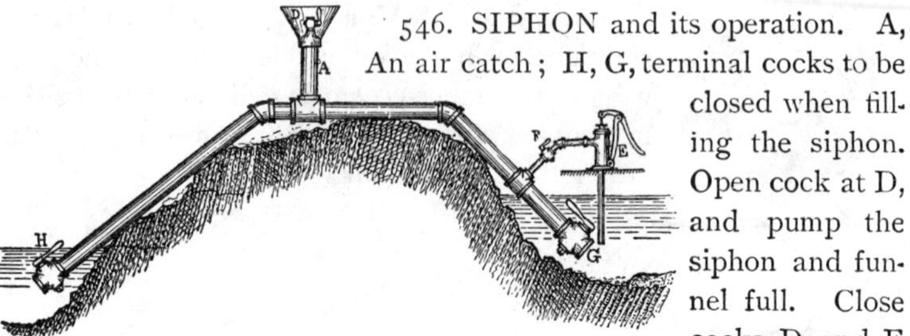

546. SIPHON and its operation. A, An air catch; H, G, terminal cocks to be closed when filling the siphon. Open cock at D, and pump the siphon and funnel full. Close cocks D and F and open H and G, when the siphon will run until chamber at A and apex of siphon are choked with air. Then close H, G, open D, and pump up again. This is very convenient for long siphons, and saves carrying of water.

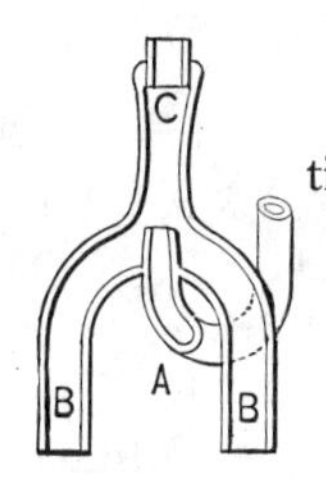

547. EJECTOR OR JET PUMP, with forked suction pipes.

548. EJECTOR OR JET PUMP.
A, the steam nozzle.
B, suction pipe.
C, the force pipe.
A crude representation of the earlier forms of the ejector.

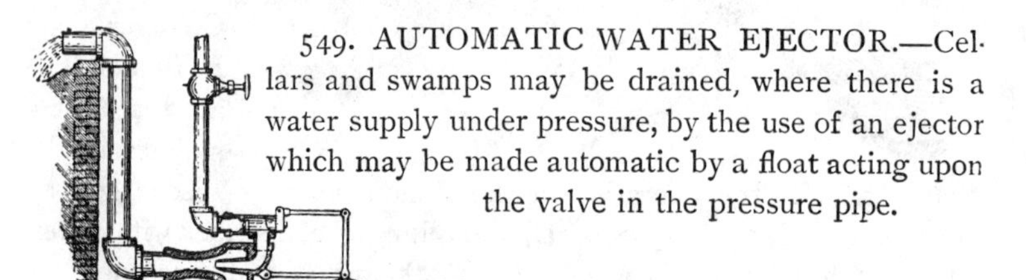

549. AUTOMATIC WATER EJECTOR.—Cellars and swamps may be drained, where there is a water supply under pressure, by the use of an ejector which may be made automatic by a float acting upon the valve in the pressure pipe.

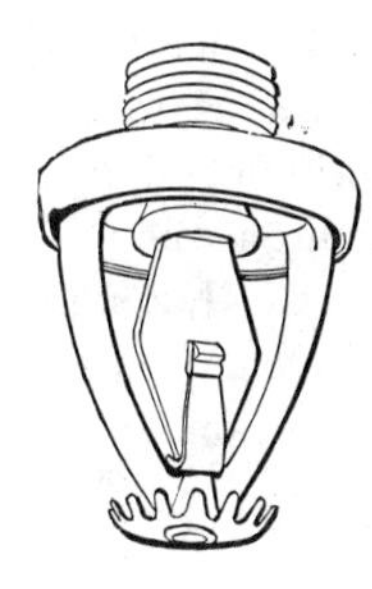

550. AUTOMATIC SPRINKLER.—The valve is held tightly closed by the diamond-shaped post resting on a bell-crank clip, which is held in position by fusible solder, melting at about 200° Fah., at which temperature the solder melts and the pressure casts the clips loose. The star washer scatters the stream.

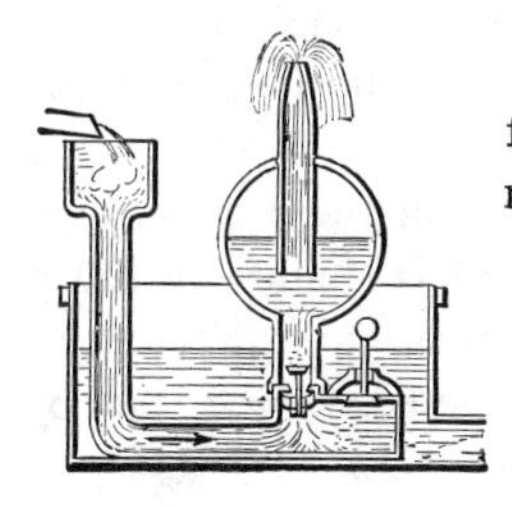

551. HYDRAULIC RAM, the "Montgolfier" idea for a fountain supplied by a water ram.

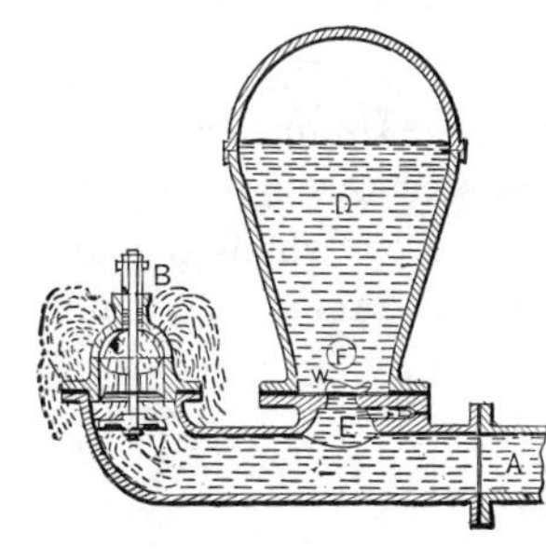

552. HYDRAULIC RAM.—A, Driving pipe; V, impact valve; C, valve-bonnet cage and spindle; W, force valve; F, outlet to force pipe; D, air chamber; E, snifting hole, sometimes furnished with a small ball valve which allows air to draw in at each rebound of the drive-water column, and thus to keep the air chamber supplied with air.

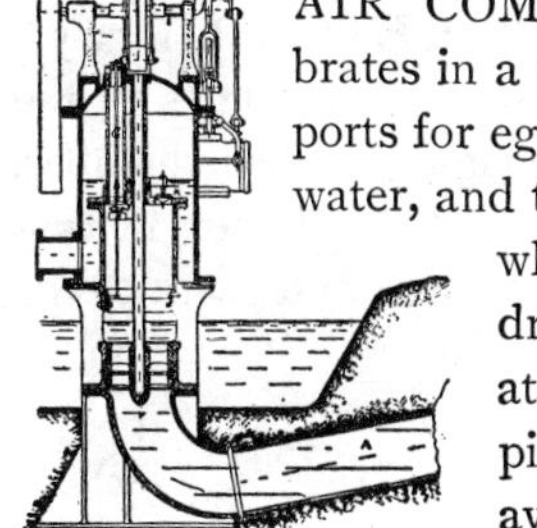

553. "PEARSALL'S" HYDRAULIC RAM AND AIR COMPRESSER.—A hollow or open piston vibrates in a cylinder, perforated all around with escape ports for egress of water. An air chamber receives the water, and the air which is drawn in through the ports, which becomes compressed. A small air motor drives a crank shaft and fly-wheel, which operate the piston. By the sliding motion of the piston in closing the ports, water hammer is avoided, thus enabling the use of a ram of very large dimensions.

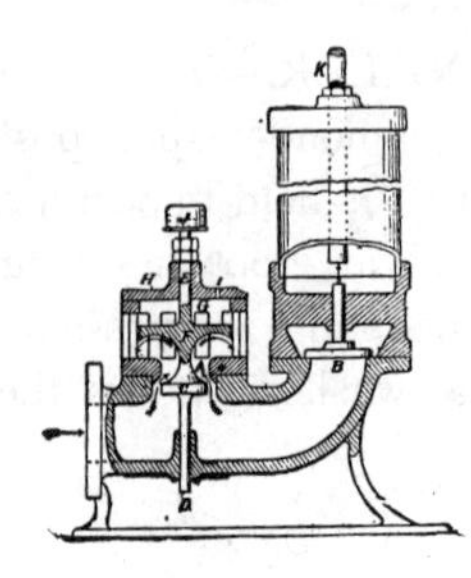

554. SILENT HYDRAULIC RAM.—The curved reaction disc, F, serves to lift the piston valve, C, quickly without shock. The air cushion at G stops the lift at the moment of closure of piston valve, C. J, a stop set-screw; H, valve cage; B, force valve; K, force pipe; I, vent hole to air cushion.

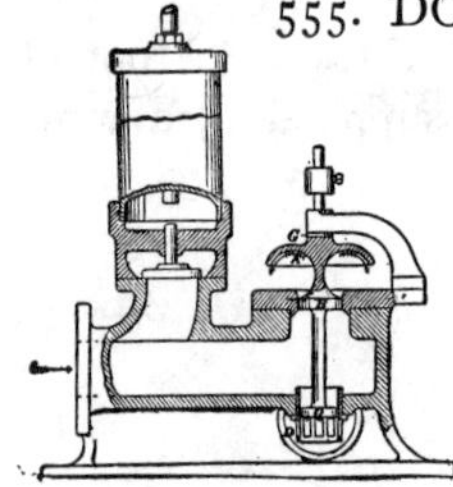

555. DOUBLE-PISTON REACTION HYDRAULIC RAM.—The two pistons, B and O, are on the same spindle with curved reaction disc, A. G is a leather washer to soften the contact with guide yoke. The cage at D guides the lower piston and serves to increase greatly the freedom of water-flow from the drive pipe, thereby increasing the duty of the ram.

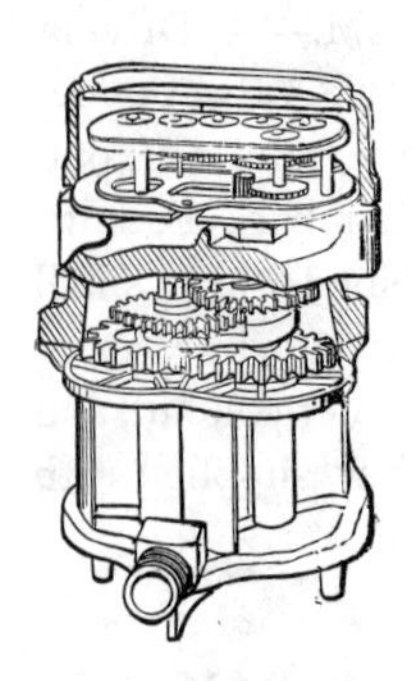

556. WATER METER.—"Union" water meter model. The water passes through a rotary motor with equalizing gear, from which the dial pointers are driven by a clock train and counter.

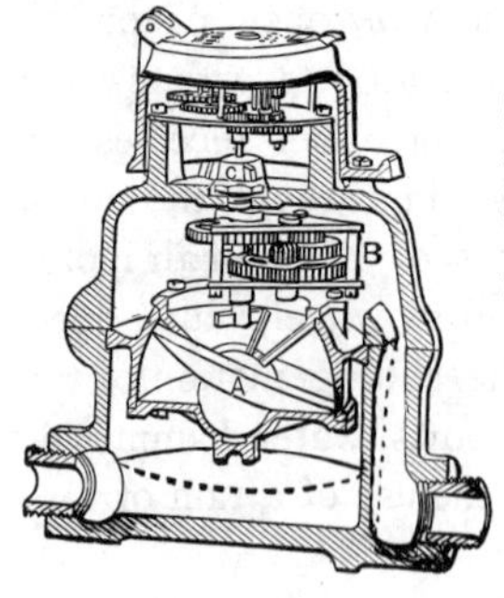

557. DISC WATER METER, "Hersey" model. The disc piston, A, oscillates by the passage of water through the disc chamber. The spindle of the disc, by its oscillating movement, rotates the crank and gears of the index-wheel train.

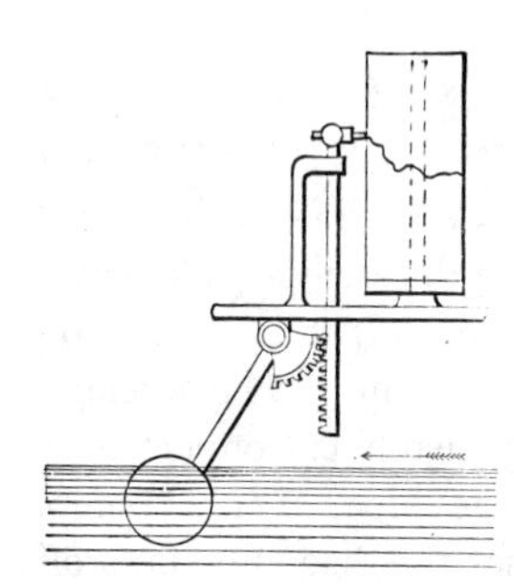

558. WATER METER, "Thompson" model. A swinging disc movement on ball socket, operated by a flow of water, rotates a vertical crank spindle and gear train with index hand above the dial.

559. WATER-VELOCITY INDICATOR AND REGISTER.—Variations in velocity of a stream varies the position of the float, which is registered on a traverse card by a pencil.

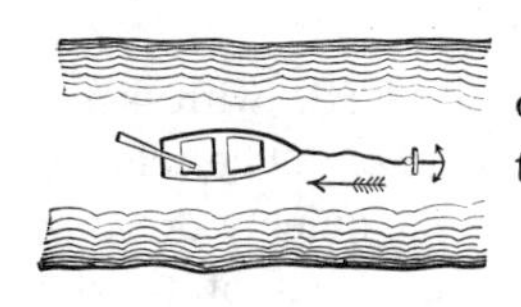

560. ANCHORED FERRYBOAT.—One of the few methods of crossing a stream by the action of the current.

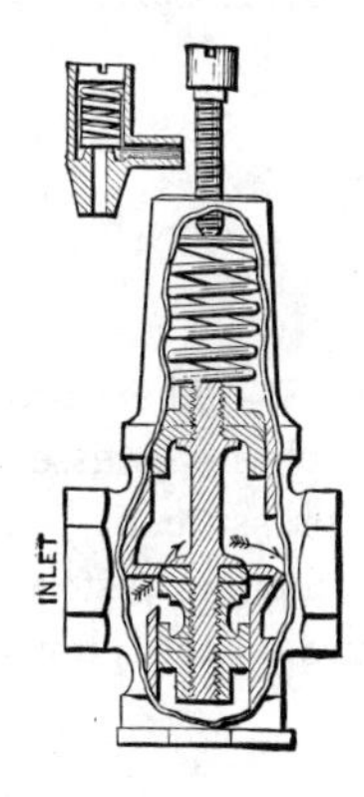

561. "MUELLER" WATER PRESSURE REGULATOR, for reducing a high-pressure works to any required pressure in the service pipe. A spindle with one disc valve, two cupped leather piston valves, and a regulating spring. The high pressure in the house service pipe is relieved by the closure of the inlet valve, due to the differential area of the piston valves. When water is being drawn, the valve opens wide by the relief from pressure at the upper piston valve.

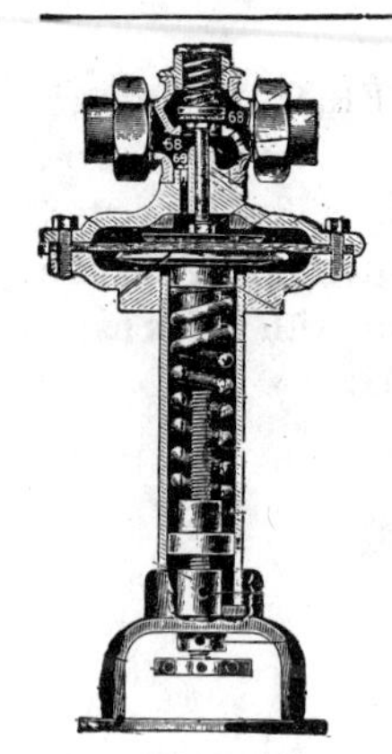

562. "MASON" WATER PRESSURE REGULATOR.— Over-pressure on the low-pressure side depresses a diaphragm and draws the valve to its seat. Adjustment for difference of pressure is made by compressing or releasing the spring pressure under the elastic diaphragm, by the screw and nut at the bottom.

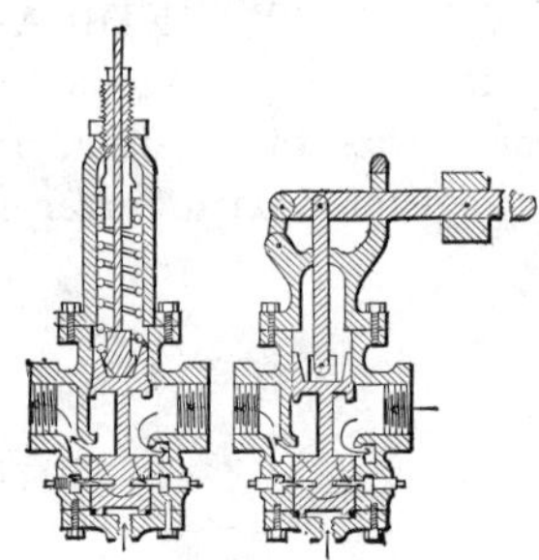

563. PUMP WATER PRESSURE REGULATING VALVE.— A balanced piston valve, with a differential balance by spring or lever and weight, is placed on the steam pipe to a pump. The opening beneath the lower piston is connected to the water discharge pipe of the pump. Over-pressure raises the disc and shuts off steam.

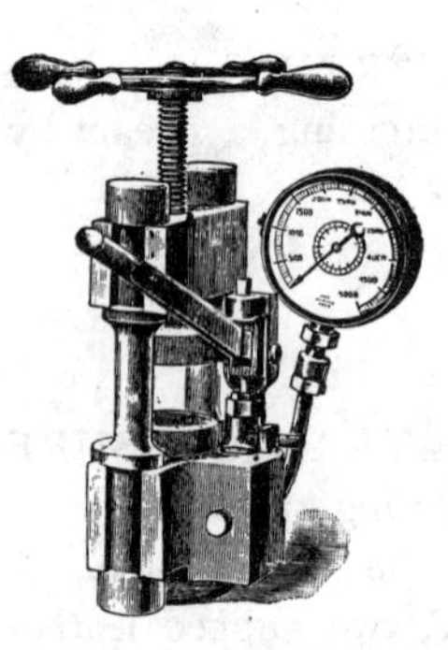

564. HYDRAULIC PRESS, with screw adjustment of upper platen. The closing down of the upper platen is quickly done by the screw, when a small movement of the hydraulic piston is required for the pressure.

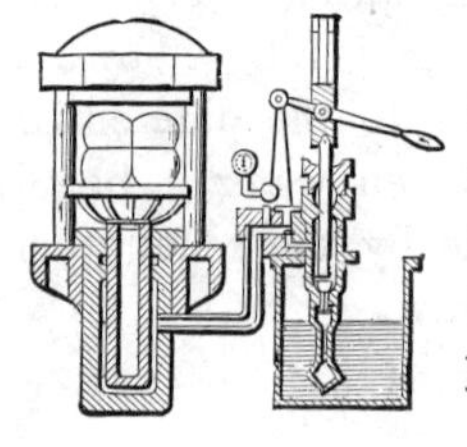

565. HYDROSTATIC PRESS.—There are many modifications of this principle for presses and elevator lifts. The gross pressure of the ram is as the areas of the ram and pump pistons multiplied by the pounds pressure on the pump piston.

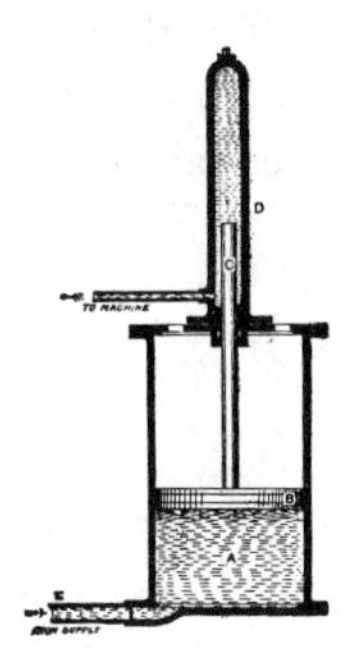

566. HYDRAULIC INTENSIFIER.—High pressure obtained from low pressure by differential pistons. A, Low-pressure cylinder; D, high-pressure cylinder and plunger.

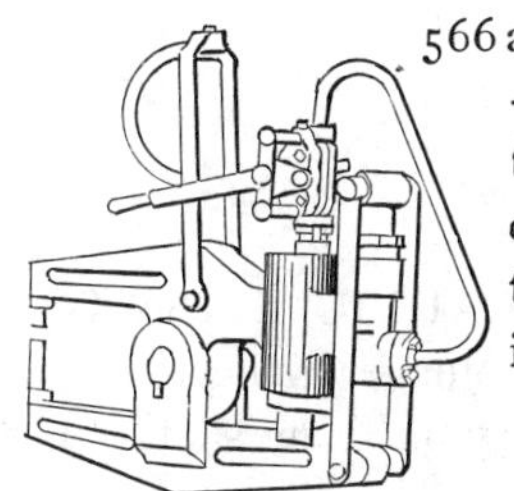

566 a. PORTABLE HYDRAULIC RIVETER.—An inverted hydraulic ram is operated by the small pump and lever attached to the top of the ram. The return stroke is made by the small reverse ram at the rear of the driving ram.

567. HYDRAULIC RAIL BENDER.—The plunger is moved with great force by the pressure from a small piston plunger operated by a hand lever, on the same principle as with the hydraulic jack. It is suspended by the eyes, and can be used for straightening or bending rails on the track.

568. HYDRAULIC RAIL PUNCH, constructed in the same line as the rail bender and hydraulic press. The loops are for suspending and to allow the punch to be easily handled in any position.

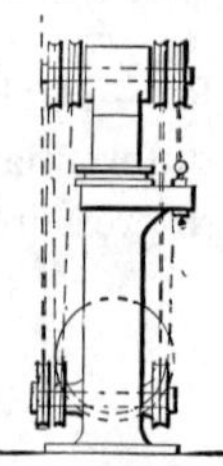

569. HYDRAULIC ELEVATOR LIFT with multiplying cable gear. The cable is carried under and over cross-head sheaves on each side to equalize the pressure on both sides of the plunger.

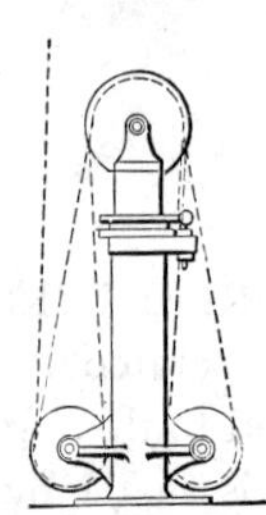

570. HYDRAULIC ELEVATOR LIFT with pulley sheaves central over plunger.

571. HORIZONTAL HYDRAULIC ELEVATOR LIFT, with central-plunger pulley. Cable winds on small pulley on drum shaft. For light lift.

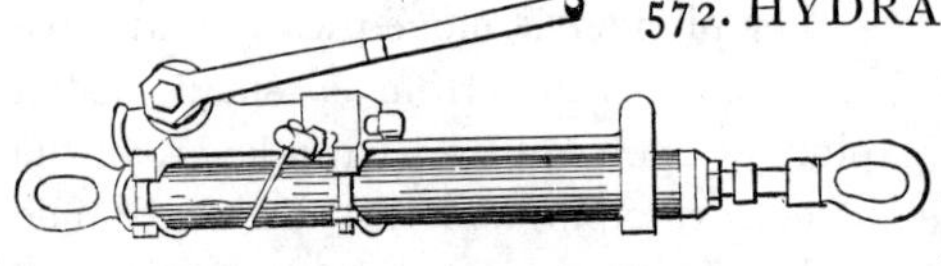

572. HYDRAULIC PULLING JACK. — The lever operates a small pump which forces water to the upper side of the piston and draws the piston rod and ring. The small screw and handle is the relief valve to return the water below or to the opposite side of piston for return.

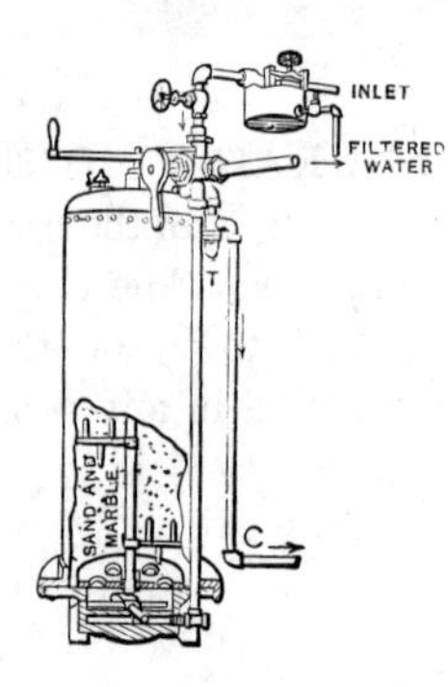

573. WATER PURIFYING FILTER, "N. Y. Filter Mfg. Co." pattern. A diaphragm near the bottom holds the gravel and sand filtering material. There is a shaft through the middle of the tank, with arms for stirring the sand while cleaning by a back-waterflow. The water is fed at the top with a small portion of alum at the rate of one pound to 7,000 gallons of water. The small tank at the top is the alum dissolver with the regulating valves.

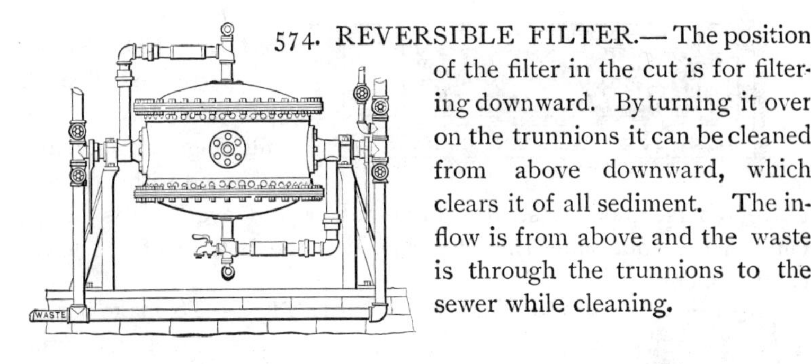

574. REVERSIBLE FILTER.— The position of the filter in the cut is for filtering downward. By turning it over on the trunnions it can be cleaned from above downward, which clears it of all sediment. The inflow is from above and the waste is through the trunnions to the sewer while cleaning.

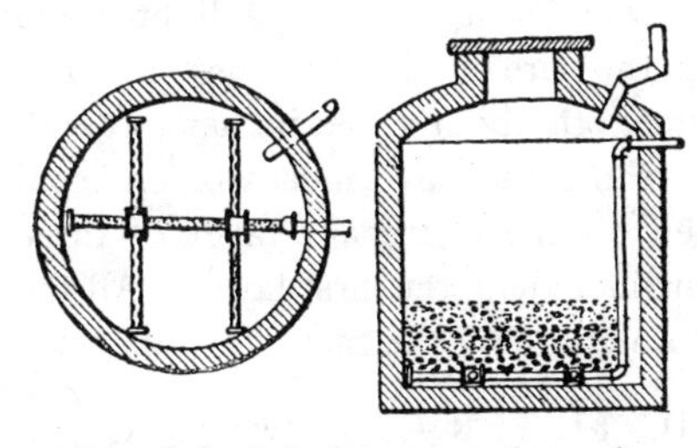

575. FILTERING CISTERN, plan.

576. Section. The pump pipe extends to the bottom of the cistern and across, with lateral branches. The pipes on the bottom to be perforated with one-sixteenth inch holes, enough to give a free flow of water to the pumps. Cover the pipes with sifted gravel larger than the holes in the pipes to a depth of six inches, then a layer of sharp, clean sand six inches thick, a layer of charcoal four inches thick, and a final layer of sand six inches thick.

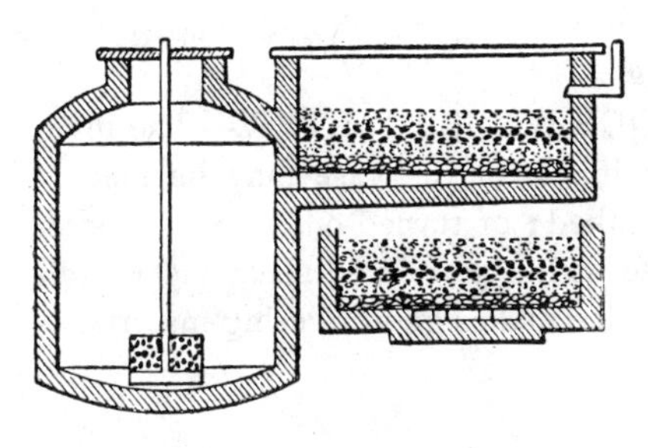

577. FILTERING CISTERN.—The rain-water is caught in a flat filter basin with gravel and sand spread on a perforated floor and drained into the cistern. The pump pipe is fixed to the perforated diaphragm of a two-chambered metal cylinder, the upper section of wihch may be filled with a bed of sand and charcoal in layers.

578. Cross-section of basin.

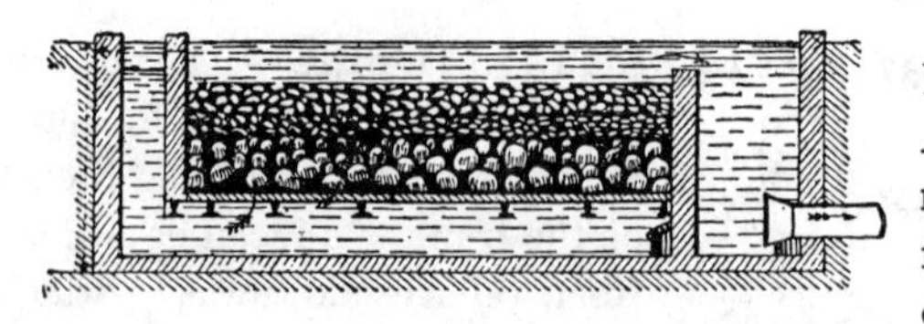

579. UPWARD-FLOW FILTER.—A perforated floor is made of any desired filtering capacity and charged with layers of gravel, coarse and fine sand, with an inflow and overflow, as in the cut. A wash-out outlet should be made in the bottom of the lower compartment.

580. DOMESTIC FILTER.—To make a filter with a wine barrel, procure a piece of fine brass wire cloth of a size sufficient to make a partition across the barrel. Support this wire cloth with a coarser wire cloth under it and also a light frame of oak, to keep the wire cloth from sagging. Fill in upon the wire cloth about three inches in depth of clear, sharp sand, then two inches of charcoal broken finely, but no dust. Then on the charcoal a layer of three inches of clear, sharp sand, rather finer than the first layer. All the sand should be washed clean before charging the filter.

581. DOMESTIC FILTER.—Use two stone pots or jars, the bottom one being a water jar with side hole; if no faucet can be used, the top jar can be removed to enable the water to be dipped out. The top jar must have a hole drilled or broken in the bottom, and a small flower-pot saucer inverted over the hole. Then fill in a layer of sharp, clean sand, rather coarse. A layer of finer sand, a layer of pulverized charcoal with dust blown out, then a layer of sand, the whole occupying one-third of the jar.

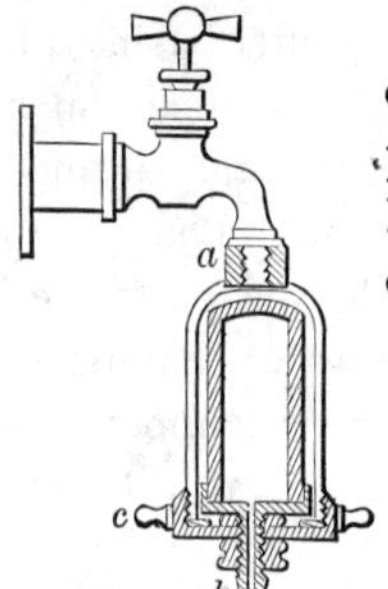

582. POROUS WATER FILTER.—The inverted cup on the inside of the case may be made of potters' clay, baked; or turned out of porous stone. Fibre, enclosed within perforated sheet metal walls, or wire gauze also makes good filtering material.

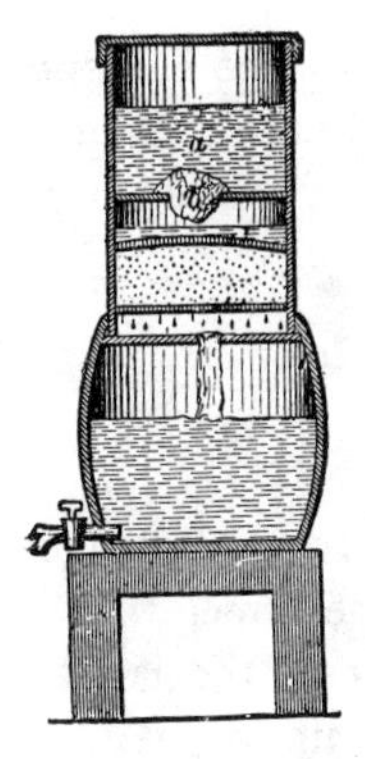

583. STONEWARE FILTER for household use. The lower jar for storage of filtered water. The upper jar has a hole filled with sponge that filters the dirt out; beneath, a bed of charcoal on a porous stone or earthen plate.

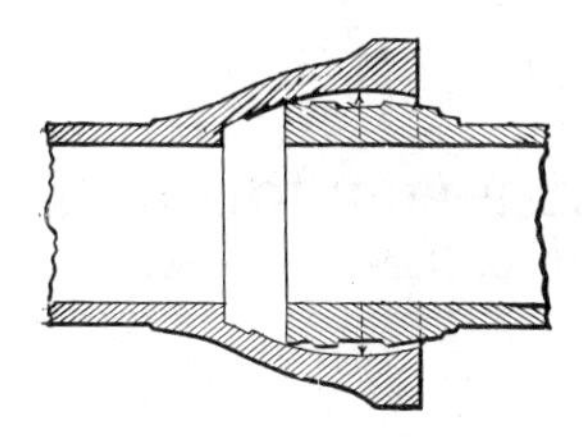

584. "WARD" FLEXIBLE PIPE JOINT.—The internal surface of the hub is made spherical. The corrugated pipe end is inserted and the space filled with lead and calked.

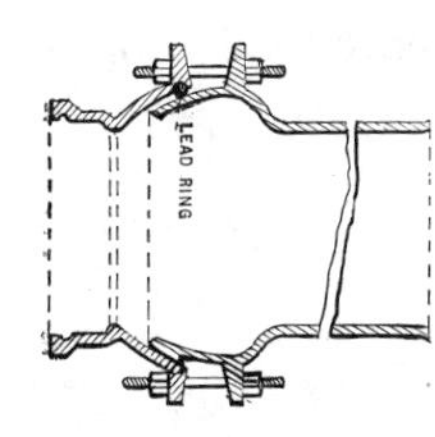

585. FLEXIBLE BALL JOINT.—Flanges are cast upon the spherical ends of the pipes. The joint is packed with a lead ring and drawn together with bolts at any angle within its limit.

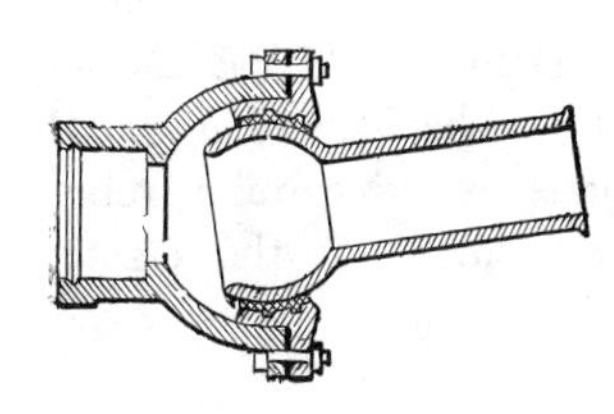

586. FLEXIBLE PIPE JOINTS, for submarine pipe lines. The head joint is first made up in the gland. The flange joint is bolted when the pipe is laid in line ready for lowering.

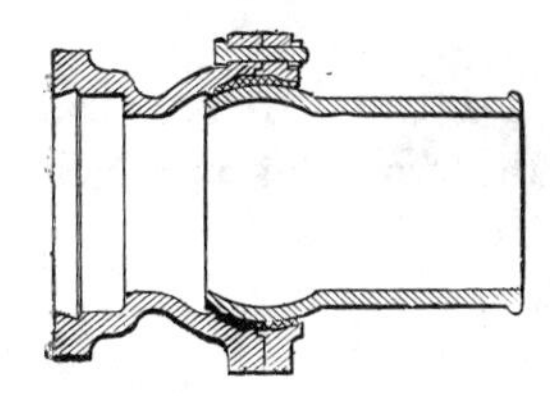

587. FLEXIBLE PIPE JOINT, in which the lead joint is made between a divided socket, which does not require the pouring of melted lead; a lead ring is used.

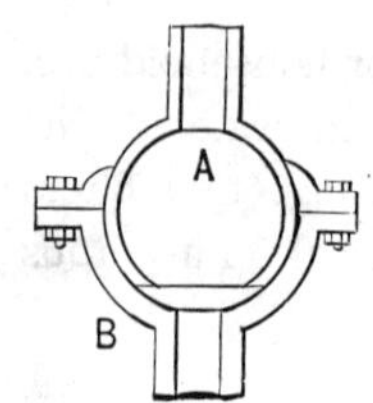

588. FLEXIBLE PIPE JOINT.—The ball end, A, of a pipe is ground to a tight fit in the socket, B, of another pipe and held in place by a bolted flange.

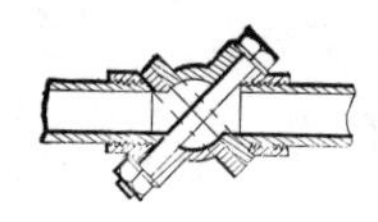

589. UNIVERSAL PIPE JOINT.—The flanges are faced at 45° to the line of the pipe, with a through bolt at right angles to the faces of the flanges. The joint may be made at any angle up to 90°.

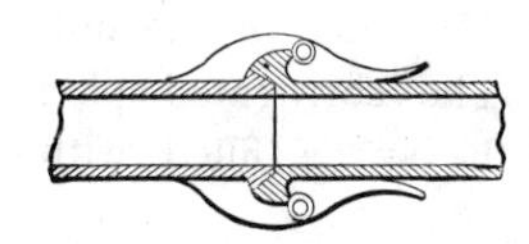

590. TOGGLE CLIP PIPE JOINT.—A quick connecting joint for hose.

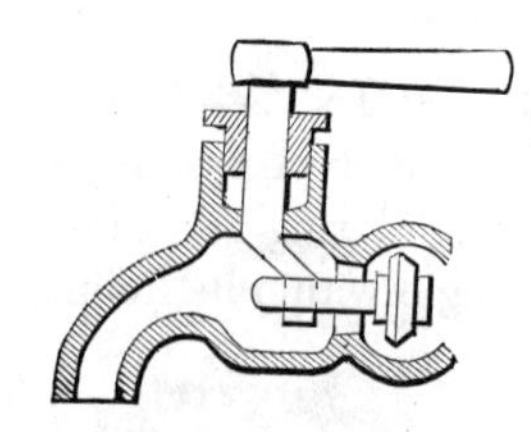

591. BIBB, with crank-moved valve opening against the pressure.

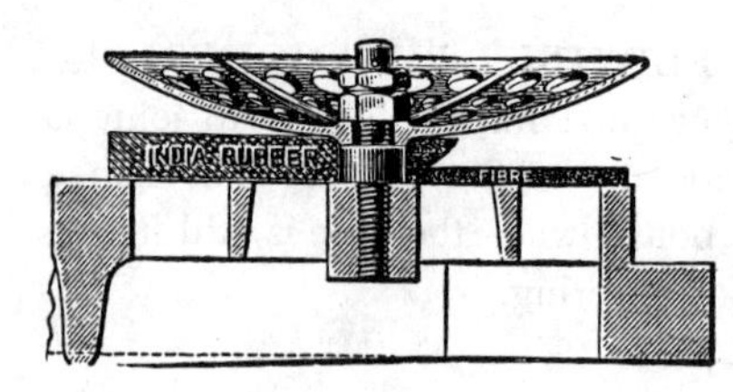

592. DISC VALVE AND GUARD.—The spherical guard is perforated to give quick relief to the movement of the elastic disc.

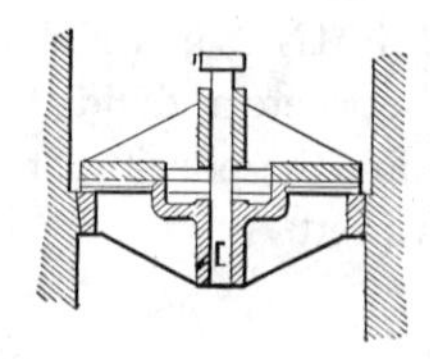

593. DOUBLE BEAT DISC VALVE.—The central seat is borne by the cross bar in which the guide pin of the valve is set.

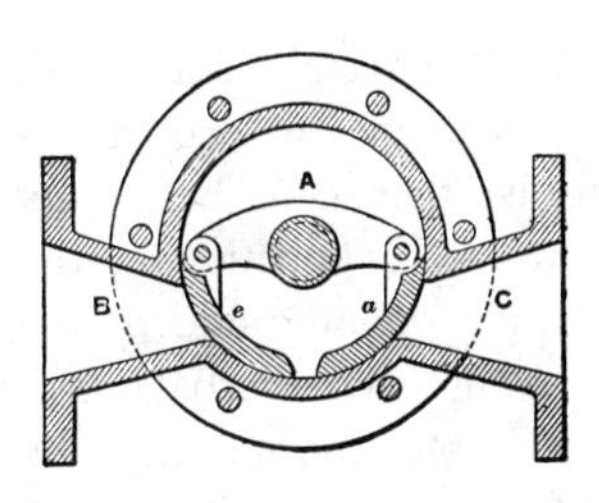

594. HYDRAULIC VALVE, used on elevators. Cylindrical in form, the valves move across the ports by a rock shaft and arms.

A, pressure chamber. B, C, to elevator cylinder.

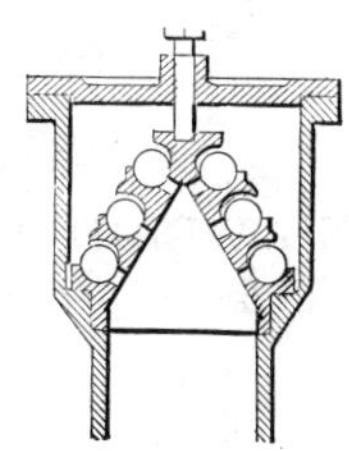

595. MULTIPLE BALL VALVE.—The cone-valve seat is in two parts; the cover or cage is held in place by the screw in the cap.

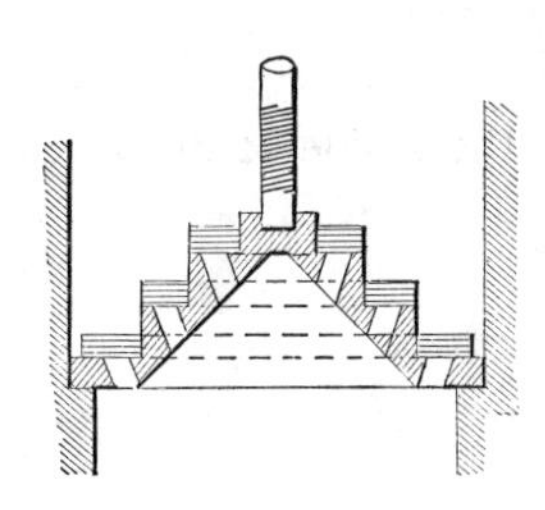

596. MULTIPLE RING VALVE, for enlarged valve area with small lift.

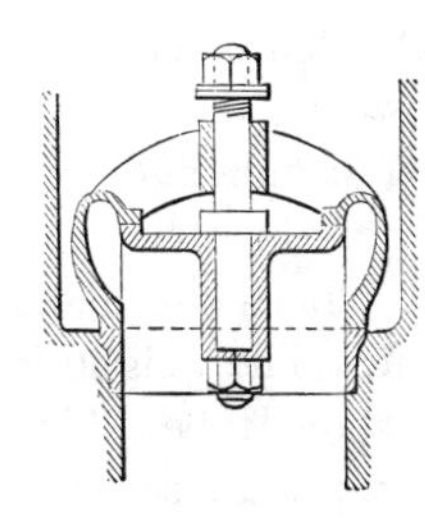

597. DOUBLE-BEAT PUMP VALVE, Cornish model. The upper seat is supported by a cross-bar, in which is fixed the guide-pin that carries the valve.

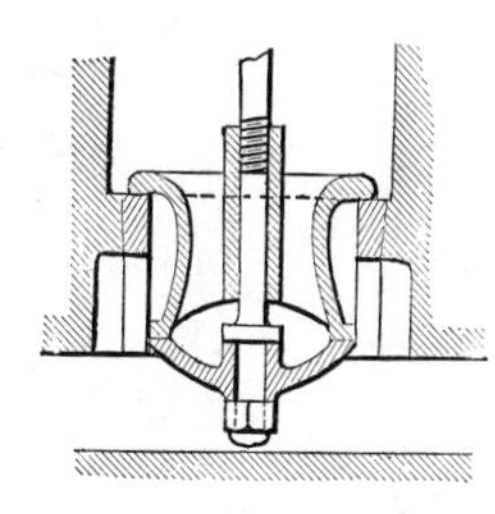

598. DOUBLE-BEAT PUMP VALVE or relief valve.— The valve spindle may be loaded by weight or spring.

599. VIBRATING MOTION of a trough discharging water alternately in two directions. The trough is balanced below its centre of gravity, and has a partition at the middle. The water falls on one side of the partition until the trough is overbalanced, when it turns and discharges the water. The partition is thrown over and the other end of the trough is then filled. A crude form of water meter.

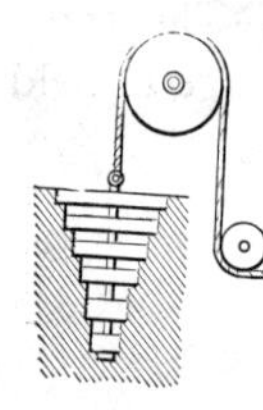

600. VARIABLE COMPENSATING WEIGHTS for a hydraulic lift. The weights are picked up one after the other.

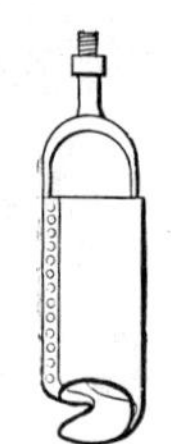

601. SAND AUGER.—Used on the inside of deep well pipes with open bottom.

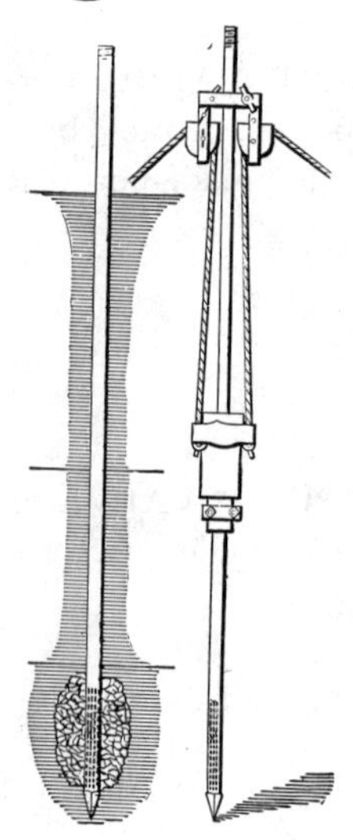

602. DRIVEN WELL.—A clamp strongly bolted to the well pipe on which the weight strikes to drive the tube. A clamp and two sheaves are bolted at the top of the tube with ropes rove through the sheave blocks and made fast to the weight for raising it. The weight is hollow, and rides loosely over the tube. The clamps are raised as additional pipes are screwed to the well pipe.

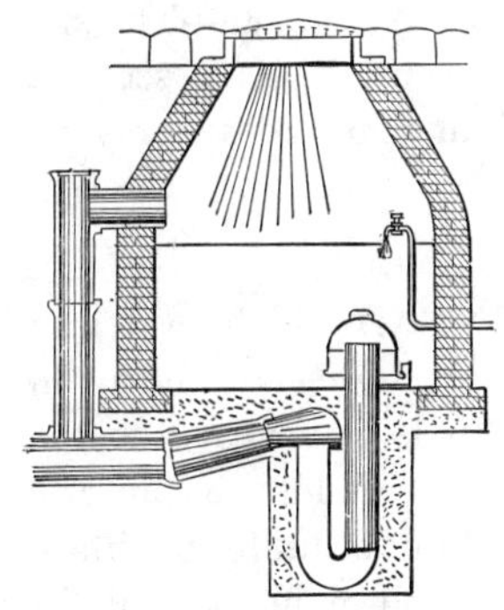

603. AUTOMATIC FLUSH SEWER TANK, "Miller" model. In this form the siphon is inverted, holding the water seal to balance the water head in the tank in the uptake of the siphon. The cap over the long end of the siphon is to seal the air in the siphon until the sewage pressure is equal to the water-balanced leg.

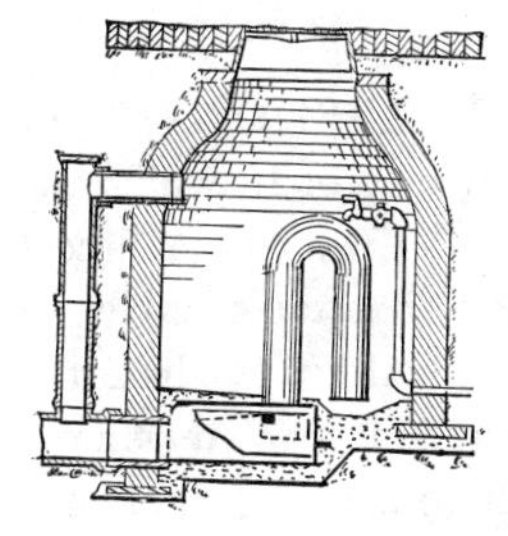

604. AUTOMATIC FLUSH SEWER TANK, "Van Vraken" model. The inverted siphon opens into a tipple pan which seals the outlet of the siphon until the sewage in the tank reaches the level of the bend, when a general discharge takes place.

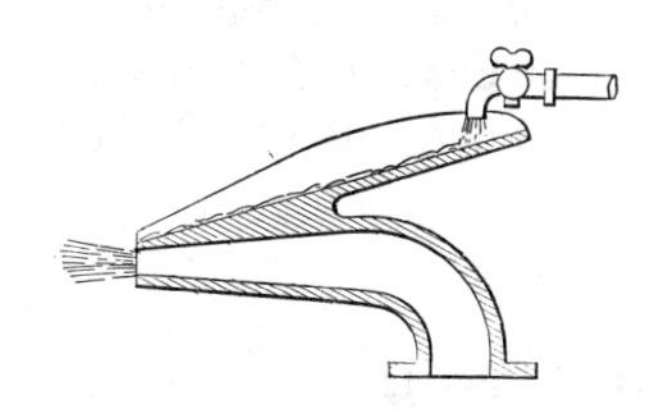

605. ATOMIZER.—A small stream running down an incline is atomized at the nozzle by a blast of air.

606. BALL AND JET NOZZLE.—The ball is held in contact with the jet by the adhesion of the water to the rolling surface. The ball should be very light. The principle is the same for an air jet, only that a very light ball must be used. With the low ball in the conical nozzle the ball can lift no higher than to give vent to the water or air under the same area as the neck of the nozzle.

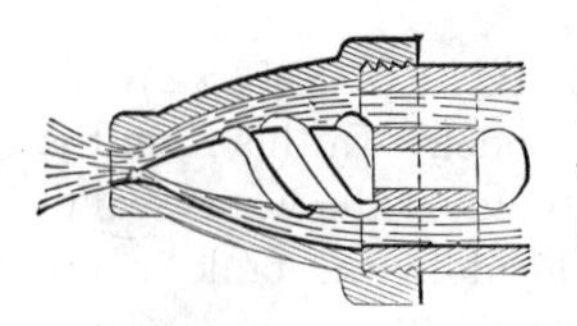

607. SPRAY JET NOZZLE.—The spiral wings on the central cone set the water into a whirl, and induce a spray by centrifugal action.

608. HERO'S FOUNTAIN.—The water in the upper basin exerts a pressure upon the air in the lower receptacle, which is transferred to the surface of the water in the middle basin and forces it up in the jet. Many beautiful modifications of this principle are shown in modern devices.

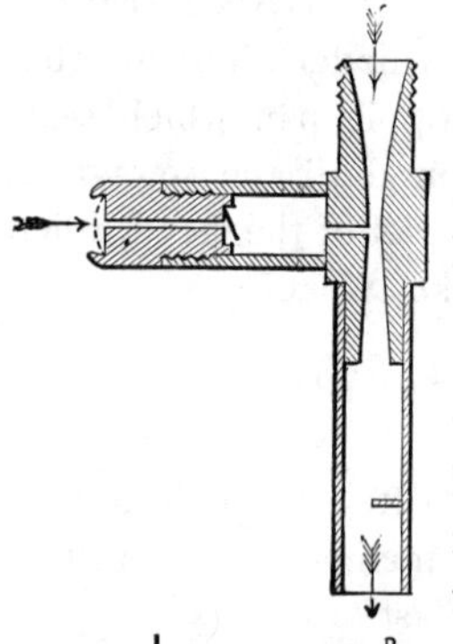

609. "CHAPMAN" ASPIRATOR or vacuum pump. A water ejector in which the propelling power may be derived from a faucet of any town water-works, or a tank having a head of seventeen feet, equal to one-half the static water-head of a vacuum. Water enters at the conical end. There is an elastic check valve in the branch tube or vacuum connection. It will produce a vacuum equal to the barometric height, less the height due to the tension of the vapor of water.

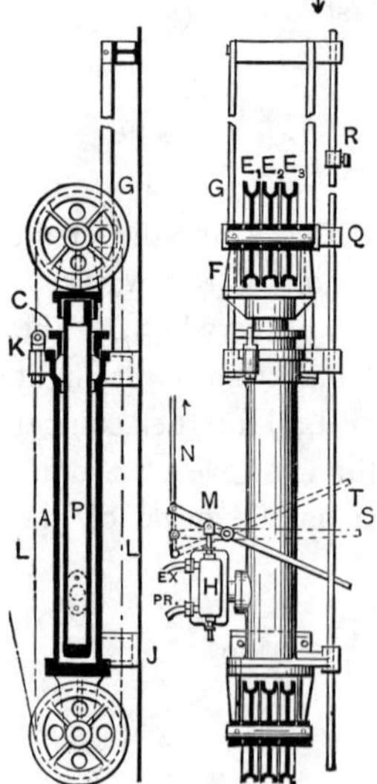

610. HYDRAULIC LIFT for a crane or elevator. Section showing cylinder plunger and sheaves.

611. Plan, showing position of valve chamber and valve lever in three directions for stop, start, and reverse. The side rod limits the extreme movement of the plunger by automatically operating the valve lever.

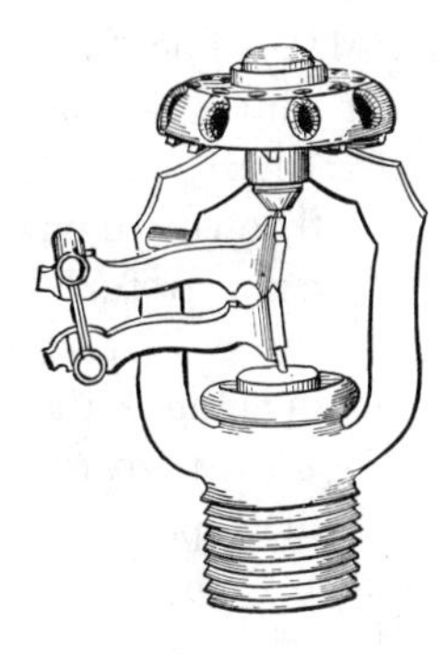

611*a*. HYDRAULIC SPRINKLER HEAD.—The levers of the toggle joint are held in place by a strip of fusible alloy that melts at about 212 deg. and allows the levers to fly open and release the water spray valve. Pressure on the valve is adjusted by the screw and nut at the top.

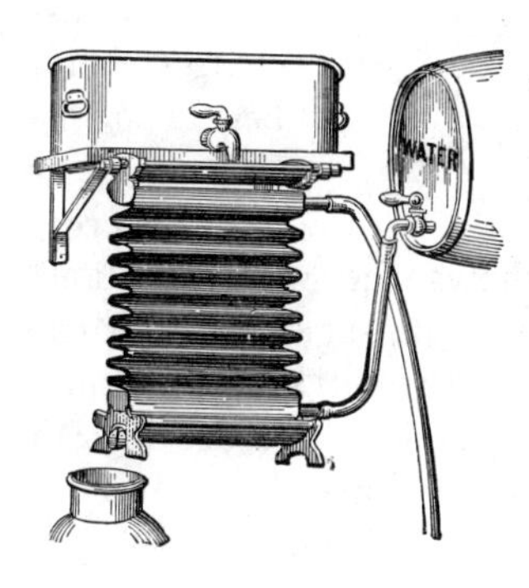

611*b*. MILK COOLING DEVICE.—Milk is fed from a tank over the surface of a hollow copper pan, corrugated to retard its flow, while cold water is circulated through the pan in the opposite direction.

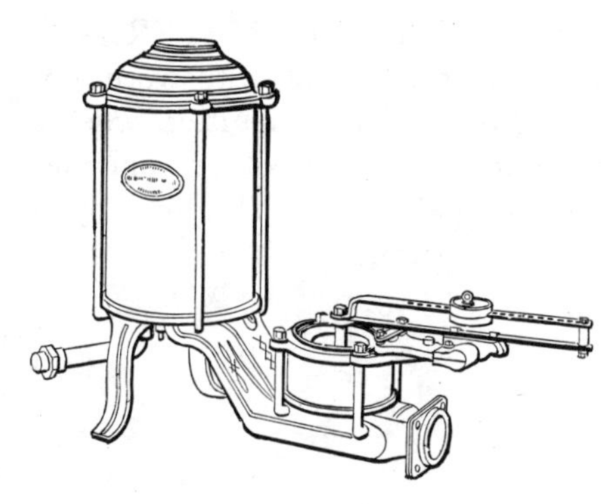

611*c*. HYDRAULIC IRRIGATION ENGINE.—Capacities up to 750,000 gallons per day. They are made adjustable for the best conditions of operation by the sliding weight on the valve lever. The double acting type will pump pure spring water by the use of impure water of streams. Will run on 2 feet fall. Rife Engine Company, Roanoke, Va.

611*d*. FOUR STAGE CENTRIFUGAL PUMP.—The four volute wheels are fixed to the shaft. The water enters to the wheel A, is thrown out and returns through the opening in the stationary partition to the center of the next wheel, and so on to the discharge chamber D at the right.

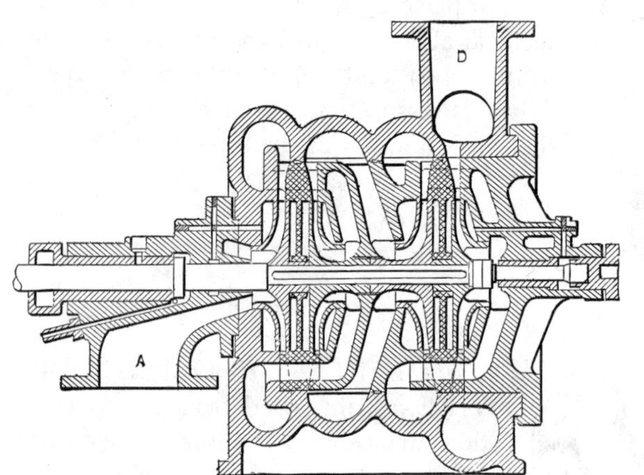

In the cross section *d* shows the volute sections and *h* the cross passages in the partitions. At 900 revolutions per minute it sustains a forcing pressure of 240 pounds per square inch.

Made in Switzerland.

611*e*. CURRENT METER. — A propeller on a spindle with worm operating two geared register wheels, graduated to 1,000 revolutions. Stopped and started for time by a string and spring pawl.

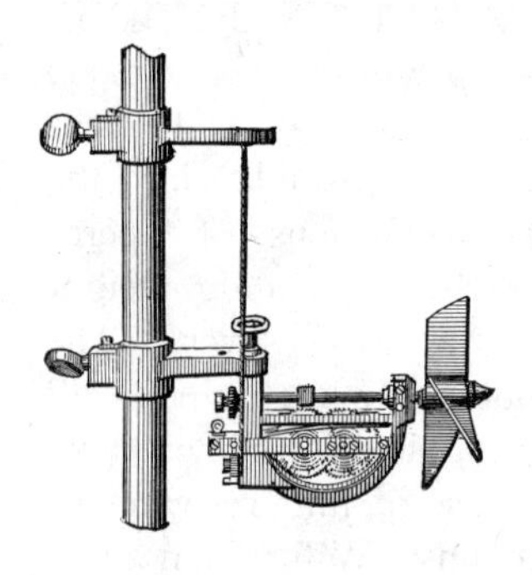

Section VIII.

AIR POWER APPLIANCES.

Windmills, Bellows, Blowers, Air Compressors, Compressed Air Tools, Motors, Air Water Lifts, Blowpipes, Etc.

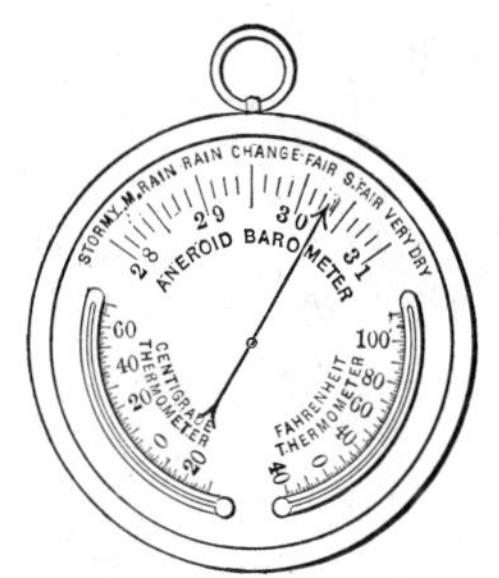

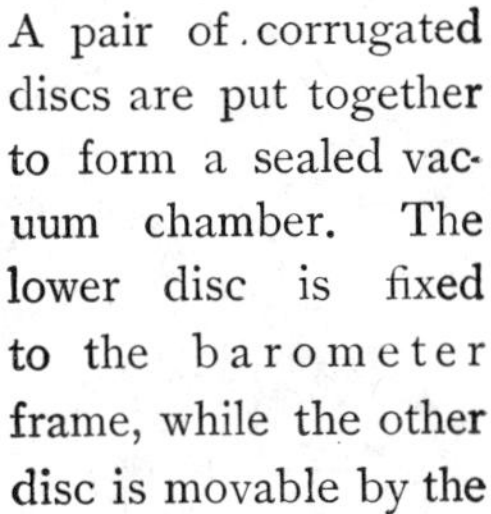

612. ANEROID BAROMETER.—A pair of corrugated discs are put together to form a sealed vacuum chamber. The lower disc is fixed to the barometer frame, while the other disc is movable by the difference in air pressure, and, through a gear to increase the motion, moves the index hand on the graduated dial.

612 a. Corrugated disc and gear.

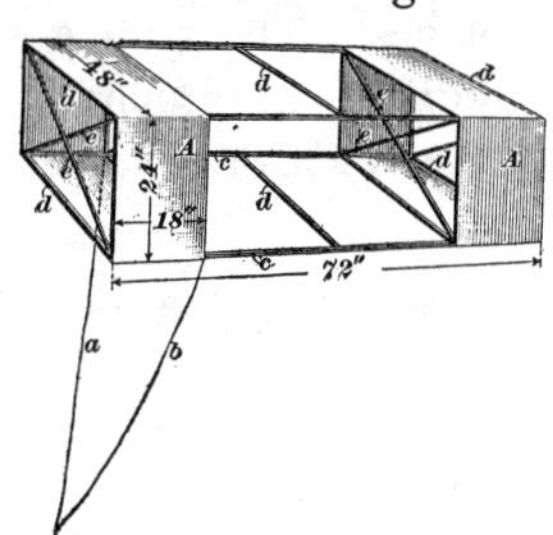

613. BOX KITE.—A light frame of pine, spruce, or bamboo is braced as shown in the cut. Fine, light cambric is stretched over each end, all in proportion to the figures in the cut. The bridle is attached one-quarter of the length of the box from the front on the bottom frame.

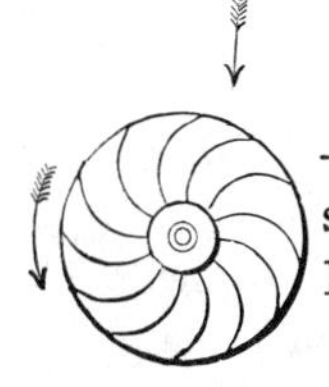

614. CURVED VANE WINDMILL OR MOTOR.—The wind pressure is greater against the hollow side of the curved blades than against the other side. Hence the motor motes.

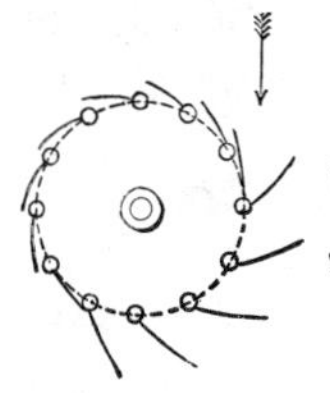

615 FEATHERING WINDMILL.—The light jointed blades are forced out when their edge catches the wind, and the mill goes.

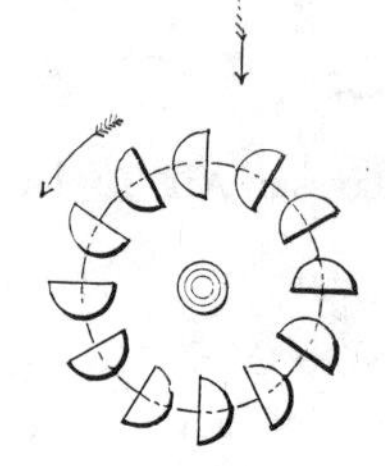

616. HEMISPHERICAL CUP WINDMILL.—The pressure of the wind is greater against the hollow side of the cups than against the spherical side, and the mill rotates. Also used for anemometers.

617. WINDMILL OF OUR GRANDFATHERS, with reefing sails. A few still in use in the United States.

618. WINDMILL AND STEEL TOWER.—Mill with a single series of blades. The tail-piece is pivoted to the mill-head, and is swung around to turn the face of the mill from the wind by a governor.

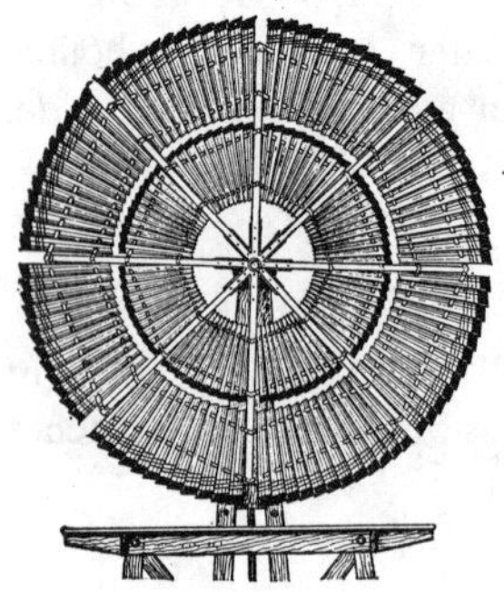

619. MODERN WINDMILL.—Two series of concentric blades fastened to the purlines of a braced radial frame. The blades are fixed at an angle of about 35° to the plane of the wheel. A peculiarly constructed mechanism turns the wheel edgewise to the wind to stop it, or to regulate its position in a high wind.

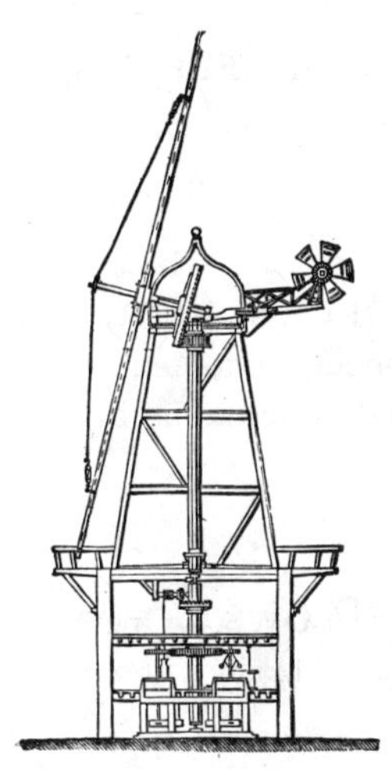

620. ANCIENT WINDMILL and gearing for a two-stone flour mill. The windmill is turned toward the wind by a small windmill at right angles on the tail frame, with pinions and shaft connecting with a circular rack around the revolving dome. These mills, used for grinding grain, are the principal source of power in Eastern countries.

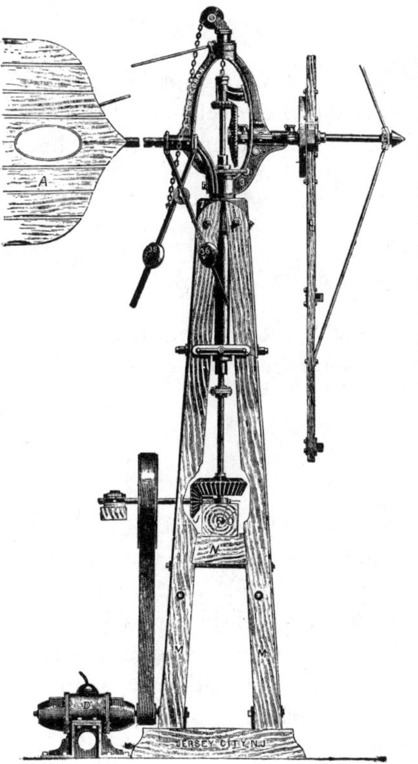

621. ELECTRIC WINDMILL PLANT, "Corcoran" model. The windmill-driven dynamo charges a storage battery, which has an automatic cut-out when the mill runs too fast or too slow. The mill has also a regulator throwing it out of the direct course of the wind when running too fast, or for stopping the mill.

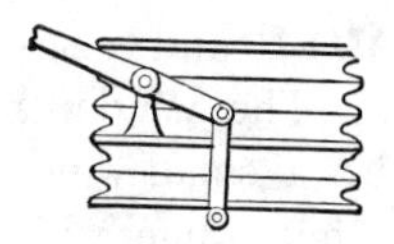

622. SMITH'S CIRCULAR BELLOWS, in two parts for uniform blast.

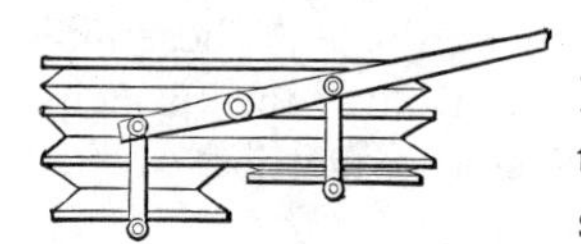

623. DOUBLE ORGAN BLOWING BELLOWS.—The upper section equalizes the air pressure from the alternating blower sections.

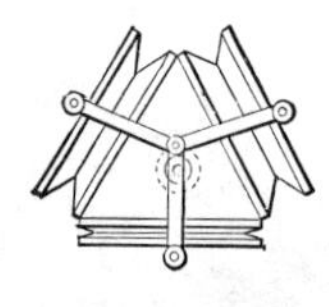

624. THREE-THROW BELLOWS.—Operated by a crank, and gives constant blast without an equalizer.

625. FOOT BELLOWS, for a blowpipe. A spring raises the top of the bellows. The rubber bag is confined to the netting to prevent bursting. The step at the left is for the foot.

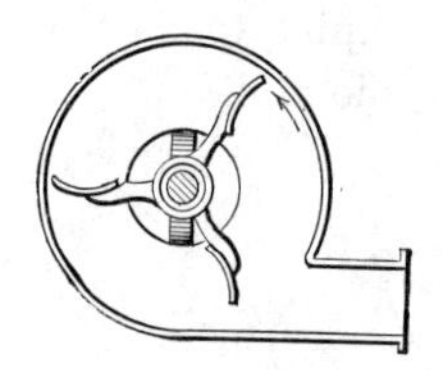

626. FAN BLOWER.—An ordinary model as used for blowing forge fires

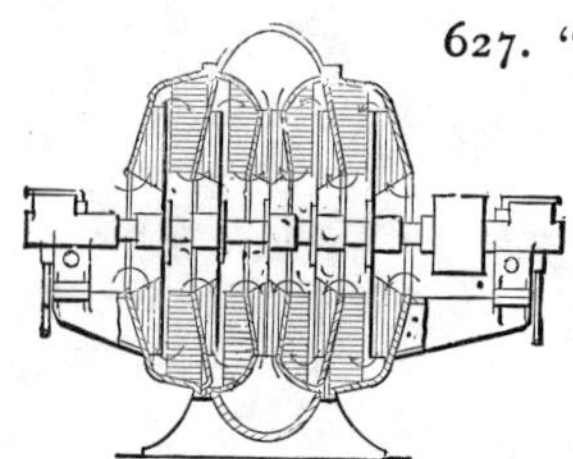

627. "HODGES" COMPOUND BLOWER.—The action is a triple effect. The air is drawn in at each side of the blower and thrown out at increasing pressure successively by the fans on each side, and returned successively by the stationary partitions, with a final discharge at the central annular chamber.

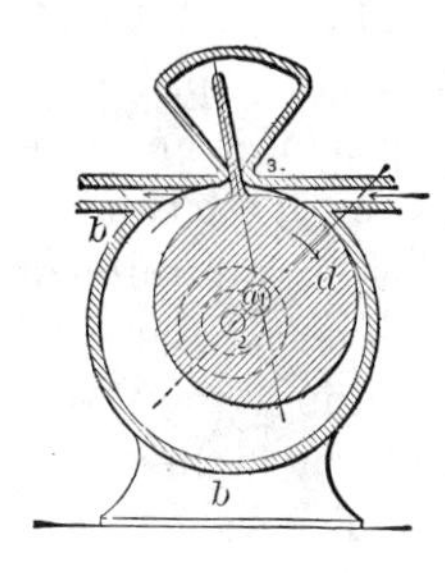

628. "WEDDING" ROTARY BLOWER.—A swinging winged cylinder moving in contact with an outer shell. The wing rides in a slot in the shell with a cavity to give it freedom of motion. The central cylinder is driven by a crank-pin or eccentric on a shaft central with the shell.

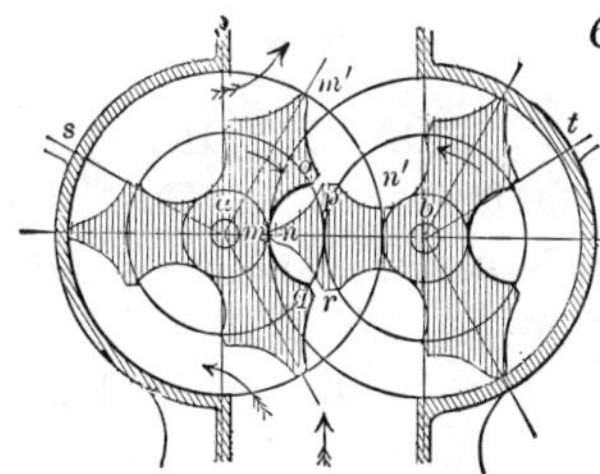

629. "FABRY" ROTARY BLOWER.—Two wheels of three teeth each rotate in a two-part cylindrical case. The teeth on and near the line joining the axis mesh alternately for a part of a revolution, so as to make a continuous closure to the passage of air between the wheels.

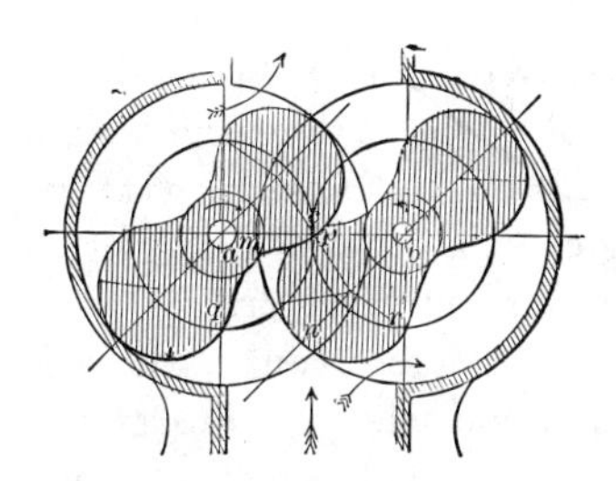

630. "ROOT" ROTARY BLOWER. An early form. Has been also used as a pump.

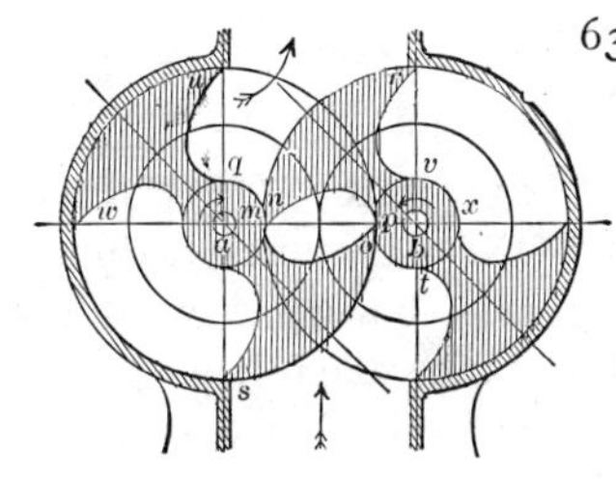

631. "ROOT" ROTARY BLOWER.—Present design. The extended surface of the periphery of the wheels allows them to run loosely in the shell without friction, and with very small loss by air leakage.

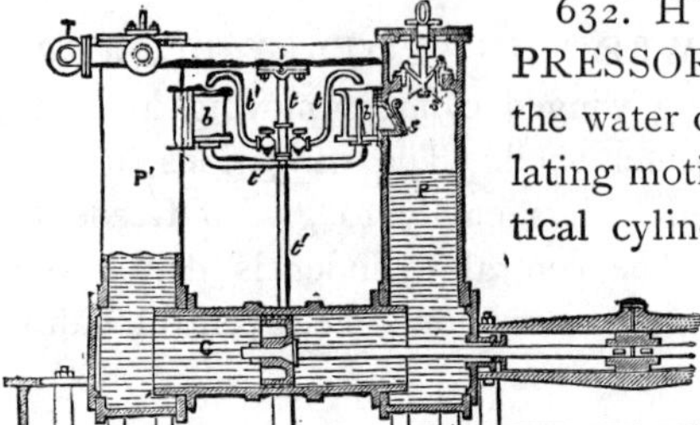

632. HYDRAULIC AIR COMPRESSOR.—A reciprocating piston in the water cylinder, G, produces an oscillating motion in the water of the two vertical cylinders, drawing in air through the flap valves at the side, and discharging the compressed air through the valves at the top. The water pipes, *t*, *t*, *t*, are to supply the place of water ejected through the air valve by delivering all the air compressed at each stroke of the piston.

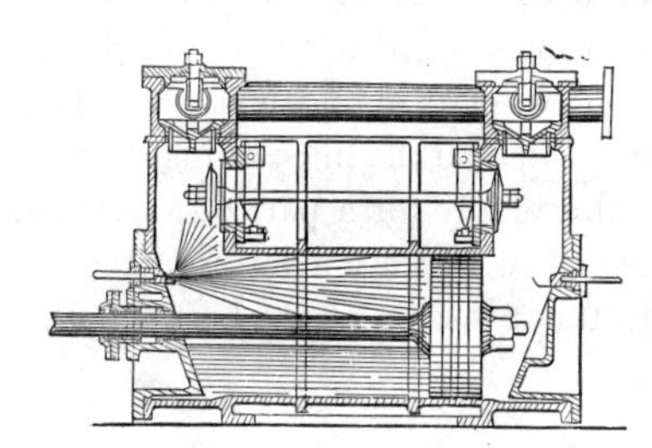

633. PISTON HYDRAULIC AIR COMPRESSOR, "Dubois & Francois" model. Water was constantly injected into the cylinder to cool the air, the excess being discharged through the air valves. An early type.

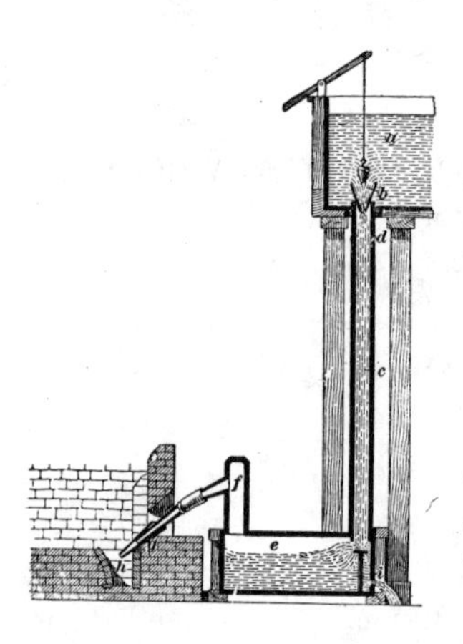

634. TROMPE OR HYDRAULIC AIR BLAST.—One of the early devices for furnishing an air blast to a forge. The falling column of water draws in air through the small inclined orifices at *d*, carrying it into the reservoir *e*, where it separates, and is discharged through the tuyère pipe at *b*. The outlet at *i* discharges the water through an inverted siphon, carried high enough to balance the air pressure.

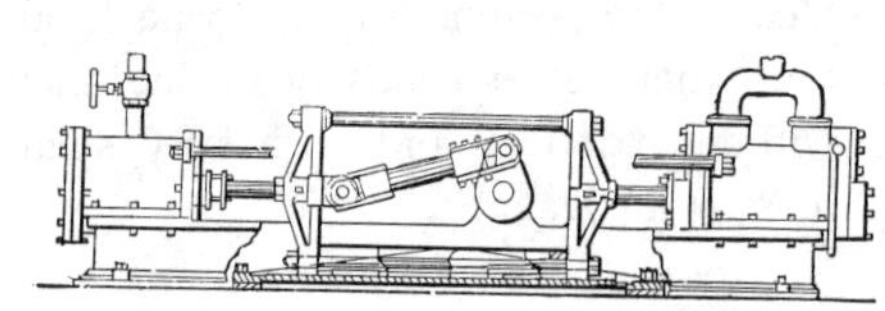

635. AIR COMPRESSOR.— Elevation of duplex type, showing connecting rod and yoke frame.

"Clayton" model.

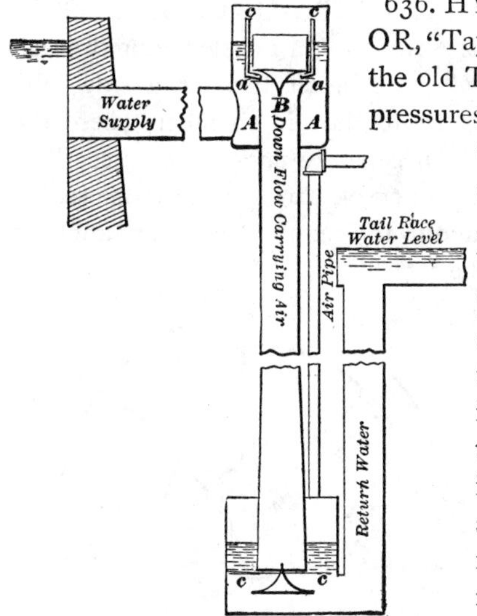

636. HYDRAULIC AIR COMPRESSOR, "Taylor" system. The principles of the old Trompe blower extended for high pressures. A number of air tubes, *c*, *c*, terminate at the conical entrance of the down-flow pipe, B, at *a*, *a*. A supply of water to the chamber A, A, and its flow down the pipe, draws air through the small pipes, carrying it down to the separating tank, *c*, *c*, where it is liberated at the pressure due to the hydrostatic head. The air is delivered through a pipe, as shown in the cut, and the water rises through a pipe to the tail race.

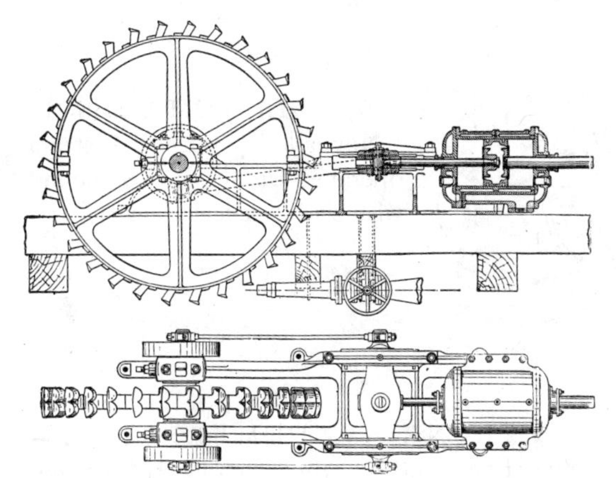

637. AIR COMPRESSOR. — Pattern of the "Ingersoll-Sergeant Drill Co." Operated by a Pelton wheel. Vertical section.

638. Plan.

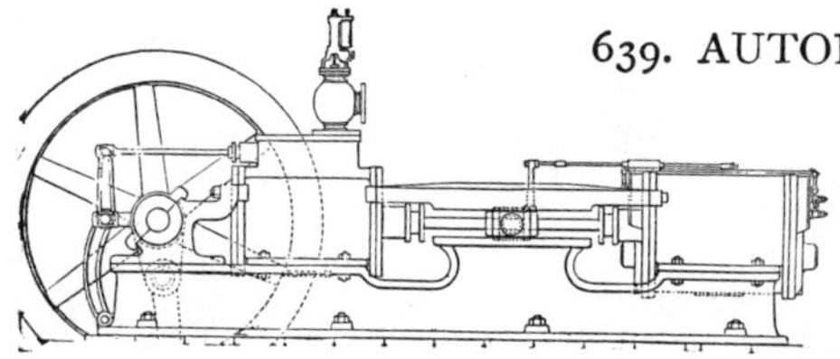

639. AUTOMATIC AIR COMPRESSOR, "Bennet" model. Showing the valve gear of a simple lever connected by link to the eccentric.

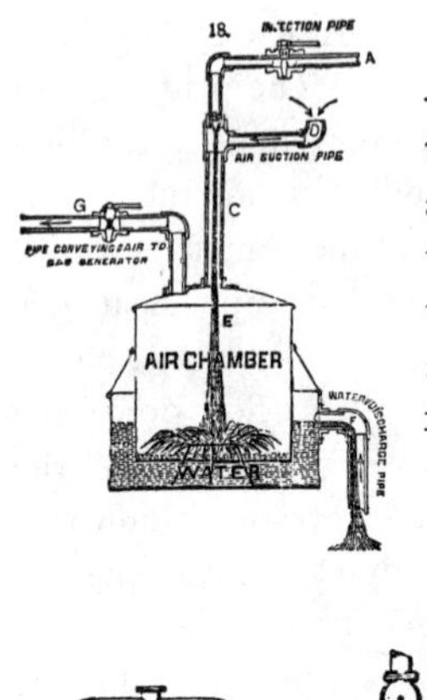

640. WATER JET AIR COMPRESSOR. — A jet of water from a nozzle falling through the tube C draws in air through a side tube and forces it into the air chamber, where the water and air separate under pressure. The water is siphoned off through the water seal at a height due to the required pressure and the force of the jet.

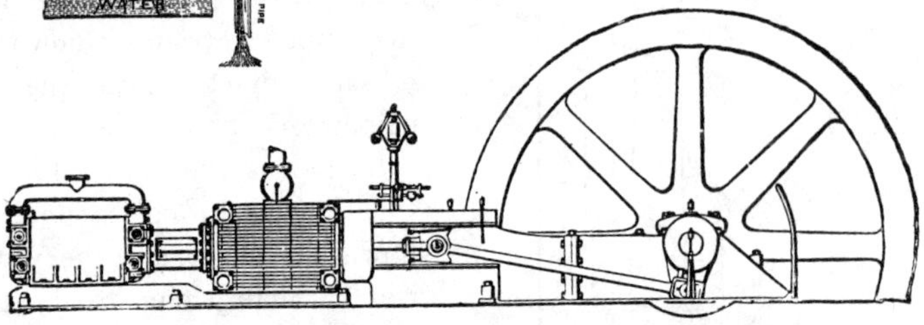

641. AIR COMPRESSOR.—Driven by a Corliss engine, direct connected.

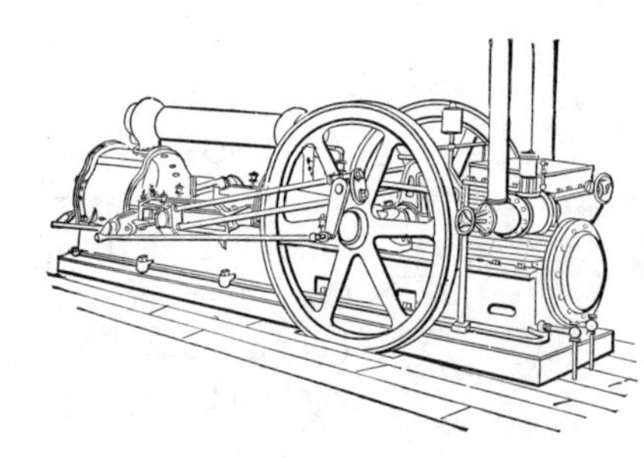

642. AIR COMPRESSOR, "Norwalk" pattern. A steam operated tandem compound with an intercooler.

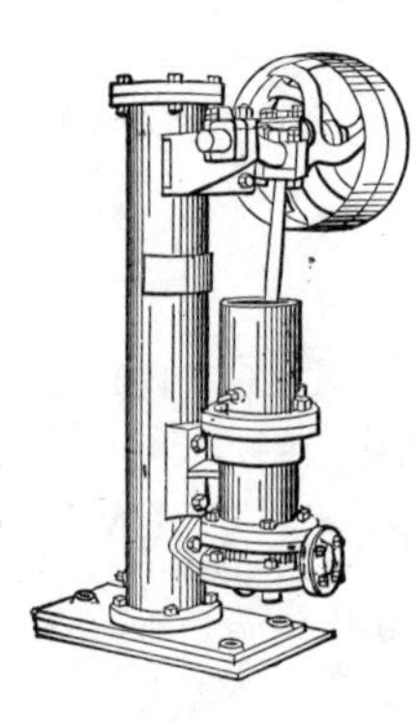

643. TRUNK AIR COMPRESSOR.— Mounted on receiver. Single-acting, belt driven. A very compact model.

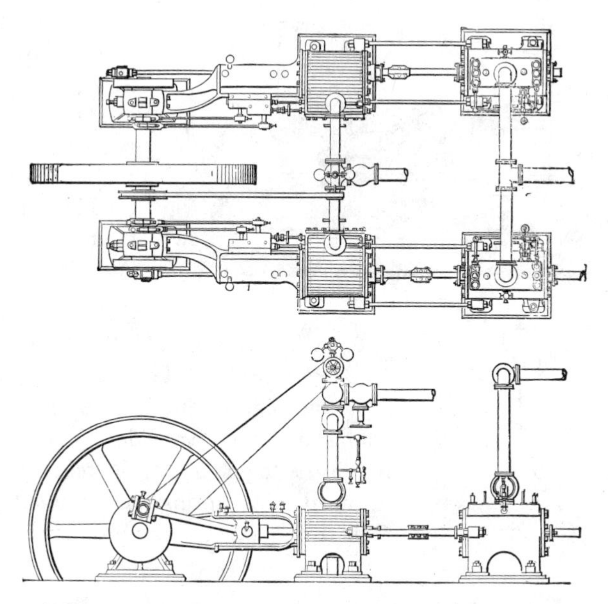

644. DUPLEX STEAM ACTUATED AIR COMPRESSOR, "Ingersoll-Sergeant" model. The air cylinders are tandem to each steam cylinder with steam and air governors.

645. Elevation.

646. COMPOUND AIR COMPRESSOR.— Air is drawn in through the ports A, passes through the annular valve in the large

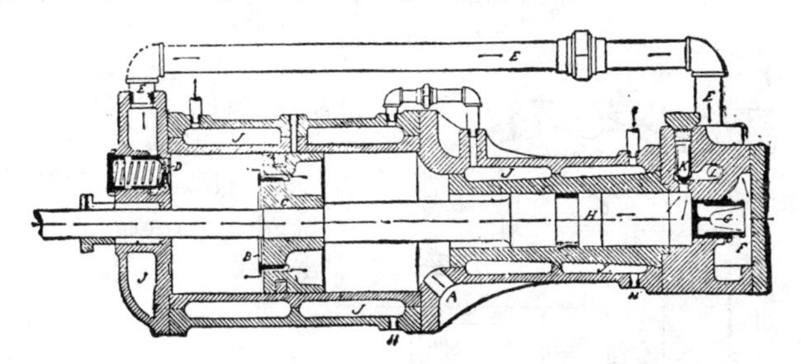

piston, and is forced through the valve D and pipe to the high-pressure inlet valve G; it is further compressed and delivered through the valve A′, and passage L. Both pistons are single, acting in opposite directions.

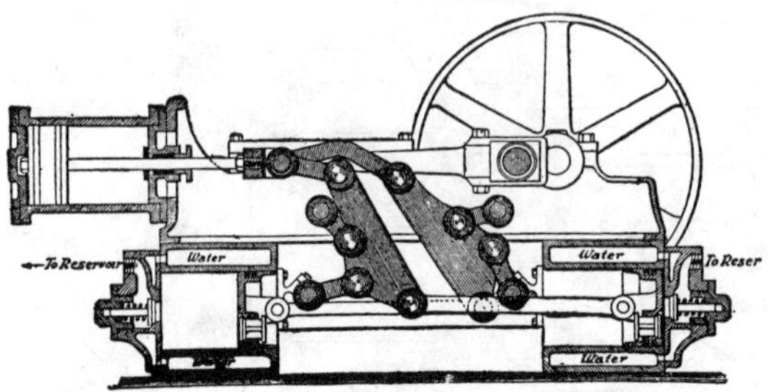

647. DUPLEX AIR COMPRESSOR, with parallel motion beams to two single-acting air cylinders from a double-acting steam cylinder. "N. Y. Air Brake" model.

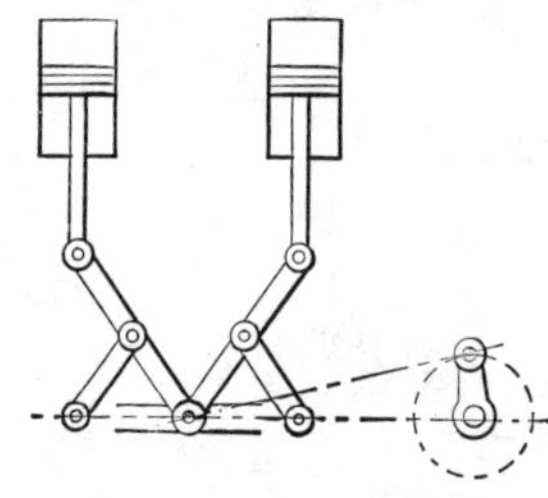

648. TOGGLE-JOINT DUPLEX AIR COMPRESSOR.—The crank moves the common joint of the long arms in a horizontal direction on a slide. The straightening of the toggle greatly increases the power of the pistons during the terminal part of their stroke, when the air pressure is greatest.

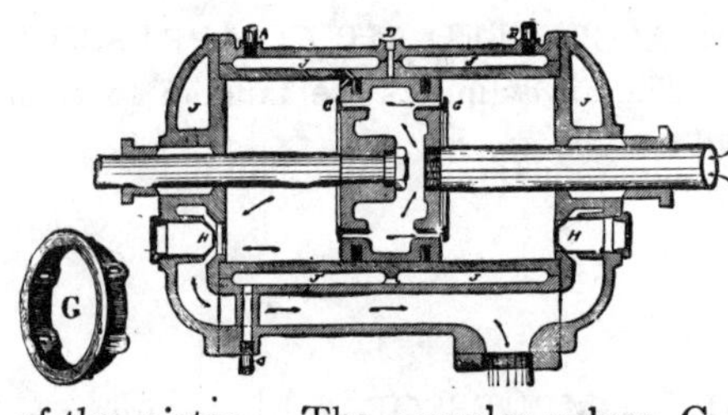

649. AIR COMPRESSOR CYLINDER, PISTON AND VALVES.—Pattern of the "Ingersoll, Sergeant Drill Co." Takes its air through a hollow piston rod at E to the interior of the piston. The annular valves, G, G, open and close by their momentum. H, H, discharge valves closed by springs; J, J, water jacket.

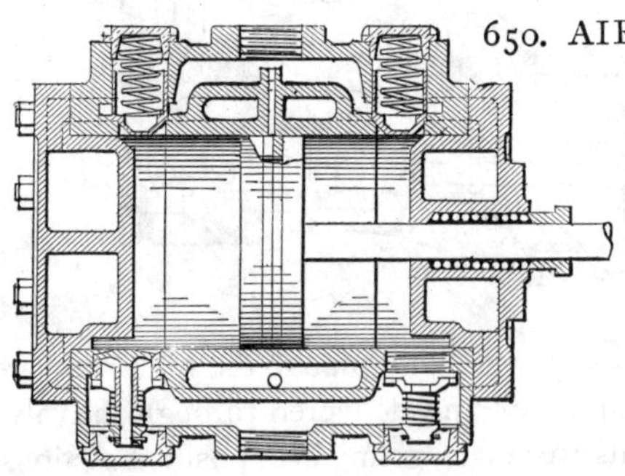

650. AIR COMPRESSING CYLINDER, with vertical lift valves, water-jacketed cylinder and heads. "Ingersoll-Sergeant" model.

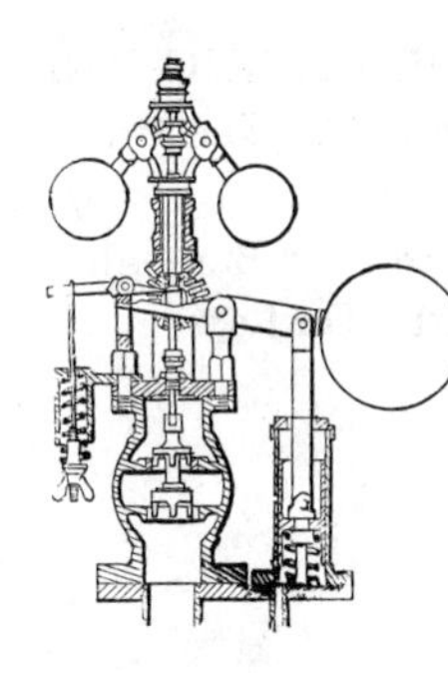

651. AIR COMPRESSOR GOVERNOR.—Controlling the speed by the ordinary action of the governor balls, and also reducing the compressor to minimum speed when the air pressure becomes excessive. The ball and lever at the right are lifted by the air pressure in the small piston, and force the valve rod and throttle down to give the smallest motion to the compressor. "Clayton" model.

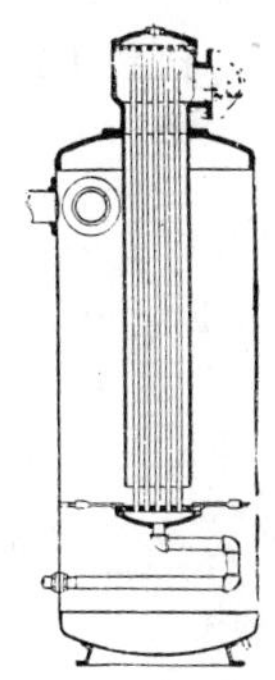

652. AIR COOLING RECEIVER, for cooling the air from a compressor. A series of tubes between headers with water circulation cools the air and condenses the excess of moisture. "Ingersoll-Sergeant" model.

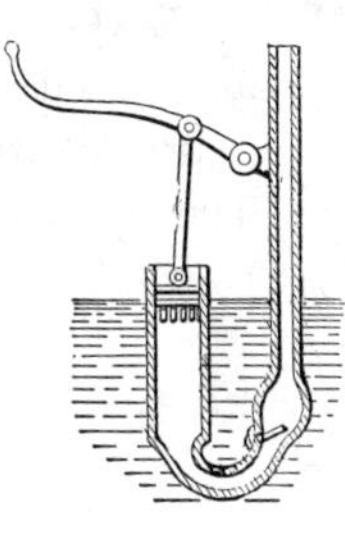

653. SINGLE VALVE AIR PUMP.—The upper part of the cylinder is perforated, so that the piston when drawn up produces a partial vacuum, and when past the perforation the air or gas rushes in to fill the cylinder. The one valve holds the pressure in the delivery pipe.

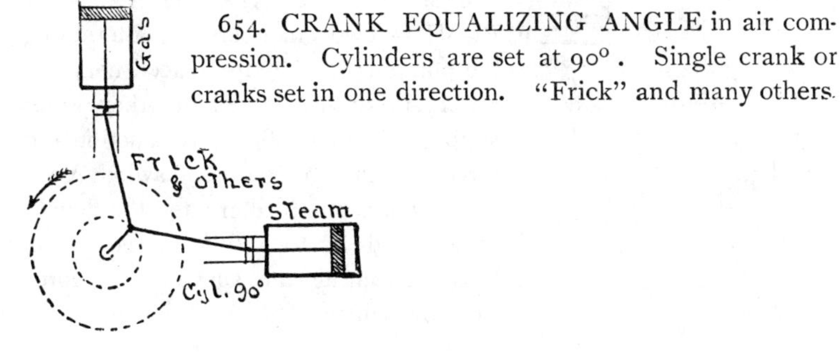

654. CRANK EQUALIZING ANGLE in air compression. Cylinders are set at 90°. Single crank or cranks set in one direction. "Frick" and many others.

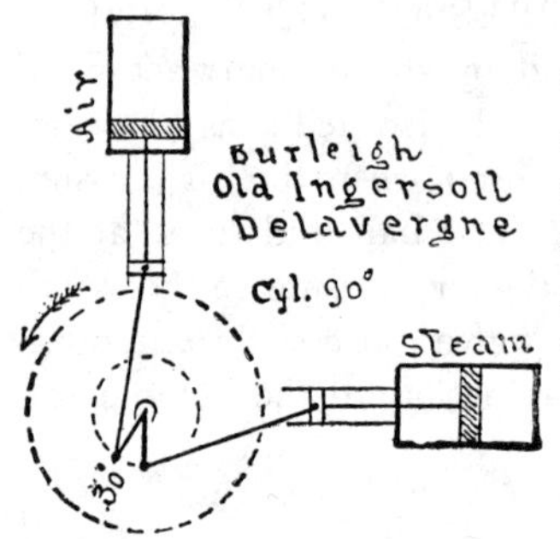

655. CRANK EQUALIZING ANGLE in air compression. The cylinders are set at an angle of 90° and two cranks are set at 30°. "Burleigh," early "Ingersoll," and "De Lavergne" system.

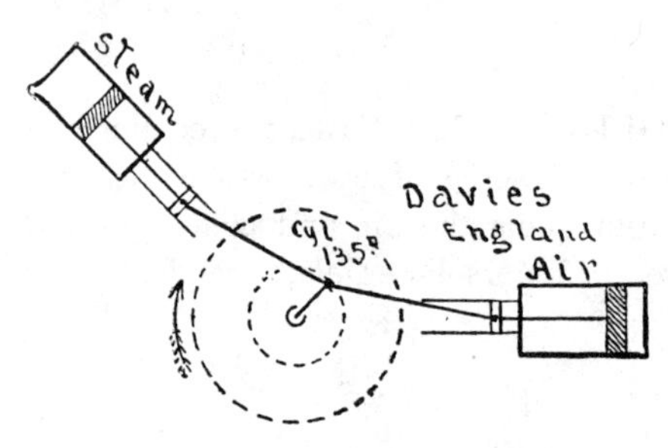

656. CRANK EQUALIZING ANGLE in air compression. The cylinders are set at an angle of 135°. "Davies" system in England.

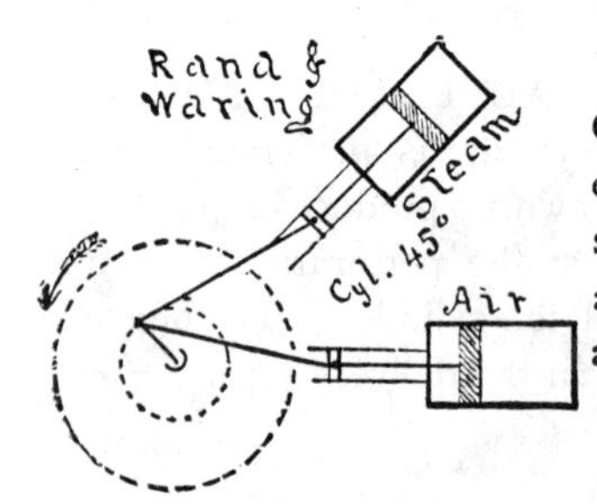

657. CRANK EQUALIZING ANGLE in air compression. Used to equalize the mean pressure of the steam and air pistons. The cylinders are set at an angle of 45°. "Waring" and "Rand" system.

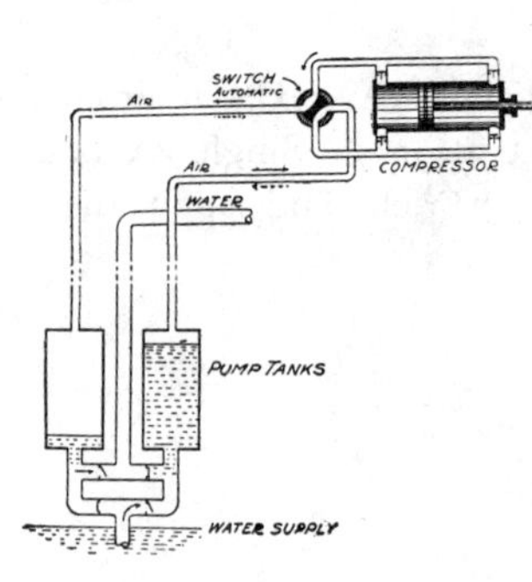

658. DIRECT AIR PRESSURE PUMP.—Two chambers for alternating the pumping action are placed near the water surface in a well or other water supply. The chambers have suction and force valves. A four-way switch cock near the air pump alternates the flow of compressed air to and from the pump, thus alternating the suction and force from the tanks.

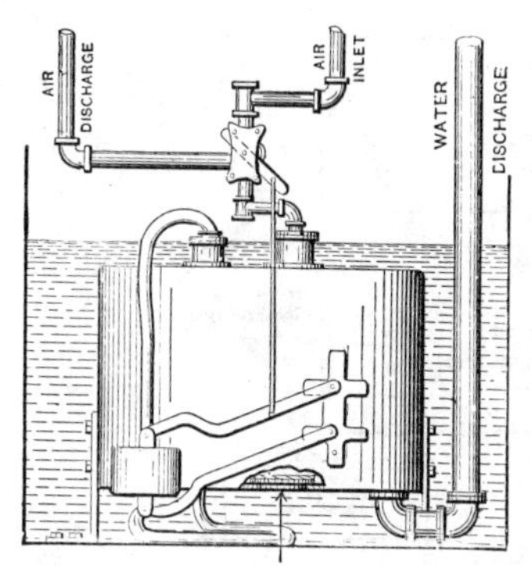

659. COMPRESSED AIR WATER ELEVATOR.—A tank is submerged in which there is a pivoted float that, by its raising and falling, operates a double-ported air valve for filling the tank, by discharging the air, and for discharging the water by the admission of compressed air. A single-flap valve at the bottom of the tank admits the water. The valve is thrown only at the top and bottom of the float stroke.

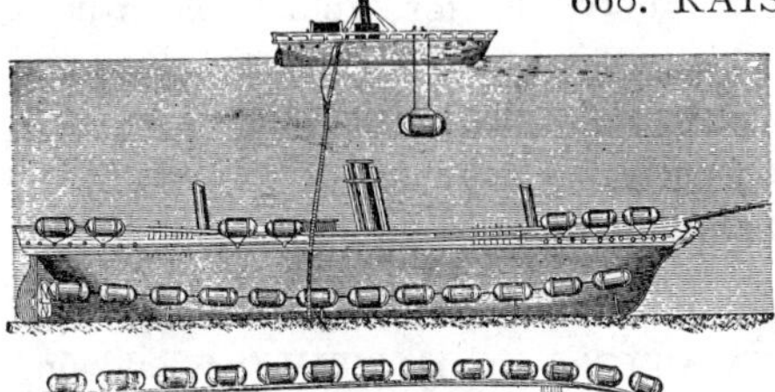

660. RAISING SUNKEN VESSELS by compressed air. Casks or bags fastened to the sides or placed inside of a vessel, and inflated with air under pressure, are used for raising sunken vessels.

661. COMPRESSED AIR LIFT SYSTEM of pumping water from deep wells. The pressure in the air pipe must be greater than the hydrostatic pressure of the water at the bottom of the pipe, and in quantities sufficient to make the ascending column of air and water in the flow pipe lighter in its total height than the weight of an equal column of solid water of the depth of the well from the surface of the water to the bottom of the pipe.

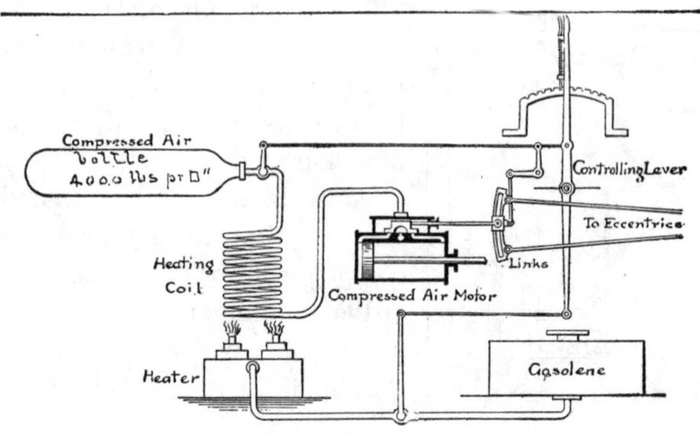

662. COMPRESSED AIR POWER for automobile trucks. Compressed air at about 4,000 lbs. per square inch is stored in steel bottles. Reheated in a coil over a burner under reduced pressure, and made a power factor in a compound engine. Controlled by link valve gear and a reducing pressure valve.

663. COMPOUND PNEUMATIC LOCOMOTIVE, "Baldwin" type. Two high-pressure air receivers. An intermediate pressure receiver fed automatically from which the high-pressure cylinders are operated. The low-pressure cylinders receive the exhaust from the high-pressure cylinders, and exhaust at almost atmospheric pressure.

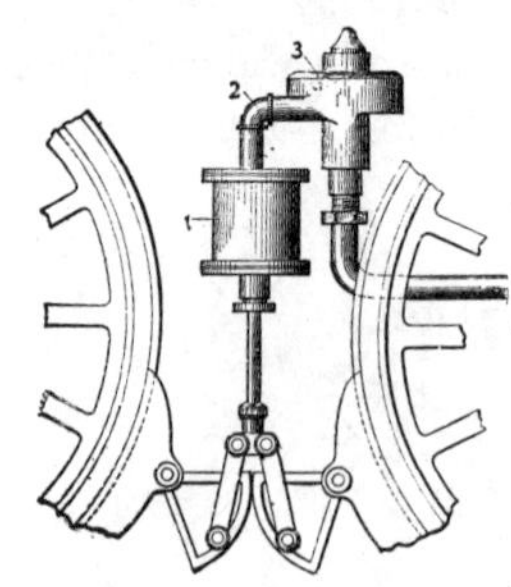

664. LOCOMOTIVE AIR BRAKE.—1, Air cylinder; 3, reducing valve. The piston is directly connected by links to the cam sectors, which press the brake shoes.

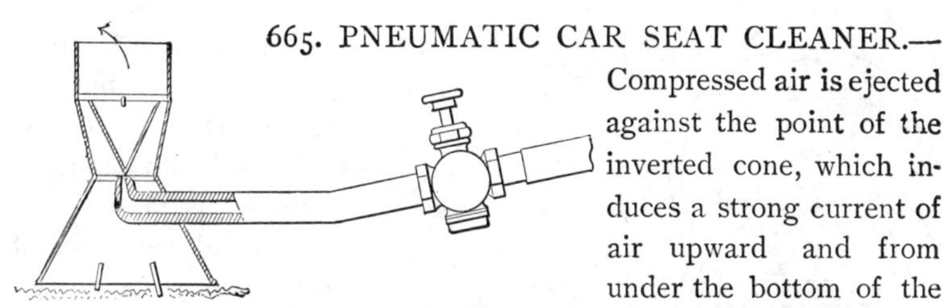

665. PNEUMATIC CAR SEAT CLEANER.—Compressed air is ejected against the point of the inverted cone, which induces a strong current of air upward and from under the bottom of the inverted funnel, drawing the dust from the fabric and projecting it through a hose out of the windows.

666. AIR SPRAY NOZZLE for dusting with compressed air. A broad, thin nozzle from which a blast of compressed air penetrates fabrics, clearing them of dust. A good cleaner of plain and carved woodwork.

667. PNEUMATIC PAINT SPRAYER.—An ejector nozzle for compressed air, with a side feed for the paint. An inverted conical nose-piece is flattened to a thin opening to project the spray paint in a thin sheet.

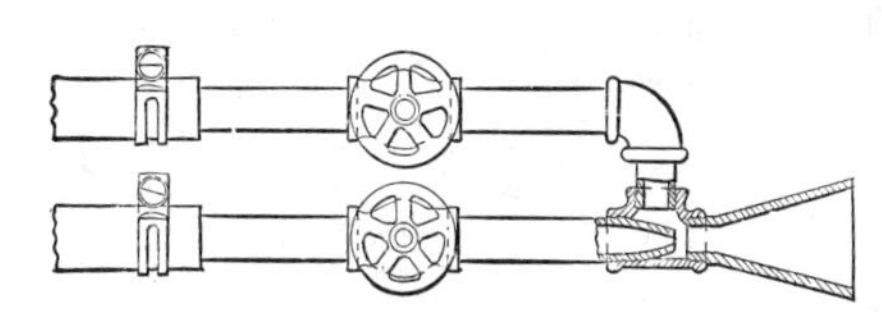

668. PORTABLE FIRE EXTINGUISHER.—The tank is nearly filled with a saturated solution of carbonate of soda and water. The glass cup is filled with acid and sealed by the cap. To use it, turn the tank quickly, top down, when the ball falls and breaks the acid cup, producing pressure by the liberation of gas.

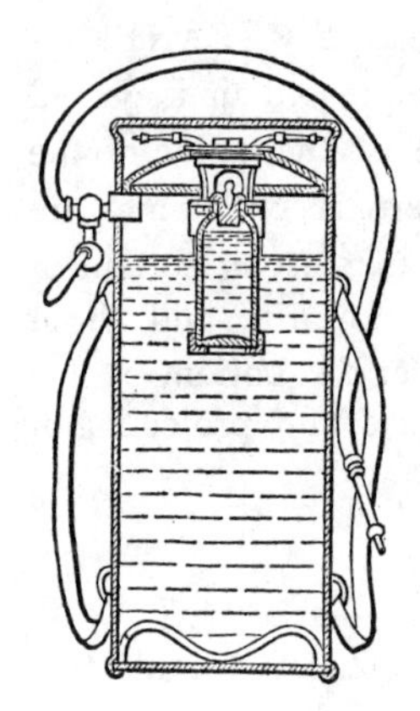

669. FIRE EXTINGUISHER.—The tank is filled with a saturated solution of bicarbonate of soda in water to five-sixths of its capacity. A small glass bottle filled with sulphuric acid, with a loose lead stopper, is placed in a cage at the top of the tank, and the cover of the tank fastened. To use, turn the tank over, which spills the acid, generating pressure by liberating carbonic acid gas.

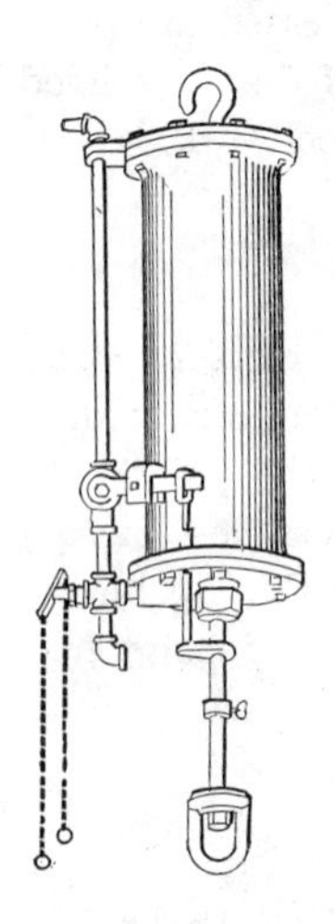

670. COMPRESSED AIR LIFT, "Clayton" model. Showing safety stop on the piston rod, which automatically stops the lift at any set point by closing the air valve.

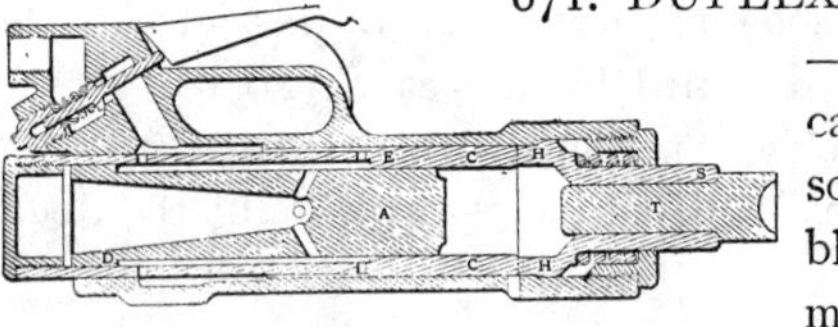

671. DUPLEX PNEUMATIC RIVETER. — The striking piston, A, is encased in a striking cylinder, C, so that the tool, T, receives a blow alternately from the hammer piston, A, and from the cylinder, C, on the tool socket, H. The method of operation is shown by the differential piston areas. The hand is relieved from jar by this operation.

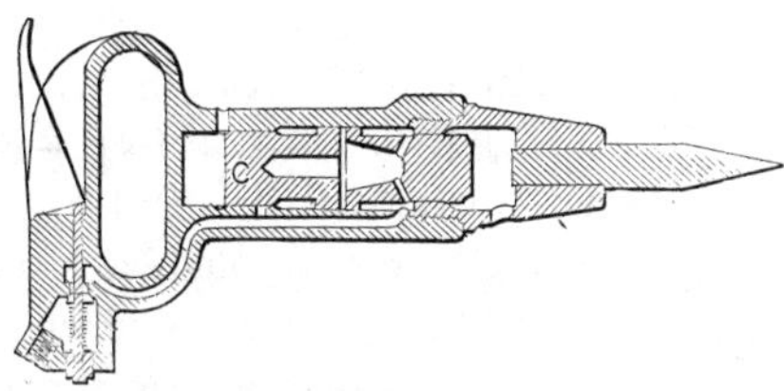

672. PNEUMATIC HAMMER.—Constructed on similar lines with No. 673, with the addition of a counterbalance piston, C, which, by its reaction and cushion, relieves the body of the tool and the hand from excessive jar.

673. PNEUMATIC HAMMER.—F is the flexible hose connection. When T is pressed, compressed air enters through the piston valve and ports P*o*, into the cylinder, as indicated by the arrows in the cut. The piston will first move to the top. The effective pressure is that due to the area of the piston. When P has given the blow, exhaust takes place through S and E, and the piston P is brought back by means of the pressure in the annular space B, acting only on the collar at D.

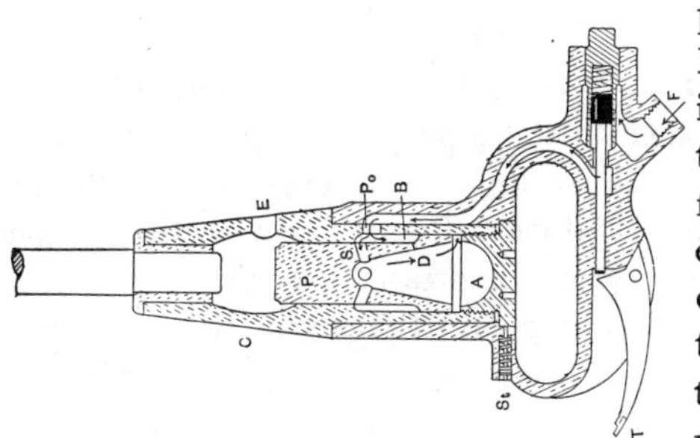

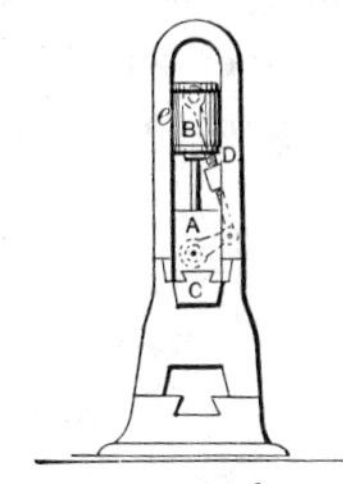

674. "HOTCHKISS" ATMOSPHERIC HAMMER.—The hammer-head, A, is connected directly with the piston within the vibrating cylinder, by a piston rod. The cylinder is connected to the crank by an outside rod, vibrating vertically by the motion of the crank, which also carries the piston and hammer with a cushioned stroke, due to compression of the air within the cylinder.

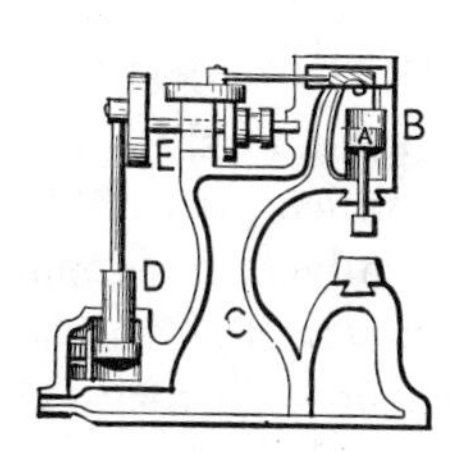

675. "GRIMSHAW" COMPRESSED AIR HAMMER.—A belt-driven air compressor, D, furnishes compressed air to drive the piston, A, and hammer. A variable friction pulley on the belt shaft, E, regulates the stroke of the hammer by varying the admission of compressed air to either side of the piston. The friction-valve driving pulley slides on the feathered shaft by the action of the foot lever.

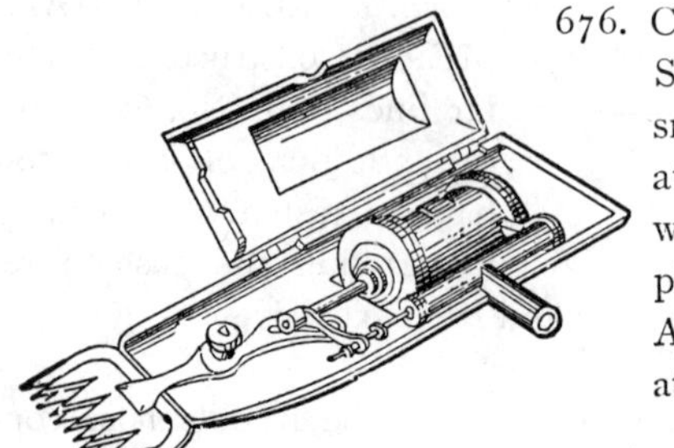

676. COMPRESSED AIR SHEEP-SHEARING MACHINE.—A small piston vibrates and operates the cutters through a lever with a diagonal slot in which a pin in the piston-rod head slides. An arm on the piston rod operates the valves.

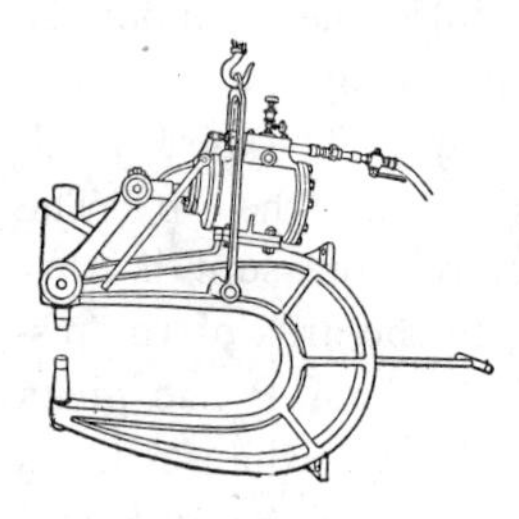

677. PORTABLE RIVETER, "Allen" model. The toggle joint is pivoted to a cam and also within the trunk piston. By the differential trunk form, the return stroke economizes the compressed air, the large piston area giving great power to the riveting stroke.

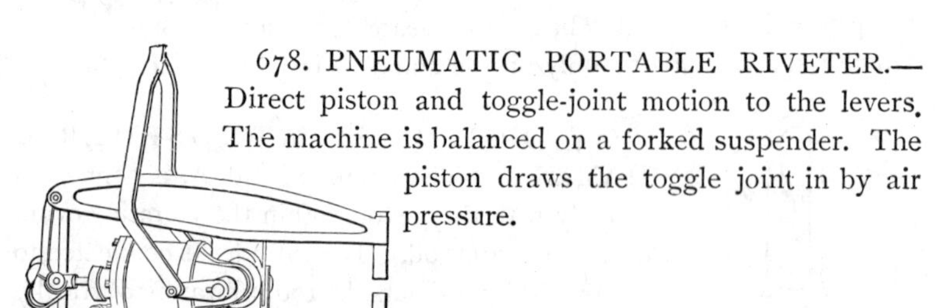

678. PNEUMATIC PORTABLE RIVETER.—Direct piston and toggle-joint motion to the levers. The machine is balanced on a forked suspender. The piston draws the toggle joint in by air pressure.

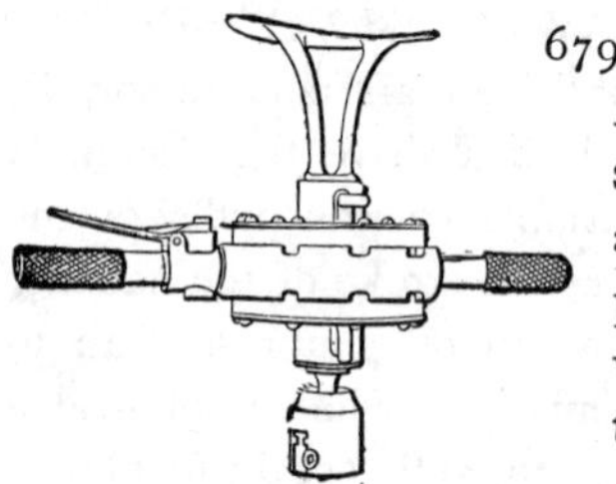

679. PNEUMATIC BREAST DRILL.—A rotary air motor is fixed to the drill-spindle, in a case to which the handles and breast-plate are attached. Compressed air enters through the handle with the valve lever and is exhausted through the opposite handle.

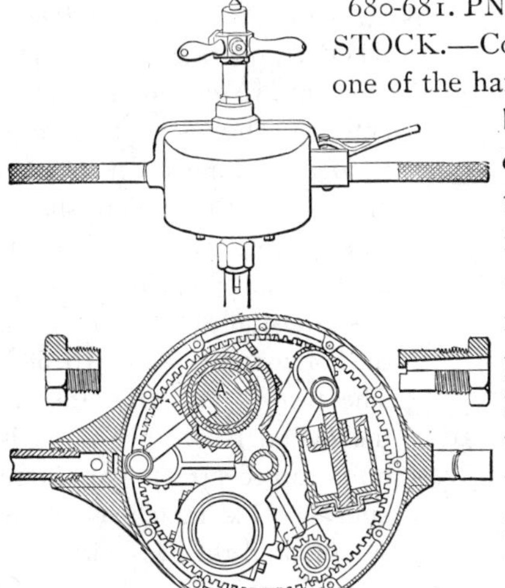

680-681. PNEUMATIC MOTOR DRILL STOCK.—Compressed air enters through one of the handles with its flow controlled by a lever and valve. The exhaust enters the case from the port in the oscillating cylinder trunnions. The three double-acting pistons are directly connected to cranks and pinions which mesh with an internal spur gear, which is fast to the outer shell. The spider which carries the cylinders and pinions is fast on the central spindle and revolves with it. The inlet and exhaust ports are shown in the horizontal section of the top trunnion at A, No. 681.

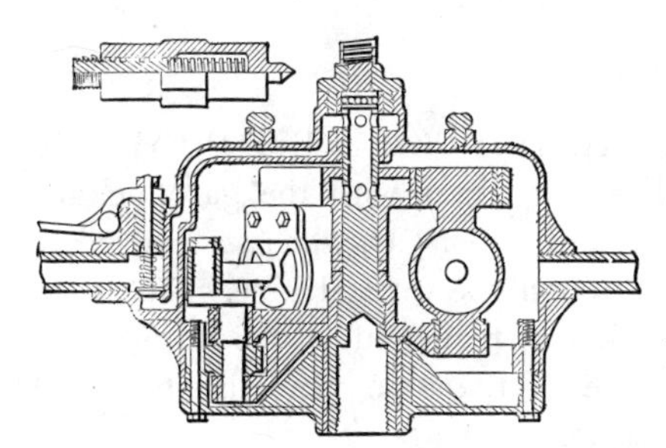

682. Is the vertical section, showing the compressed air valve and port passages opening into a cavity in the central spindle and to the trunnion ports.

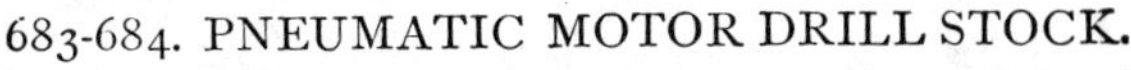

683-684. PNEUMATIC MOTOR DRILL STOCK. —A horizontal rotary motor, over the centre of the spindle, carries on one end of its shaft a bevel pinion, which drives a bevel gear attached by the lower section of the case to the drill spindle. The inlet and exhaust ports and valve are shown in the vertical section, No. 684.

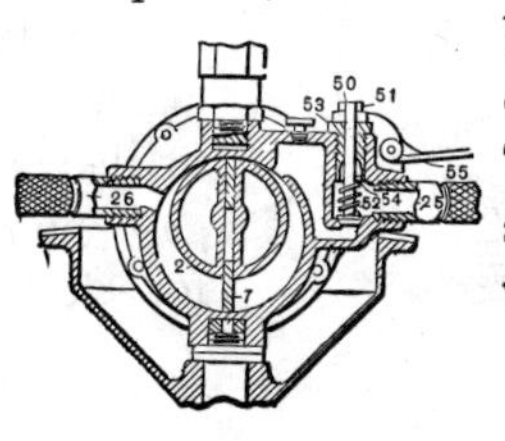

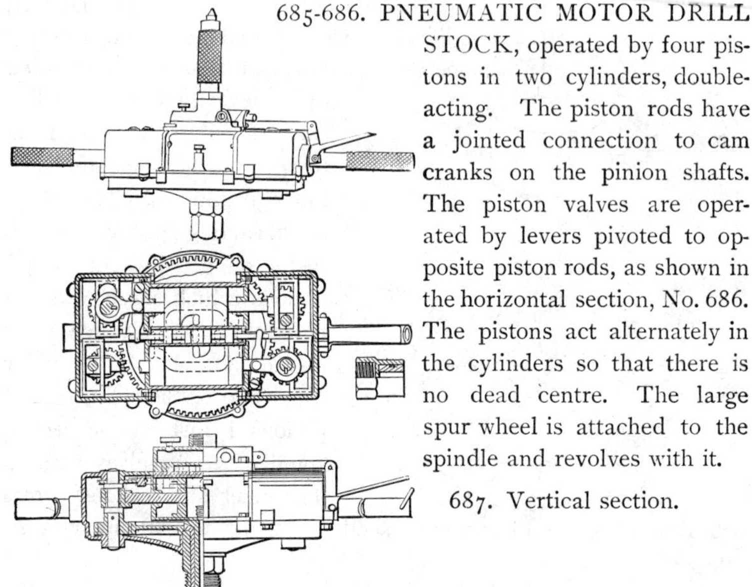

685-686. PNEUMATIC MOTOR DRILL STOCK, operated by four pistons in two cylinders, double-acting. The piston rods have a jointed connection to cam cranks on the pinion shafts. The piston valves are operated by levers pivoted to opposite piston rods, as shown in the horizontal section, No. 686. The pistons act alternately in the cylinders so that there is no dead centre. The large spur wheel is attached to the spindle and revolves with it.

687. Vertical section.

688. AIR AND GASOLINE TORCH.—Air is pumped into the tank with the gasoline, and forms a saturated air and vapor gas, which is carried to the Bunsen burner through the vertical pipe. The additional air for combustion is regulated at the burner, and the vapor at the valve in the pipe near the tank. A gauge shows the pressure.

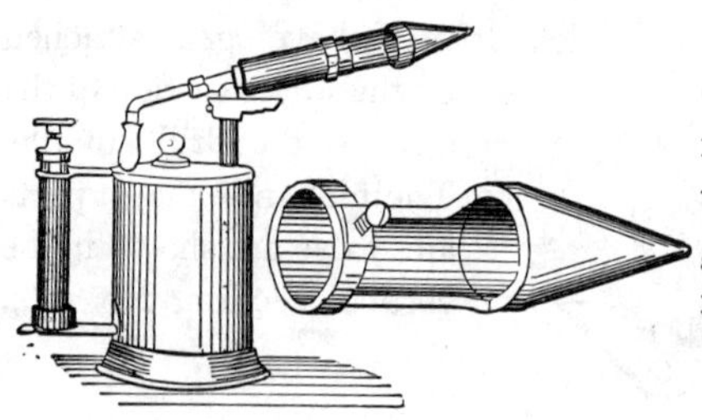

689. TORCH SOLDERING COPPER.—The conical tip is made of copper, and slips on to the nozzle of a plumber's gasoline torch. Used largely for electric wire connections.

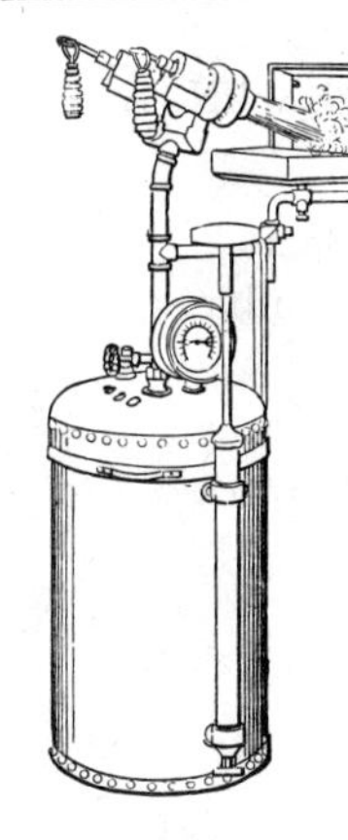

690. AIR AND GASOLINE VAPOR BRAZER, double flame. The pressure of vapor to the Bunsen burners is regulated by a valve near the top of the tank. The valve handles hanging from the stems regulate each burner.

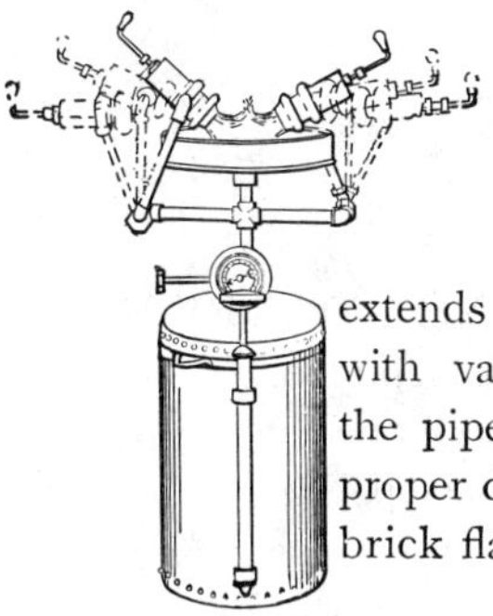

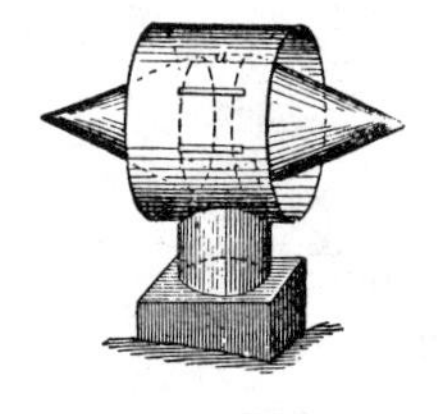

691. AIR AND GASOLINE BRAZING APPARATUS.—A small attached pump forces air into a tank holding a small quantity of gasoline. A gauge shows the air pressure. From the top of the tank a pipe extends to two oppositely placed Bunsen burners with valves for regulating the flame. Swivels in the pipe allow the burners to be adjusted to the proper distance from the piece to be brazed. Fire-brick flame plate.

692. DOUBLE CONE VENTILATOR.—The up-take enters between the cones. The smoke has its exit around the edge of the leeward cone.

693. SPIRAL VANE OR COWL, for a chimney top. The wind catching in the wings causes it to revolve and increase the draught.

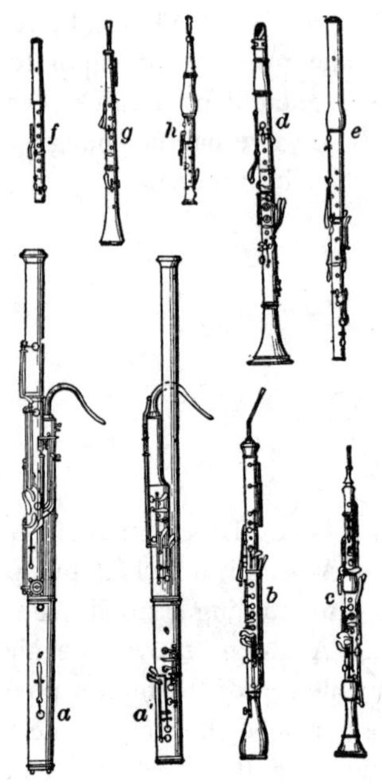

694. WIND INSTRUMENTS.

a, *a'*, bassoons.
b, cors Anglais.
c, oboe, or hautbois.
d, clarionet.
e, flute.
f, octave, or piccolo.
g, musette.
h, flageolet.

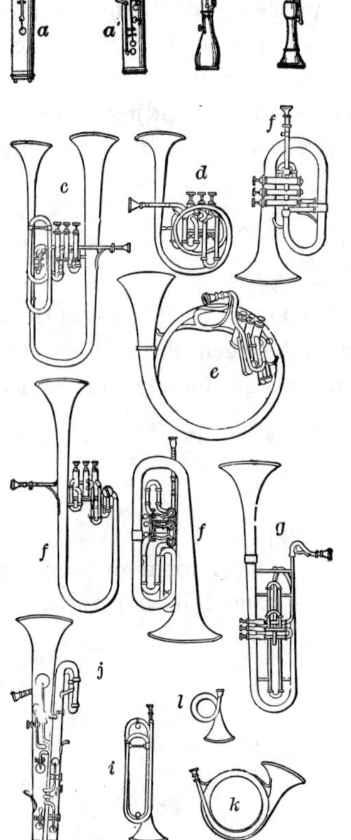

695. WIND INSTRUMENTS

a, Sarrusophone.
b, Saxophone.
c, Duplex pelitti.
d, *d*, Cornets *à pistons*.
e, Helicon *à pistons*.
f, *f*, *f*, Saxhorns.
g, Clavicor.
h, Trombone.
i, Trumpet.
j, Ophicleide.
k, Hunting horn.
l, Post horn.

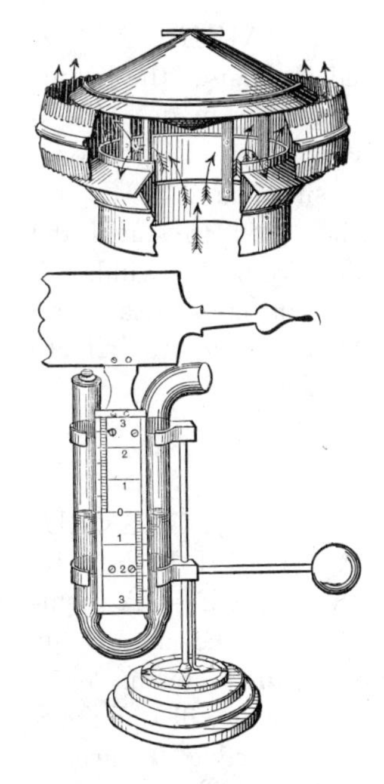

696. VENTILATOR OR COWL FOR A CHIMNEY TOP.—The corrugated edges of the outside guard ring intensify the draft by directng the wind in a vertical direction.

697. A WIND GAUGE for obtaining the force of the wind in inches height of water, from which the wind pressure per square foot may be obtained from the measured hydrostatic pressure of the water.

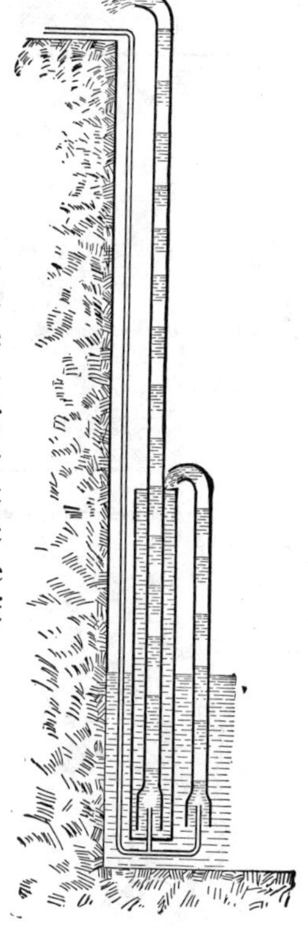

698. COMPOUND POHLE AIR LIFT.—In compounding an air lift for mine drainage but one-half the depth of sump is required as for a single lift and by still further duplicating the lifts, shallower sumps may be utilized with economy. Air pressure must be greater than the total hydrostatic pressure of the receiving tank. The sump should be one-half the height of the receiving tank in depth.

699. THE PRAIRIE WINDMILL, called in Kansas the "Jumbo Mill," generally made with 6 arms on an axle placed in a north and south position and lower half covered with a box to shield the lower paddles from the wind. A simple crank connection to a pump supplies sufficient water to irrigate 6 acres of garden land.

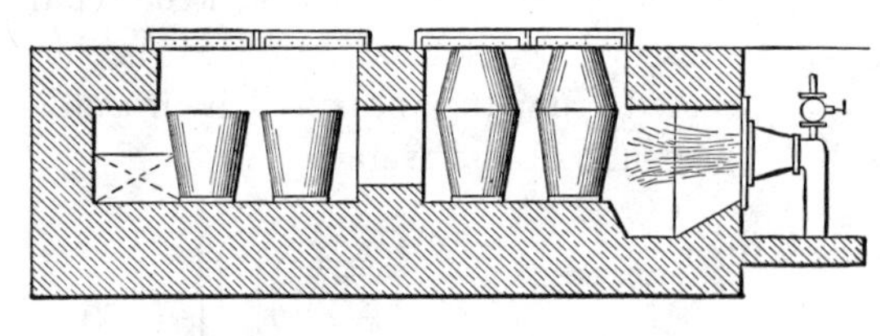

700. GAS CRUCIBLE FURNACE. — A compressed air or air and steam blow pipe will operate a crucible furnace for melting metals.

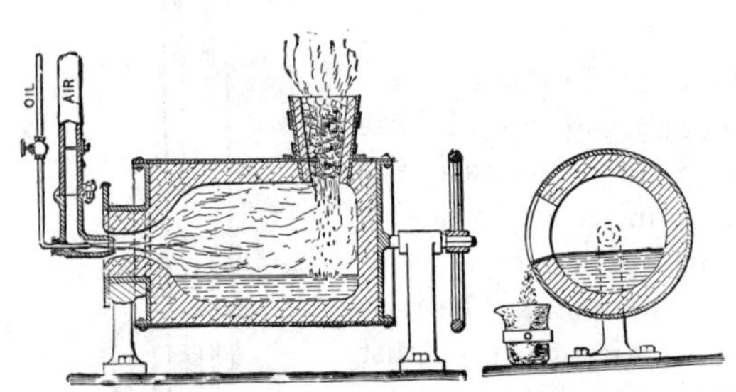

701. OIL BURNING MELTING FURNACE. — A cylindrical casing lined with fire brick. An oil burner. The feed hopper is removed to pour the metal as shown in the right-hand section.

702. MECHANICAL FLYER. — A small windmill with two or four blades, when quickly revolved by the string and forked spindles makes a pretty illustration of a flying machine.

Section IX

ELECTRIC POWER AND CONSTRUCTION.

GENERATORS, MOTORS, WIRING, CONTROLLING AND MEASURING, LIGHTING, ELECTRIC FURNACES, FANS, SEARCHLIGHTS AND ELECTRIC APPLIANCES.

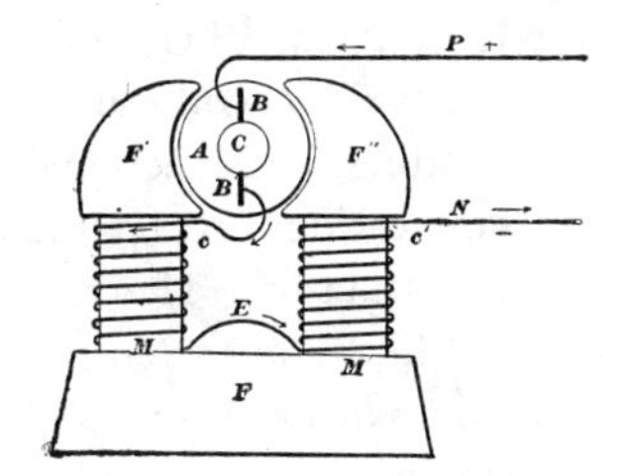

703. SERIES WOUND MOTOR OR GENERATOR.—A motor if the current is supplied through the wires P and N, and a generator if the armature is rotated, when the current can be taken from the wires P and N

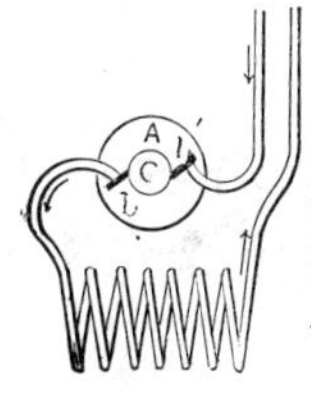

704. ELECTRIC GENERATOR CONSTRUCTION.—Series winding in which the armature, field winding, and external circuit are in series or one continuous line. Best for arc lighting. A, armature; C, commutator; *b* and *b′* brushes; the coil showing the field winding.

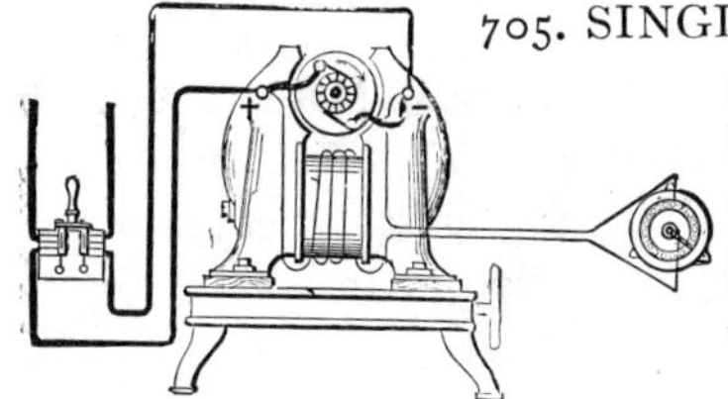

705. SINGLE-POLE SHUNT GENERATOR, showing the shunt-winding connection with the brushes and branch wiring to a rheostat controller. The heavy lines are the main current with a switch.

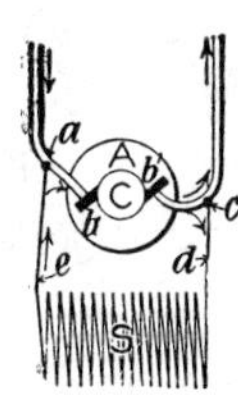

706. ELECTRIC GENERATOR CONSTRUCTION.—Shunt winding, in which the field winding is in parallel with the armature winding and connected with the circuit at the brush holders. A, Armature; C, commutator; *b* and *b′*, brushes; *a*, *c*, field connections; S, field winding.

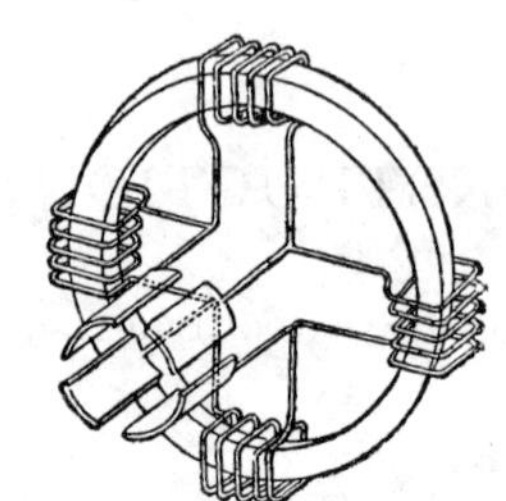

707. FOUR-POLE RING ARMATURE, showing intermediate connections with the commutator bars from a continuous winding or closed coil.

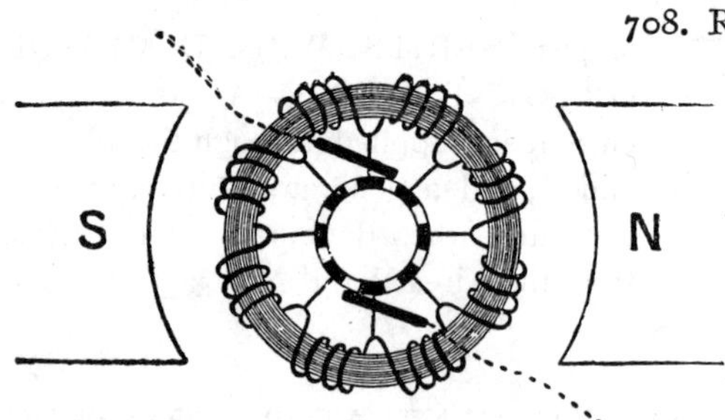

708. RING ARMATURE.—Method of continuous winding and sectional connections with the commutator. The dotted lines are the circuit connection with the brushes.

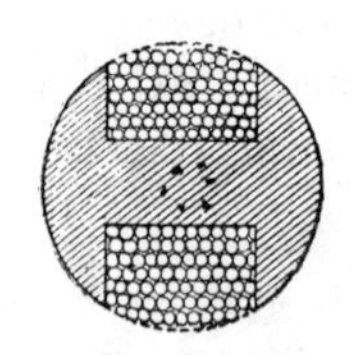

709. TWO-POLE OR SHUTTLE-SPOOL ARMATURE.—Section of spool with end over winding; usually made of cast iron.

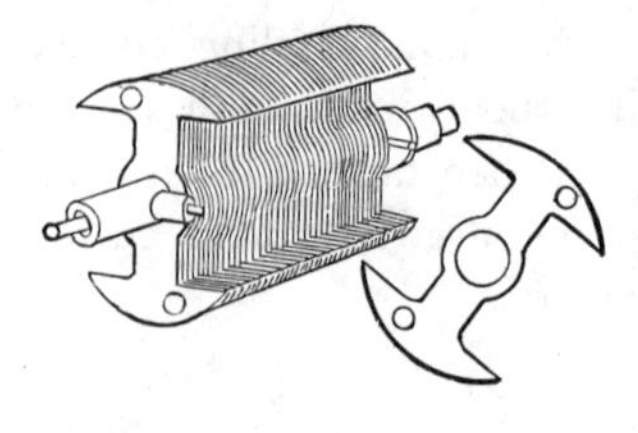

710. SHUTTLE ARMATURE, made with soft sheet-iron plates riveted together. The strongest current armature for small two-pole generators.

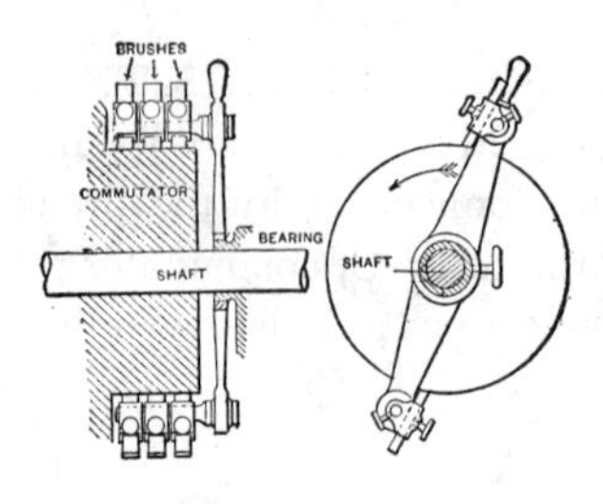

711. MULTIPLE BRUSH COMMUTATOR.—The brushes are adjustable on the pivots of the handle bar, and are given an even pressure on the commutator by springs.

712. Front view.

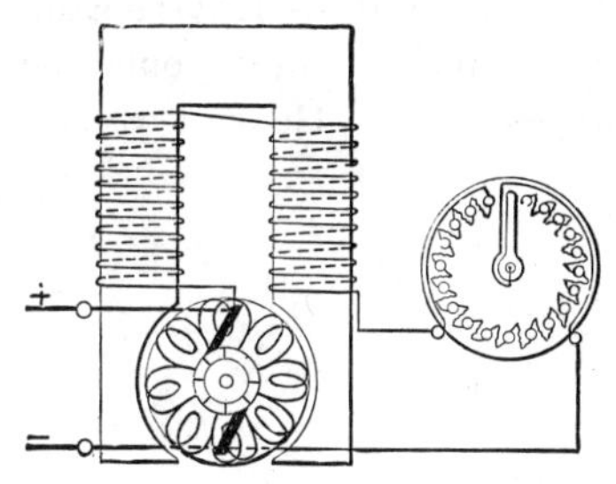

713. BIPOLAR SHUNT GENERATOR, showing the shunt winding on both fields and its connection to the brushes, with intervening rheostat controller.

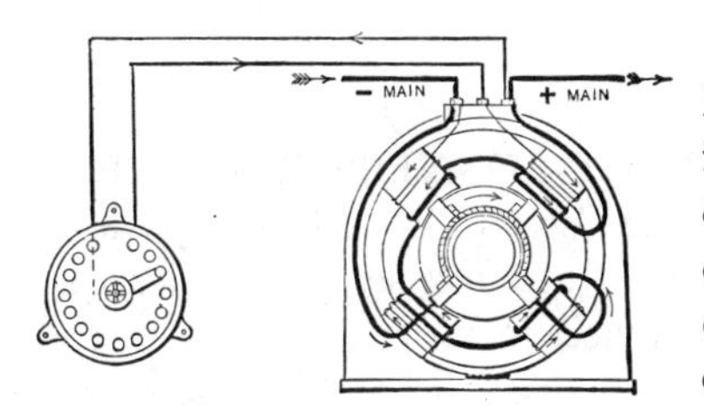

714. FOUR-POLE COMPOUND GENERATOR, showing shunt winding and rheostat connection. Wiring is successive on each pole in the opposite direction for both shunt and current.

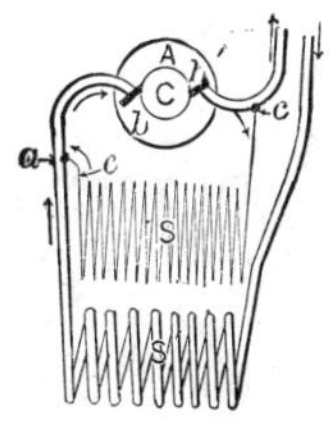

715. ELECTRIC GENERATOR CONSTRUCTION.—Compound winding, in which a winding of the field magnets is in shunt with the armature, and a second winding of the field magnets is in series or direct connection with the outer circuit. The shunt winding should be small wire. S′, Shunt connected with armature brush holders; S, large wire field winding in main circuit.

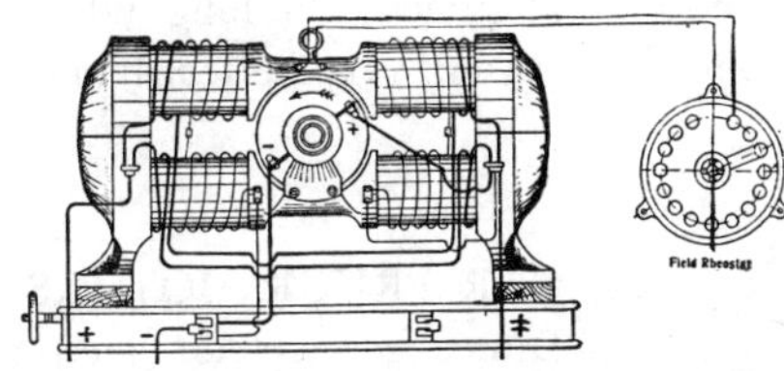

716. CONSEQUENT-POLE COMPOUND GENERATOR.—The opposite field pieces are wound in opposite directions and have opposite polarity in the same piece at the centre. The shunt winding is in the same direction as the field winding and connected to the brushes with an intervening rheostat.

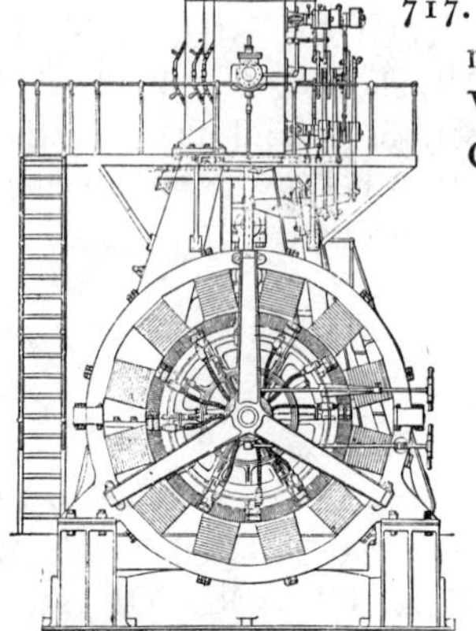

717. TRIPLE-EXPANSION ENGINE and multipolar dynamo. Direct-connected. Vertical types of the General Electric Company.

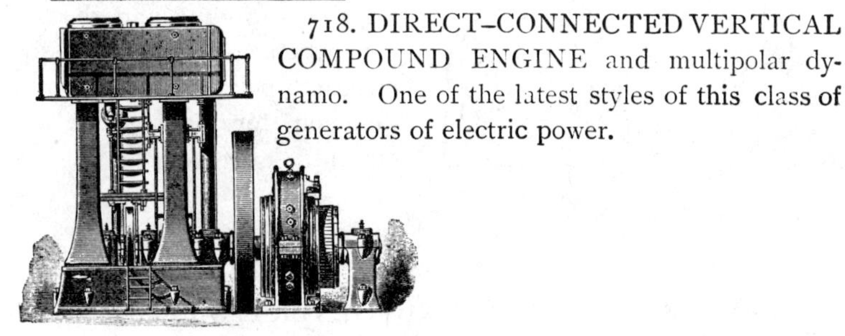

718. DIRECT-CONNECTED VERTICAL COMPOUND ENGINE and multipolar dynamo. One of the latest styles of this class of generators of electric power.

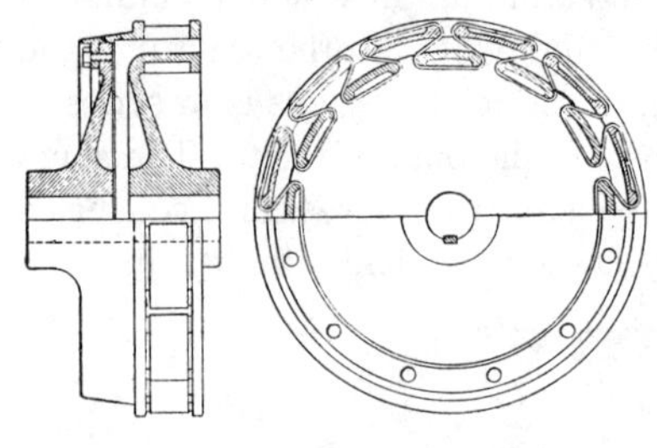

719. FLEXIBLE COUPLING for engine and generator direct connection.

720. Plan. The "Zodel" coupling. A flange on each shaft with overhanging crowns interlapping. A continuous belt over the outside and under the inside crowns allows of considerable variation in alignment and longitudinal vibration in the shafts. If a rubber belt is used, very perfect insulation may be obtained.

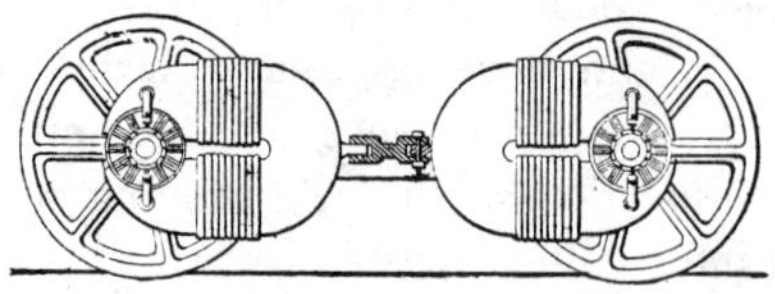

721. CAR TRUCK MOTORS. — Direct-connected electric motors on street-car axles.

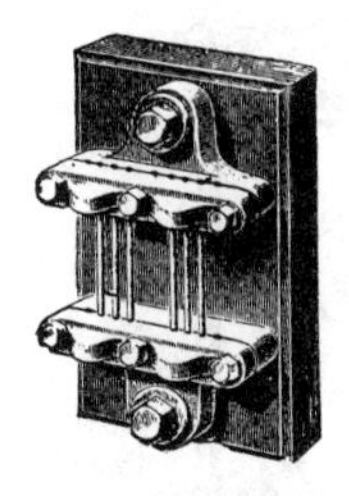

722. ELECTRIC FUSIBLE CUT-OUT.—The fuse wires or strips are connected to the circuit on insulated porcelain blocks. They are made of resisting metal or alloy of tin and lead of sufficient capacity for the required current without excessive heat. Overcurrent melts the wire or strips and opens the circuit.

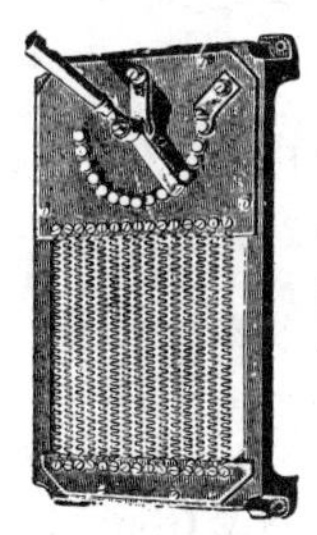

723. RHEOSTAT OR RESISTANCE COILS, with variable switch. Coils are made of iron, platinum, or German silver wire. The switch connections are so made that the coils may be made to connect the line with one or any number in series.

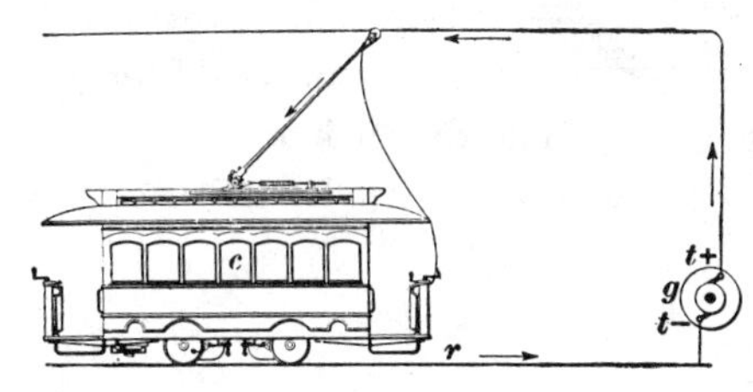

724. TROLLEY CAR, showing the circuit from the generator *g*, through the line wire to car and return by rail circuit.

725. SECTIONAL FEEDER SYSTEM for electric railways. The trolley wire line is divided into a convenient number of sections for feeders from a long main line, or divided into several feeder lines, as shown in the cut.

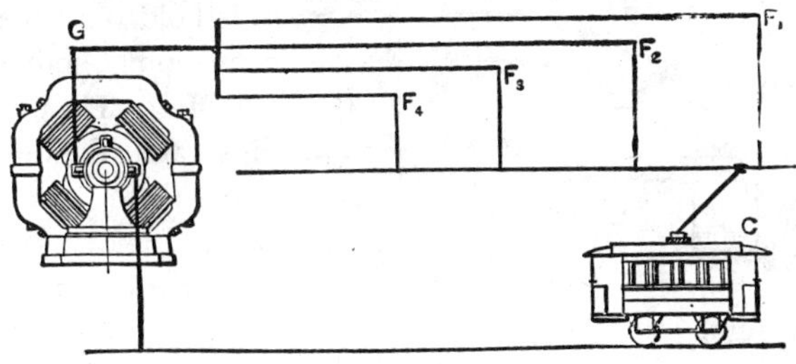

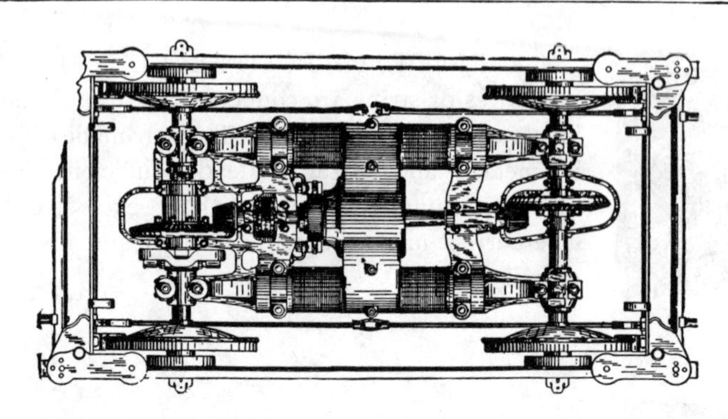

726. STREET RAILWAY SINGLE MOTOR geared to both axles. "Rae" system. The motor is carried on a frame and is journaled to both axles.

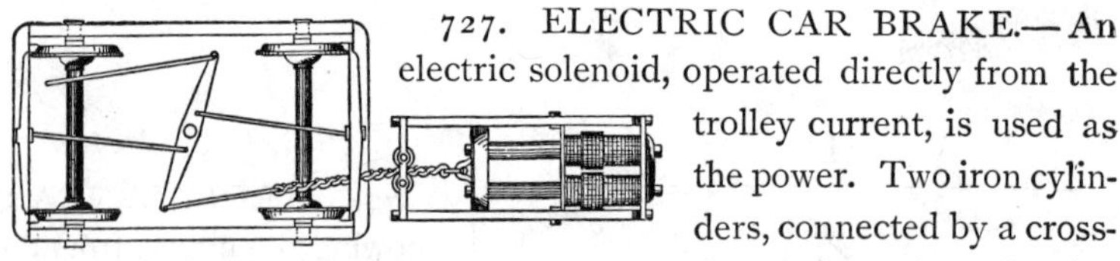

727. ELECTRIC CAR BRAKE.—An electric solenoid, operated directly from the trolley current, is used as the power. Two iron cylinders, connected by a cross-head, form a U-shaped magnet, which is drawn into the solenoids when the current is turned into the coils. Regulation is made by switches and rheostat.

728. ELECTRIC STREET-CAR BRAKE.—A solenoid, operated by the trolley current, pulls up the brake levers. The springs around the piston rods hold back the connections, acting as buffers. The pistons are divided into three parts each, to soften the jerk when turning on the electric current.

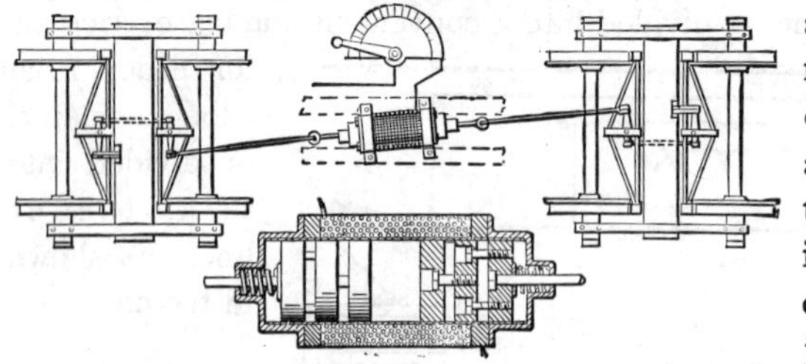

729. Section of solenoid, with the take-up pistons.

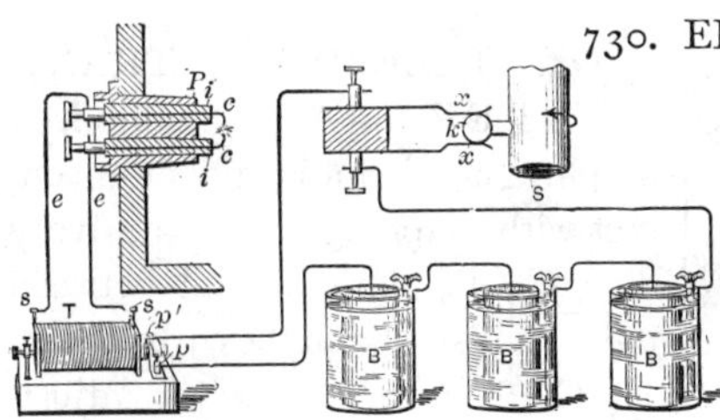

730. ELECTRIC IGNITER, used on explosive motors. The batteries, B, B, B, in series; a sparking coil, T; a braker, *k*, revolving on the shaft, the insulating plug, P, and the platinum electrodes, *c*, *c*, with the wiring, are the principal parts in this device.

731. SPARKING DYNAMO, or generator for a marine gasoline engine. Permanent horseshoe magnets, with an armature revolved by a belt from the fly-wheel of the engine. With a true rim on the fly-wheel, the pulley of the generator may be covered with leather or rubber and pressed lightly against the rim of the fly-wheel.

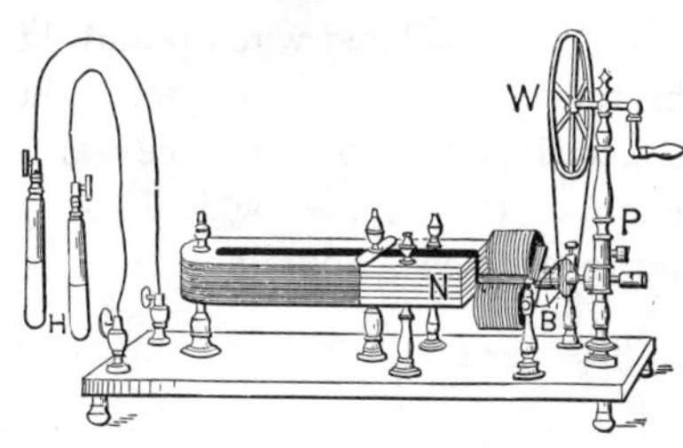

732. MAGNETO-ELECTRIC MACHINE.—The revolution in the field of a permanent magnet of an iron armature wound with an insulated conductor, terminating in a commutator or pole-changing device, from which the conducting wires extend through the base of the instrument to the posts and handles, H.

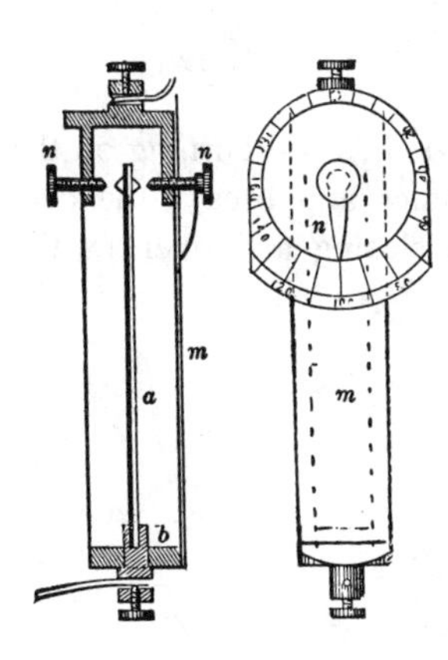

733. ELECTRIC THERMOSTAT.

734. Two strips of thin sheet steel and brass are fastened together by soldering or riveting, and to a base with binding-post in an insulated frame. A cap, with binding post and adjusting screw and index plate, allows for electric contact of the spring and screw at any required temperature. By making a double-wiring, a damper may be made to open or close within a small range of temperature.

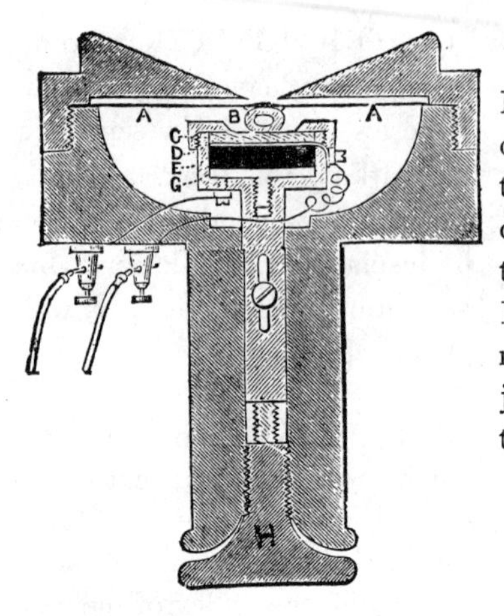

735. TELEPHONE TRANSMITTER.—A, A, thin iron diaphragm ; B, india rubber in contact with diaphragm and the ivory disc, C ; D, platinum foil between the ivory disc, C, and the carbon disc, E ; G, disc and screw for adjustment of carbon contact ; H, adjusting screw for diaphragm contact.

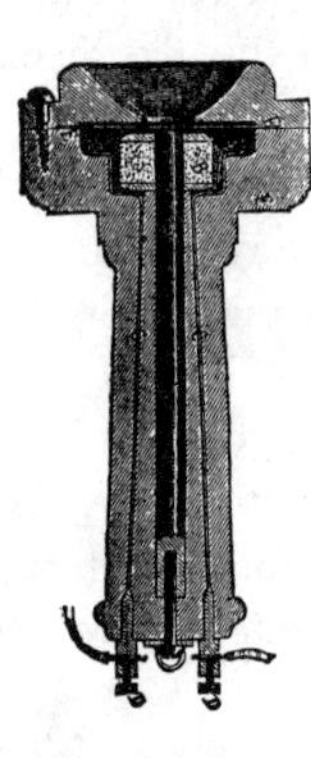

736. TELEPHONE RECEIVER.—A central magnet, with a coil of fine insulated wire around the end, next the vibrating plate or diaphragm. The variations in the electrical current produce variations in the intensity of the magnet, which set up vibrations of sound in the iron diaphragm.

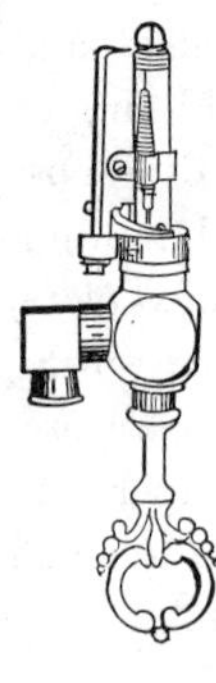

737. ELECTRIC GAS LIGHTER.—Turning on the gas brings the electrodes in contact, and breaks the contact, which produces a spark by closing and opening the battery circuit.

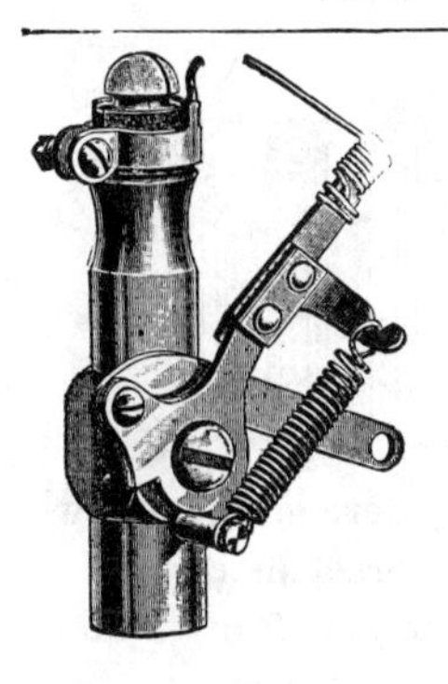

738. ELECTRIC GAS LIGHTER.—Non-short-circuiting. The wiping spring is insulated, and there is no electric current except at the instant of lighting.

739. POCKET ELECTRIC LIGHT.—A dry battery, with a small incandescent lamp connected with it by a break-piece operated by the thumb. A small lens at the front protects the lamp and concentrates the light. Gives a constant light for several hours. Battery easily renewed.

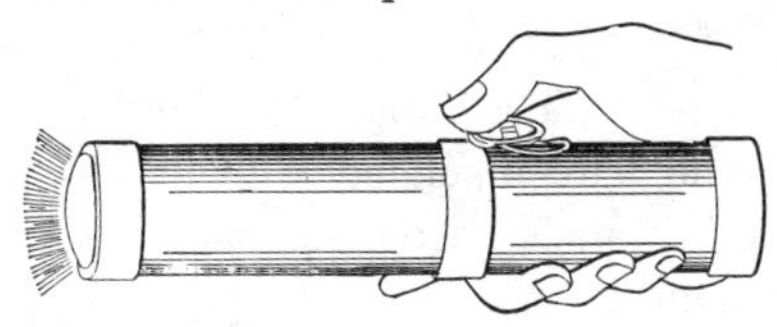

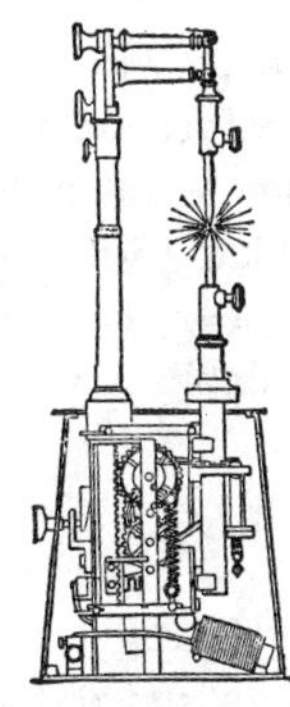

740. ARC LIGHT AND REGULATING GEAR, "Faucault" model. The upper carbon runs down by a rack and gear governed by a fly, which is stopped or let go by variations in the current.

741. LUMINOUS FOUNTAIN.—The lower end of the jet nozzle is fitted with a strong disc of plate glass. A concave mirror, placed in the focus of an arc light just below the glass disc, brilliantly illuminates the water jet.

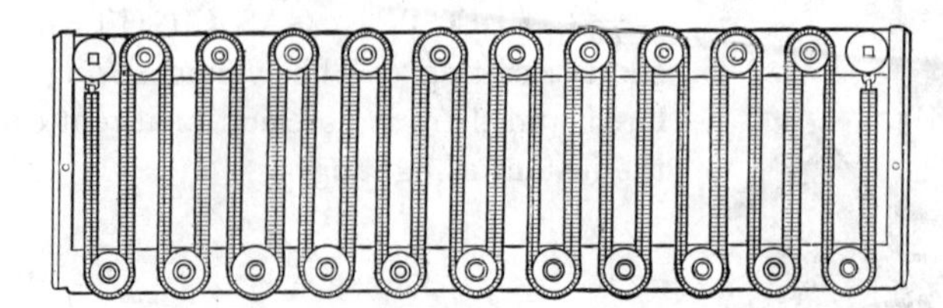

742. ELECTRIC HEATER.—Coils of German silver wire wound around asbestos cords and rove over porcelain buttons for insulation. The buttons may be fastened to a frame of any required form.

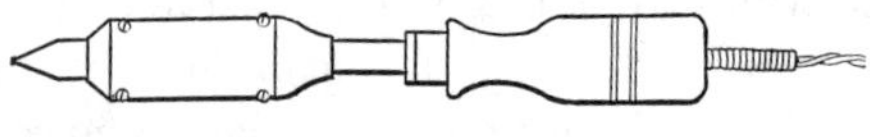

743. ELECTRIC SOLDERING COPPER.—The copper is wound with a coil of resisting material; platinum wire insulated with asbestos, and the coil covered with a protecting shell. Connections are insulated and pass through the hollow handle.

744. ELECTRIC SAD IRON.—The iron is a shell frame with a smooth face on the bottom. A resistance coil made of iron, German silver, or platinum, insulated with asbestos, is wound in spirals as near the bottom plate as can be made available for the greatest amount of heat.

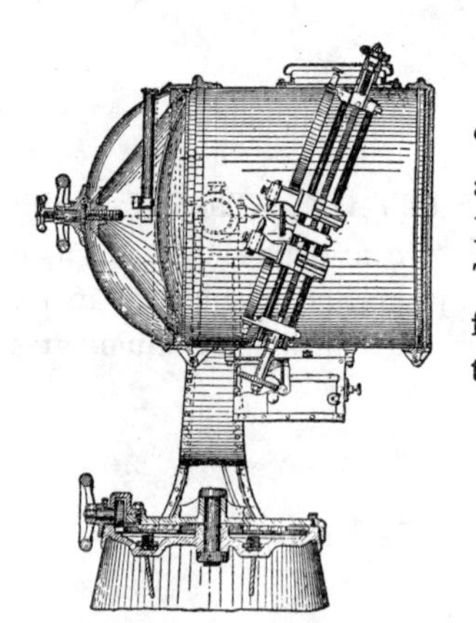

745. ELECTRIC SEARCHLIGHT, "Edison" model. An arc light in front of and in the focus of a concave reflector. It gives a beam of light nearly parallel. The front of the case has a plane glass for protection. It swivels in all directions.

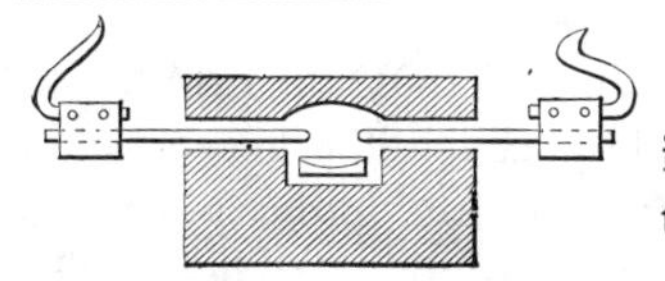

746. ELECTRIC FURNACE, showing the recess and flat crucible. Electrodes of hard carbon and connections.

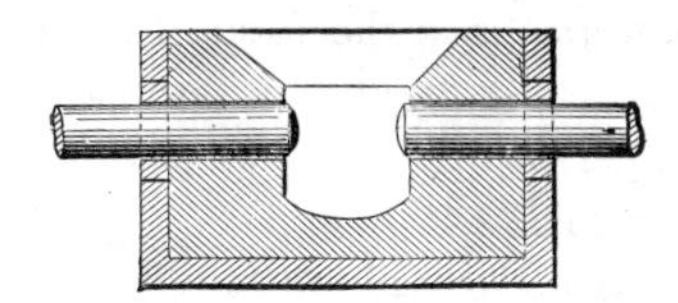

747. OPEN TOP ELECTRIC FURNACE.—A cavity in a box of refractory material with holes on each side through which the insulated carbon electrodes are inserted.

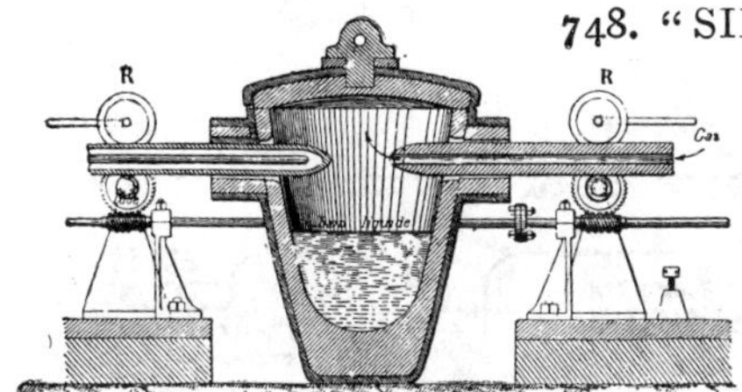

748. "SIEMEN'S" ELECTRIC GAS FURNACE.—Gas enters the crucible through a hollow carbon electrode. The opposite electrode is a copper tube closed at the end with an inner tube for circulation of water to keep the end of the copper electrode from burning. The electrodes are adjusted by the rollers.

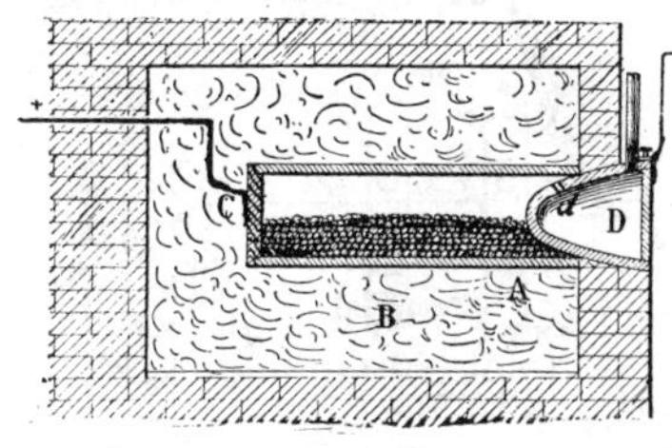

749. "COWLES" ELECTRIC FURNACE.—A cylinder, A, is made of silica or other heat-resisting material. A carbon plug, C, is connected with the positive wire, and a graphite crucible, D, answers as the negative electrode and stopper, also as an exit for gases generated in the retort; B, a bed of insulating material.

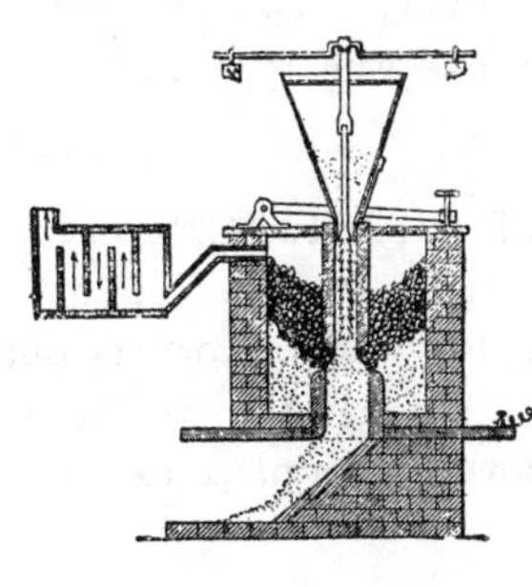

750. ELECTRIC FURNACE, "Cowles" hopper model. The upper electrode is a vertical carbon tube fixed to the hopper. The lower electrode is a larger carbon tube fixed to the furnace floor. The tubes are banked with carbon and lime. The charge is fed down from the hopper by a barbed rod, reciprocated by a crank. The gases generated are drawn off through a condenser.

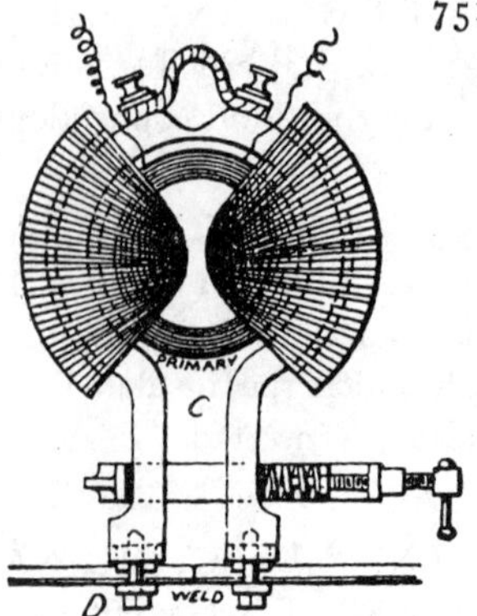

751. ELECTRIC WELDING PLANT.—The secondary coil is the heavy bar of copper enclosing the primary coil to which the clamps are attached. The magnetic material is in the form of coils of iron wire wound around the primary coil and copper hoop.

C, clamp arms.

D, pieces to be welded.

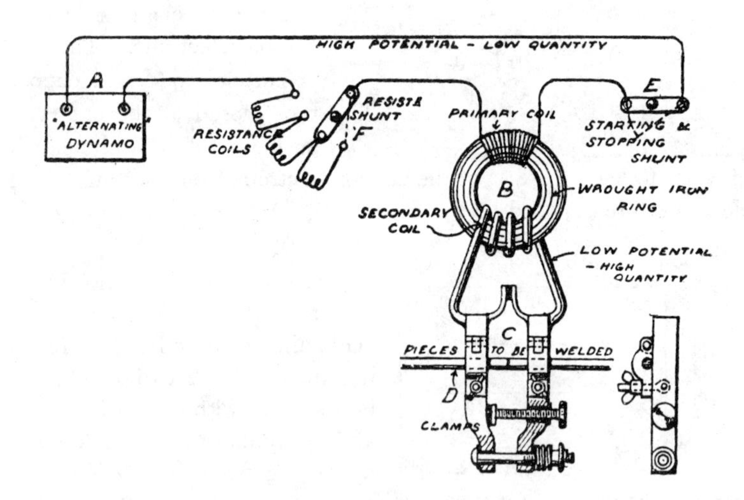

752. ELECTRIC WELDING PLANT.—A, Alternating dynamo; F, resistance coils and switch; B, transformer; C, clamping jaws; D, rods or pieces to be welded; E, switch in the primary circuit.

753. PORTABLE ELECTRIC MOTOR DRILL PLANT, with a stow flexible shaft. A spool on the motor winds up or lets out the electric wires, so that the apparatus may be quickly moved from place to place.

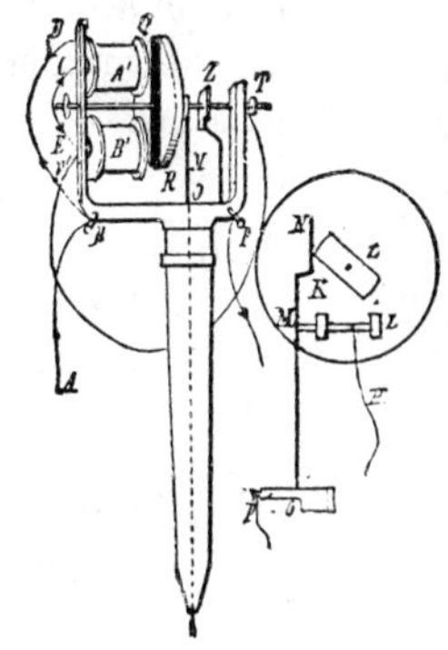

754. ELECTRIC PERFORATING PEN, "Edison" model. Consists of a small pointed tube with a perforating needle on the inside vibrated by a small electro-magnetic motor fixed on top of the pen. A′, B′, Armature coils on iron studs fixed to frame; Q, R, revolving arm and fly-wheel; Z, commutator; N, M, O, spring current breaker. The pen produces a stencil of fine perforations on a glazed sheet of paper from which many copies may be made by a brush and ink.

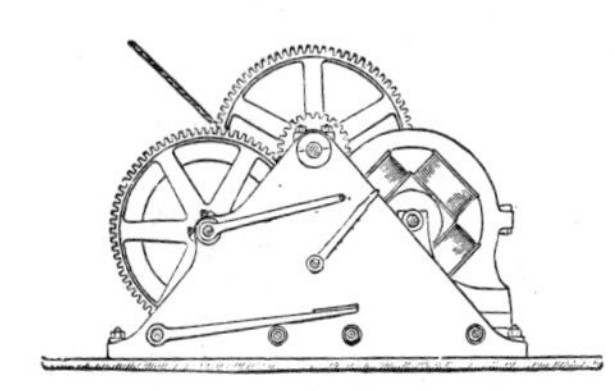

755. ELECTRIC HOIST.—The foot lever is the friction brake. The left-hand lever is for release, the right-hand lever is the starter.

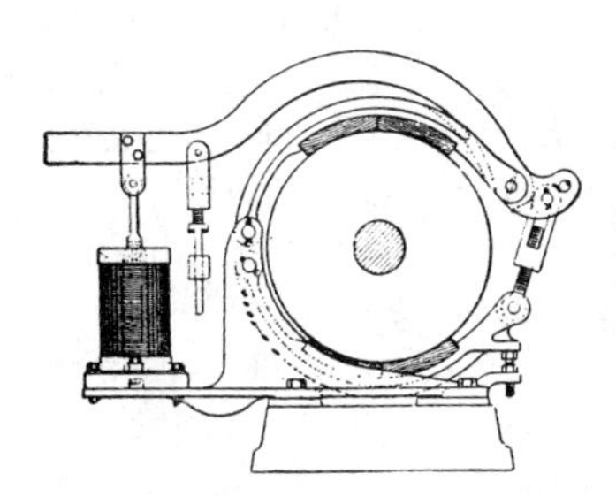

756. ELECTRIC BRAKE.—The brake shoes are fixed to two adjustable curved levers and an operating lever — a solenoid magnet being the operating power.

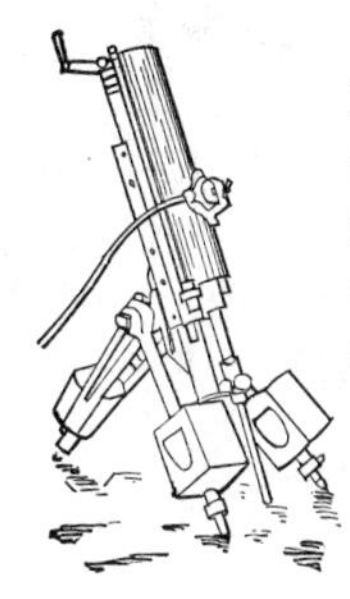

757. ELECTRIC ROCK DRILL, "General Electric Co.'s" model. A series of electric coils are fixed along the cylinder. The iron plunger traverses the interior of the coils, which are charged successively by the electric current through traverse brushes on a straight commutator.

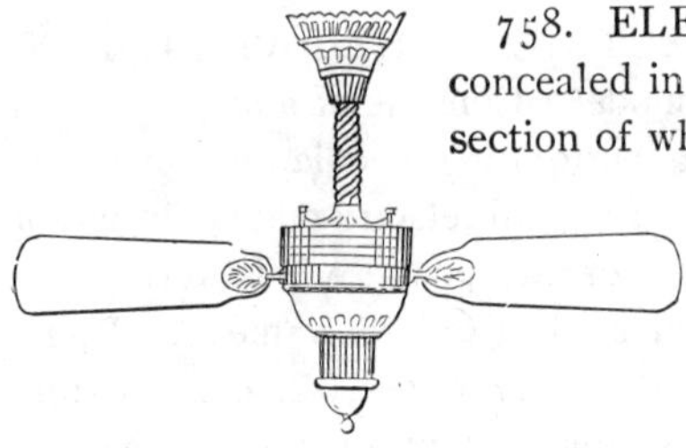

758. ELECTRIC FAN.—The motor is concealed in the central chamber, the middle section of which revolves with the arms.

759. ELECTRIC-DRIVEN FAN, "Edíson" model. Fan on same shaft with the armature. Ball bearings. Runs with four ordinary batteries.

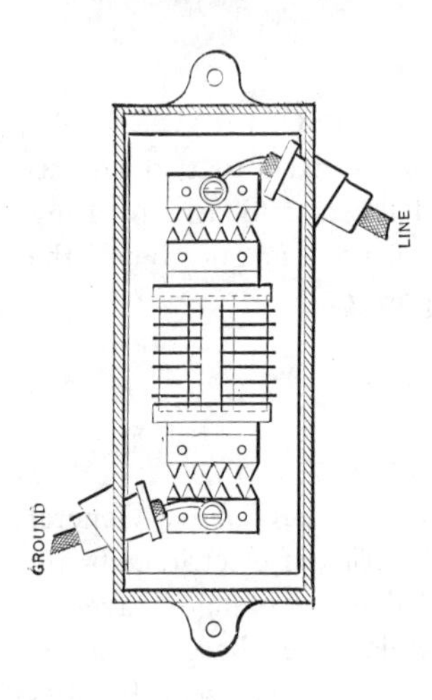

759*a*. NON-ARCING LIGHTNING ARRESTER.—This arrester is made up of a number of circular discs of non-arcing composition, separated from each other by thin sheets of mica. A discharge jumps readily through these thin sheets of mica, as their combined resistance is much less than that of a single sheet of aggregate thickness. The discharge also divides itself into hundreds of little fine sparks through the mica, which do not have sufficient body to pull an arc.

This device thus does away with the necessity of fuses, magnets or moving parts.

759*b*. AMPEREMETER.—Simple form showing the principle of operation. The attraction of the electric current in the solenoid coil, draws the iron core within the coil against the gravity of the core and frame which swings on a frictionless bearing.

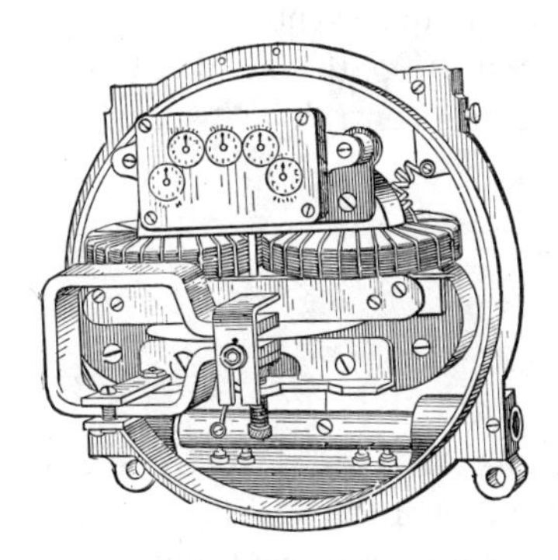

759*c*. RECORDING WATT-HOUR METER.—The armature is rotated under the influence of the current in the field coils. The armature spindle drives the recording gear and dial hands and is regulated by the constant retarding influence of a disk revolving on the same spindle between the poles of permanent magnets.

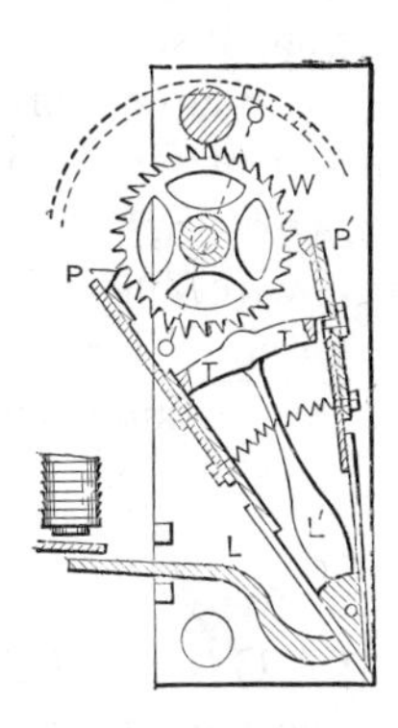

759*d*. ELECTRIC ESCAPEMENT, in which a positive motion is given to a clock from a central station by a detent and impulse action of the electric current. The bell crank lever L, L′ and its arms T, T′ are actuated by the electro magnet and alternately strike the pallets P, P′ moving the escapement one tooth each at make and break of the circuit.

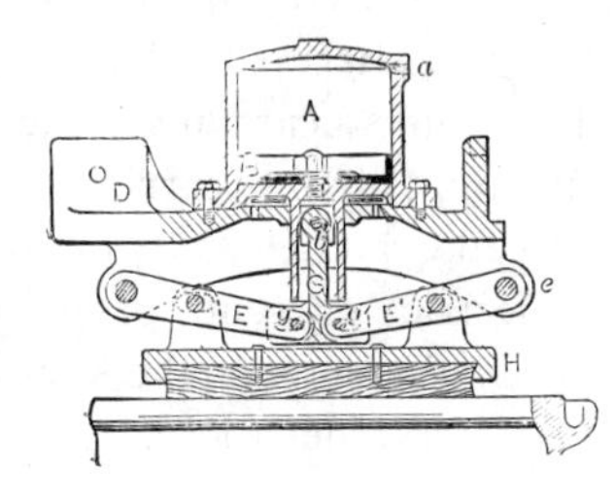

759*c*. PNEUMATIC EMERGENCY BRAKE.—Instantaneous action from air brake pressure on the piston A, which is connected by a jointed rod to the levers E, E, throws the brake H in contact with the rail with great force.

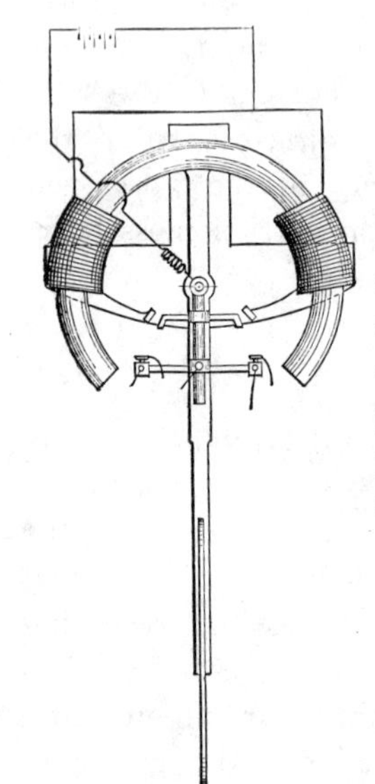

759*f*. SOLENOID ELECTRIC FAN.—A circular magnet attached to a pivoted arm and fan blade, oscillates within two electric coils by the alternate make and break of the circuit from the battery by the contact fingers and studs. Requires no switch; simply stopping or starting the fan blade puts it out and in electric control.

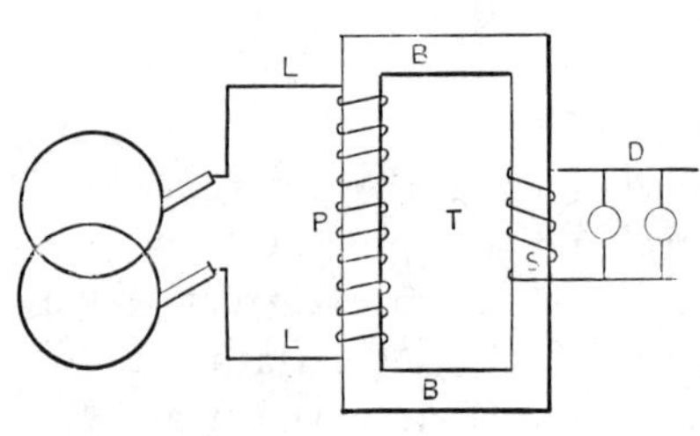

759*g*. ALTERNATING CURRENT TRANSFORMER.—L, L, main lines from dynamo A; P, primary coil of small wire and many turns; B, iron magnet core which may be made of strips of sheet iron or iron wire; S, induction coil of large wire and few turns; D, lamp circuit. Practically both coils are wound on the same core.

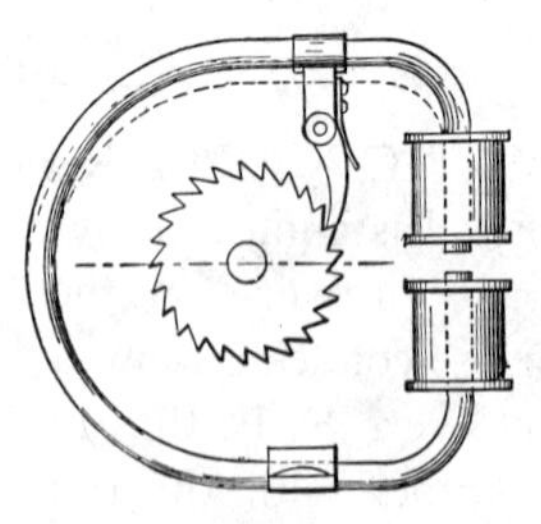

759*h*. ELECTRO MAGNETIC RATCHET DRIVER.—Used on electric clocks. The flexible iron stirrup is fixed at the bottom to the frame carrying the ratchet wheel. The upper side carries a pivoted pawl held to the ratchet teeth by a light spring. An electric current passed through the coils draws the ends of the stirrup together equal to the advance of a single ratchet tooth.

Section X.

NAVIGATION AND ROADS.

Vessels, Sails, Rope Knots, Paddle Wheels, Propellers, Road Scrapers and Rollers, Vehicles, Motor Carriages, Tricycles, Bicycles and Motor Adjuncts.

760. LEG-OF-MUTTON SAIL.—A triangular sail attached to mast and boom. 5, mainsail.

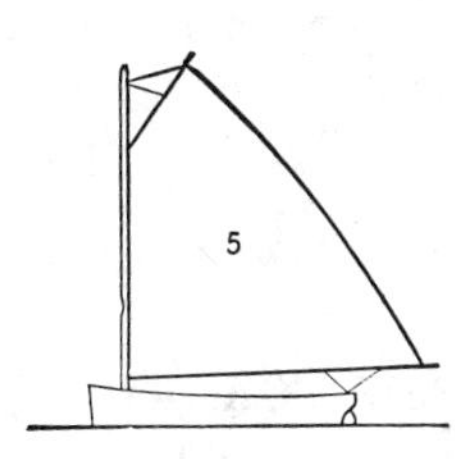

761. SKIP JACK.—A baggy sail bent to the mast and extended by a boom and gaff. The cat-boat. 5, mainsail.

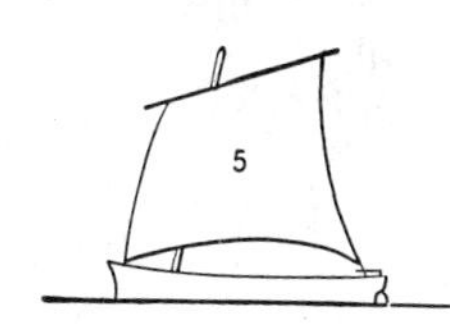

762. SQUARE OR LUG SAIL, attached to a yard. 5, mainsail.

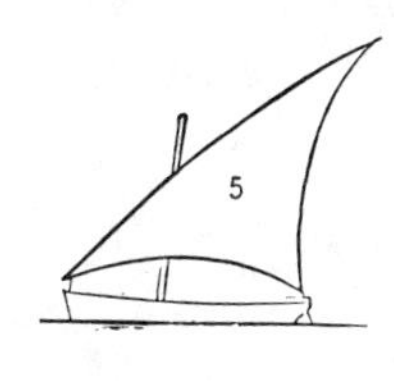

763. LATEEN RIG.—A triangular sail extended by a long yard, which is slung about one-quarter its length from the lower end, which is brought down to the tack. 5, mainsail.

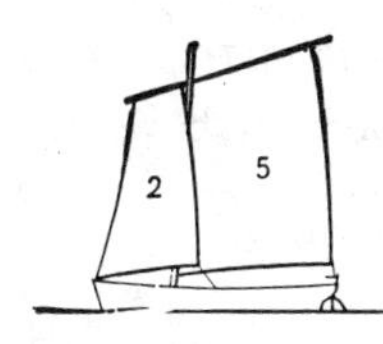

764. SPLIT LUG OR SQUARE SAIL, attached to a yard and divided at the mast, the larger portion being bent to the mast. The unequal division gives one sail the effect of a jib. 2, jib; 5, mainsail.

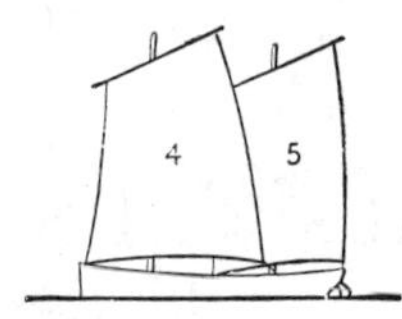

765. TWO-MASTED OR DIPPING LUG.—The sails are square, except at the top, where they are bent to yards hanging obliquely to the masts. 4, foresail; 5, mainsail.

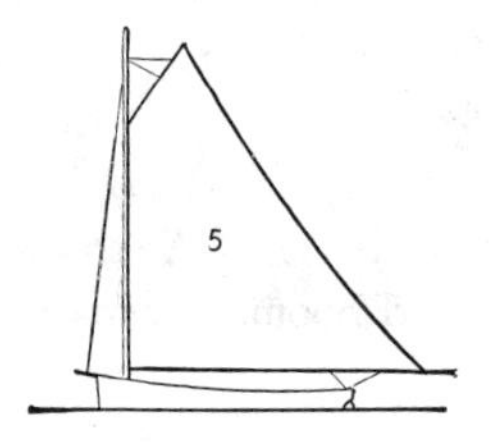

766. NEWPORT CAT-BOAT.—Sail bent to mast and extended by boom and gaff, with a fore-stay to a short bowsprit. 5, mainsail.

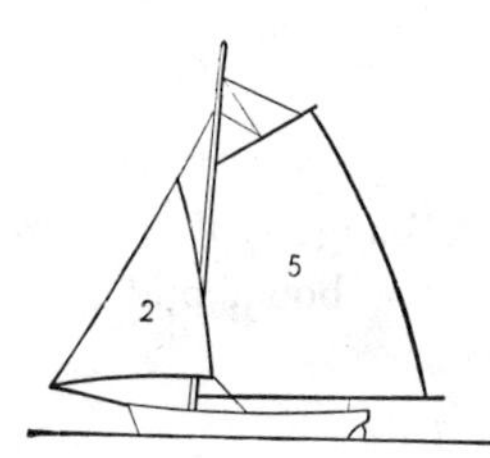

767. SLOOP.—A mainsail and jib with fore- and back-stays. 2, jib; 5, mainsail.

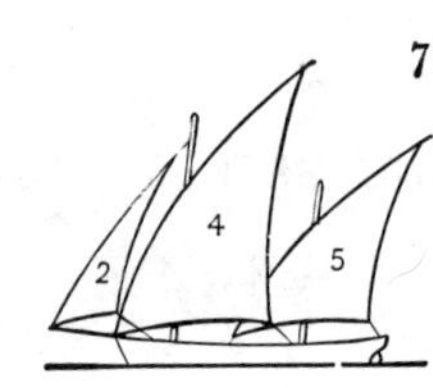

768. LATEEN-RIGGED FELUCCA.—A two-masted boat with lateen sails and a jib. 2, jib; 4, foresail; 5, mainsail.

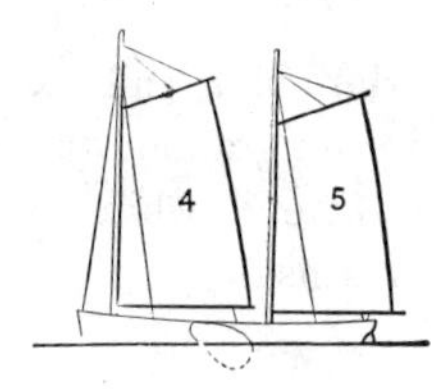

769. PIROGUE.—A two-mast schooner rig, without jib and furnished with a leeboard. 4, foresail; 5, mainsail.

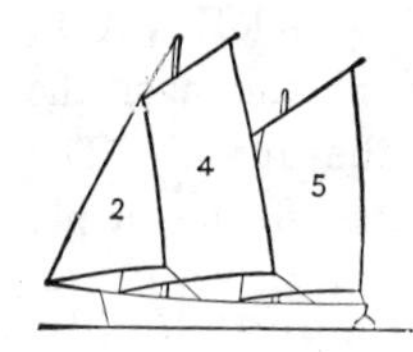

770. THREE-QUARTER LUG RIG.—Two long or lug sails with jib stayed to bowsprit. 2, jib; 4, foresail; 5, mainsail.

771. "SLIDING GUNTER," or sliding topmast. A two-masted boat, with divided masts. The triangular sails are bent to both masts, and furled by lowering the upper mast. Mainsail extended by a boom. 2, jib; 4, foresail; 5, mainsail.

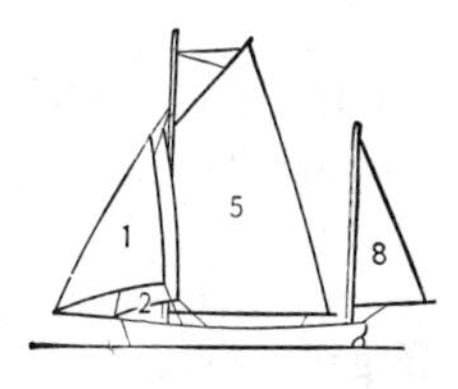

772. SKIFF YAWL RIG.—A mainsail with one or two jibs, and a small mast at the stern with a leg-of-mutton sail, extended by a boom. 1, flying-jib; 2, jib; 5, mainsail; 8, lugsail.

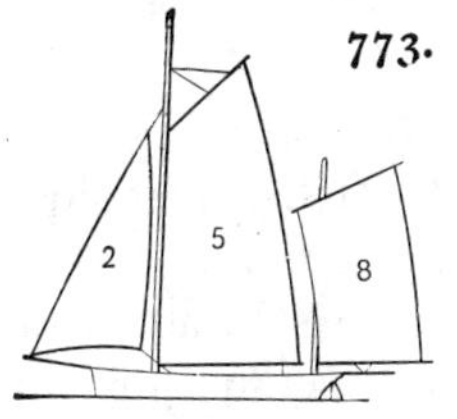

773. SLOOP YAWL.—A foremast, schooner-rig, of sheet and jib, with a lugsail and mast at the stern. Lugsail extended by a boom. 2, jib; 5, mainsail; 8, jigger.

774. JIB-TOPSAIL SLOOP.—A mainsail, two jibs and jib-topsail. The topsail is run up the topmast and extended on the gaff. Main jib-stay from masthead to bow. Fore jib-stay from topmast to bowsprit. 1, flying-jib; 2, jib; 5, mainsail; 13, gaff-topsail.

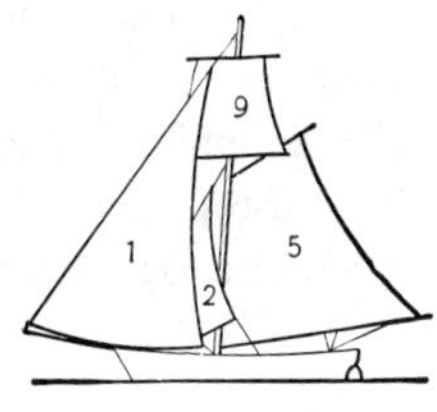

775. THE CUTTER.—A mainsail, 5; jib, 2; flying-jib, 1, and topsail, 9, are the main features of a cutter-rig.

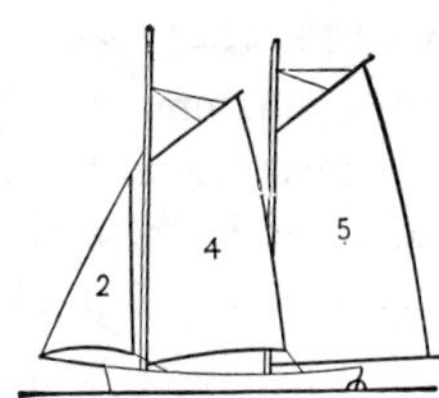

776. SCHOONER RIG.—Fore- and main sail bent to the mast, boom and gaff. Jib stayed to bowsprit. 2, jib; 4, foresail; 5, mainsail.

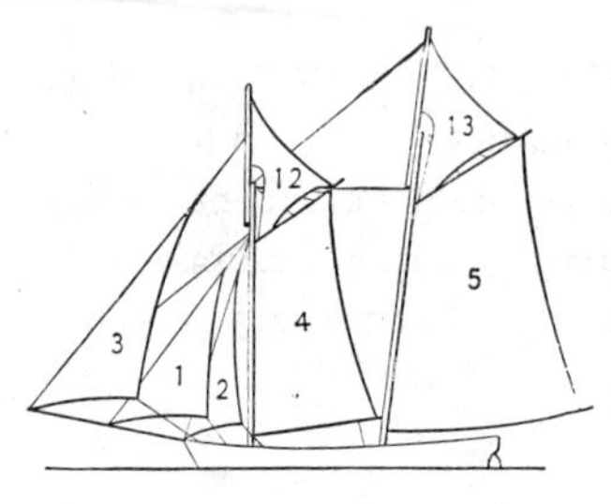

777. FULL SCHOONER RIG.—Main- and fore-sail, two or three jibs, and two topsails. 1, flying-jib; 2, jib; 3, foretop staysail; 4, foresail; 5, mainsail; 12, fore gaff-topsail; 13, main gaff-topsail.

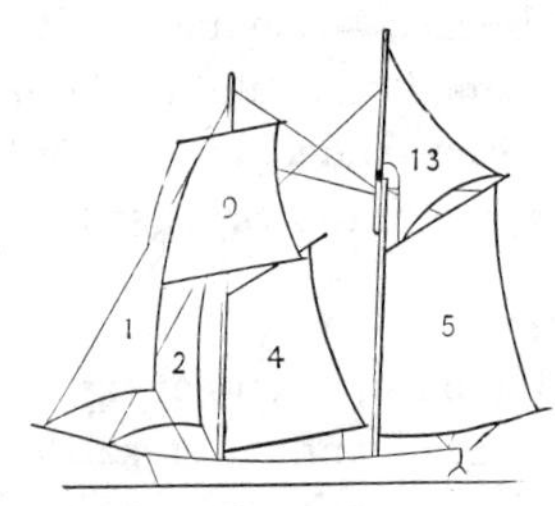

778. TOPSAIL SCHOONER.—The same rig as a schooner, except the foretop, which is a square sail bent to a yard. 1, flying-jib; 2, jib; 4, foresail; 5, mainsail; 9, fore-topsail; 13, main gaff-topsail.

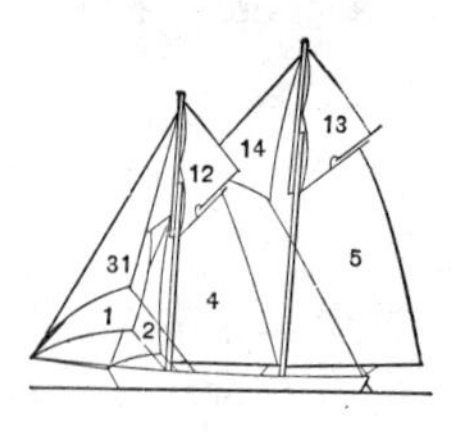

779. CLUB TOPSAIL RIG.—In addition to the full schooner rig, a club topsail is bent to a stay from the main-topmast head to the cross-trees of the foremast. 1, flying-jib; 2, jib; 4, foresail; 5, mainsail; 12, fore gaff-topsail; 13, main gaff-topsail; 14, main topmast staysail; 31, jib topsail.

780. HERMAPHRODITE BRIG.—Mainmast has a fore and aft sail, triangular topsail, and a club sail on a stay to the foremast. Foremast is square-rigged, with the addition of a fore and aft sail—hence the name half-brig, half-schooner. 1, flying-jib; 2, jib; 4, foresail; 5, mainsail; 9, fore-topsail; 13, main gaff-topsail; 14, main topmast-staysail; 22, fore-topgallant sail; 25, fore-royal.

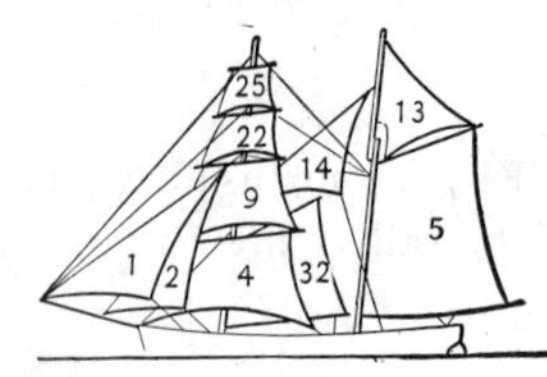

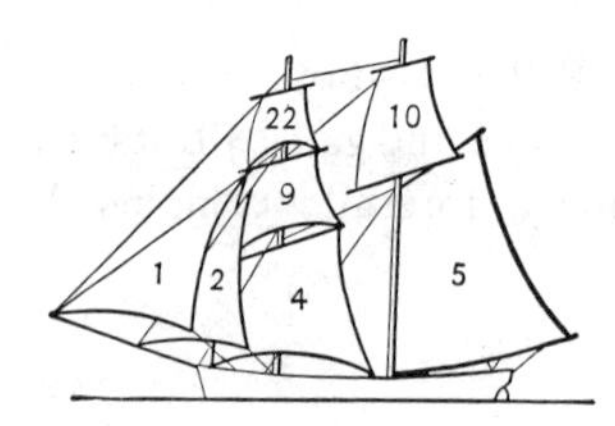

781. A BRIGANTINE.—Foremast rigged with square sails; mainmast with fore and aft sail and square-topsail. 1, flying-gib; 2, jib; 4, foresail; 5, mainsail; 9, fore-topsail; 10, main topsail; 22, fore-topgallant sail.

782. A BARKENTINE. — Schooner-rigged main and mizzen mast, full square-rigged foremast, with the addition of a fore and aft sail on the foremast. Club sails on stays from main to foremast. 1. flying-jib; 2, jib; 3, fore topmast staysail; 4, foresail; 5, mainsail; 7, spanker; 9, foretopsail; 13, main gaff-topsail; 14, main topmast-staysail; 22, fore-topgallant sail; 25, fore royal; 32, fore-trysail; 33, staysail · 34, gaff-topsail.

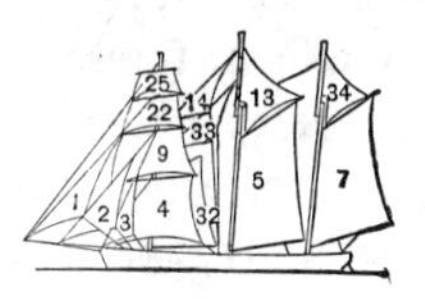

783. FULL-RIGGED BRIG.—Square sails on both main and fore mast with the addition of a fore and aft sail on the main mast. Two or three jibs. 1, flying-jib; 2, jib; 3, foretopmast-staysail; 4, foresail; 5, mainsail; 7, spanker; 9, foretopsail; 10, maintopsail; 22, foretopgallant-sail; 32, main-topgallant-sail; 20, upper maintopsail; 25, fore royal.

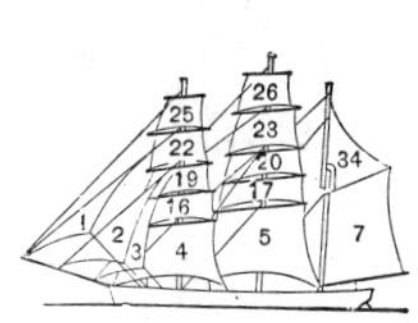

784. A BARK.—Full, square-rigged sails on fore and main masts. Schooner rig, mizzen-mast. 1, flying-jib; 2, jib; 3, foretopmast staysail; 4, foresail; 5, mainsail; 7, spanker; 16, lower foretopsail; 17, lower maintopsail; 19, upper foretopsail; 20, upper maintopsail; 22, fore-topgallant-sail; 23, main-topgallant-sail; 25, fore royal; 26, main royal; 34, gaff-topsail.

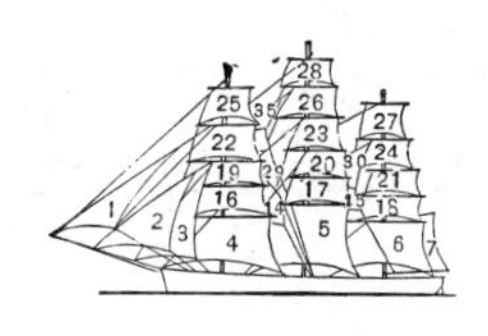

785. FULL-RIGGED SHIP, with double topsails and staysails. 1, flying-jib; 2, jib; 3, foretopmast staysail; 4, foresail; 5, mainsail; 6, cross-jacksail; 7, spanker; 14, main-topmast staysail; 15, mizzen-topmast staysail; 16, lower foretopsail; 17, lower maintopsail; 18, lower mizzen-topsail; 19, upper foretopsail; 20, upper maintopsail; 21, upper mizzen-topsail; 22, fore-topgallant-sail; 23, main-topgallant-sail; 24, mizzen-topgallant-sail; 25, fore royal; 26, main royal; 27, mizzen royal; 28, main skysail; 29, main-topgallant-staysail; 30, mizzen-topgallant-staysail; 35, main royal staysail.

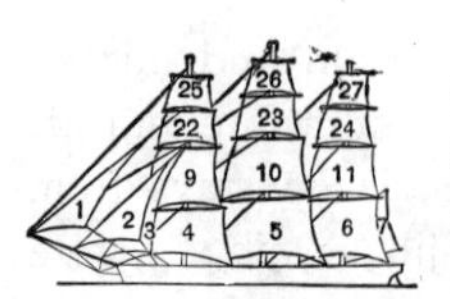

786. FULL-RIGGED SHIP.—Square sails on fore, main, and mizzen mast, with a fore and aft sail on mizzen mast. Three jibs. 1, flying-jib; 2, jib; 3, foretopmast-staysail; 4, foresail; 5, mainsail; 6, cross-jacksail; 7, spanker; 9, foretopsail; 10, maintopsail; 11, mizzentopsail; 22, foretopgallant sail; 23, maintopgallant-sail; 24, mizzen topgallant-sail; 25, fore royal· 26, main royal· 27 mizzen royal.

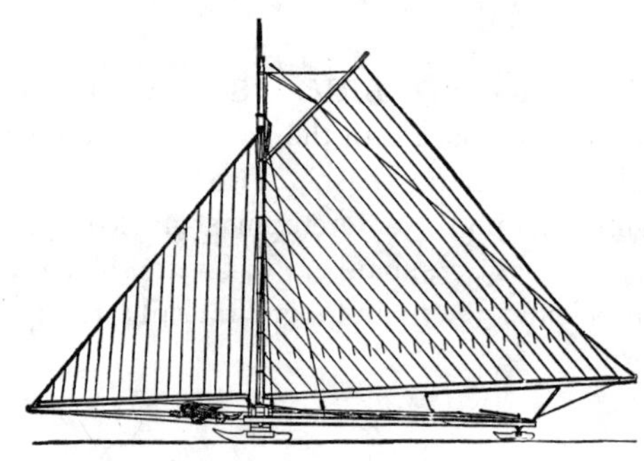

787. ICE BOAT.—A sloop-rigged frame on three runners, the rear one being the tiller runner.

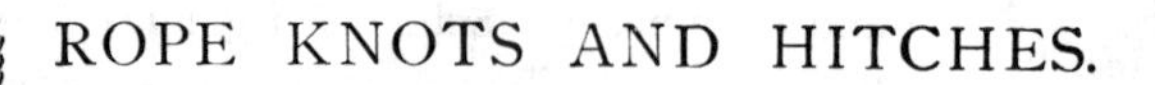

ROPE KNOTS AND HITCHES.

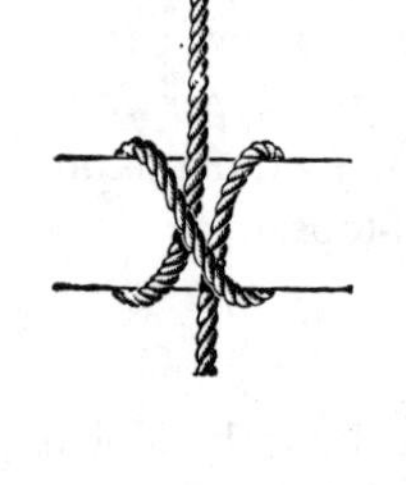

788. CLOVE HITCH.

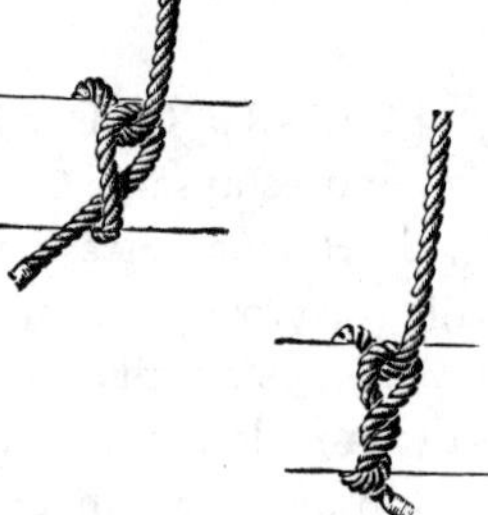

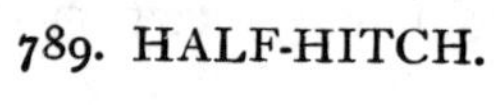

789. HALF-HITCH.

790. TIMBER HITCH.

791. SQUARE OR REEF KNOT.

792. STEVEDORE KNOT.

793. SLIP KNOT

794. FLEMISH LOO

795. BOWLINE KNOT.

796 CARRICK BEND.

797. SHEET BEND AND TOGGLE.

798. SHEET BEND. Weaver's knot.

799. OVERHAND KNOT.

800. FIGURE EIGHT KNOT.

801. BOAT KNOT.

802. DOUBLE KNOT.

803. BLACKWALL TACKLE HITCH.

804. FISHERMAN'S BEND HITCH.

805. ROUND TURN AND HALF HITCH.

806. CHAIN STOP for a cable.

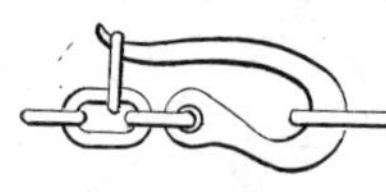

807. DISENGAGING HOOK, held by a mousing link.

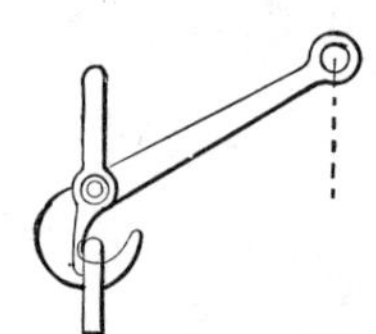

808. SLIP HOOK.—The extension of the suspension link holds the lower link in line, while a pull on the arm by a lanyard releases the load.

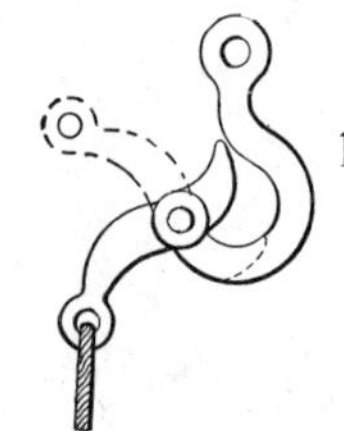

809. RELEASING HOOK.—The lever throws the link off by a pull of the lanyard.

810. BOAT DETACHING HOOK.—The standard is fastened to the boat. A tongue is pivoted to its upper end and passes through the hook of the tackle-block. A lever with an eye to catch the tongue is pivoted to the upright standard, with a lanyard attached at the bottom. A simultaneous pulling of the two lanyards detaches both ends of a boat at once.

811. SWINGING OAR LOCK.—The hook C of the oar lock is swivelled on a post, D. which is fastened to the gunwale by a flange staple and latch or by extending the swivel through the gunwale.

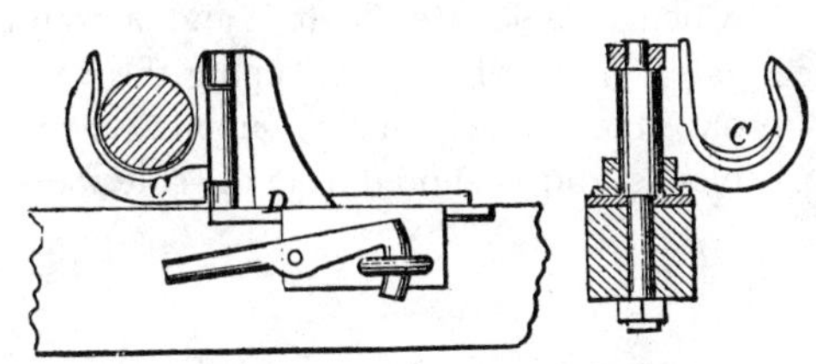

812–813. PIVOTED STEPS for a boat-landing. One edge of each step is pivoted to the lower stringer, the other edge to the upper stringer by a hanger. On a level the steps form a floor, as the end of the ladder falls with the tide the hangers lift the forward edge of the step to keep it level. The shore posts are fixed and vertical. Stringers are pivoted to posts.

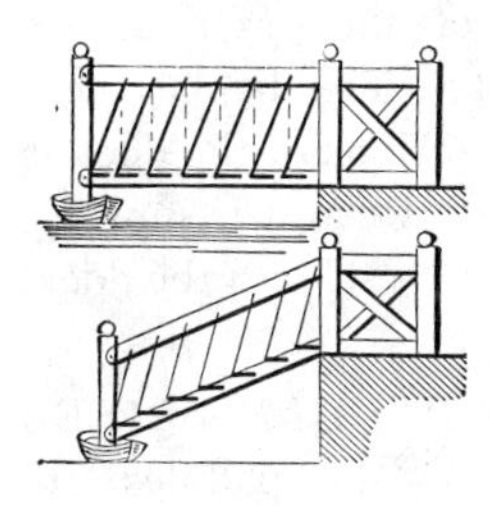

814. SCREW ANCHOR for buoys. Is screwed to the required depth in the sand by a long box wrench.

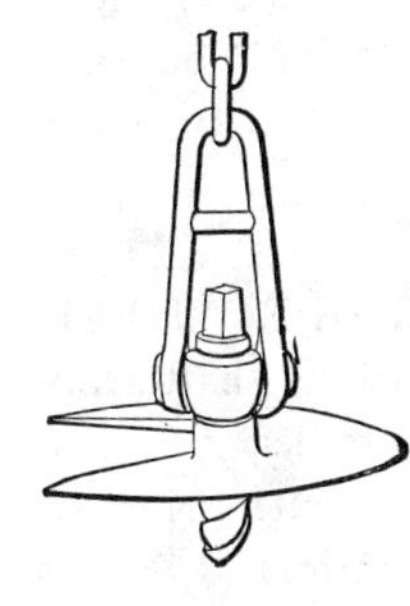

815. FLOATING LIGHTHOUSE.—A floating buoy filled with compressed gas (Pintsch system). Supplies a constant light of high power in the lantern for several days.

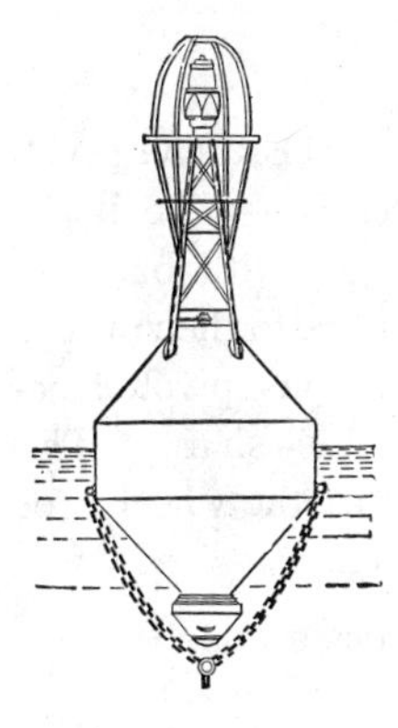

816. STONE DRY-DOCK, into which vessels are floated and a water gate closed, when, by pumping the water out, the vessel settles upon bearing blocks, and is shored from the side walls.

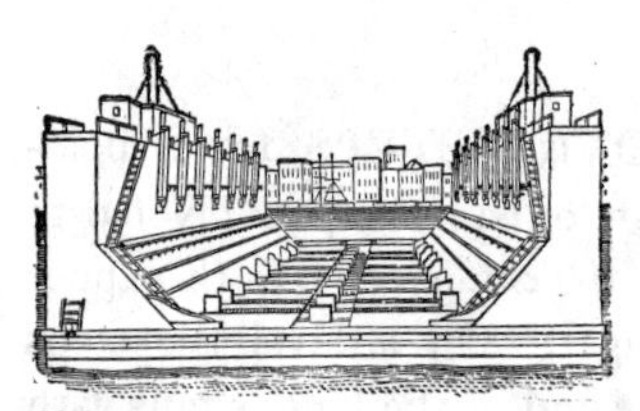

817. FLOATING DRY-DOCK, in which the lifting power is derived from the displacement of the water in the interior of the dock. The displacement area of the side extensions of the dock is sufficient to balance it when it is sunk, by filling the lower part with water in order to float a vessel into the dock.

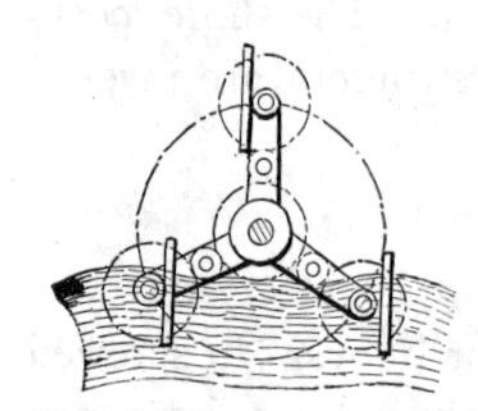

818. FEATHERING PADDLE WHEEL OR WATER MOTOR.—The paddles are kept in a vertical position by a planetary gear. The central gear is fixed. The pinions and gear on the arms keep the paddles in a vertical position in the water.

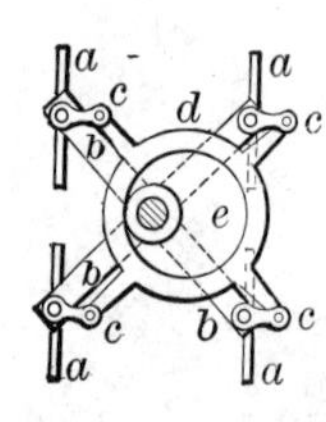

819. VERTICAL BUCKET PADDLE WHEEL.—The buckets, *a*, *a*, *a*, *a*, are pivoted to the shaft arms, *b*, *b*. To the pivots are attached cranks, *c*, *c*, *c*, *c*, which are pivoted to the arms of an eccentric ring revolving with the shaft on a fixed eccentric, *e*. By this arrangement the buckets are kept vertical.

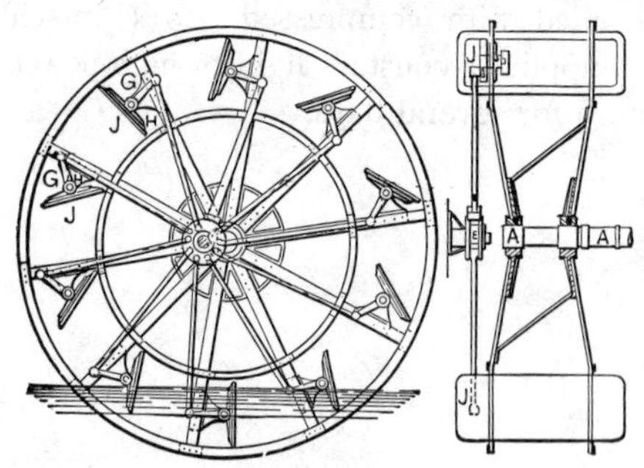

820. FEATHERING PADDLE WHEEL.—The buckets are hinged with back levers and turned to their proper position by arms pivoted eccentric to the shaft. The framework of the wheel is of iron or steel.

821. Cross section.

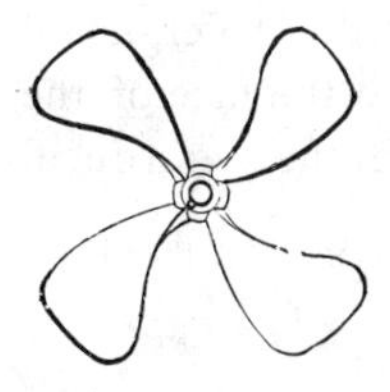

822. OUTWARD THRUST PROPELLER WHEEL.—The blades pitch forward to throw the water outward as well as backward, to increase the thrust or power of the wheel.

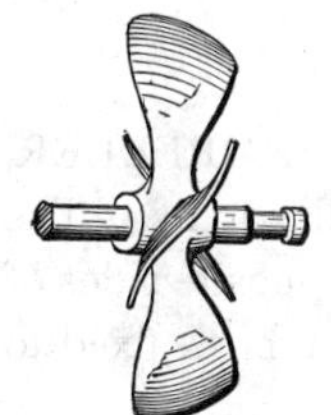

823. SCREW PROPELLER. Four blades. Ordinary form for heavy draft tugs and tow-boats.

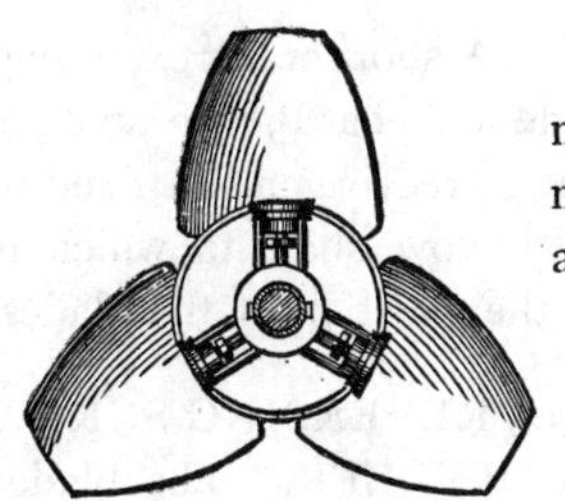

824. SCREW PROPELLER, "Griffith" model. The inclination of the blades is made adjustable, and they are attached to a rim outside from the hub.

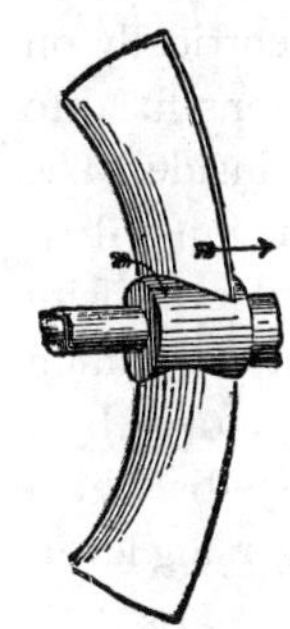

825. SCREW PROPELLER, "Hodgson's" model. The blades are curved backward to prevent the centrifugal direction of the water when passing the blades. Claims on fore-and-aft direction of blades by inventors, are not in harmony with the best practice in propeller design.

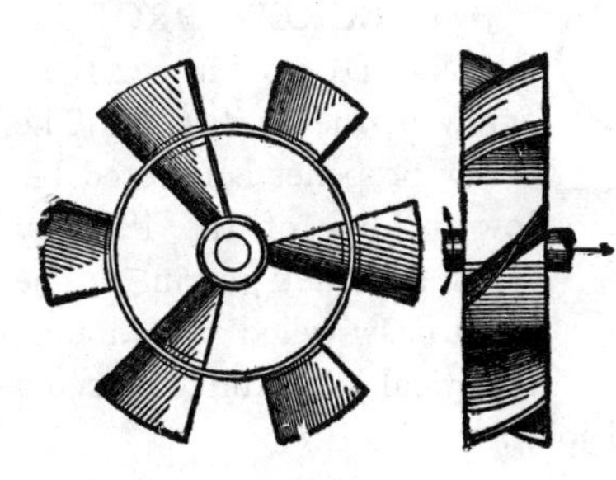

826. SCREW PROPELLER, the "Ericsson" model. A rim connecting all the blades, supposed to counteract the centrifugal tendency of the water.

827. Side view.

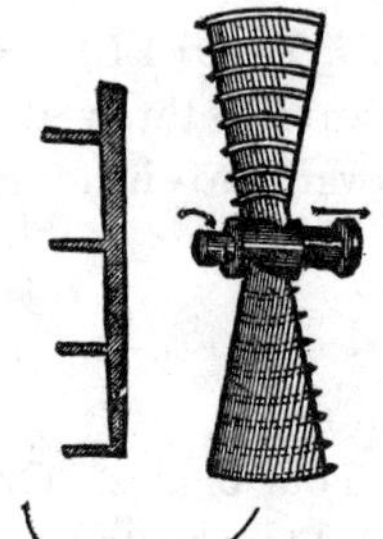

828. SCREW PROPELLER, "Vergne's" model. The projecting ribs from the face of the blades are intended to neutralize the centrifugal action of the water.

829. Section of blade.

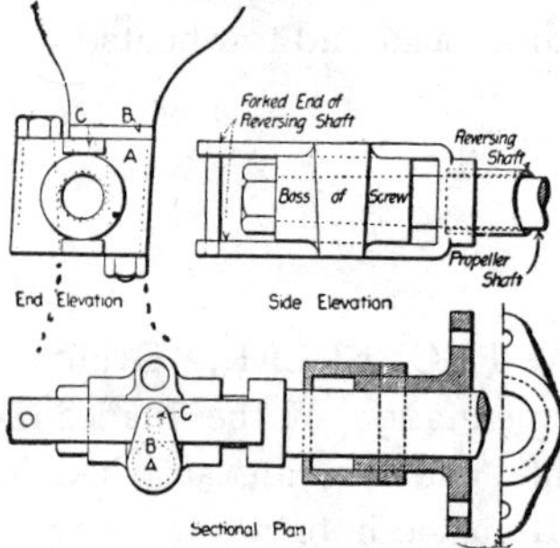

830. REVERSING PROPELLER, for launches and small yachts. The blades are socketed on opposite sides of the shaft and through a boss fixed to the driving shaft.

831. Plan. A short crank extending from the blade socket at B, with an elongated hole at C, receives a pin fixed to a yoke and hollow shaft to which is given a fore-and-aft motion for changing the position of the blades.

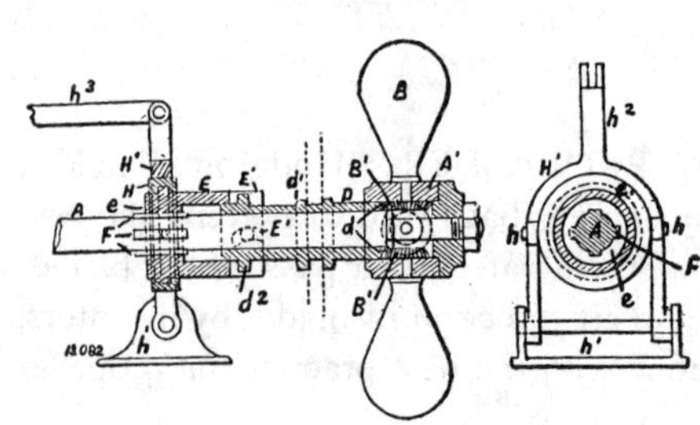

832. REVERSING SCREW PROPELLER.—The blades are pivoted concentrically on the hub, with pinions fixed to the shanks on the inside. The hub is fixed to the inner driving shaft. A sleeve, with gear-cut end to fit the pinion teeth, revolves with the shaft. An inclined slot-sleeve E, moved by a yoke lever, gives a slight rotary motion to the geared sleeve by which the four blades are reversed. 833. Section of shaft and reversing lever.

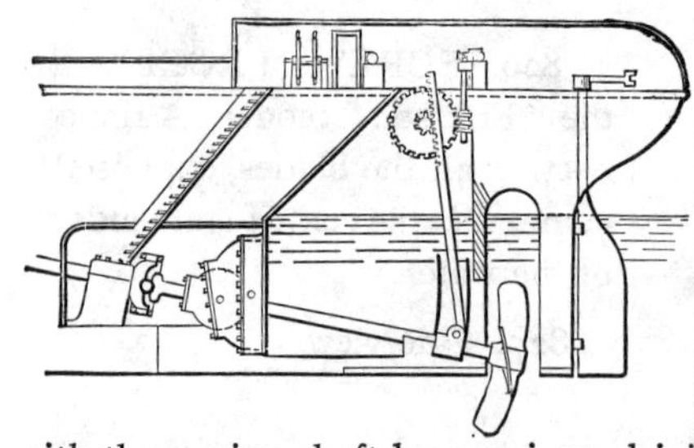

834. SCREW PROPULSION.—Deep immersion screw of the *Britannic.* The screw propeller is lowered below the line of the keel by worm and rack gearing. The shaft is swivelled by a double spherical joint and connected with the engine shaft by a universal joint.

835. REVERSING SCREW PROPELLER. — The central shaft is the driver, and has a small longitudinal motion by a clutch and lever to shift the position of the blades. The outer-end sleeve is fast on the driving shaft, and carries the blades in sockets on each side of its centre. A hollow short shaft, free on the driving shaft, but fixed longitudinally, turns in a socket on the stern post. A rack on each side meshes in a gear sector attached to each blade socket, so that the blades are reversed by the fore-and-aft movement of the driving shaft.

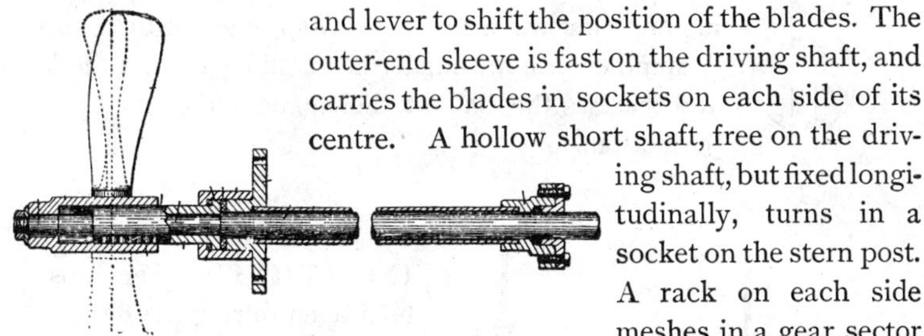

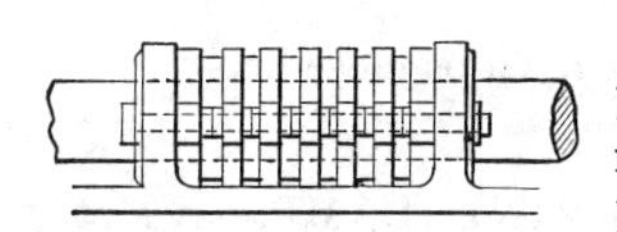

836. THRUST BEARING for a propeller shaft. The collar brasses are set in mortices in the frame; they are made in halves and bolted together.

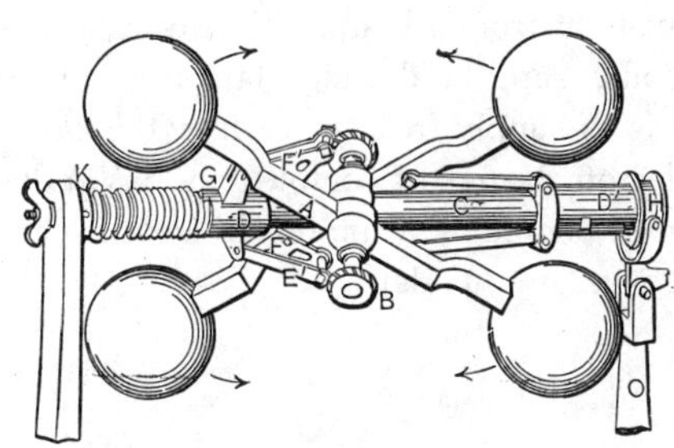

837. "SILVER'S" MARINE GOVERNOR.—The two pair of balls are pivoted to the revolving shaft at the centre of their connecting arms. Their centrifugal tension is held and adjusted by the helical spring I, and thumb-screw. The opening of the balls moves the sleeve, D, for controlling the valve gear.

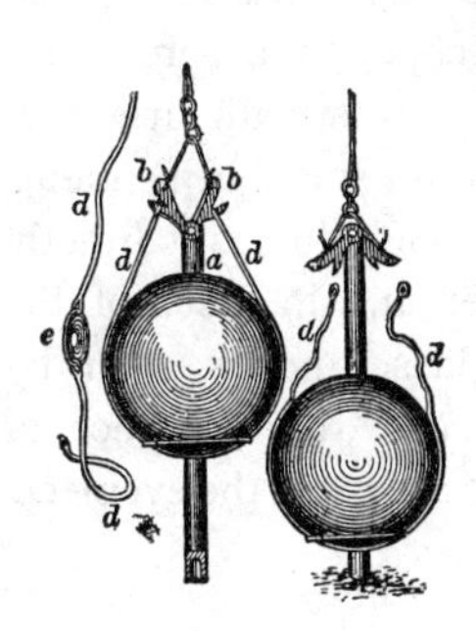

838. DEEP-SEA SOUNDING BALL.—The sounding line is held by the pivoted horns *b*, *b*, which are thrown down when the rod passing through the ball touches bottom; this releases the wire sling *d*, that holds the ball, when the rod and line can be easily drawn up. Has been used in four-mile depths of the ocean.

839. Release position.

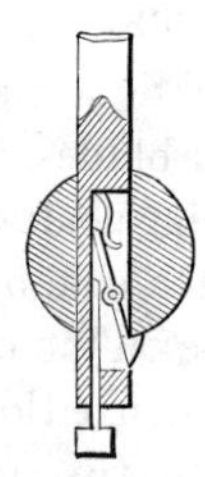

840. SOUNDING WEIGHT RELEASE for deep-sea sounding. A hollow spindle attached to the sounding-line encloses a hook lever, sprung out by a spring. A spindle, with an impact head, slides behind the lever and releases the ball at the moment the head strikes the bottom.

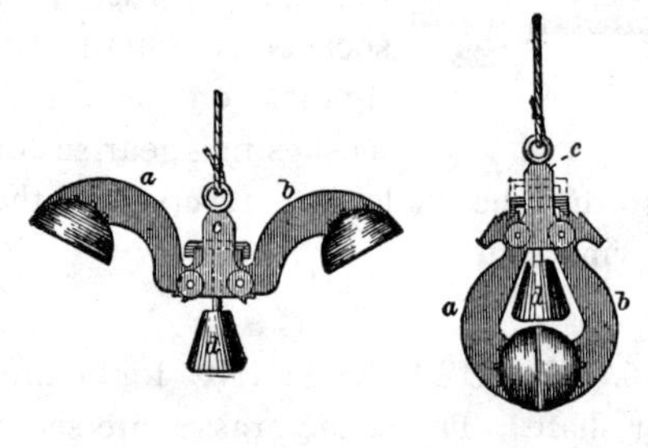

841.. SAMPLER SOUNDING WEIGHT.—The cups are held open during the descent by a clip, which is disengaged when the bob strikes the bottom. The cups spring together by the release of the catch.

842. Cups closed.

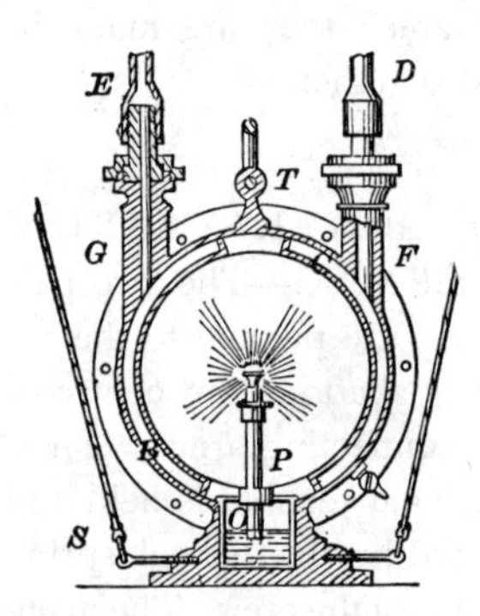

843. SUBMARINE LAMP.—A strong iron case with convex lenses. An ordinary bright light from a lamp, with two hose connections, sling and guide lanyards. One hose is to supply fresh air, while the other carries off the gas of combustion. "Vander Weyde" model. A powerful electric arc light is a later model.

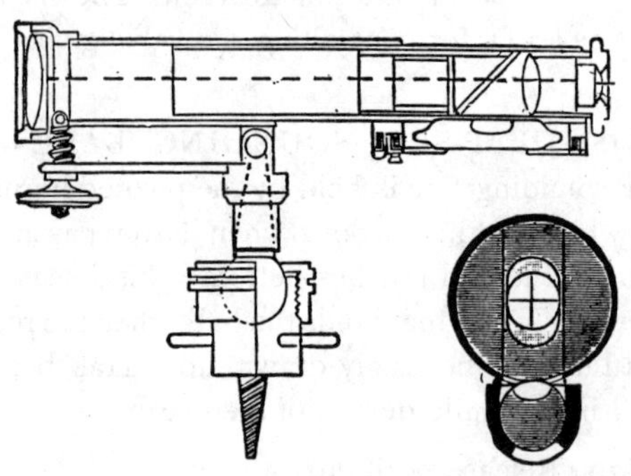

844. ROAD BUILDERS' LEVEL. — A draw telescope, on a screw and swivel base, with arm and screw for small adjustment. The bubble is directly under and in focus of the eyeglass, and is seen by reflection from a piece of glass at 45° in the eyepiece.

845. Section through reflector and level.

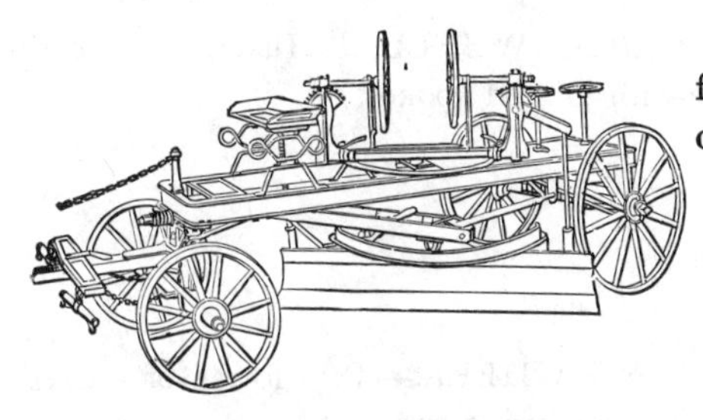

846. ROAD MACHINE, for scraping and levelling common roads.

847. REVERSIBLE ROAD ROLLER.—The tongue is attached to the frame that carries the driver's seat, and is balanced by the weight on the rear arm. By unlocking the catch the horses wheel around the roller with the tongue and seat frame, and the tongue is relocked on the other side of the wheel frame.

848. ROAD ROLLER.—Steam-driven. One of the heavy class now improving our roads.

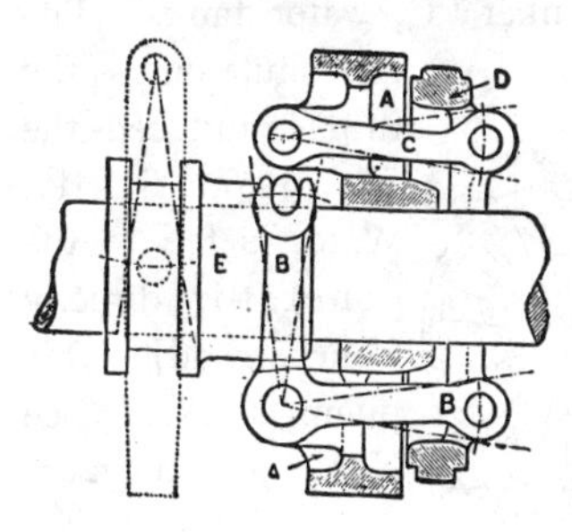

849. SINGLE ECCENTRIC REVERSING GEAR.—Used on traction engines. A is a wheel keyed on the crank shaft; D is the eccentric; C, a link; B, B, bell crank, connected to sleeve and eccentric. The movement of the sleeve E by the lever throws the eccentric D to the centre and to the opposite position for reversal.

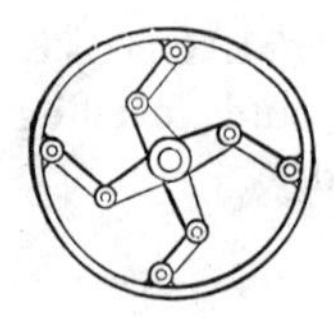

850. ELASTIC WHEEL, "Huxley." A steel spring tire with jointed spokes.

851. SPRING WHEEL.—Two forms of curved spring spokes and spring rim.

852. ELASTIC WHEEL, with steel tire and spring spokes.

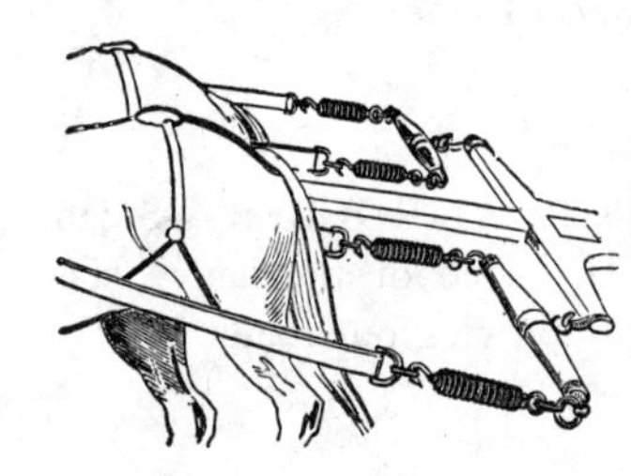

853. APPLICATION OF TRACE SPRINGS for trucks and heavy wagons. Saves the shoulders of horses from fatigue and abrasion.

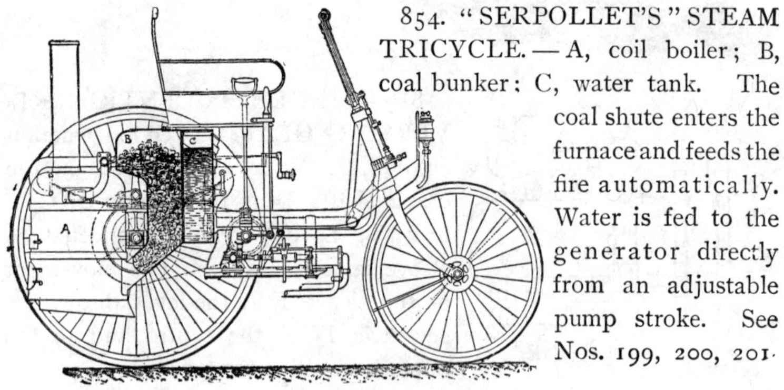

854. "SERPOLLET'S" STEAM TRICYCLE.—A, coil boiler; B, coal bunker; C, water tank. The coal shute enters the furnace and feeds the fire automatically. Water is fed to the generator directly from an adjustable pump stroke. See Nos. 199, 200, 201.

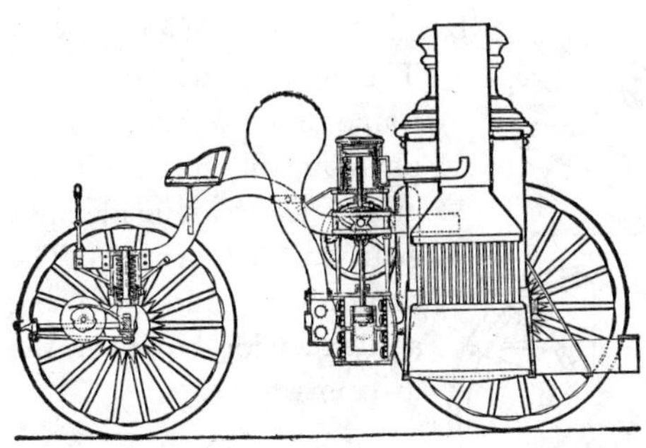

855. STEAM FIRE ENGINE. Vertical tubular boiler. Vertical steam pump, with yoke connection to fly-wheel crank. "Gould" pattern.

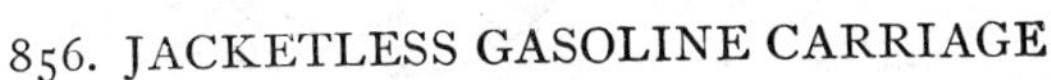

856. JACKETLESS GASOLINE CARRIAGE MOTOR, with two cylinders in line on two cranks at opposite points. Four-cycle type. Explosion in cylinders simultaneously, reducing vibration. Cylinder cooled by air circulation over the radial ribs.

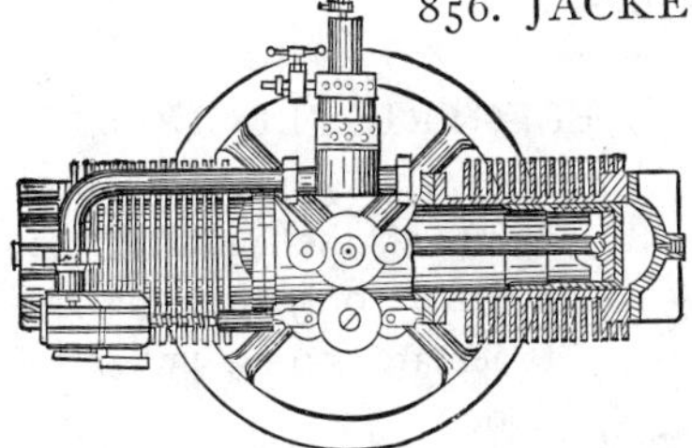

857. GASOLINE MOTOR CARRIAGE.—Two full seats and single seat for driver. The middle seat turns over to get at the motor and gear.

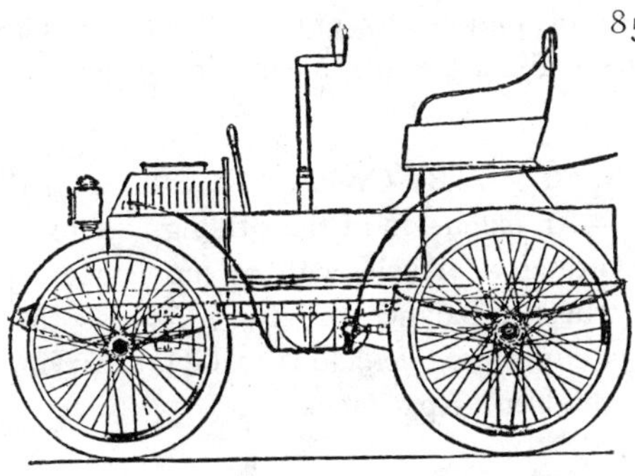

858. LIGHT ELECTRIC CARRIAGE, with single seat. The motor is attached to the frame and geared to a speed shaft, and by sprocket and chain to the wheel axle.

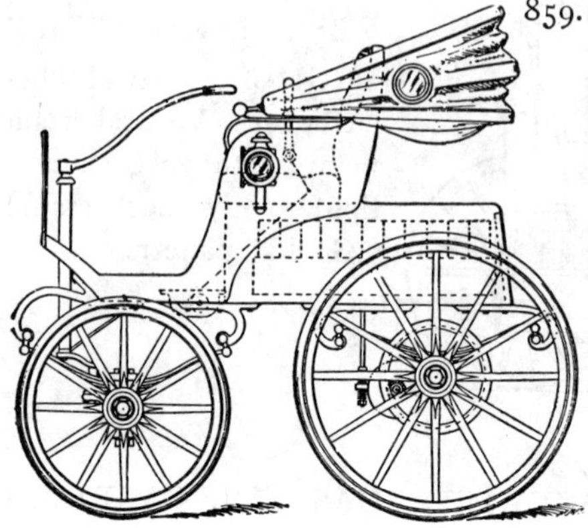

859. ELECTRIC PHAETON.—The motors are fixed to a frame under the floor of the phaeton, with their pinions meshing with an inside spur gear on each wheel. The batteries are under the seat and extension box over the driving wheels.

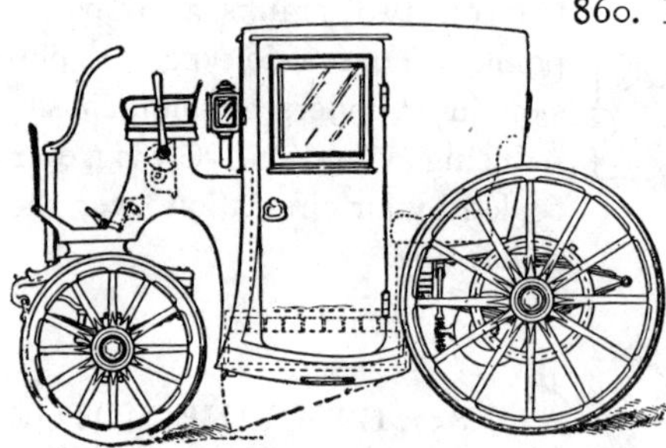

860. ELECTRIC BROUGHAM. The same general arrangement of the motor as in No. 859, only that the batteries are stored under the floor.

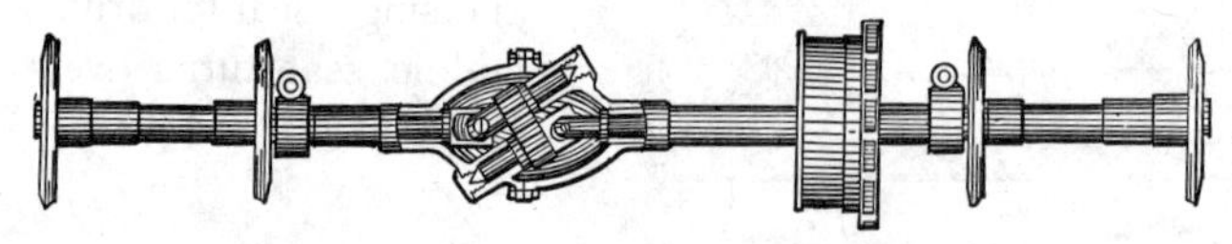

861. DIFFERENTIAL GEAR for a tricycle. The bisected shaft is connected to a pair of pinions by universal joints. The pinions are pivoted at an angle of about 30° in a free-moving sleeve box.

862. BABY-CARRIER TRICYCLE.—An extension of the driving axle of an ordinary bicycle, with a supplementary wheel to balance and for safety, so that a convenient vehicle is made for carrying children or packages

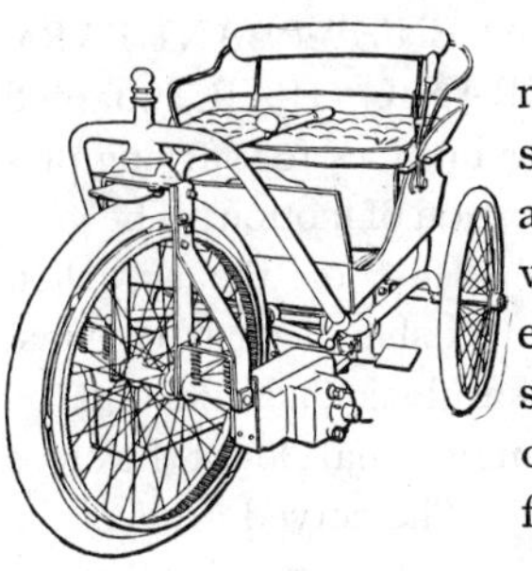

863. ELECTRIC TRICYCLE, "Barrow" pattern. The single forward wheel is swivelled to the vehicle frame for steering and is also the driving wheel. It has a spur wheel on the inside of the rim in which the electric-motor pinion meshes. The motor swings with the steering-wheel frame, and is connected to the battery under the seat by flexible wiring.

864. ICE BICYCLE.—An attachment of a runner and a toothed rim for any bicycle; making bicycling a winter sport on the ice.

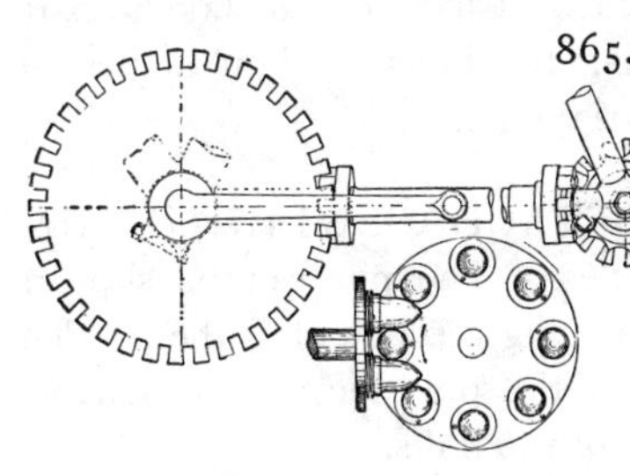

865. BICYCLE GEAR.—Transmission by fore and aft shaft with pin-tooth gearing. "Sagar" model.

866. Pin-tooth wheel and pinion.

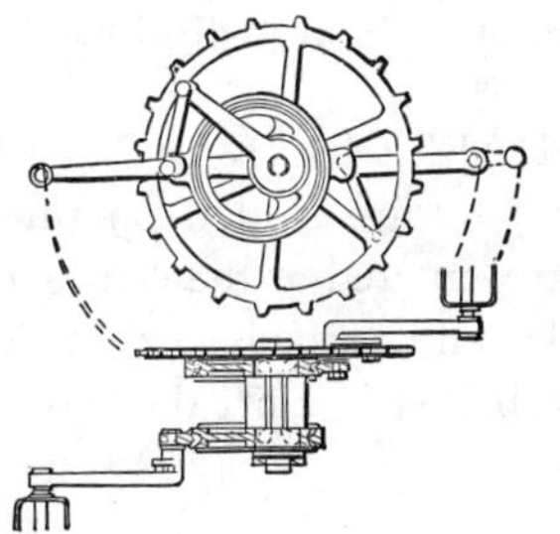

867. BICYCLE CRANK.—A device for shortening the up-crank stroke. The eccentrics are fixed to the frame. The cranks and eccentric straps revolve on ball bearings, carrying by link connection the secondary crank shaft and sprocket wheel.

868. Horizontal plan.

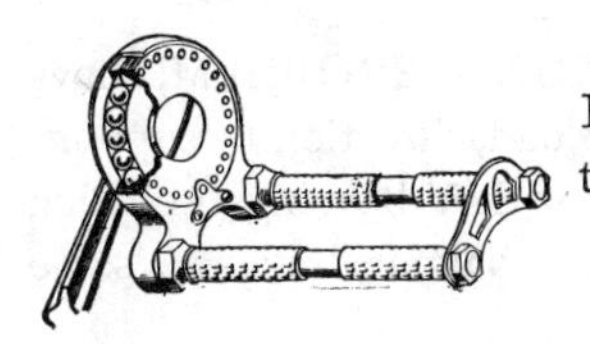

869. SWINGING BALL-BEARING BICYCLE PEDAL.—Carries the feet close to the ground.

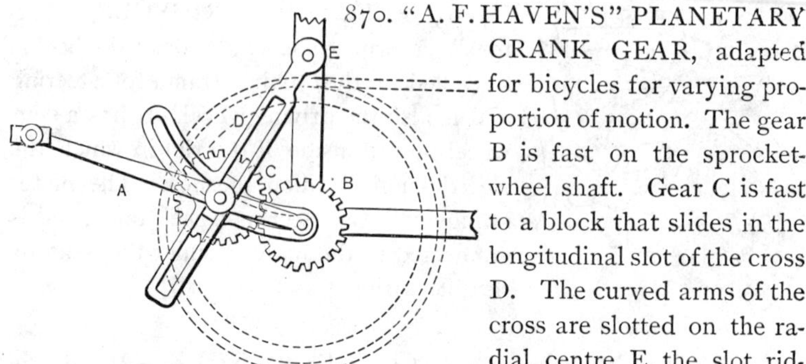

870. "A. F. HAVEN'S" PLANETARY CRANK GEAR, adapted for bicycles for varying proportion of motion. The gear B is fast on the sprocket-wheel shaft. Gear C is fast to a block that slides in the longitudinal slot of the cross D. The curved arms of the cross are slotted on the radial centre E, the slot riding over the sprocket shaft, allowing the radial arm D to pass the shaft. The crank A is pivoted to the shaft and the sliding block. With equal gears, the sprocket wheel makes two revolutions for one of the crank.

871. DETACHABLE LINK CHAIN for bicycles. Chain can be taken apart by turning the links at right angles to the run of the chain.

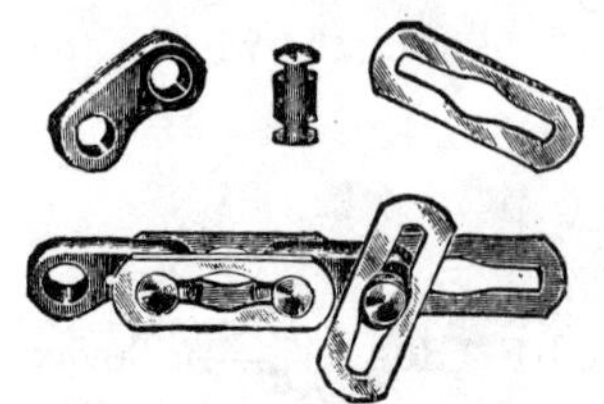

872. DETACHABLE LINK CHAIN for bicycles. The pin can be slipped out by drawing the links together. The grooves in the pins lock in the narrow slot ends of the links.

873. Centre link pin and slip link.

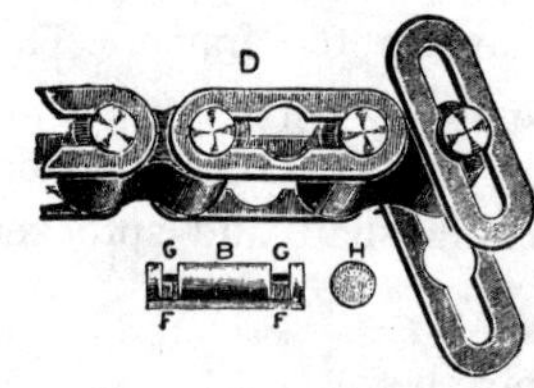

874. DETACHABLE LINK BICYCLE CHAIN.—The pins are slotted on three sides at G, G, are entered at the centre of the outside links and turned so that the straight back will rest against the end of link slot.

875. Pin showing slots.

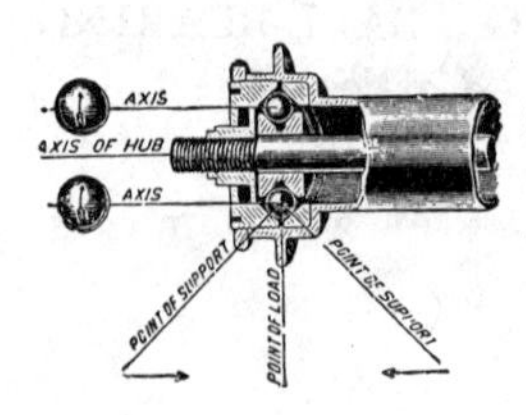

876. BALL-BEARING PROBLEM, showing the direction of load, direction of support, and axis of rotation with V bearings in which the angular thrust is balanced in the same journal.

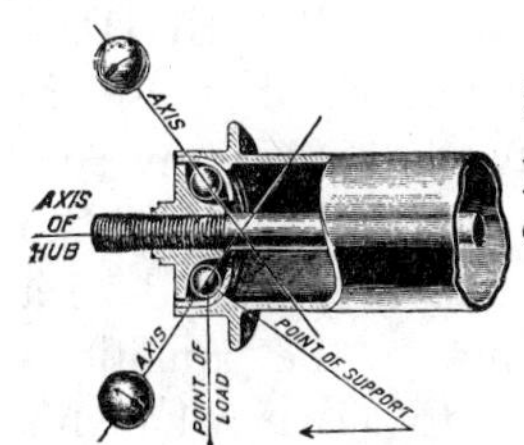

877. BALL-BEARING PROBLEM, showing the direction of load, the direction of support, and the axis of rotation with angular quarter-curve bearings and angular thrust.

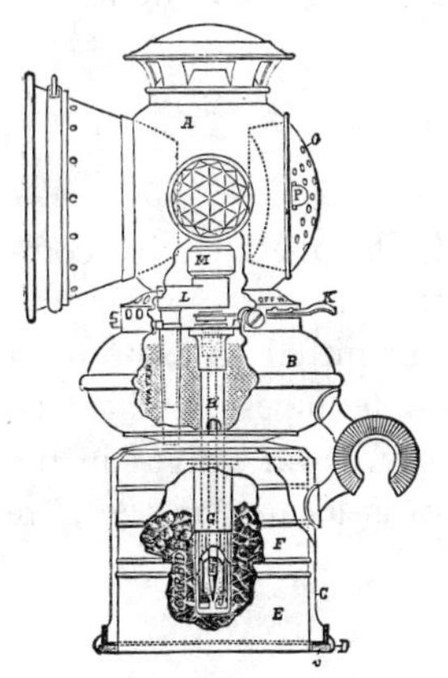

878. ACETYLENE BICYCLE LAMP. Gas is generated in the lower compartment by admission of water in small quantities from the compartment B, through a needle valve operated by the handle K. L, gas tube; M, burner. The gas pressure is regulated by the hydrostatic head of water in the reservoir B. If gas is generated too fast, the water is held back by the gas pressure.

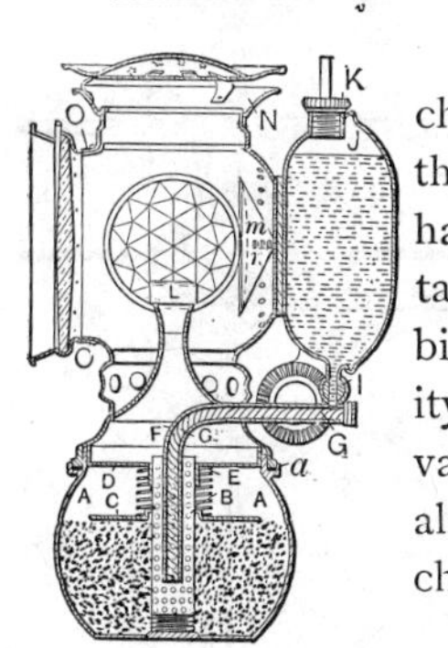

879. ACETYLENE BICYCLE LAMP.—A charge of pulverized calcium carbide is placed in the lower chamber. A charge of water of one-half the weight of the carbide is placed in the tank, J. The wick G carries water to the carbide by capillary action and pressure from gravity. The gas is aerated in the burner. The valve at I regulates the flow of water, which is also retarded by the gas pressure in the carbide chamber.

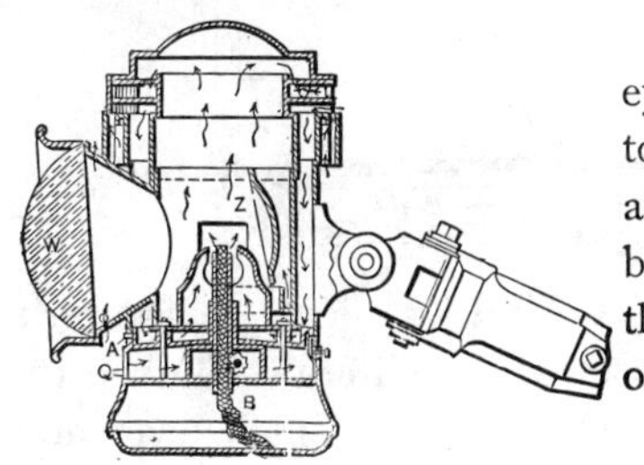

880. BICYCLE LAMP.—W, bull's-eye lens. Air enters at O, and passes to the flame between the wick tube and guard, and flickering is prevented by air's exit through small passages in the shell of the lamp. Z, reflector; B, oil chamber.

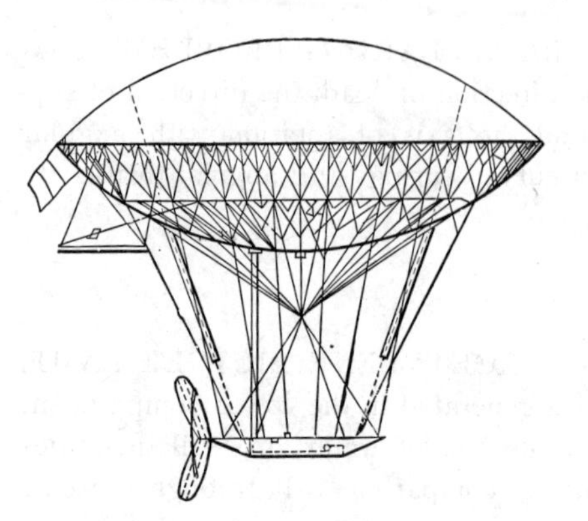

880*a*. AIR SHIP.—General form of those in use that have had any success. Too many come to grief and they should be a warning to the ambitious soarer, to learn what his predecessors have done.

880*b*. RAILROAD GATE operated by compressed air.—A hand pump in the gate house compresses air which is transmitted to pistons at the gate bar posts by double pipes so that the bars are raised or closed by the air valves in the gate house. The air pump is located in the gate house.

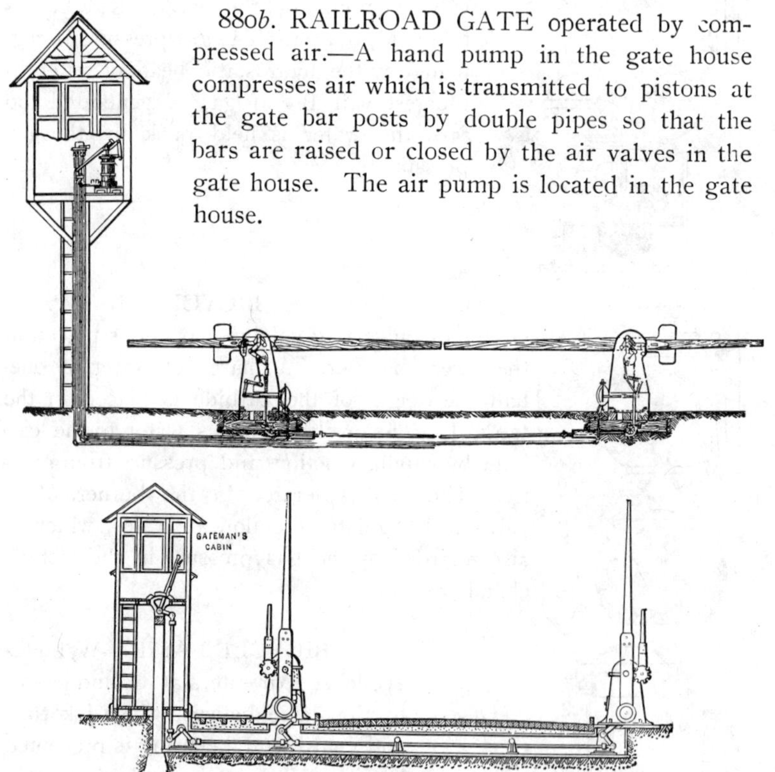

880*c*. RAILROAD CROSSING GATE.—A lever in the gate house operates the gate arms by a series of bell cranks linked together and to sectors in the gate arm boxes. A small lever and sector at each arm to close the foot walk.

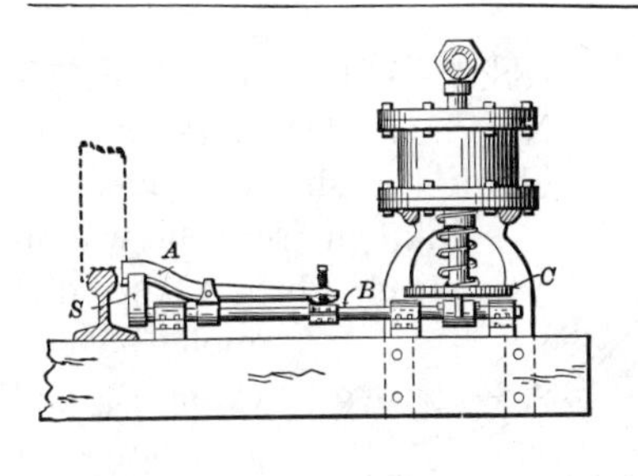

880*d*. RAILWAY PNEUMATIC SIGNAL.—An air cylinder near the rail is operated by the wheels passing over the lever A, pressing it against the cam S, on a rocking shaft B, lifting the plate C, and the connected piston. Elevation and plan.

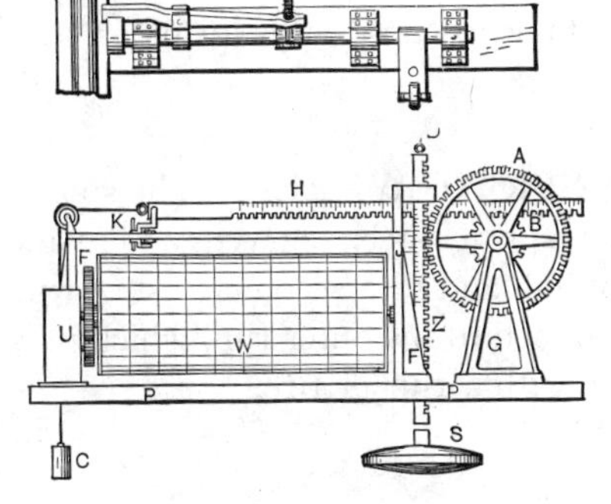

880*e*. A SELF REGISTERING TIDE GAUGE.—S is the float in the tide well; Z the rack meshed in the wheel A; B is a pinion meshed in the horizontal rack H, which carries the marking point K. The barrel W carries a paper marked by the hours of the day and driven by a clock. D and C are tension weights to take up any looseness in the gear.

880*f*. NOVEL STEERING GEAR on Emperor William's yacht "Meteor."—A right and left screw shaft with links between the nuts and rudder post cap. Rudder buffer rings at each end of the shaft to take up the jar.

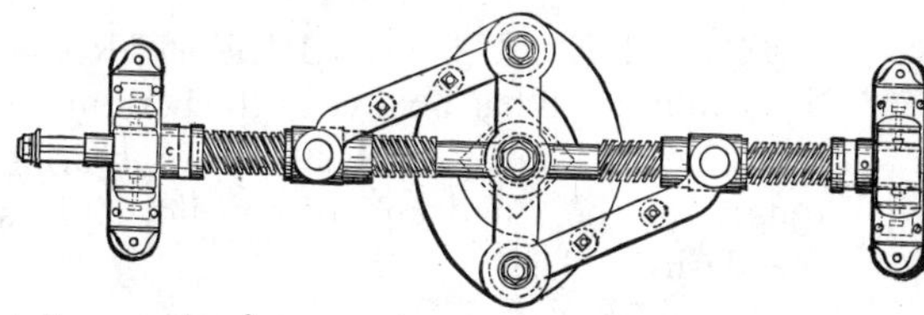

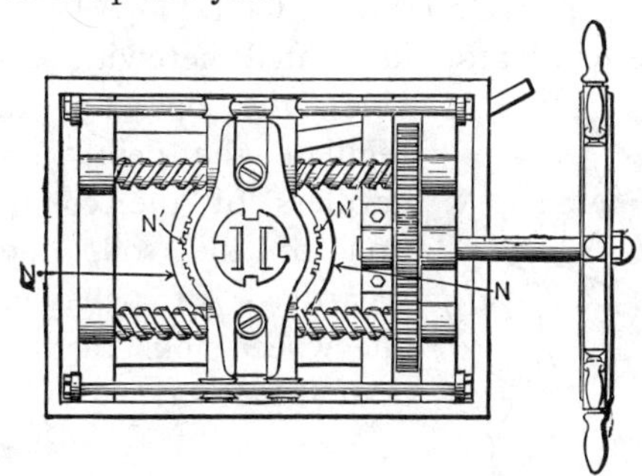

880*g*. SHIP'S STEERING GEAR.—A slotted cross head fixed to the rudder post, in the slot of which two nut bearings traverse; the nuts being carried in opposite directions by right and left-hand screws operated by the steering wheel and gears.

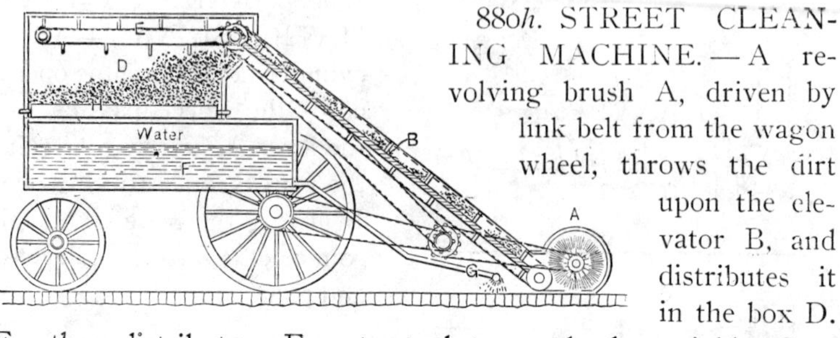

880*h*. STREET CLEANING MACHINE.—A revolving brush A, driven by link belt from the wagon wheel, throws the dirt upon the elevator B, and distributes it in the box D. E, the distributer; F, water tank to supply the sprinkler G.

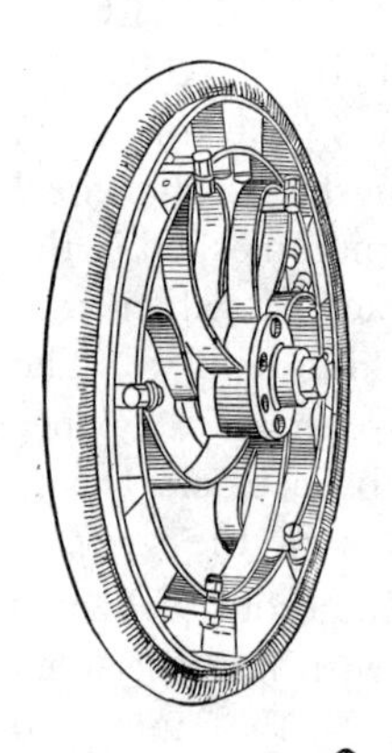

880*i*. A SPRING WHEEL.—The spokes are of flat spring steel, curved and made fast to the hub and to a secondary steel rim, which is clipped to the outer rim, which may be solid or with a rubber tire.

880*j*. THE AUTOMOBILE HORN.—The rubber ball has a valve at the bottom for charging with air. A whistle or a vibrating tongue at the small end of the horn gives the desired blast.

880*k*. ADJUSTABLE CULTIVATOR.—The cultivator illustrated herewith enables the operator to regulate the depth regardless of the condition of the soil, and regardless of whether one wheel sinks farther than the other.

Section XI.

GEARING.

RACKS AND PINIONS; SPIRAL, ELLIPTICAL, AND WORM GEAR; DIFFERENTIAL AND STOP-MOTION GEAR; EPICYCLICAL AND PLANETARY TRAINS; "FERGUSON'S" PARADOX.

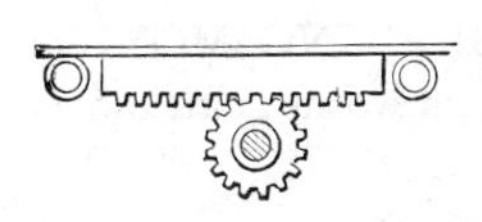

881. ORDINARY RACK AND PINION. —Reciprocating motion, from circular or rectilinear motion as desired.

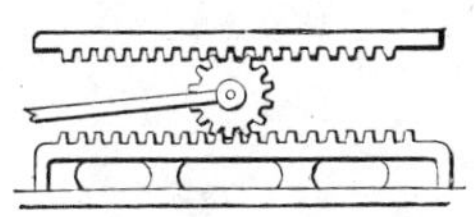

882. DOUBLING THE LENGTH OF A CRANK STROKE by a fixed and a movable rack. The crank rod connects with a pinion, which rolls on a fixed rack, carrying a reciprocating rack to double the distance of the crank throw.

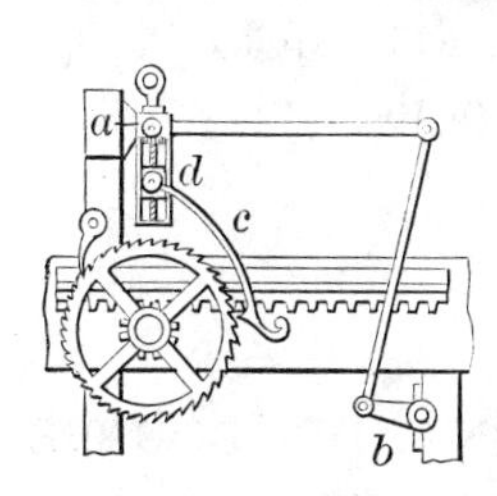

883. SAWMILL FEED.—By the revolution or rocking of the crank *b*, the adjustable bell-crank lever *a* is vibrated, which gives the hook pawl *c* the desired motion to turn the ratchet wheel and pinion which, meshing in the log bed-rack, feeds the log to the saw. The rate of feed is adjusted by the screw and traverse block *d*.

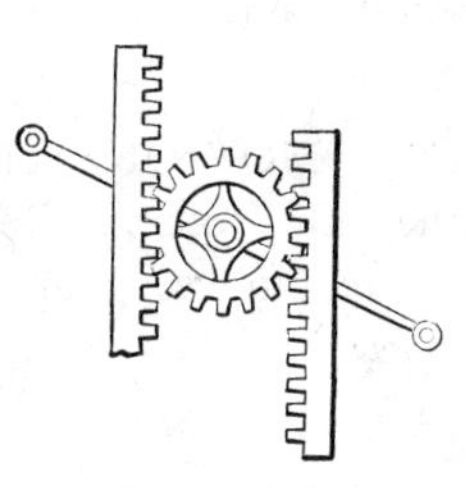

884. RACK MOTION used for air pumps. The racks are directly connected with the pistons of a single-acting air or other pump, and operated by a brake lever.

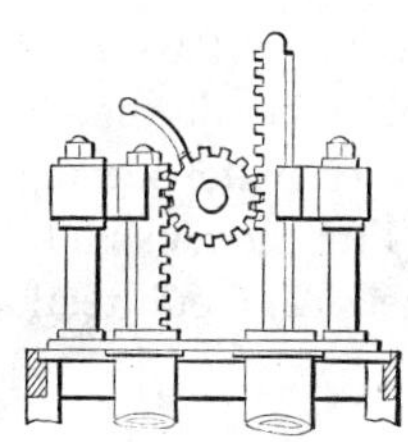

885. AIR-PUMP MOVEMENT.—Two racks connected directly with the pistons, with guides, are operated by a pinion and lever.

886. CIRCULAR RACK and pinion gear. A variable thrust bearing.

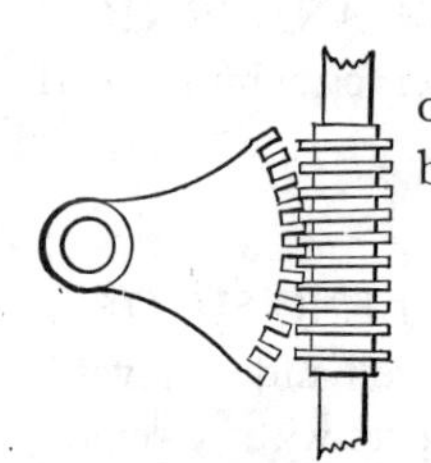

887. RECTILINEAR VIBRATING MOTION of a spindle having an endless worm gear, moved by a spur-gear sector.

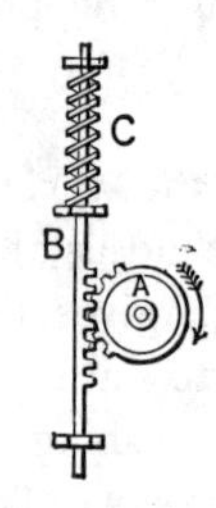

888. VERTICAL DROP HAMMER or impact rod, in any position. Continual motion of sector pinion lifts or draws back the rack-rod B, which quickly drops or springs forward on the release of the teeth.

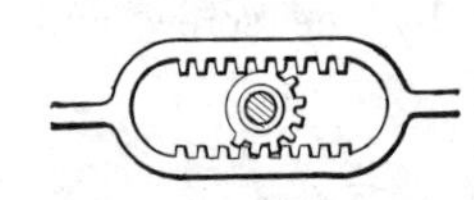

889. SECTOR PINION AND DOUBLE RACK.— Rectilinear reciprocating motion from the continual motion of a sector pinion.

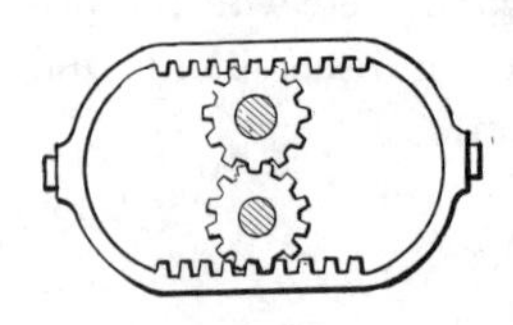

890. RECIPROCATING MOTIONS of two pinions, geared together and to opposite racks, producing rectilinear reciprocating motion to the racks, or *vice versa*.

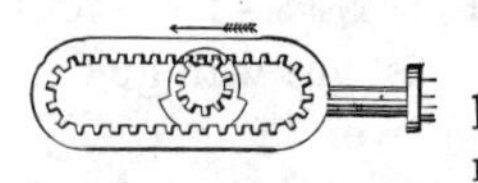

891. CRANK SUBSTITUTE, "Parson's" patent. A reciprocating double rack alternately meshing in a pinion. A cam face plate running in smooth ways in the racks and fast to the pinion lifts the racks into and out of gear alternately at the end of each stroke. The end teeth keep the pinion in mesh.

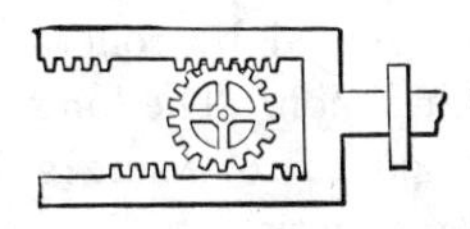

892. ALTERNATE CIRCULAR MOTION of a spur pinion from rectilinear motion of a mutilated rack gear.

893. CRANK SUBSTITUTE. Two loose pinions with reverse ratchets attached to shaft, with pawls on pinion ratchets. Each rack meshes with reverse pinion for continual motion of shaft. Many variations of this device are in use.

894. QUICK BACK MOTION given to a rack slide by a sector gear and slotted arm; operated by a pin in a revolving face plate.

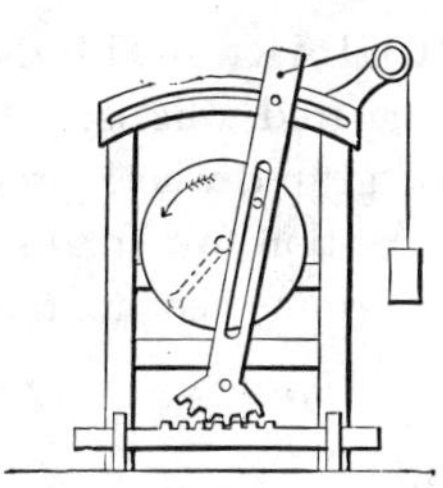

895. ALTERNATE RECTILINEAR MOTION from a swinging lever with sector and rack. The lever has a quick return motion, operated by a wrist pin on a face plate, and free from backlash by the weight and lanyard attached to end of lever.

896. RECIPROCATING RECTILINEAR MOTION of a double rack; gives a continuous rotary motion to the central crank. Each stroke of the rack alternates upon one or the other of the sectors. A curved stop on the centre gear is caught on the pins in the rack, to throw it into mesh with the opposite sector.

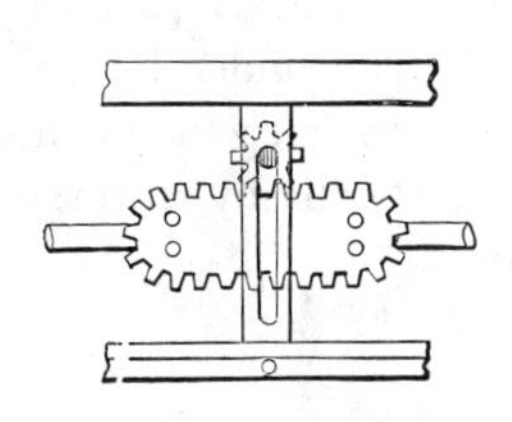

897. RECIPROCATING RECTILINEAR MOTION of a bar carrying an endless rack. A mangle device. The pinion shaft moves up and down the slot, guiding the pinion around the end of the rack.

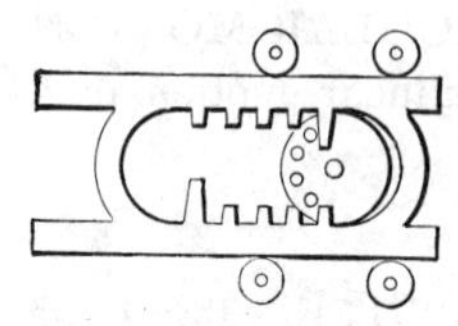

898. MANGLE RACK, guided by rollers and driven by a lantern half-pinion. The long teeth in the rack act as guides to insure a tooth mesh at the end of each motion.

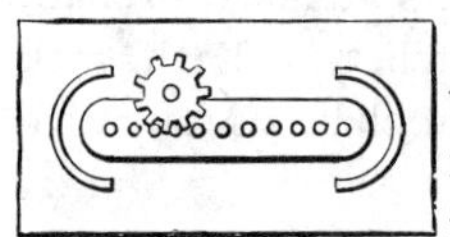

899. MANGLE RACK.—A reciprocating motion of a frame to which is attached a pin-tooth rack, the pinion being guided by the shaft riding in a vertical slot, not shown.

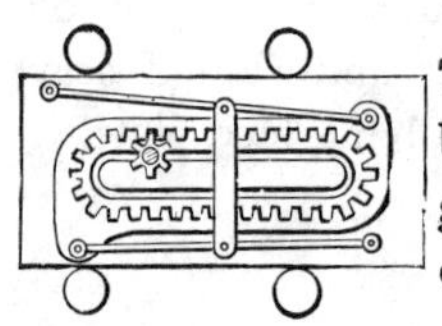

900. MANGLE RACK with stationary pinion. The rack and slot frame are jointed to the mangle box, riding in mesh with the pinion by the slot guide, leaving the mangle box free to ride and tip on the rollers.

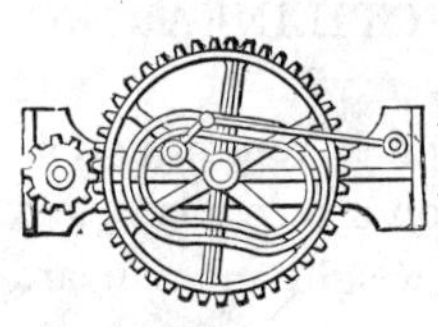

901. ALTERNATE CIRCULAR MOTION from continuous motion of geared wheels. A grooved cam revolving with a geared wheel produces a variable or alternate motion to a crank, through a pin in the groove connected to the crank and to a fixed point by a connecting rod.

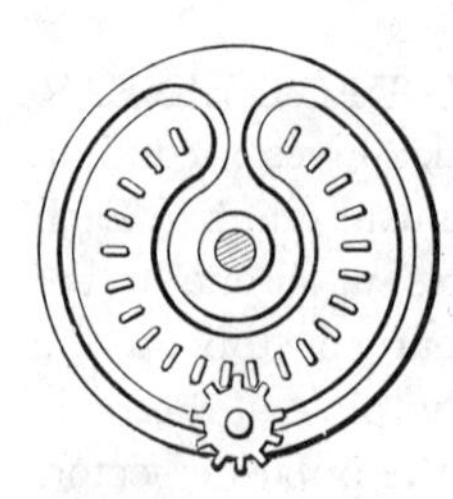

902. MANGLE WHEEL with equal motion forward and return. The pinion moves over the same teeth in both motions. The pinion moves vertical in a guide slot, not shown. The end of the shaft is guided vertically by the groove keeping the pinion teeth in mesh.

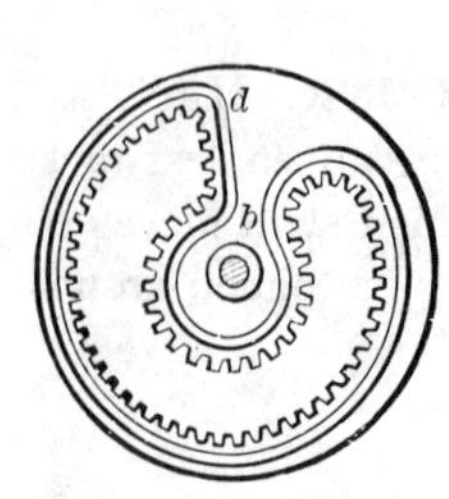

903. "MANGLE WHEEL" GEAR in the operation of which the speed varies in every part of its revolution. The pinion shaft is guided by the groove in the face of the wheel to keep the teeth in mesh, but rises and falls vertically by traversing a slotted guide, not shown.

904. CONTINUOUS ROTARY MOTION of a pinion producing reciprocating motion of the double-geared wheel carrying drum of a mangle. The slotted stand allows the pinion shaft to rise and fall, its end guided by the slot in the return-gear wheel to give the mangle drum a quick return.

905. MANGLE WHEEL with grooved guides, uniform motion through nearly a revolution, and quick return.

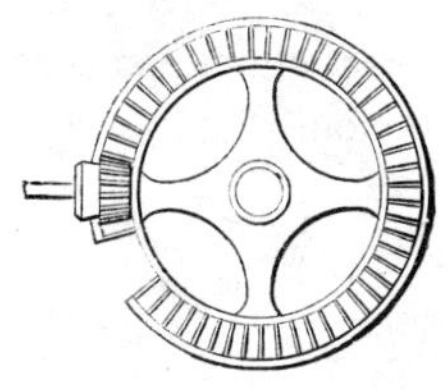

906. MANGLE MACHINE GEAR.—Large wheel is toothed on both faces. The pinion traverses from one side to the other of the geared wheel through the open space.

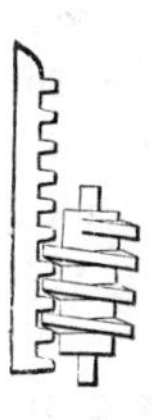

907. WORM SCREW RACK.—Continued motion of a worm screw meshed in a rack to produce motion in the rack from a fixed position of the worm, or with a fixed rack; the worm. sliding over a feather-key shaft, will drive sliding nuts holding a hoisting car or platform.

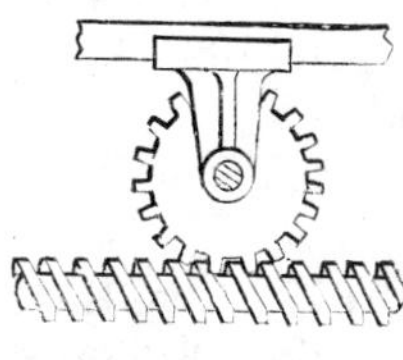

908. ROTARY MOTION of worm gear from an ordinary screw, or when the screw has great pitch, rotary motion of the screw may be obtained from the rotation of the worm-gear wheel.

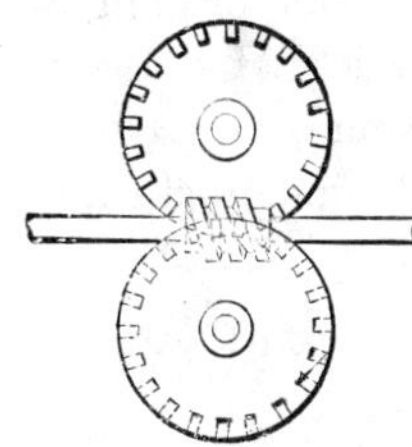

909. ADJUSTABLE FEED ROLLS driven by worm gear. The roll gears have elongated teeth on their face meshing with the screw on each side, which allows of considerable variation of the depth of feed.

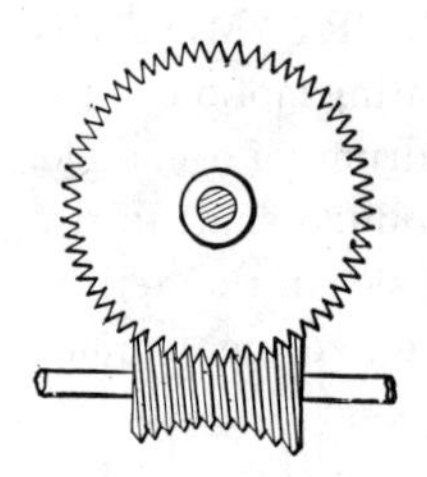

910. SAW-TOOTH WORM GEAR.—By the saw-tooth form of the teeth of both wheel and worm, and the concave pitch lines of the worm, a large area of contact is given to the teeth.

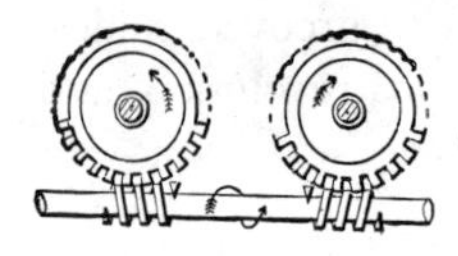

911. RIGHT- AND LEFT-HAND WORM GEAR for feed rolls or drums.

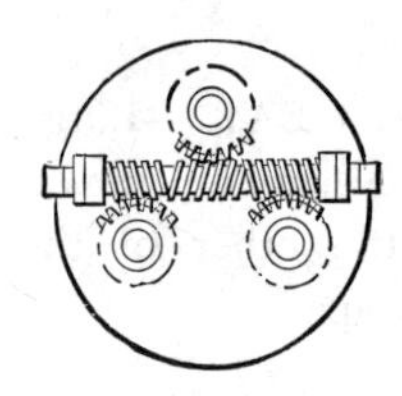

912. THREE-PART WORM SCREW, for operating three screw gears for a chuck, so that the jaws close in the same direction.

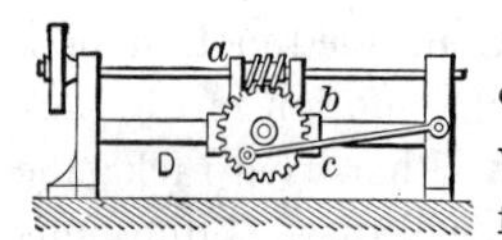

913. TRAVERSING MOTION from circular motion of a worm gear. The worm wheel and spur gear are relatively held by the frame *b*, and slide freely on shaft *a* and guide bar D. The feathered key on shaft *a* allows the worm to turn with the shaft, while the connecting rod *c*, by having one end fixed to the frame and the other end attached to a crank pin on the spur gear, gives the sliding frame with spur gear and worm a reciprocating motion equal to the throw of the crank pin.

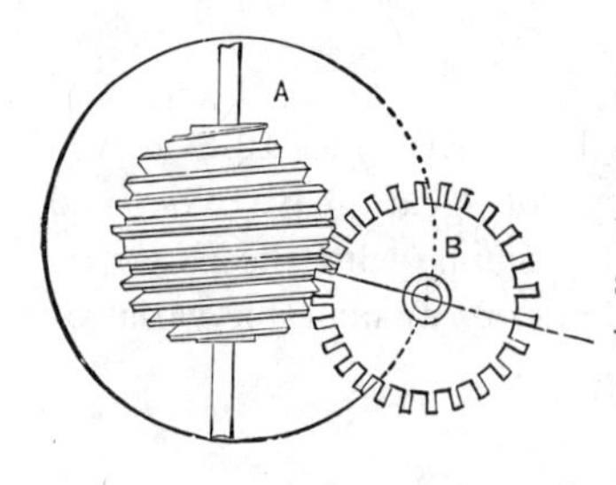

914. GLOBOID SPIRAL GEAR WHEELS.—The revolution of the globoid gear A gives a variety of differential motions to the spur gear B as it swings between the limits practicable with the globoid teeth.

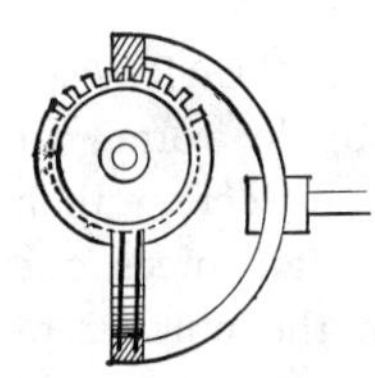

915. INTERNAL WORM-GEAR WHEEL for driving a spur-gear pinion.

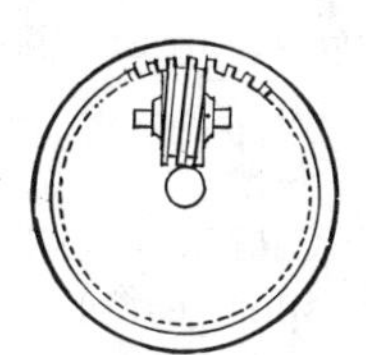

916. WORM-GEAR PINION to drive an internal spur-gear wheel.

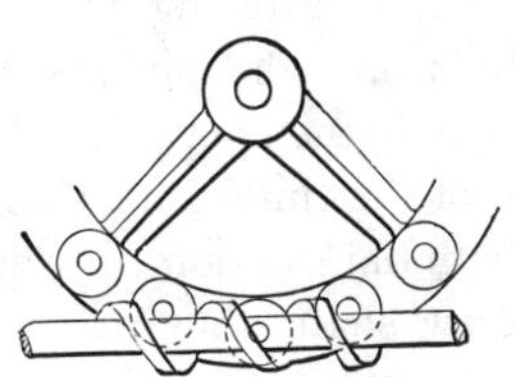

917. ANTI-FRICTION WORM GEAR.—The worm-wheel bearings are on friction rollers running on pins.

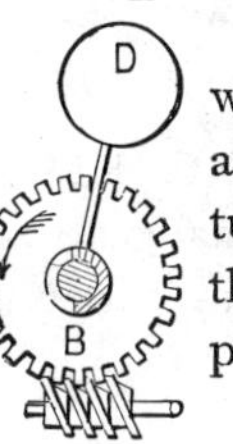

918. RELEASE ROTARY MOTION.—A worm wheel B, fast on a shaft to which is attached a loose arm and weight D, that carries the arm quickly over a half-turn, more or less, as required. The worm wheel lifts the arm and weight to beyond the vertical position by a pin in the shaft. See 919.

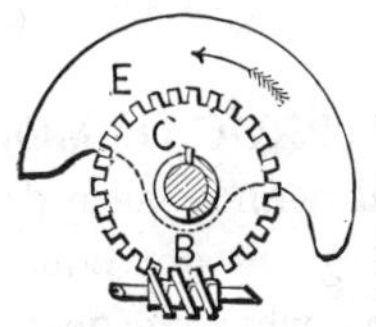

919. RELEASE ROTARY MOTION.—A sector weight E, moving loose on a shaft to which is fixed a worm wheel driven by a screw. The weighted sector is lifted by a pin resting in the half-section of the hub of the worm wheel until it reaches the point at which gravity carries it over a half-turn, more or less, as required.

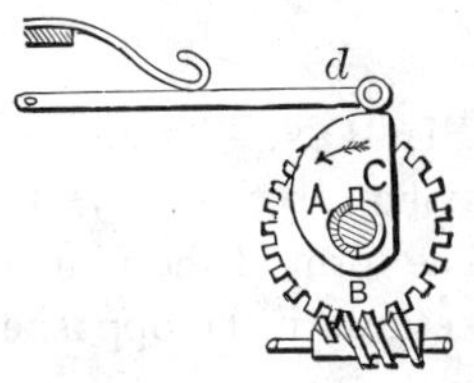

920. RELEASE CAM.—Uniform motion is communicated to the gear wheel, B, fixed on its shaft with a pin at C. The cam is loose on the shaft, with a stop section to meet the pin at C. The lever *d* has a spring and a roller on the cam. The lever *d* is raised by the motion of the cam until its straight face reaches the roller, when the lever falls suddenly, throwing the cam forward.

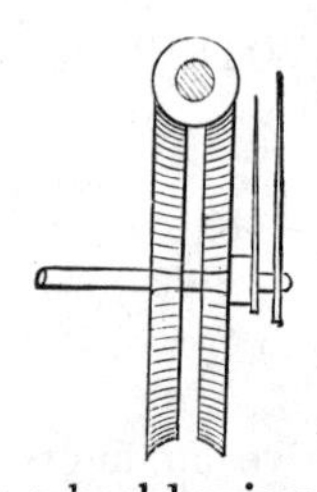

921. HUNTING TOOTH WORM GEAR, used for planetary or clock motion. The double worm-gear wheel may have one or more teeth in one section than in the other. The motion of the worm advances one wheel in proportion to the difference in the number of teeth. If the difference is as 100 to 101, the worm will make 10,100 revolutions for one revolution of the wheel having 101 teeth, over the wheel having 100 teeth.

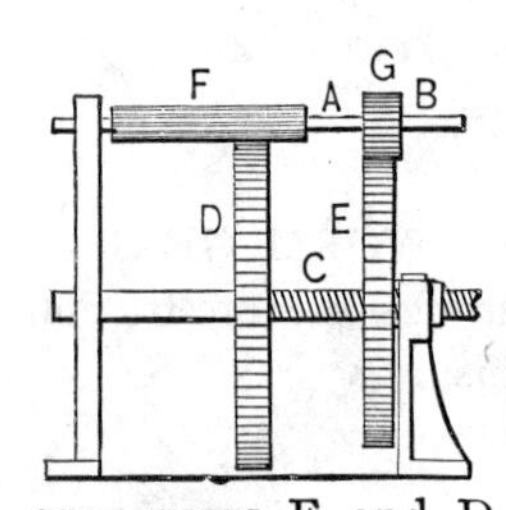

922. DIFFERENTIAL SCREW AND GEAR MOVEMENT.—The spur gear E is fixed to a screw hub or nut, revolving in the head of the short standard. The pinions F and G vary in size to match the spur gears D and E. The revolution of the pinions and shaft A, B produces a differential motion in the spur gears E and D. D is fixed to the screw shaft, thus driving the screw shaft forward at a very slow rate and great power.

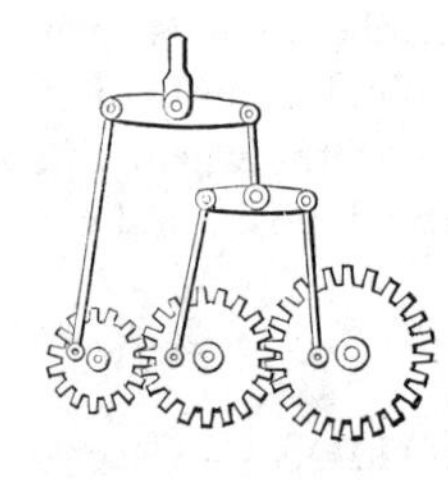

923. COMPLEX ALTERNATING RECIPROCAL MOTION from three unequal gears and two walking-beams giving an endless variety of motions to the terminal connecting rod.

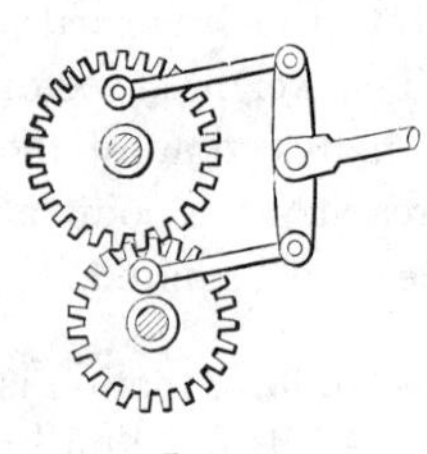

924. ALTERNATING RECIPROCAL MOTION from two crank gears and connecting rods to a walking-beam. When the gears are equal the motion of the rod is uniform; when the gears are unequal the motion of the rod is proportionally a varying differential one.

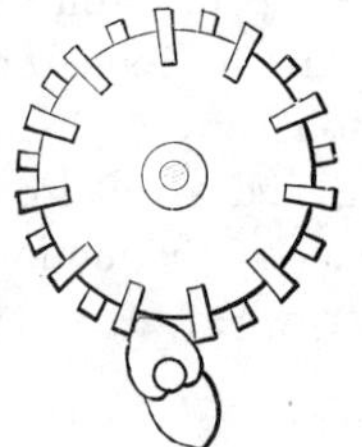

925. TWO-TOOTHED PINION.—Transmission of motion to a wheel having a series of teeth alternating on each side. The form of the pinion cam teeth locks the wheel teeth until the opposite cam catches its wheel tooth.

926. PIN WHEEL AND SLOTTED PINION, by which a change of speed is obtained by shifting the pinion along its shaft.

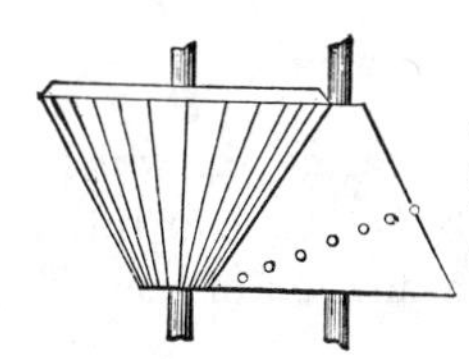

927. VARIABLE ROTARY MOTION from cone gears. A toothed cone is matched to an inverted cone with pin teeth to gear with the variable pitch of the cone teeth.

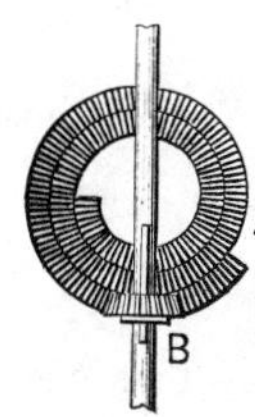

928. SCROLL GEAR.—Increasing velocity is obtained by a geared scroll plate with a sliding pinion on a constant speed shaft.

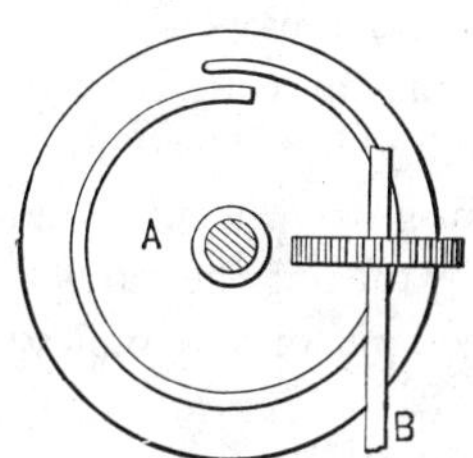

929. SPIRAL HOOP GEAR for special and slow transmission of power and motion to a shaft at right angle. One revolution of wheel A moves shaft B one tooth of its gear.

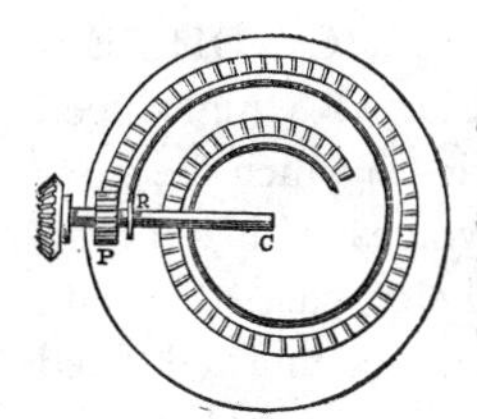

930. ACCELERATED CIRCULAR MOTION by a volute gear. The pinion P and guide disc R move along the feathered shaft C, following the rail guide, and returns by reversal of the motion of the driving shaft C.

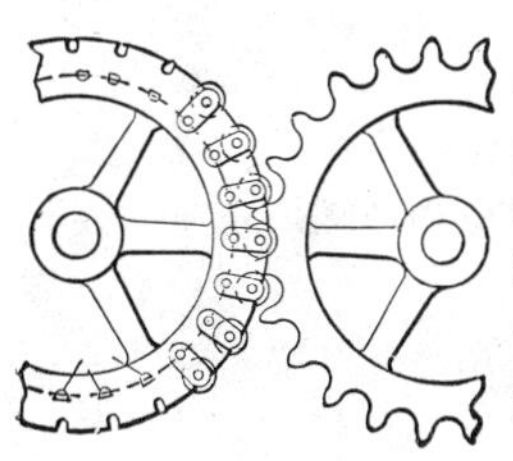

931. ROLLER-BEARING GEAR TEETH. —A double-flanged wheel with roller-bearing notches cut to the pitch of the wheel. The rollers are held in place by straps bolted to an inner circle of the flanges. The meshing wheel has its teeth skeletoned to make room for the roller teeth.

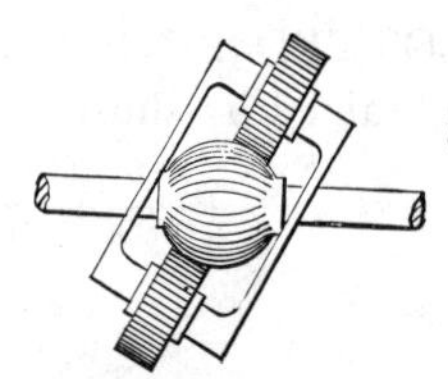

932. BALL GEAR with traverse pinions. Has a very limited traverse of the pinions.

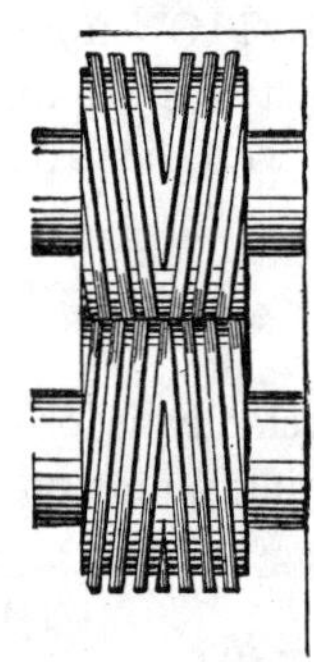

933. SPIRAL GEARING.—V gearing, in which the teeth are at a small angle with the plane of rotation, makes a perfectly silent transmission of power.

934. EXPANDING PULLEY.—The sectional rim pieces with their arms have a radial sliding joint on the hub arms, and are moved out or in by pins projecting into the spiral slots on the central spur-gear wheel. The movement of the wheel *c*, by turning the ratchet pinion *d*, moves all the sections of the pulley equally.

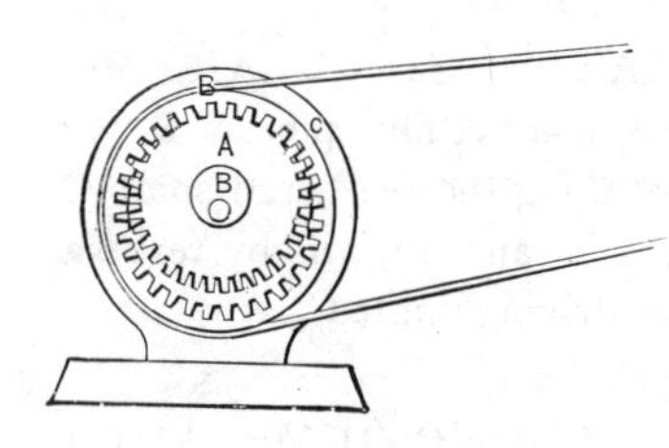

935. CONCENTRIC DIFFERENTIAL SPEED.—B, high-speed shaft and eccentric on which the slow-speed gear A revolves with a differential motion by being carried around in mesh with the larger internal fixed gear C, giving a slow motion to the belt pulley B.

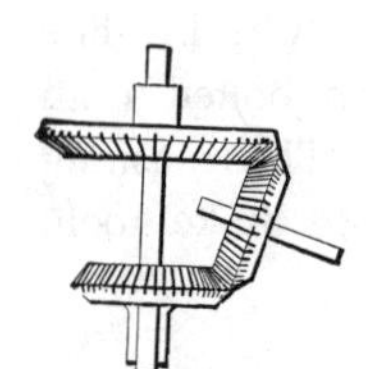

36. DIFFERENTIAL MOTIONS on concentric shafts by bevel gear.

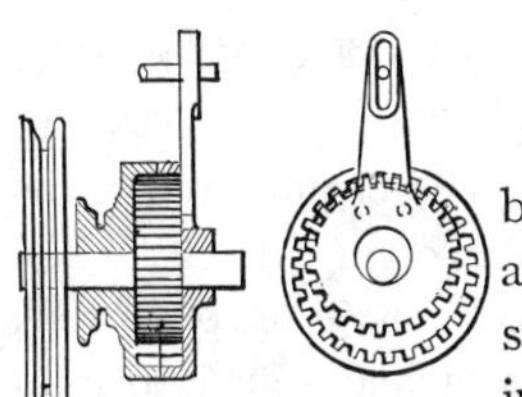

937. DIFFERENTIAL GEAR, section.

938. Plan. Used in differential pulley blocks. The cam and large grooved pulley are fixed on the shaft, the revolution of which swings the small gear in mesh with the larger internal gear, and rotating the large gear, shell, and the chain lift pulley, with a speed due to the difference in the number of teeth in the gears.

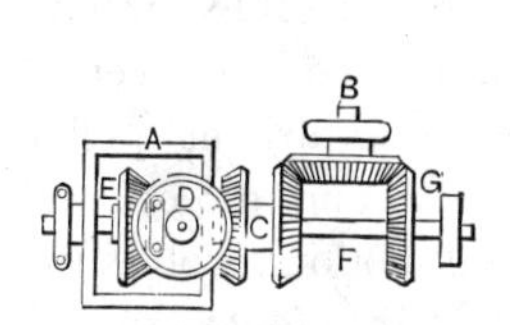

939. DOUBLING THE NUMBER OF REVOLUTIONS on one shaft. B, driving shaft and bevel wheel; G, bevel wheel fast on shaft F; C, two bevel wheels on hollow shaft running on shaft F; A, frame fast on shaft F, and carrying bevel wheel D; E, bevel wheel running loose on shaft F. Revolution of B gives contrary and equal motions to shaft F and double-bevel wheel C. Frame A and its bevel wheel D, revolving in contrary direction to C, doubles the speed of bevel wheel E.

940. MULTIPLE GEAR SPEED in line of shaft. Pinion E is fast on small shaft. B and C are fast together and pivoted on the *y* sleeve which runs loose on an extension of the small shaft gear; D is fast on the large shaft, and gear A is fixed to the bearing. Speed may thus be increased or decreased on a continuous line of shafting by the relative number of teeth in the different bevel gears. When the multiple of the teeth in A and C is less than the multiple of the teeth in B and D, the gear D and the large shaft will revolve forward or in the same direction as the pinion E. When the multiple of A and C is greater than the multiple of the teeth in B and D, the gear D and large shaft will revolve backward or in the opposite direction from the pinion E. The "Humpage" reducing gear.

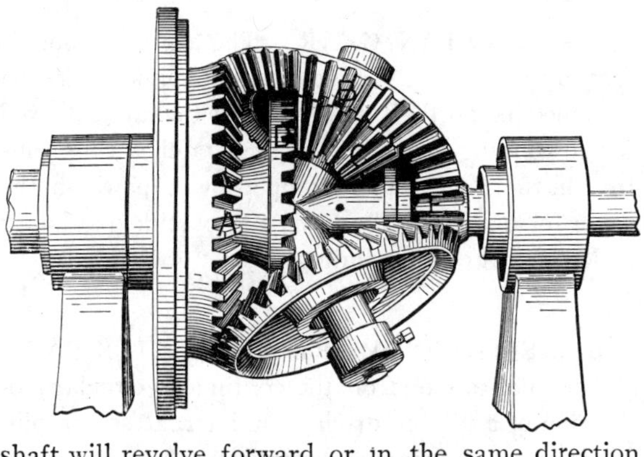

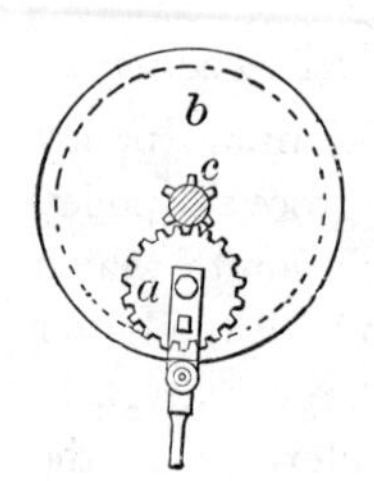

941. VARIABLE THROW TRAVERSING BAR, used in silk spooling. The spur gear *a*, to which is affixed a crank and jointed guide rod, turns freely on a pin fixed in the revolving disc *b*. The pinion *c* is fixed on a central shaft or otherwise, allowing the disc *b* and its attached spur gear *a* to revolve around the pinion *c*, thereby producing a varying throw of the guide rod for each revolution of the disc *b*.

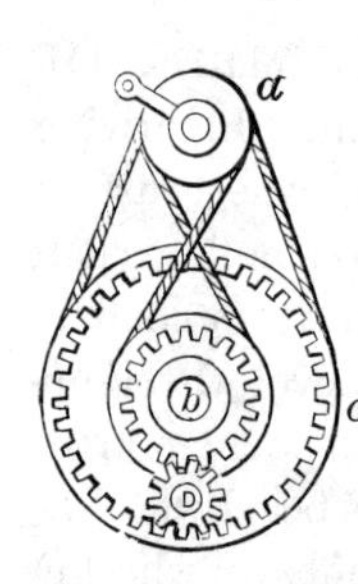

942. REVOLUTION OF A PINION around its own centre and also around the common centre of two externally centred gears. *a*, driving pulley with cross band to gear pulley *b*, and direct band to gear pulley *c*. The differential motion revolves the pinion D around its own axis and around its external axis *b*. A planetary motion.

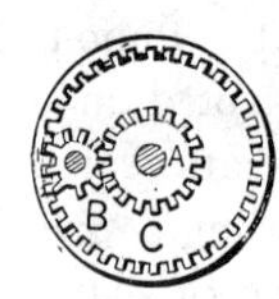

943. DIFFERENTIAL SPEED of two gears in different directions on the same shaft. A, driving pinion; B is geared to the shaft pinion A and to the internal spur gear C, and runs on a fixed journal.

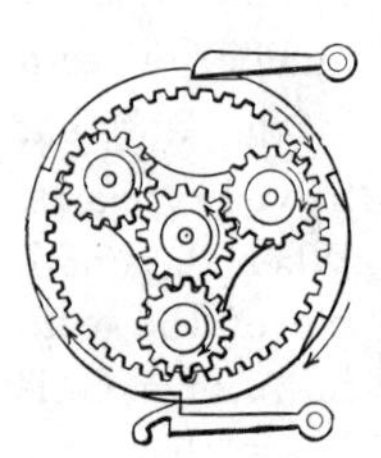

944. CAPSTAN GEAR.—The central pinion is fast to the shaft. The intermediate pinions are on a frame free on their own axes, but the frame is fixed to the winding drum. The gear ratchet ring runs free on the shaft, but is stopped by a pawl on the drum for quick speed and by the outside pawls for a slow speed of the winding drum.

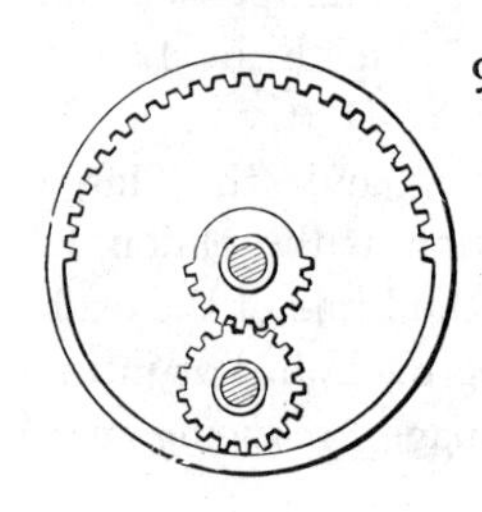

945. SLOW FORWARD AND QUICK BACK circular motion from the continuous circular motion of a pinion, driving an internal sector pinion and an external sector gear.

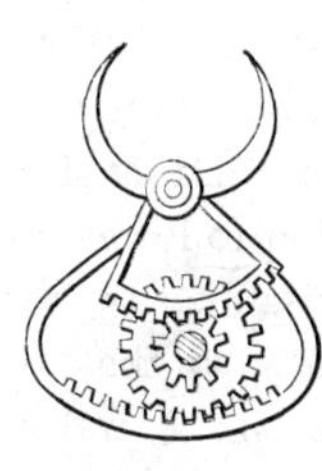

946. GEARED GRIP TONGS.—The radial distances of the sectors are in proportion to the diameters of the two pinions, which gives the jaws an equal motion, closing them with a strong grip by the action of the pinions.

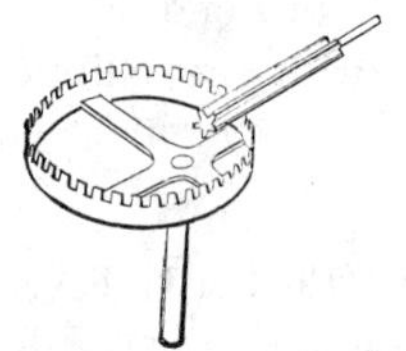

947. VARIABLE CIRCULAR MOTION by a pinion driving an eccentric crown wheel.

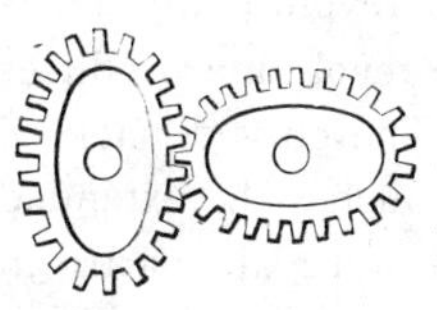

948. ELLIPTICAL SPUR GEAR for variable speed, the amount of which is governed by the relative lengths of the greater and lesser axes of the pitch lines of the elliptical gears.

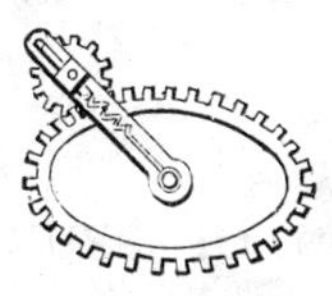

949. ELLIPTICAL GEAR WHEEL and pinion for variable motion of a pinion from uniform speed of an elliptic gear. The pinion shaft is carried in a box in a slotted arm and held in contact by a spring or other means.

950. IRREGULAR CIRCULAR MOTION from a circular gear train. A, the driver, with a spur gear B, attached eccentrically; C, a pinion, and D, the driven wheel. The three pinions are connected with pivoted arms; then the swinging of the spur wheel B around its eccentric axis will give a variable motion to the wheel D.

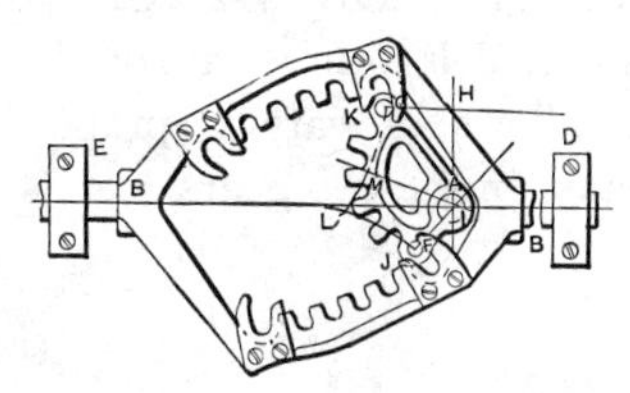

951. VARIABLE RECIPROCATING MOTION from a rotating spiral spur sector meshed in racks inclined to the line of motion. The pitch lines of the racks are curved to match the pitch line of the spiral sector. The pins F on the sector mesh with the stop jaws J, K, on the rack frame, alternately at each half revolution.

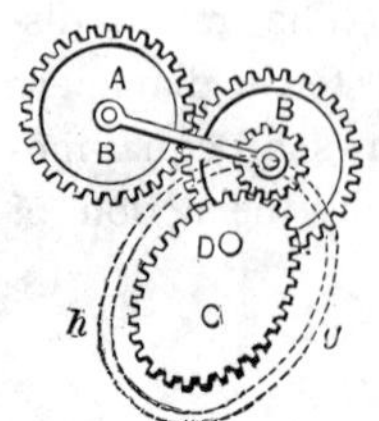

952. IRREGULAR CIRCULAR MOTION from an elliptically eccentric gear train. C is the elliptic driving wheel turning with the shaft at D. B is the intermediate gear with a pinion follower to the eccentric gear C. A and B are attached by an arm pivoted on their respective shafts, so that B rises and falls to keep the gear in mesh; *h* and *g* is an elliptical slot in a plate attached to C, in which the end of the shaft of B traverses to keep the pinion B in gear with the elliptic wheel C.

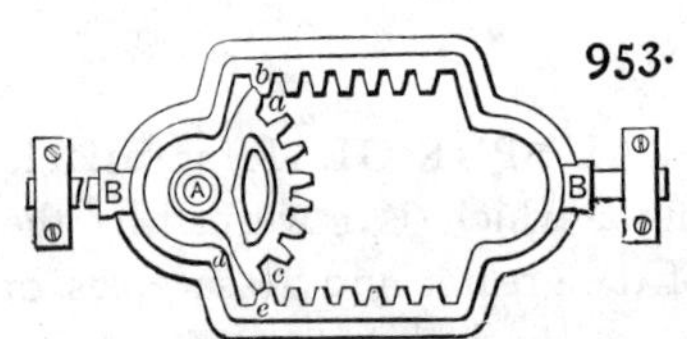

953. ALTERNATING RECTILINEAR MOTION by the revolution of a sector by which one revolution produces both motions. The curved back of the sector just touches the extended tooth of the rack frame at *d*, while the teeth at *e* and *b* are partly in mesh with the enlarged sector end teeth, thus preventing back-lash or locking of the teeth.

954. INTERMITTENT MOTION OF SPUR GEAR.—A is the driver. The pin J and the dog L are on the front side of the gear; the pin R and dog P are on the back. This class of gears may be made in varying proportion to suit the required stop motion of the gear B, A being the driver.

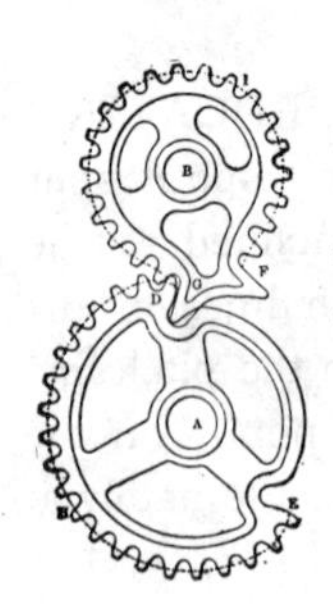

955. INTERMITTENT MOTION OF SPUR GEAR, in which the dogs G and F form a part of the driven gear B. This form allows of varying proportions of stop and speed motion in the two gears. A is the driving gear.

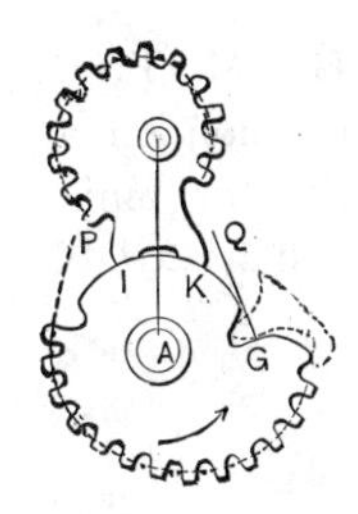

956. SPIRAL STOP-MOTION GEAR. — In this form a variable motion, in addition to the stop, is given to the driven wheel B. The dotted section at G shows the mesh of the spur, K, of the stop wheel. A is the driving wheel.

957. FAST AND SLOW MOTION SPUR GEAR, or a quick return when operating a slide motion by a crank. The driving gear B is composed of gear sectors of differential radius to correspond with the sectors of the driven gear A. The horns and studs M, L are back of the face of the gears and make contact with the studs N and O, on the sector wheel A, guiding the wheels to mesh in the other pair of sectors.

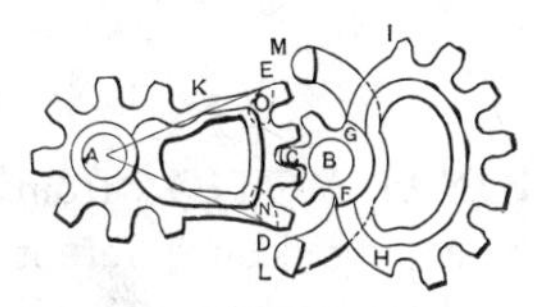

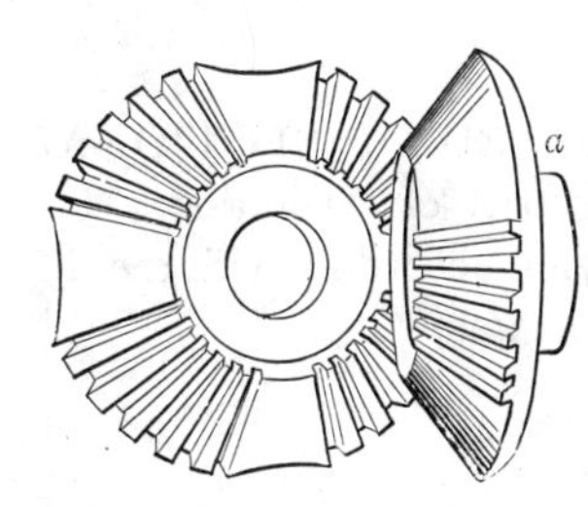

958. MITER INTERMITTENT GEARS.—The driver makes one revolution to one-quarter of a revolution of the driven gear. The blank part of the driving gear is milled down to the pitch line, and runs in the corresponding concave of the four-part driven gear.

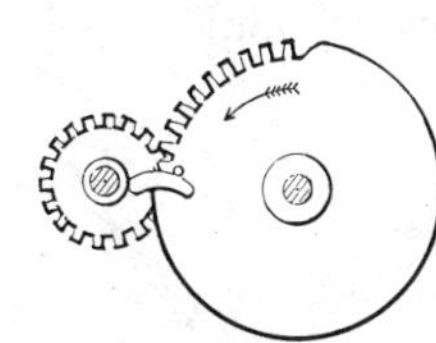

959. INTERMITTENT ROTARY MOTION, from continuous rotary motion of a sector-toothed wheel. Part of the pinion is cut out of the same curve as the smooth part of the wheel, and acts as a stop until the pin on the wheel strikes the arm on the pinion and guides the contact of the teeth.

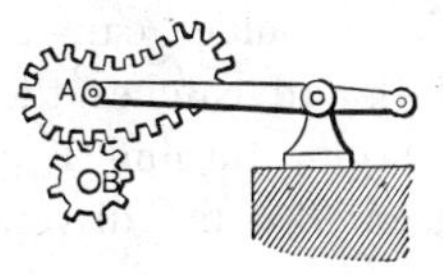

960. IRREGULAR VIBRATORY MOTION of an arm, A, from the rotary motion of a pinion, B

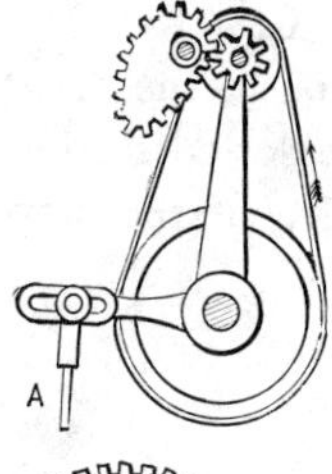

961. VARIABLE VIBRATING MOTION given to a rod, A, by the rotation of a pinion on an irregular-toothed wheel on a fixed axis; the pinion being carried by a bell-crank lever, with a variable slot adjustment.

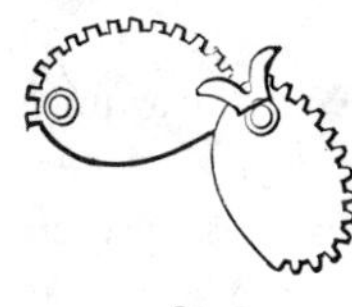

962. MOTION BY ROLLING CONTACT of elliptical half-geared wheels. The fork serves as a guide to enter the teeth into mesh.

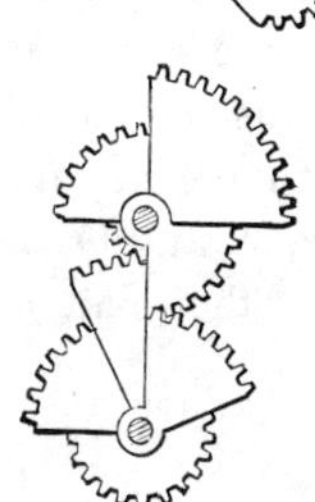

963. VARIABLE SECTIONAL MOTION from sector gears. The sectors are arranged on different planes, so that each pair shall be matched and all so adjusted that their teeth will mesh at their proper periods.

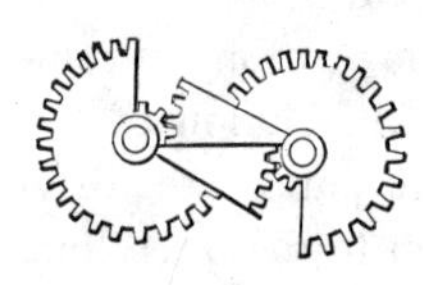

964. UNIFORM SPEED OF SECTIONAL SPUR GEAR during part of revolution. The motions varying suddenly according with the differential radii of the sectors.

965. SCROLL GEARING.— For increasing or decreasing the speed gradually during one revolution.

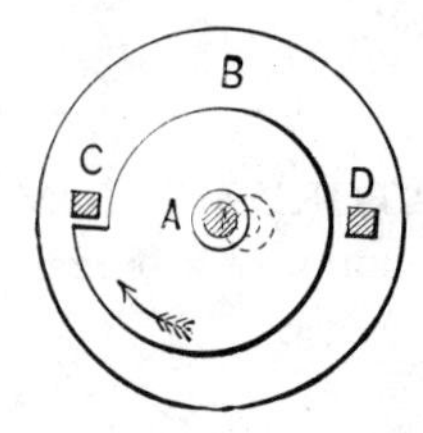

966. INTERMITTENT ROTARY MOTION from eccentric circular motion. C and D are pins concentric with wheel B. The shoulder cam A runs eccentric to the shaft of B, and catches the pin C or D at every revolution, turning B a half-revolution, and the reverse if B is the driver.

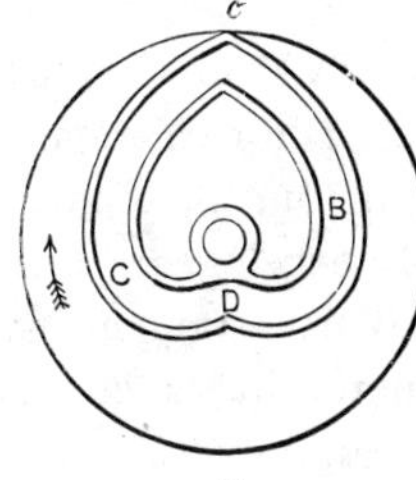

967. STOP ROLLER MOTION, used in wool-combing machines. The heart-shaped slot B, in the driving disc D, carries a roller stud, giving it a forward, backward, and stop motion. A pin on the back of the disc at *c* lifts the pawl G (Fig. 968), allowing it to pass over one of the spaces between the notches, and at the next half-revolution carrying the roller shaft forward one notch. The roller is attached to the shaft F, and by the action of the heart-shaped cam makes one-third of a revolution backward, and two-thirds of a revolution forward.

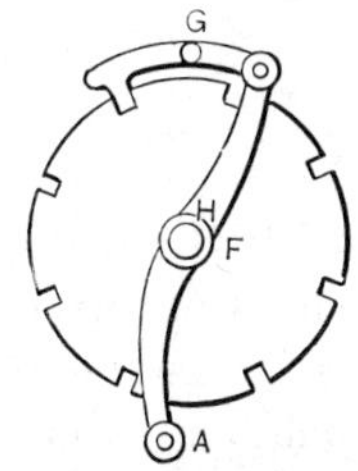

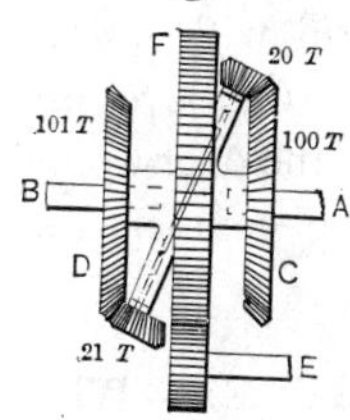

969. CHANGE GEAR MOTION.—The loose sleeve revolving freely on the concentric ends of the shafts A and B carries a diagonal shaft, with bevel pinions fast on each end; also a spur wheel, driven by the governing shaft and pinion E. Any motion given to the spur wheel F, by the pinion E, varies the speed of shaft B—A being the driving shaft.

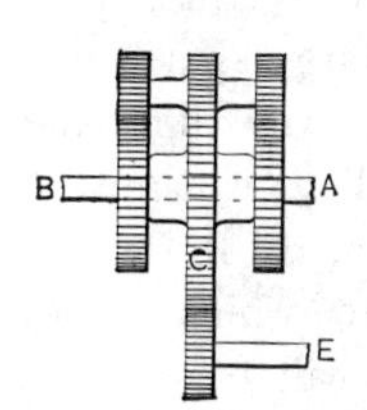

970. CHANGE GEAR MOTION, with spur gearing only. The spur wheel C moves freely on the disconnected shafts A and B. A short shaft and two fast pinions have a free motion near the periphery of the spur wheel C. The fast spur wheel on the shaft A is the driver. Any motion of the central spur wheel given by the shaft and pinion E varies the motion of the shaft B greater or less than the driving shaft, according to the direction of the governing motion.

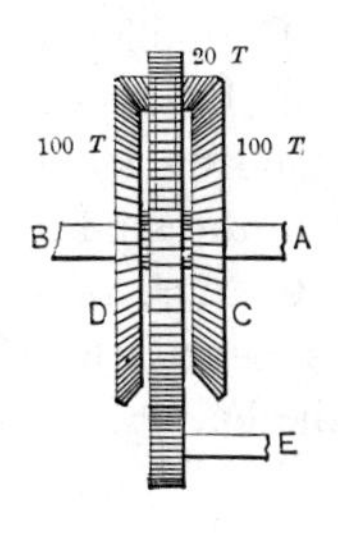

971. CHANGE GEAR MOTION.—The shafts A and B are disconnected, and carry a loose hub and spur wheel in which is pivoted the bevel pinion T. The bevel wheel C is fast on shaft A, and D is fast on shaft B. Any motion given to the central spur gear either way by the pinion shaft E varies the speed of the driven shaft B either faster or slower than the driving shaft A.

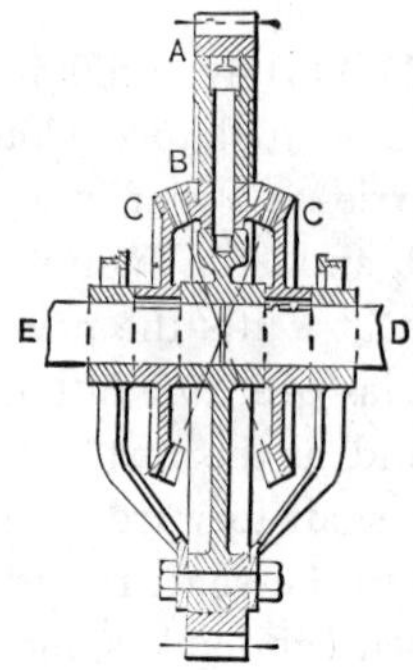

972. DIFFERENTIAL DRIVING GEAR. — Used on the driving shaft of motor carriages. A, is the driven gear from the motor; B, a bevel pinion pivoted laterally; C, C, bevel gears fast on the divided shaft E, D. This arrangement allows one wheel to advance in turning a curve, and at the same time to receive an equal impulse with the other wheel.

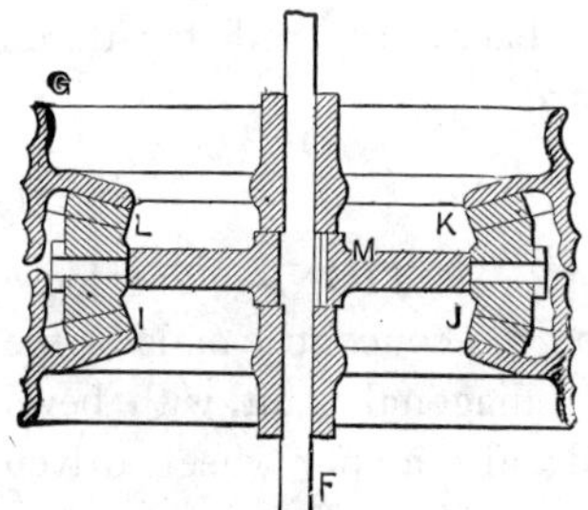

973. EQUALIZING PULLEY for rope transmission. The arm carrying the small bevel gears is fast on the shaft. The divided pulley runs loose on each side of the arm with its two bevel gears meshed with the bevel pinions. Any variation in the over-wound rope by tension will be compensated by the pinions.

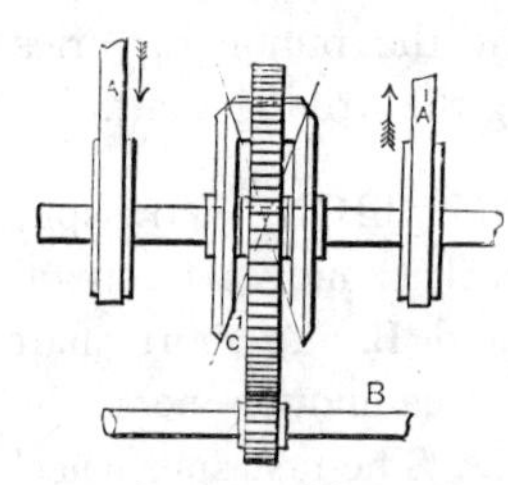

974. EQUALIZING GEAR. — When driven by the belts A, A′, with equal speed in opposite direction, the large spur wheel and shaft B do not move. Any difference in the speed of the belt pulleys will revolve the large spur wheel and shaft B forward or backward, according to which pulley runs fastest. The velocity of the large spur wheel will be one-half the difference of the pulley velocities. If B is the driving shaft, A and A′ may be the wheels of a vehicle.

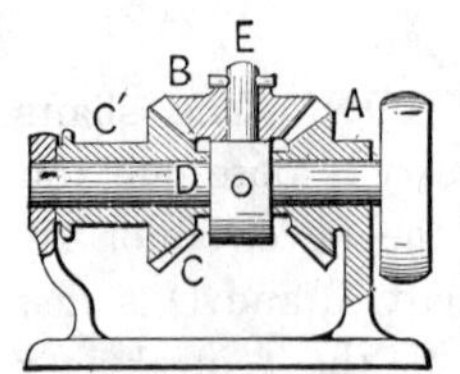

975. DOUBLING A REVOLUTION on the same shaft, "Entwistle's" patent. The pulley at A is the driver on the shaft D. The bevel gear at A is fixed. The stud E is fast on the shaft. The bevel wheel B revolves freely on the stud E. The bevel wheel C and its pulley C′ runs loose on the shaft. The revolution of the stud E with its bevel wheel around the fixed bevel wheel A doubles the speed of the bevel wheel C and pulley C.

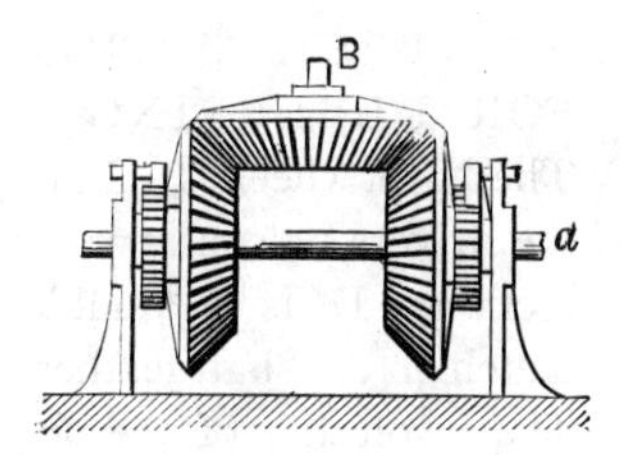

976. CONTINUOUS SHAFT MOTION from an alternating driving shaft. The ratchets fixed to the bevel gears on the shaft *a* are operated by pawls fixed to the shaft, the rocking of which revolves the bevel gear and shaft B in one direction.

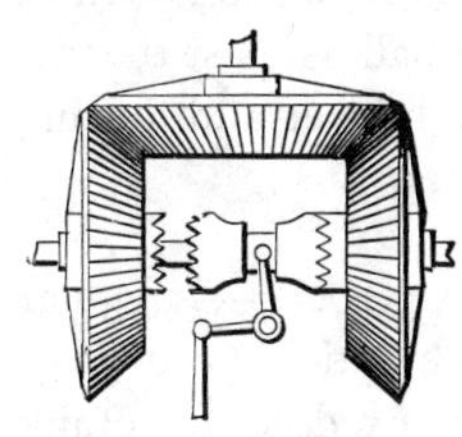

977. ALTERNATING MOTION of a shaft at right angles to a driving shaft by three bevel gears and double clutch. Bevel gears on clutch shaft run loose. Clutch slides on a feather or key, and is operated by a Y-lever and groove in clutch.

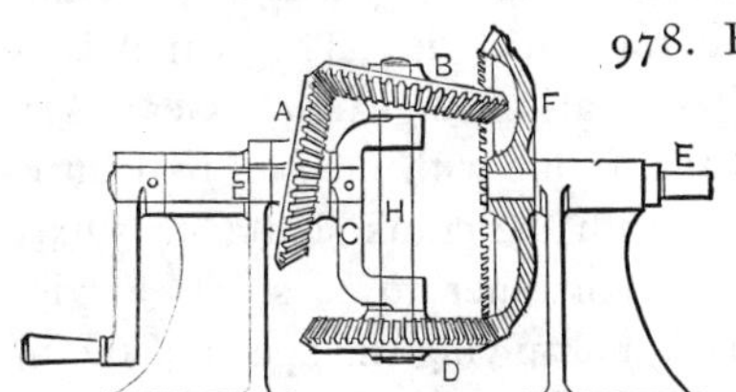

978. ECCENTRIC WHEEL TRAIN.—The elliptical bevel gear A is fixed to the crank shaft bearing at an angle to allow the elliptical bevel wheel B to clear the bevel wheel F. The arm C is fixed to the crank shaft; B and D are fixed to the shaft H, giving to the shaft E an irregular reversed motion from the motion of the crank shaft.

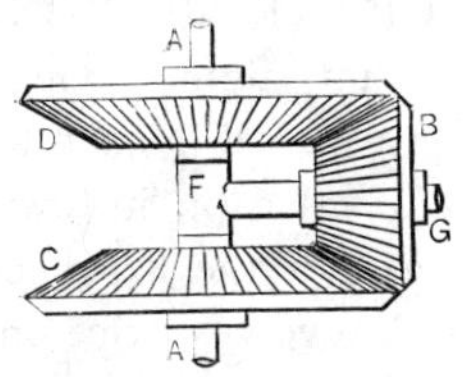

979. EPICYCLIC GEAR.—The arm F G is fast on the shaft A A. The bevel wheel is loose on the arm. The bevel wheels D and C are loose on the shaft A A. Differential motions of the two wheels C D will produce a rotation of the arm F G, around and with the shaft A, or, by making the arm loose on the shaft, a differential motion may be made by shaft and arm.

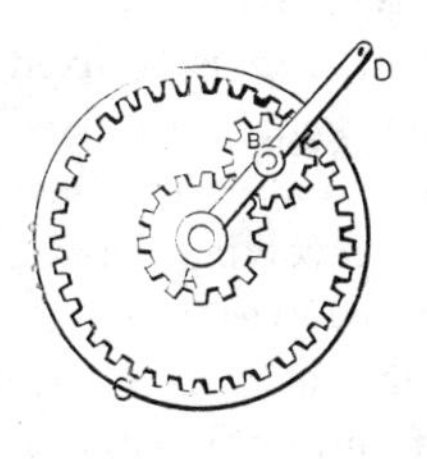

980. EPICYCLIC TRAIN.—If gear wheel C is fixed, and the arm D moved around its axis at A, the gear wheel B will have a retrograde motion, and the gear wheel A a faster motion in the direction of the motion of the arm. If wheel A is fixed, B and C will have unequal forward motions.

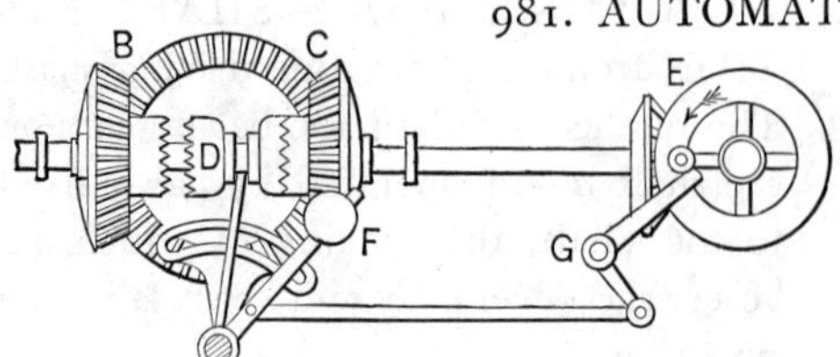

981. AUTOMATIC CLUTCH MOTION FOR REVERSING.—The bevel wheels B, C are the drivers in contrary direction; D is a double clutch on the shaft feather. The revolution of the pin on bevel wheel E moves the weighted ball F through the action of the bell-crank lever and connecting rod until the ball is past the vertical centre, when it falls over, striking the clutch lever and moving the clutch to the opposite or reverse wheel, and *vice versa*.

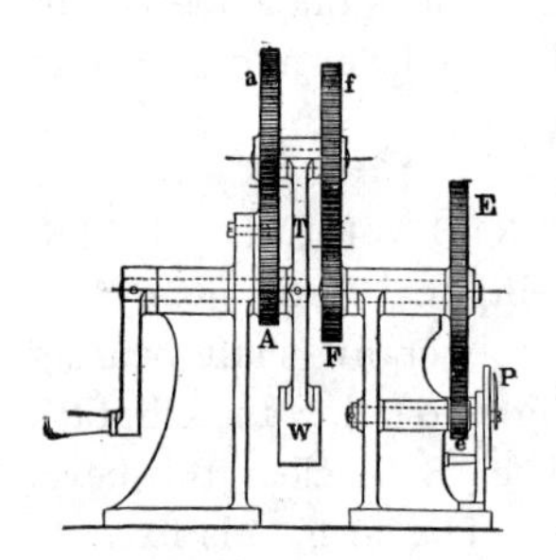

982. ECCENTRIC GEAR.—Irregular inverse motion from elliptic speed gear. The balanced arm T is fixed to the crank shaft and turns with it. The gear A is elliptical, as is also the gear *a*. Gear A is fixed to the frame with one of its centres coincident with the crank shaft; *a*, is fixed in the same manner to a shaft carrying the gear F, multiplying the speed of the index pointer P with a differential velocity, due to the eccentricity of the elliptical gears.

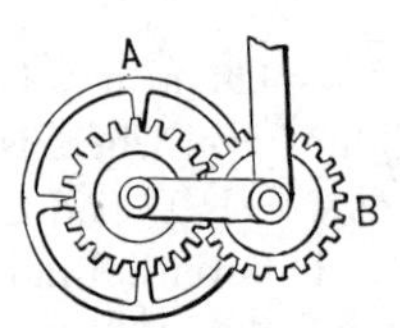

983. SUN AND PLANET CRANK MOTION, used by James Watt on the steam engine. Gear centres are held by connecting arm. B is fixed to connecting rod, and does not revolve on its own centre, but moves around the axis of the fly-wheel A with a slightly oscillating motion. The wheel A revolves twice on its axis to one circuit of B, or two strokes of the piston.

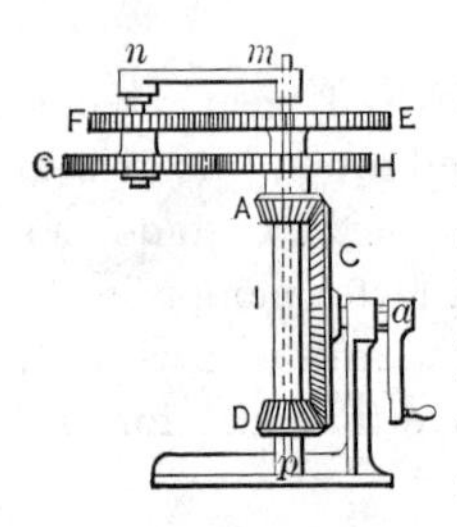

984. HIGH-SPEED EPICYCLIC TRAIN.—Bevel gear C is the driver; *m p* is a fixed shaft. Bevel pinion D and spur gear E are fixed on a hollow shaft. Bevel pinion A and spur gear H are fixed on a hollow shaft, revolving on the hollow shaft I. The arm *m n* revolves freely on the fixed shaft *m p*. The spur wheels F, G are fixed on a hollow shaft turning freely on the stud *n*.

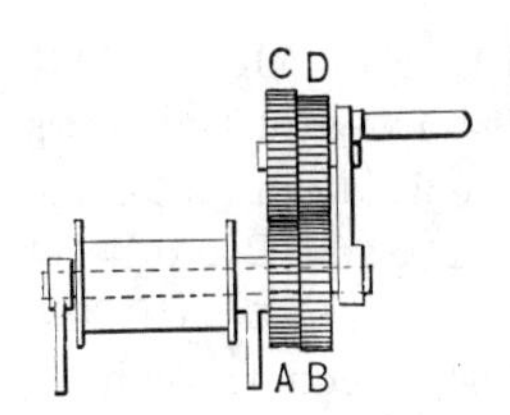

985. SUN AND PLANET WINDING GEAR.—A is fixed to the frame; B is keyed to the barrel shaft. The crank is loose on the shaft and carries a stud on which the differential gear C, D revolves.

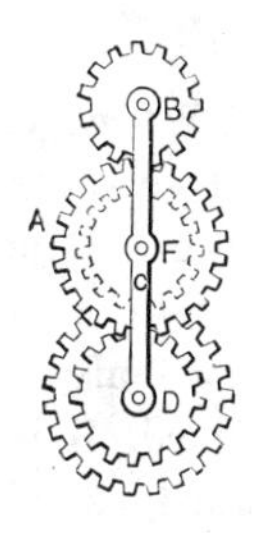

986. EPICYCLIC GEAR TRAIN.—C is the train arm which may revolve around its centre at F. The gear A is fixed. The pinion F is fast to a spindle. The gear B turns on its own axis as it revolves around the common centre. The two pinions at D are fastened together and revolve around their own axis, and also around the common centre at F. The centre spindle at F revolves with increased speed by the double gear at D. A great variety of motions may thus be made to represent planetary movement.

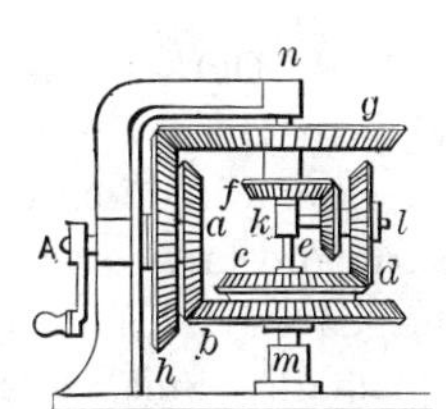

987. COMPOUND EPICYCLIC TRAIN, more curious than useful, but illustrating the changed conditions of gear motion. Gears *a* and *h* are fixed to the crank shaft. Gears *g* and *f* are fixed to a hollow shaft turning on the shaft *n m*. Gears *e*, *b* are fixed on a hollow shaft and turn on shaft *n m*. The arm *k l* is fast on and supported by shaft *n m*. Gears *e*, *d* are fixed on a hollow shaft and revolve on the arm *k l*, carrying the arm in a slow motion around the shaft axis *n m*. A variety of differential motions may be made by changing the relation of the fixed pairs.

988. PLANETARY MOTION applied to an apple-paring machine. The gear F is fixed to the crank shaft. The internal spur gear A is stationary. On turning the crank the pinion B rolls forward, carrying the arm T at half the velocity of the crank. The bevel gear A revolves with the crank, driving the spindle K with one-half the proportional speed due to the relative diameters of gears A′ and F′.

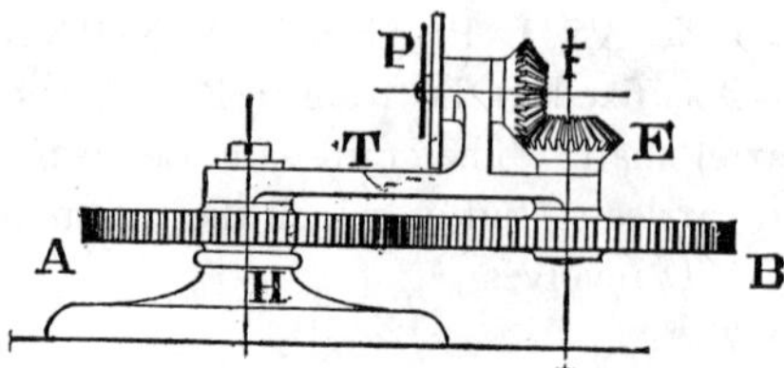

989. PLANETARY GEAR TRAIN.—The arm T revolves around the fixed gear A, on the stand H. The gear B and bevel gear E are fixed on a shaft and turn in one direction, giving a contrary motion to the bevel gear F and index hand P.

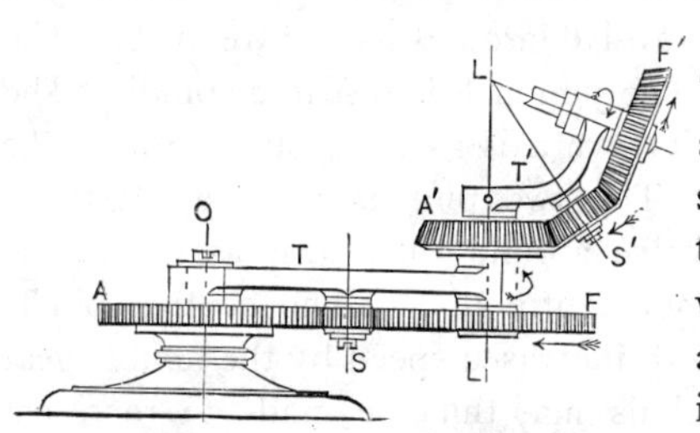

990. PLANETARY GEAR TRAIN. The arm T revolves around the fixed gear A. The small gear S reverses the motion of the gear F, to shaft of which the arm T′ is fixed. The arm T′ moves backward, carrying the pinion S′ around the bevel gear A′, which is fixed to the arm T, giving the bevel wheel F′ a forward motion, or in the same direction as the arm T.

991. "FERGUSON'S" MECHANICAL PARADOX.—The arm C revolves around the fixed gear A, carrying the gear B and train of wheels with it. The gear B revolves in the same direction as the arm and carries with it the gears I, G, E fixed to its shaft. Small differences in the number of teeth of each pair of gears gives a differential reverse motion to the gears K, H, F.

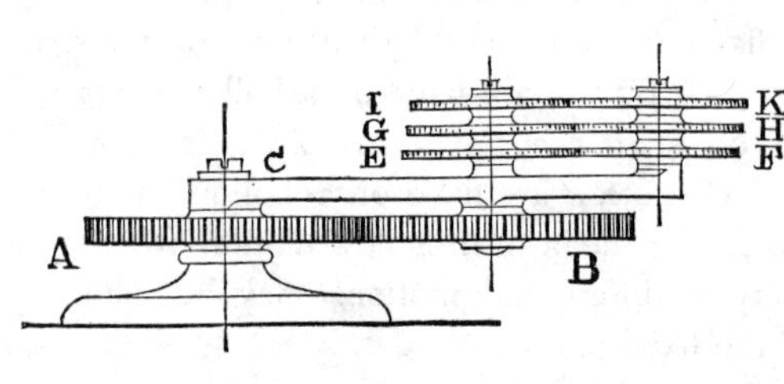

992. "FERGUSON'S" MECHANICAL PARADOX, a curious property of an epicyclic train. A is a central fixed axle and gear wheel, around which the arm C D revolves, M, a wide-gear wheel loose on a pivot set in the arm C D; N, a pivot also set in the arm and carrying three gears with a differential number of teeth, say, varying by one or two teeth. On moving the arm C D to give motion to the train, the three wheels E, F, and G will have a differential motion, which was a paradox to persons not understanding the secret.

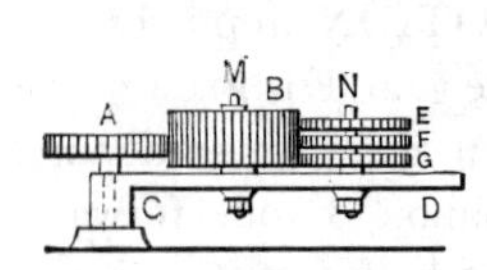

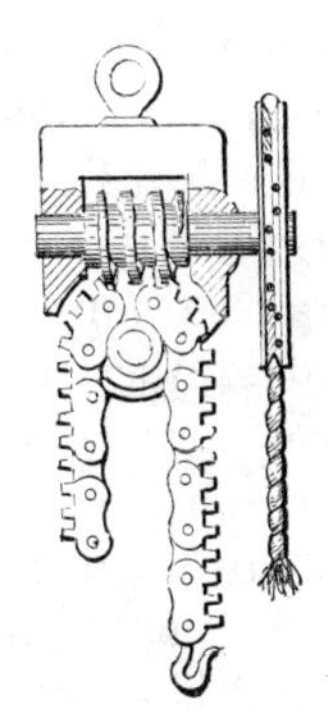

992a. LINK CHAIN HOIST.—A novel form of light weight hoists over lathes and planers. A screw gear working in a tooth chain.

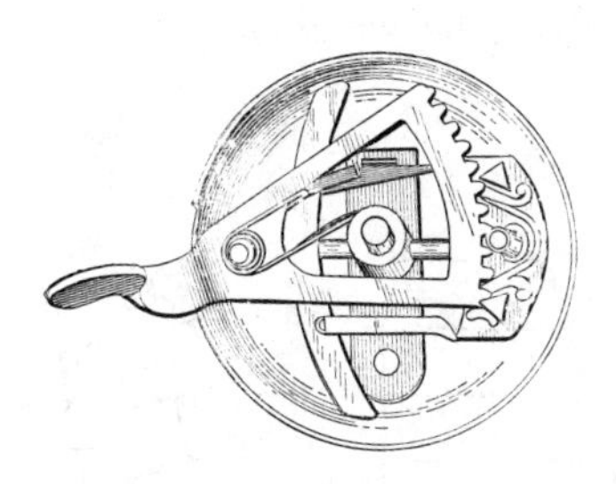

992b. BICYCLE SIGNAL BELL.—A gear sector lever operates a pallet vibrator, which is held in a central position by a spring. The sector is also held ready for a signal stroke by a spring.

992c. MULTIPLE SPEED GEAR, or paradox box.—The bevel gears A, A′, A″, are fixed to box and studs 5 and 6. C, C′ are bevel gears or shafts that carry the arms and pinions 2, 1. Each set of gears doubles the speed of the one before it, giving a final speed of 8 times the speed of the crank.

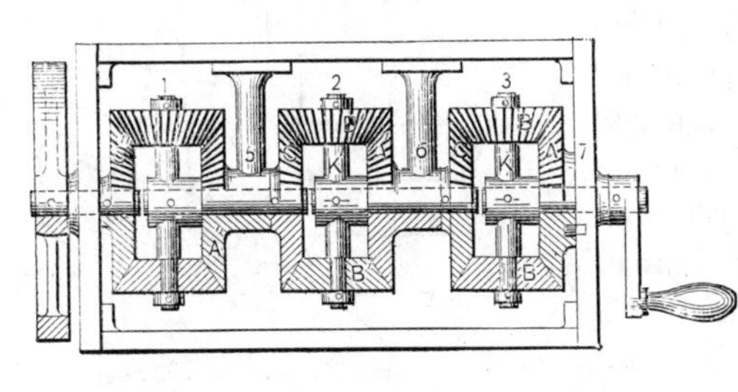

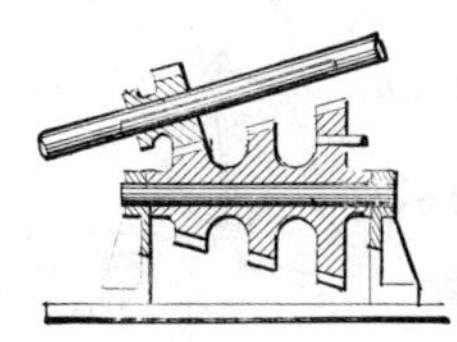

992d. CHANGEABLE MOTION GEAR.—Two or more changes of motion by bevel gear may be made by moving a pinion along a feathered shaft at an angle with the change gear shaft.

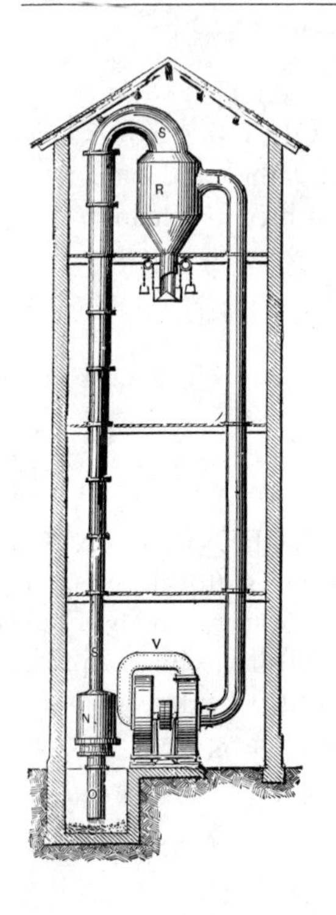

992*e*. PNEUMATIC GRAIN ELEVATOR.—V is a compound suction fan; T, pipe to receiver R; R, a receiver with a wire gauze screen to allow dust to be separated from the grain and carried off through the fan; S, lifting pipe of conical form; N, regulator with a rubber diaphragm to allow the foot nozzle O to regulate the proportion of air and grain. At the bottom of the receiver R is a conical valve to discharge the grain when it overbalances the weights.

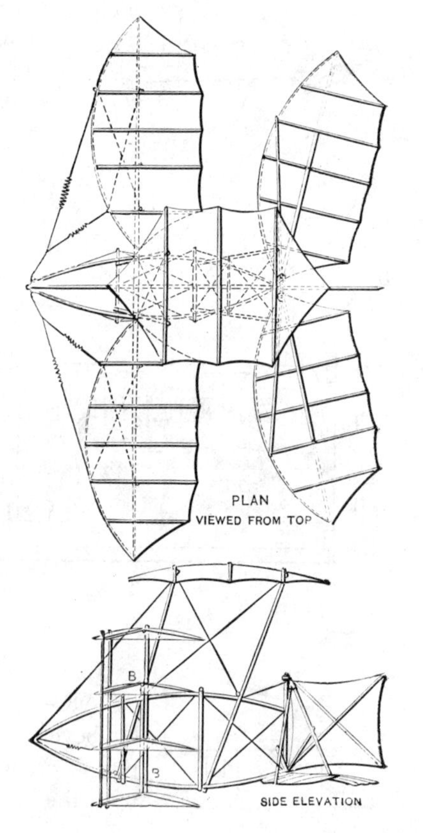

992*f*. FLYING MACHINE.—One of the many forms of experiments in aerial navigation.

The most that has yet been done in the line of human flight is to glide from a hill top or cliff and alight with possible safety.

Section XII.

MOTION AND DEVICES CONTROLLING MOTION.

RATCHETS AND PAWLS, CAMS, CRANKS, INTERMITTENT AND STOP MOTIONS, WIPERS, VOLUTE CAMS, VARIABLE CRANKS, UNIVERSAL SHAFT COUPLINGS, GYROSCOPE, ETC.

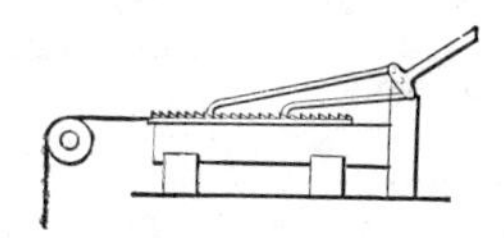

993. RATCHET BAR LIFT.—The vibration of a double-bell crank lever gives a ratchet bar and attached rope great power for lifting or tightening a binding device.

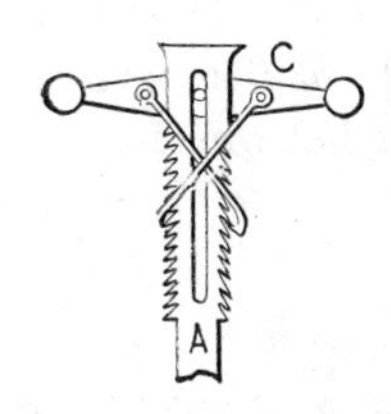

994. RATCHET LIFT.—Vibrating lever C, operates two hooked pawls on the ratchet bar A and lifts the bar. The slot serves as guide. The other member may be a suspension or standard attachment. Much used in ratchet jacks and stump-pullers.

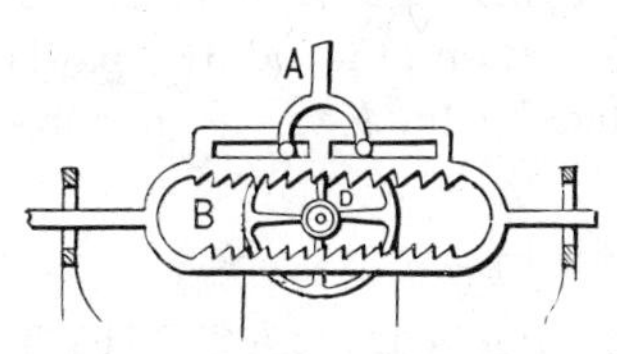

995. RATCHET GOVERNOR, for water-wheels or other prime movers. The pin cam is in constant revolution. The double-ratchet rack B, held clear of the revolving pin at normal speed, is raised or lowered by the action of the governor on the suspender A. The extension rods of the ratchet frame operate a gate or valve.

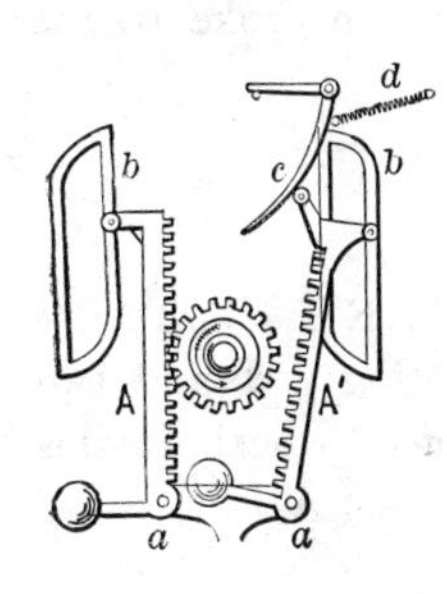

996. ROTARY MOTION, from reciprocating motion of two racks alternately meshing with a gear wheel. Racks are pinioned at *a*, *a*. The curved slots *b*, *b* guide the racks out and into gear. The bell-crank lever *c* and spring *d* serve to disengage the rack at the end of the up-stroke.

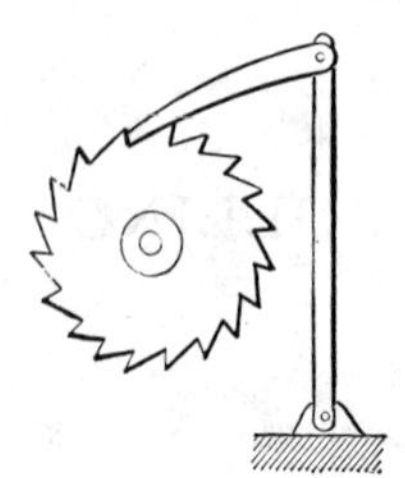

997. INTERMITTENT CIRCULAR MOTION, from a vibrating arm and pawl acting upon a ratchet wheel.

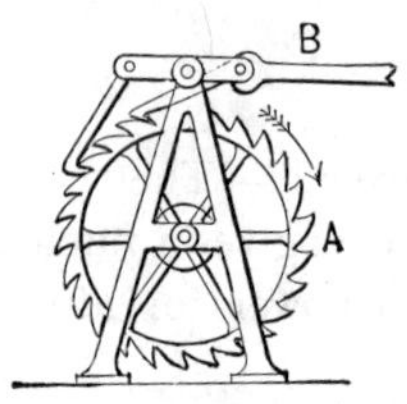

998. INTERMITTENT ROTARY MOTION of a ratchet wheel by lever and hook pawls.

B, vibrating lever.

A, ratchet wheel.

999. DOUBLE-PAWL RATCHET.—The vibration of the lever *a*, with its pawls *b*, *c*, imparts a nearly continuous motion to the ratchet wheel.

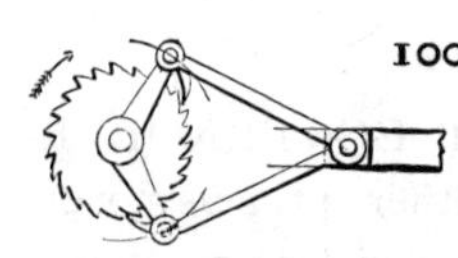

1000. CONTINUOUS FEED OF A RATCHET by the reciprocating motion of a rod, two pawls on arms, and pivoted by links to the reciprocating rod.

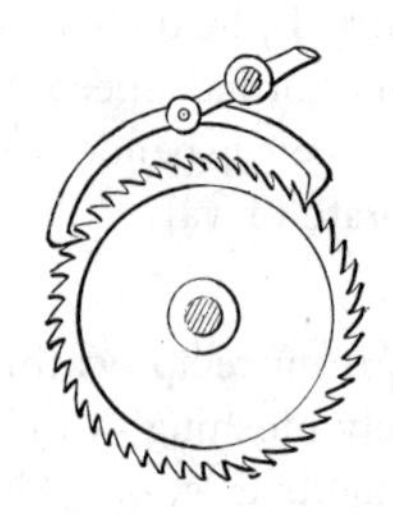

1001. DOUBLE-PAWL RATCHET WHEEL.— The lever lifts the pawls, one of which moves the ratchet wheel at up-stroke by one pawl, and again at the down-stroke by the other pawl.

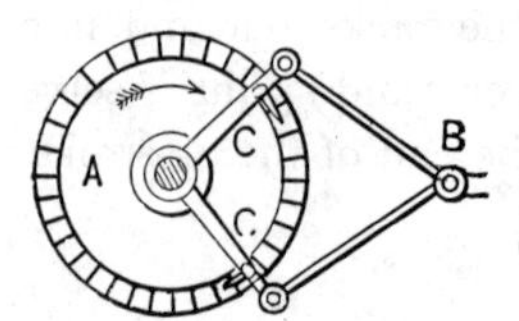

1002. INTERMITTENT ROTARY MOTION, from a reciprocating rod and two pawls, acting on a ratchet-faced wheel. Arms C, C are loose on shaft of wheel A.

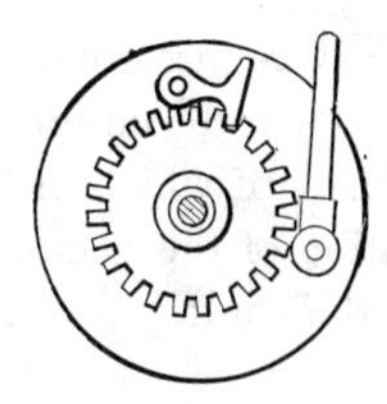

1003. INTERMITTENT CIRCULAR MOTION.—Reversible by throwing over the double pawl. Operated by a reciprocating rod attached to the disc carrying the pawl.

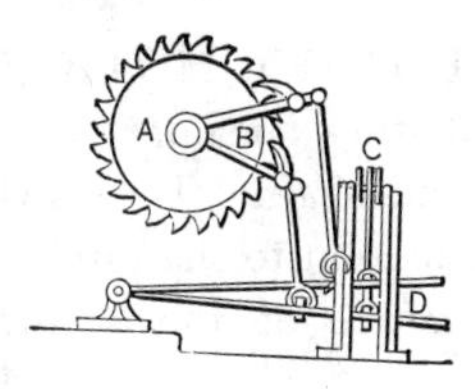

1004. RATCHET INTERMITTENT MOTION, by the operation of treadles. Pawl levers and pawls are operated through connecting rods to levers or treadles, the motion of which is made uniform by the strap and pulley attachment C.

1005. INTERMITTENT CIRCULAR MOTION—Reversible by throwing over a double pawl on the vibrating bell-crank lever. A feed motion for planing machines.

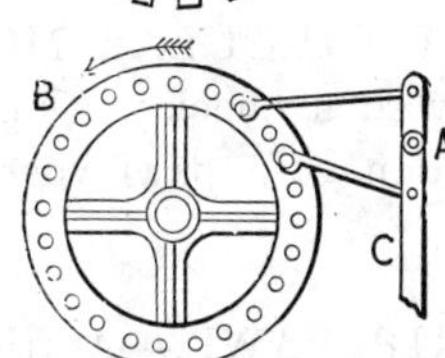

1006. INTERMITTENT ROTARY MOTION of a wheel by vibrating levers and pawls.

B, pin-tooth wheel.

A, vibrating lever.

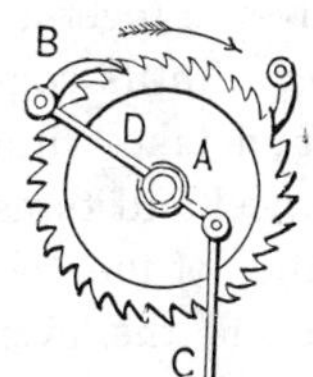

1007. INTERMITTENT CIRCULAR MOTION from a reciprocating rod. Motion varied in the ratchet wheel A by the number of teeth swept over by the pawl B.

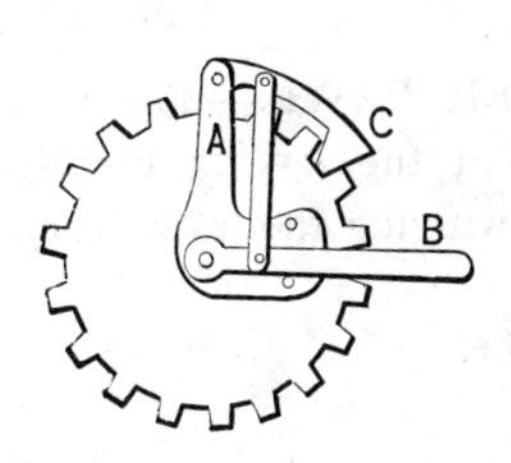

1008. PAWL LIFT.—By moving the lever between the pins in the bell-crank pawl arm, the pawl is lifted and moved to new position without dragging over the teeth of the ratchet wheel.

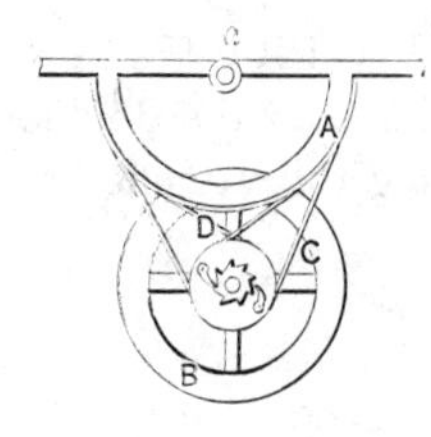

1009. OSCILLATING MOTION into rotary motion by a straight and crossed band running on two ratchet pulleys, the ratchets of which are fast on the shaft. Each oscillation of the sector lever gives a forward motion to the shaft.

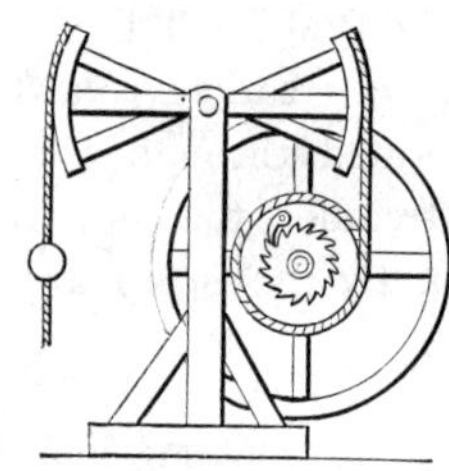

1010. CONTINUOUS ROTARY MOTION by stop ratchet and oscillating beam. The ratchet wheel is fixed on the shaft. The pawl wheel runs free and gives motion to the ratchet and shaft at every other stroke of the sector beam.

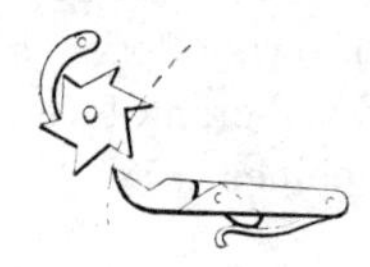

1011. INTERMITTENT MOTION of a ratchet by the oscillation of a knuckled joint tappet arm. The spring keeps the tappet extended on the forward stroke, and allows it to run over the tooth of the ratchet on its return.

1012. INTERMITTENT CIRCULAR MOTION of a ratchet wheel with a check pawl by the continuous circular motion of a pawl wheel.

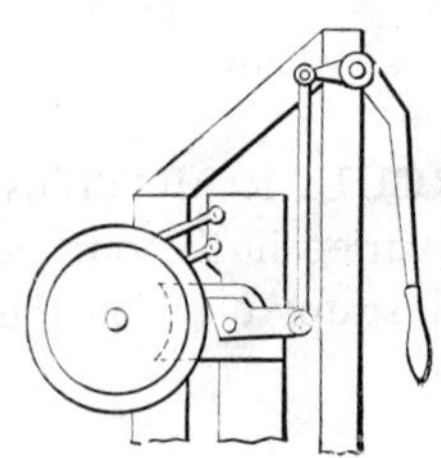

1013. WINDLASS GRIP PAWL.—A friction pawl and rim grip piece are pivoted together so that by the vibration of the lever with its connecting rod the grip pawl drops and takes firm hold of the rim of the windlass wheel and turns it with the power due to the distance of the rod attachment from the wheel centre and the lever proportions. The stop pawls act upon a separate ratchet wheel.

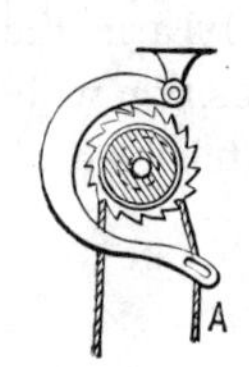

1014. RATCHET AND LEVER PAWL.—The pawl drops into the ratchet by gravity of the lever. Pulling the cord A unhooks the pawl by swinging the lever back.

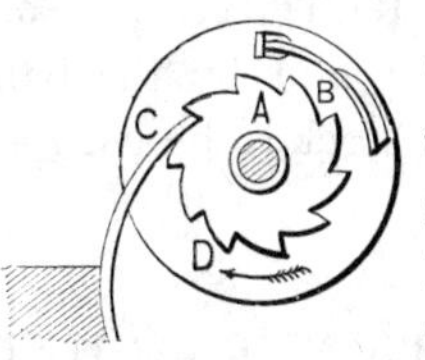

1015. INTERMITTENT ROTARY MOTION by ratchet and springs. D, driving wheel with a bent spring at B. A spring at C acts as a fixed pawl. In revolving the wheel D, the spring B lifts the spring C from the ratchet, and is itself pressed into the teeth and carries the ratchet around one tooth, when the shoulder on the spring B releases the spring C and allows it again to lock the ratchet.

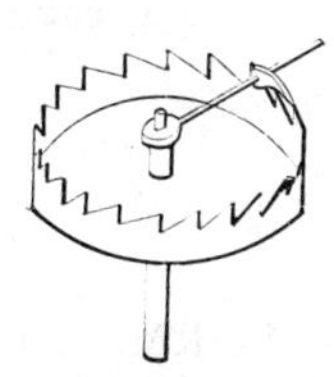

1016. INTERMITTENT MOTION of a ratchet crown wheel from the reciprocating motion of a lever and pawl

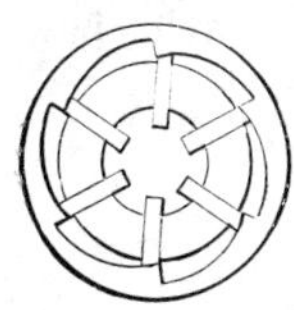

1017. INTERNAL MULTIPLE CAM for operating several slides for internal grip, or for expanding the cutters of a die stock.

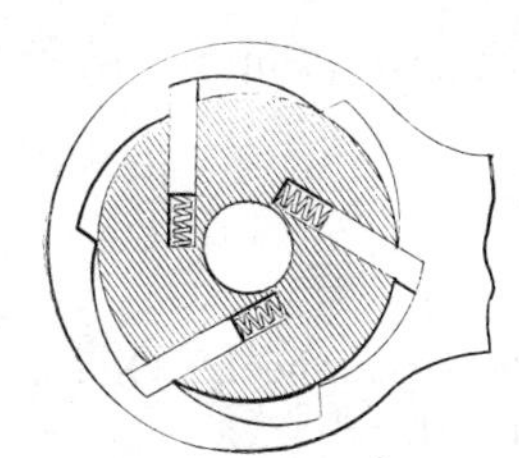

1018. RATCHET HEAD with spring pawls.

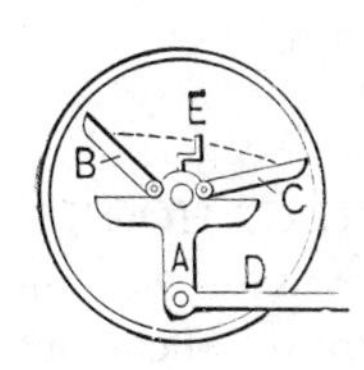

1019. INTERMITTENT CIRCULAR MOTION from oscillating motion of a lever by friction pawls. The crank E and its cord connecting with the pawls throw one or the other pawl out of lock for reversing the motion.

1020. RECIPROCAL CIRCULAR MOTION from rectilinear motion of a nut on a quick thread. The reciprocating or Persian drill stock. The screw is swivelled in the head of the stock, allowing a free movement of the drill by the motion of the nut.

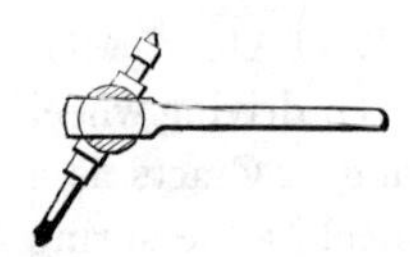

1021. BALL SOCKET RATCHET.—The pawl is within the arm socket, and by the ball ratchet form allows the drill stock to be used at an angle.

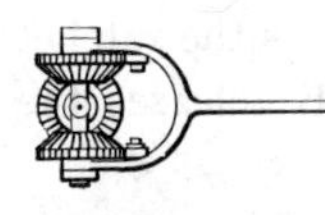

1022. CONTINUOUS MOTION RATCHET from an oscillating arm. Three bevel gears, two of which have ratchets with pawls on opposite sides, so that there is a forward motion to the spindle at each stroke of the arm.

1022 a. Elevation.

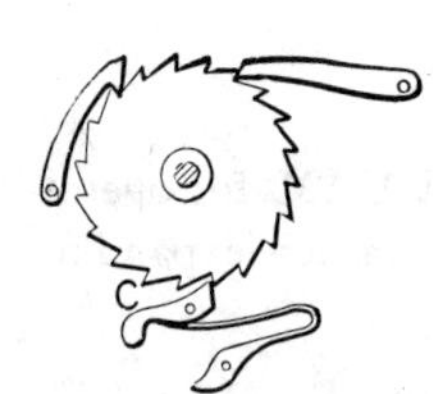

1023. STOPS OF VARIOUS FORMS for a ratchet wheel. Hook and straight gravity pawi and a spring pawl.

1024. STOPS for a spur gear. Slip pawls.

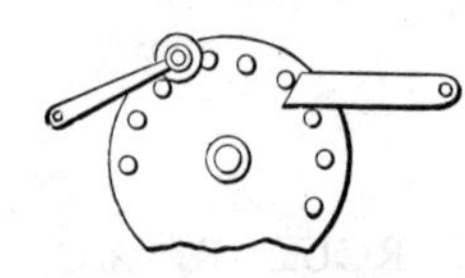

1025. STOPS for a lantern wheel. One a latch stop, the other a roller stop.

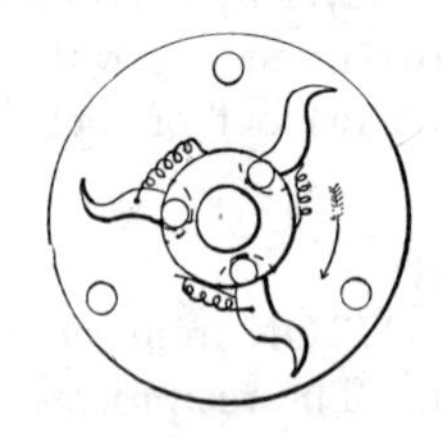

1026. SAFETY CENTRIFUGAL HOOKS.—Hooks are retained by springs until the centrifugal force of excessive speed throws them out to catch the pins in the fixed plate.

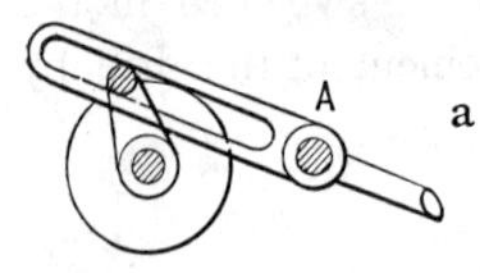

1027. CRANK MOTION for quick return of a lever. A, fulcrum of lever.

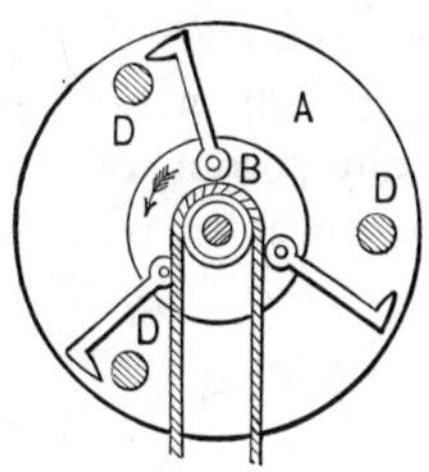

1028. CENTRIFUGAL SAFETY CATCH for hoisting drums. The studs D, D, D are fixed to the hoisting drum frame. B is a flange fast to the drum shaft and to which is pinioned the safety hooks. At ordinary speed of the drum the hooks hang back so as not to touch the studs. An unusual acceleration of speed throws out the hooks to catch on the studs.

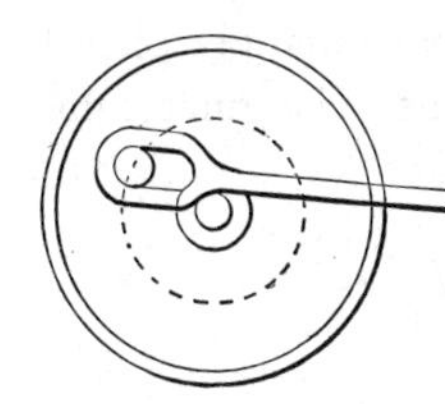

1029. STOP MOTION from a wrist or crank pin. The relative amount of stop and motion depends upon the diameter of crank-pin circle and length of the connecting-rod slot, plus the diameter of crank pin. Used in brick machines.

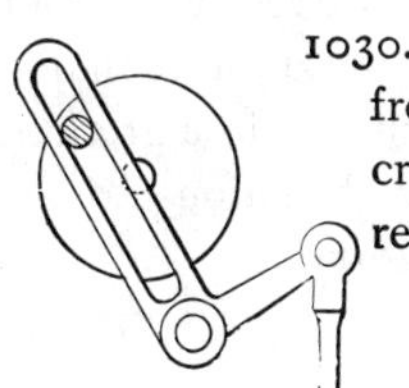

1030. VARIABLE RECIPROCATING MOTION from the circular motion of a wrist pin on a disc crank. The pin sliding in the slot makes a quick return of the bell crank and connecting rod.

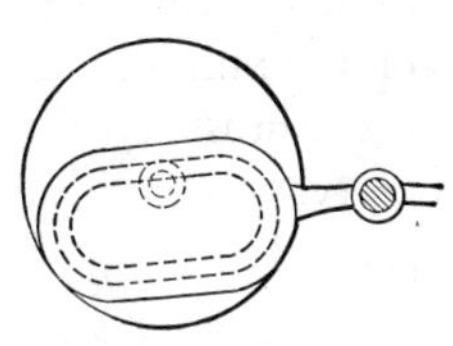

1031. IRREGULAR ROCKING MOTION in an arm having an endless groove of any required shape, with the radius of the longitudinal axis equal to the radius of the pin. Pin not shown.

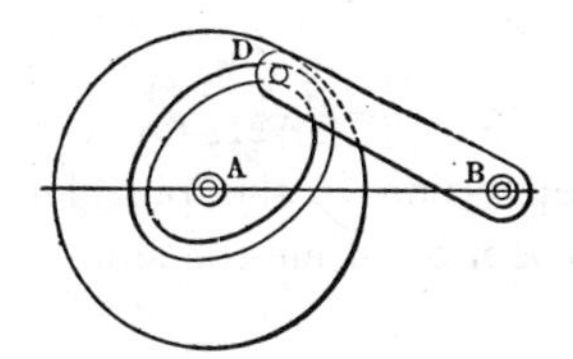

1032. ROCKING ARM by cam groove. A groove in a face plate may be so designed as to give a variety of movement to a rock shaft, with an arm and pin follower.

1033. YOKE STRAP and eccentric circular cam.

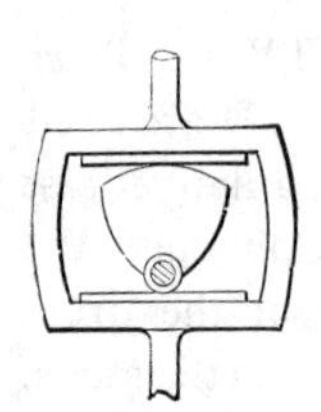

1034. TRIANGULAR CURVED ECCENTRIC, which by its peculiar form makes a stop motion at each half-revolution of the cam, for any portion of the stroke, according to the length of the concentric portion of the cam.

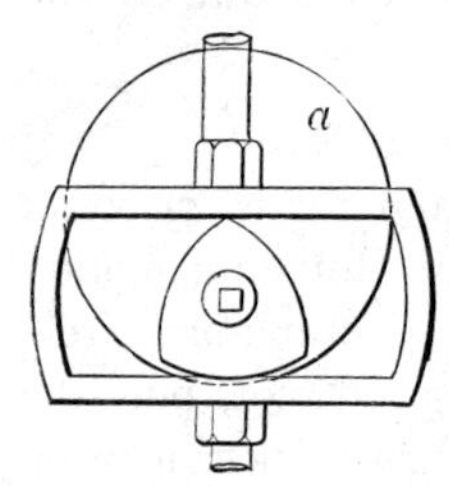

1035. TRIANGULAR ECCENTRIC for producing a stop motion at each half-revolution of the face plate *a*, by the proportional peripheral length of the outer curve of the triangular cam. Used on a French engine.

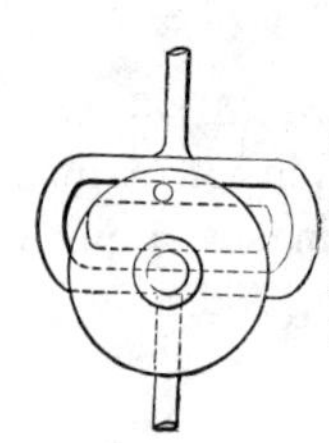

1036. RECIPROCATING MOTION with four stops, two of which are of longer duration than the others. A pin on the rotating disc, sliding in a grooved yoke, may be made to give a variety of motions to the rectilinear slide by the form of the groove.

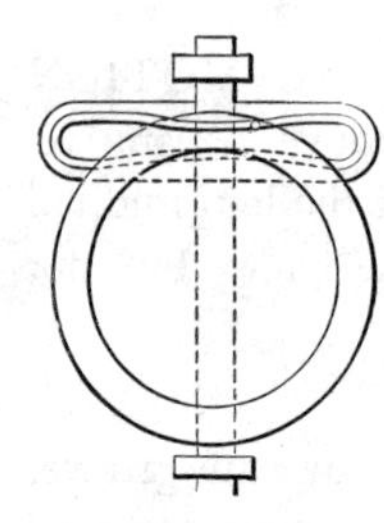

1037. UNIFORM RECIPROCATING MOTION from the circular motion of a crank or disc wrist pin. The endless groove in the cross head is made to conform in shape to the varying rectilinear motion of the wrist pin.

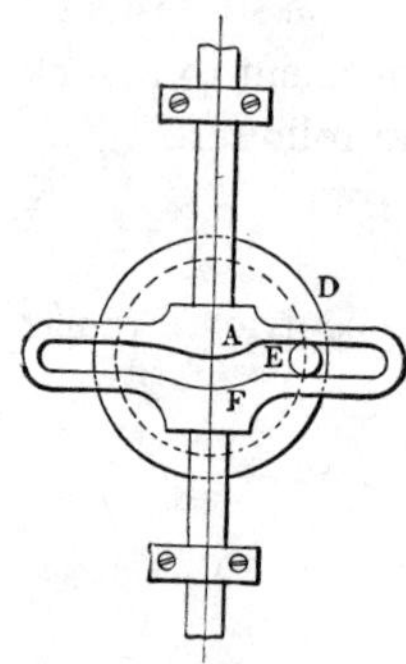

1038. NEEDLE-BAR SLOT CAM, for sewing-machines. The depression in the pin slot gives the needle a stop motion while the shuttle passes.

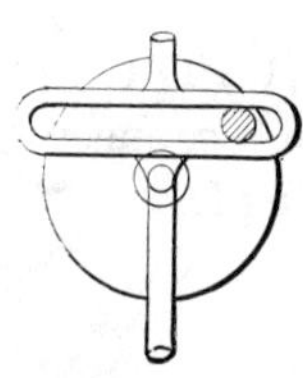

1039. SLOTTED YOKE CRANK MOTION, producing rectilinear motion of piston rod from a crank dispensing with a connecting rod.

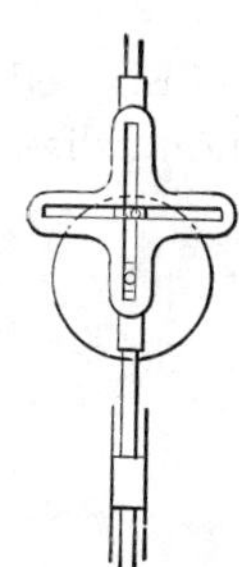

1040. TRAMMEL GEAR.—The slotted cross moves in a right line astride the shaft, while the crank pin in a block moves in the cross slot.

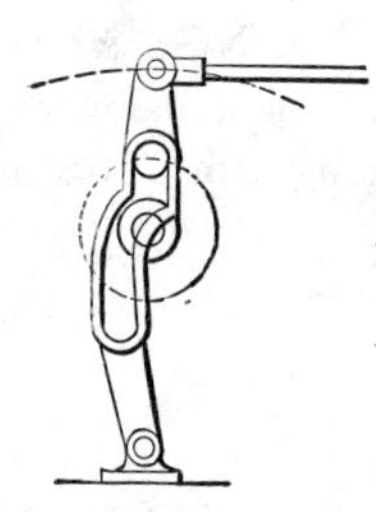

1041. SLOTTED LEVER MOTION from a crank pin. A variety of motions and stop motions may be made with this class of lever.

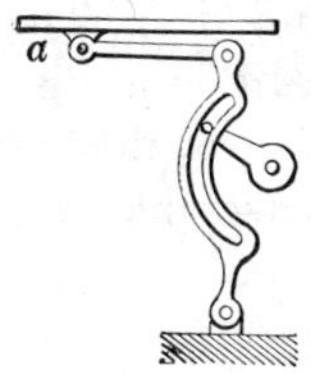

1042. INTERMITTENT RECIPROCATING MOTION from continuous circular motion. The curved slot in the lever should be radial with the crank centre for a stop. Many forms of motion may be had by variation of this device. A combination much in use for sewing-machines and printing-presses.

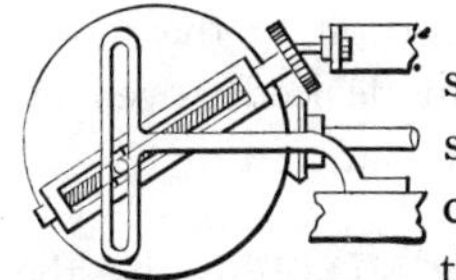

1043. VARIABLE CRANK THROW.—A screw and tappet wheel move a nut on the screw to which is fixed a wrist pin sliding in the cross slot of a carrier bar. Each revolution of the face plate brings the tappet wheel in contact with a finger, and by turning the wheel and screw moves the wrist pin to or from the centre of the wheel. Used in silk-spooling machinery.

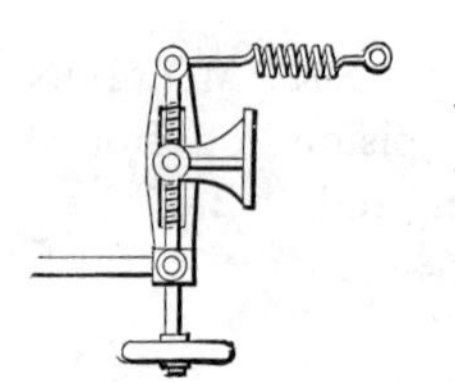

1044. VARIABLE ADJUSTMENT for the tension of a spring on the motion of a connecting rod, by varving the radii of a rocking lever.

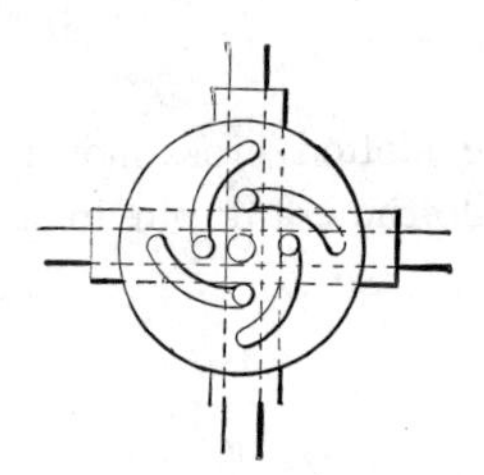

1045. FOUR-BOLT CAM PLATE, used for throwing safe bolts and for expanding dies.

1046. EQUALIZING TENSION SPRING AND LEVER.—The bell-crank lever equalizes the tension of the spring by its varying position. Its long arm is on a fixed pivot.

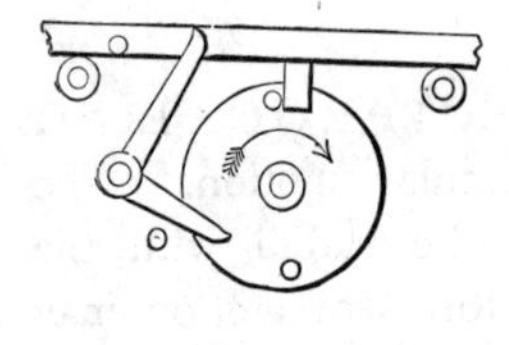

1047. ALTERNATING RECTILINEAR MOTION from studs on a rotating disc. The bar is carried forward by the stud on the disc striking the projection on the bar, and the bar returns by the movement of the bell-crank lever and opposite stud.

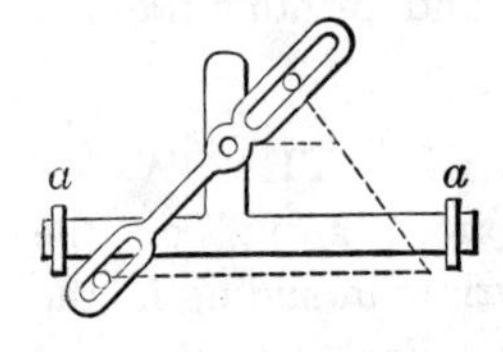

1048. TRAVERSE BAR, operated by a slotted lever. The upper pin being fixed or made adjustable for proportion to the movement of the lower pin, any desired movement of the traverse bar may be made.

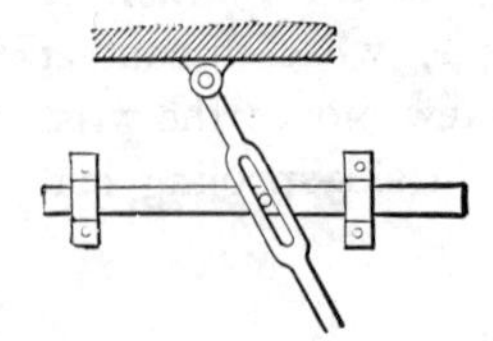

1049. RECTILINEAR MOTION by the movement of a slotted lever with one end pin-ioned. A belt shipper movement.

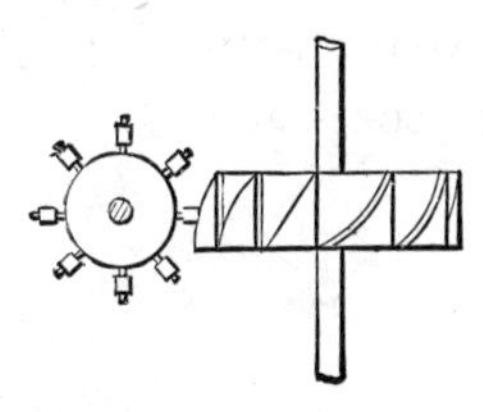

1050. INTERMITTENT ROTARY MOTION from a shaft at right angles. The friction rollers on the horizontal shaft disc move in grooves or on projections from the wheel on the vertical shaft, producing a variety of intermittent motions, due to the form of grooves or projections.

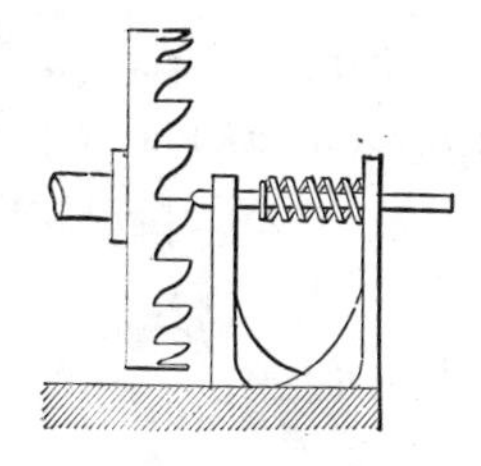

1051. VIBRATING TOOTHED WHEEL. —The rod is pressed against the teeth by the spring. A type of some electrical devices for interrupting the circuit.

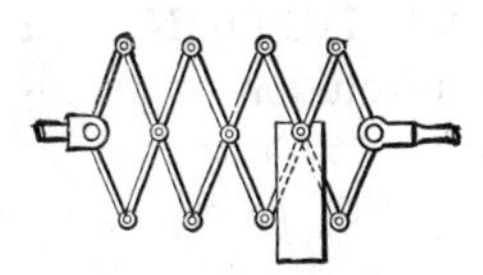

1052. "LAZY TONGS" MOVEMENT. —A system of crossed levers by which the amount of a rectilinear motion is increased by the proportional number of sections in the tongs. As a hand device it is in use as a toy, but is more useful as a reducing apparatus for a steam-engine indicator.

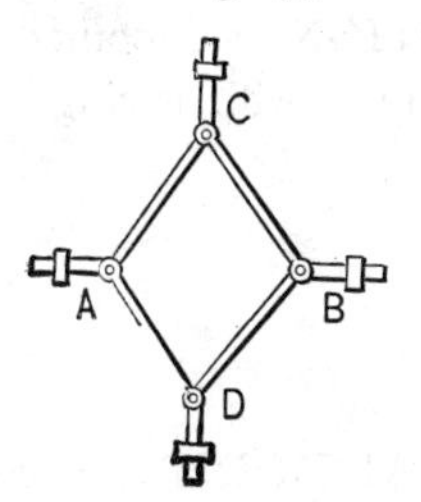

1053. QUADRANGULAR RECTILINEAR MOTION.—Rectilinear motion given to any one of the arms A, B, C, or D gives a contrary motion to its opposite arm, and a contrary motion to each of the side arms.

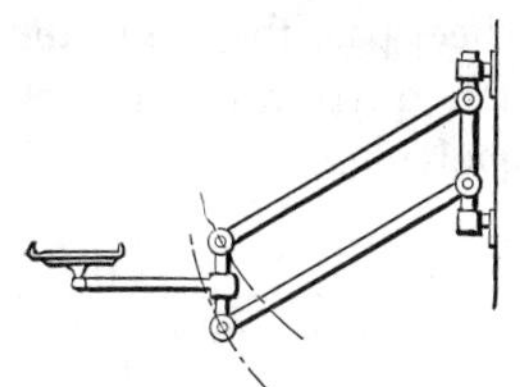

1054. PARALLEL MOTION, in a vertical line, for a swinging bracket.

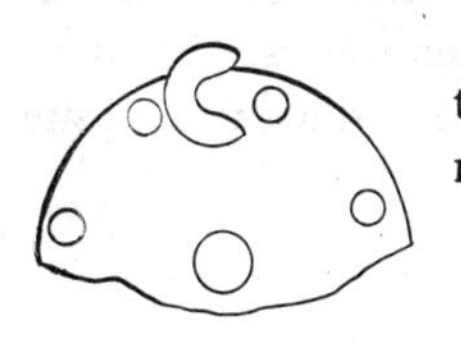

1055. INTERMITTENT MOTION of a pintooth wheel by the half-revolution of a ring segment.

1056. INTERMITTENT MOVEMENT of a pin-wheel by the vibration of a hooked arm.

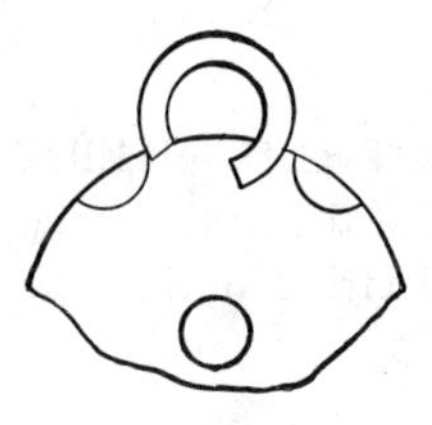

1057. INTERMITTENT MOTION of a segmental-toothed wheel by the revolution of a segmental barrel or ring.

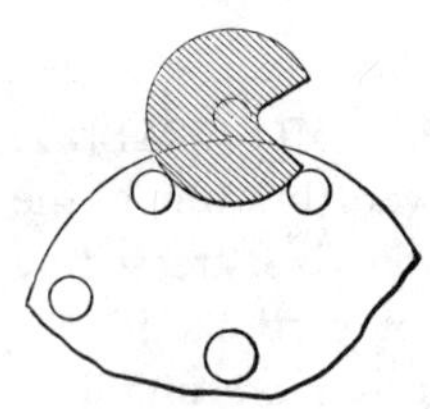

1058. INTERMITTENT MOTION of a pin-tooth wheel by the revolution of an indented tooth on a pinion.

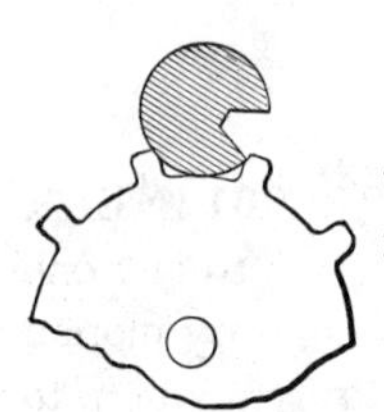

1059. INTERMITTENT MOTION of a toothed wheel by the revolution of a pinion with a single recessed tooth.

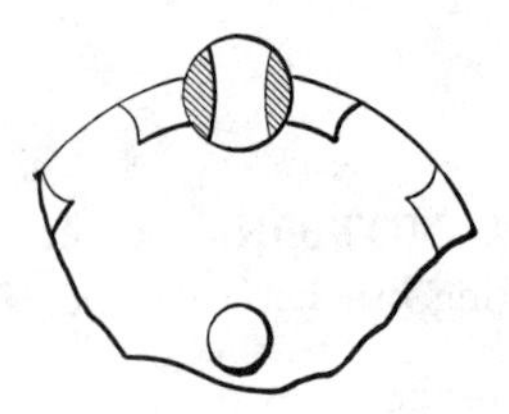

1060. ROCKING ESCAPEMENT.—The section teeth of the wheel pass the eye in the rocking cylinder at each quarter, or at each half-revolution when revolving.

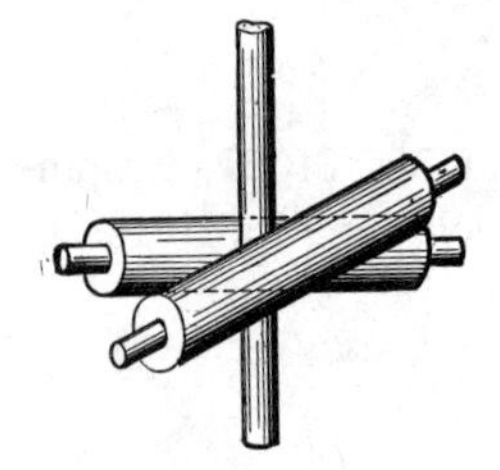

1061. ROTARY AND LONGITUDINAL MOTION of a rod between rollers, with their axes at an angle. Rollers run in opposite directions.

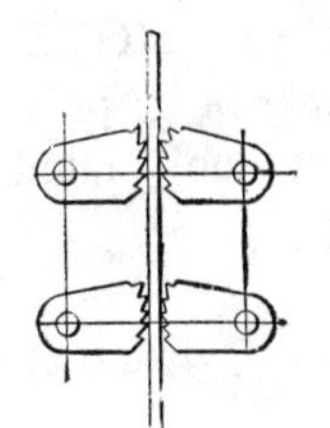

1062. RECIPROCATING FEED RATCHET.— For an intermittent feed, one pair of jaws may have a reciprocating motion. For continual feed motion both pairs of jaws should have cpposite reciprocating motions

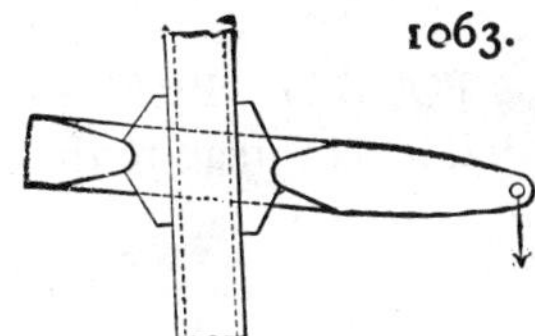

1063. FRICTION ROD FEED RATCHET.— The jaws, being pivoted in a slot in a lever, make a powerful and quick grip on a feed bar by the motion of the lever bar.

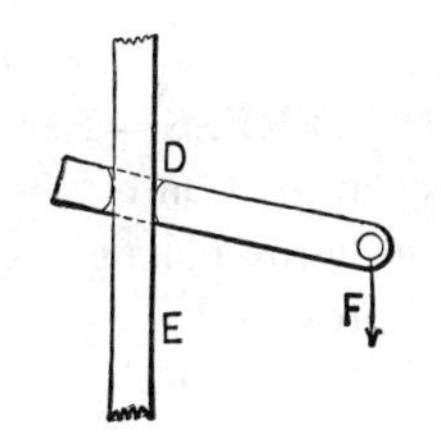

1064. FRICTION HAULING RATCHET. —A hole bored slanting through a bar D. A slot in the side of the bar, for convenience of putting on or taking off the rod or rope to be hauled, makes a handy clutching device.

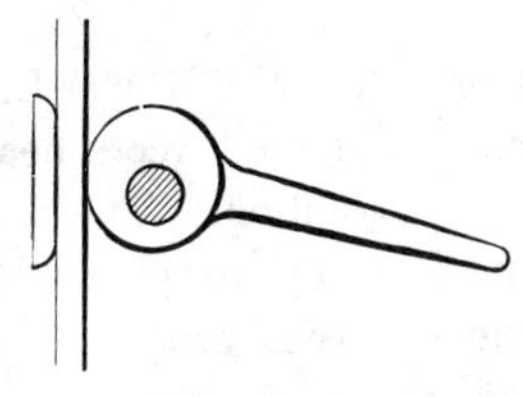

1065. CAM-LEVER GRIP for a rope or rod stop. This principle is used on safety grips for elevators.

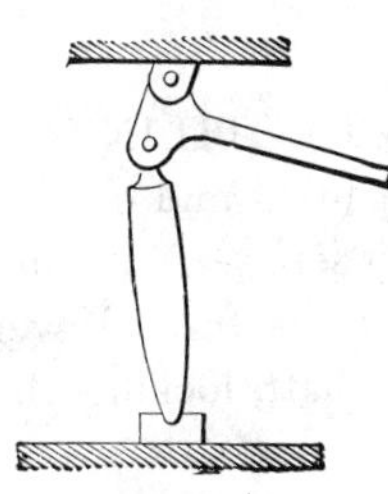

1066. LEVER TOGGLE JOINT, largely used in stamping and punching presses. This form shows great pressure when the three bearings near a linear direction.

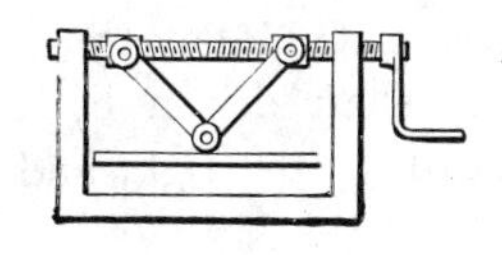

1067. SINGLE TOGGLE ARM LETTER-PRESS.—The arms are drawn together by a right and left screw.

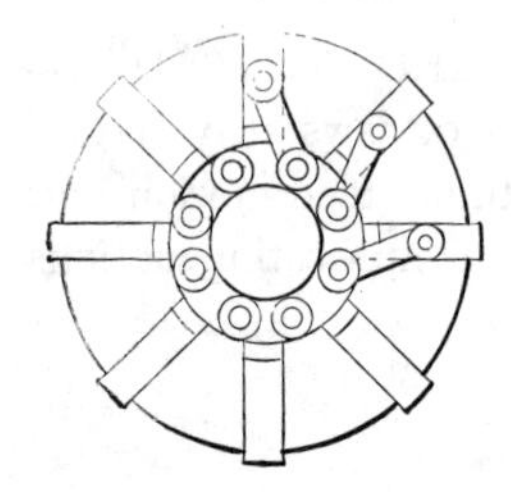

1068. TOGGLE-JOINT CAM MOVEMENT for throwing out a number of grips at once by the local movement of the jointed ring.

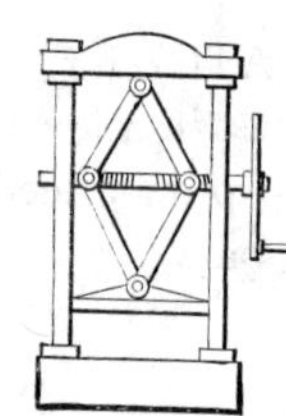

1069. DOUBLE-SCREW TOGGLE PRESS.—The screw has a right- and left-hand thread to draw the toggle joints together.

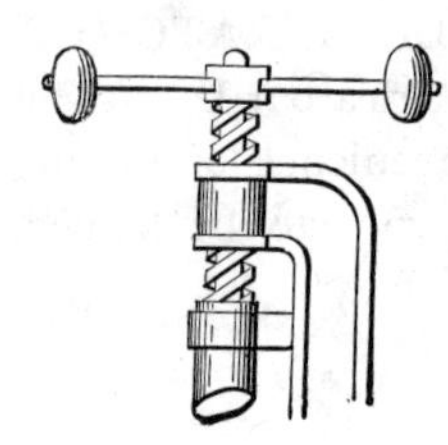

1070. SCREW STAMPING PRESS.—Rectilinear motion from the circular motion of the lever handles. The momentum of the balls gives the final power in this class of presses.

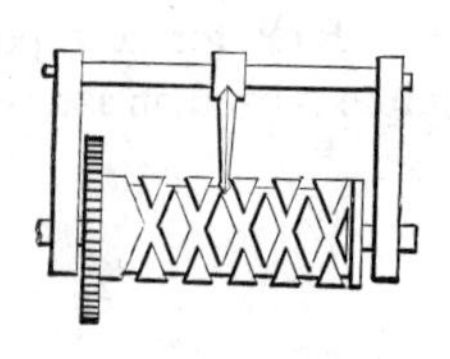

1071. MULTIPLE RETURN GROOVED CYLINDER, producing extended rectilinear motion and return by its revolution. The carrier arm has a pivoted tracer to enable a smooth passage of the opposite grooves. A spooling device.

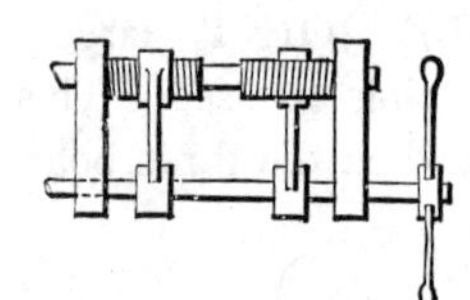

1072. RECIPROCATING RECTILINEAR MOTION by the alternate opening and closing of half nuts on a right and left screw. Nuts and arms are attached to a shaft that is thrown over by a dog on a spooling-frame shaft, locking the right- or left-threaded nut alternately.

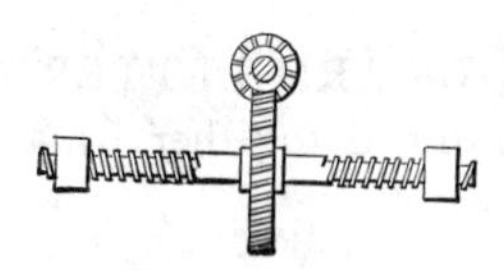

1073. RECTILINEAR MOTION by a right- and left-hand screw shaft driven by a worm gear. The nuts move on the right and left screw.

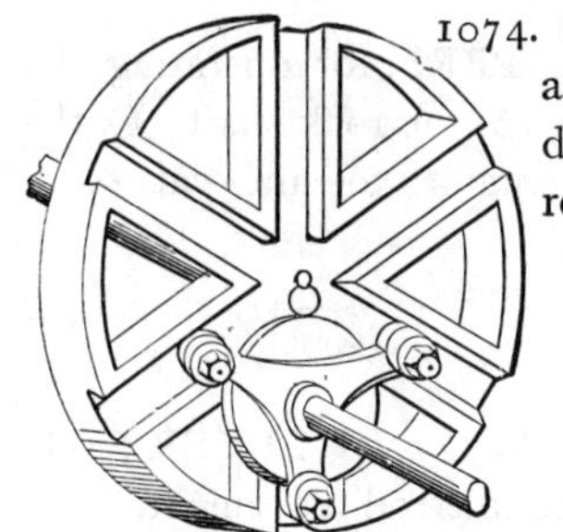

1074. SIX RADIAL GROOVED TRAMMEL and triangular shaft arms, driving or being driven by a shaft out of line. The friction rollers give freedom of motion to either gear.

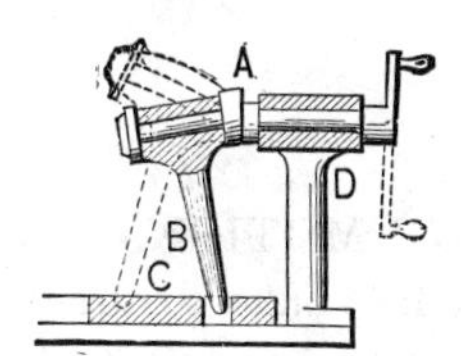

1075. RECTILINEAR RECIPROCATING MOTION of a bar, from continuous circular motion of a bent shaft.

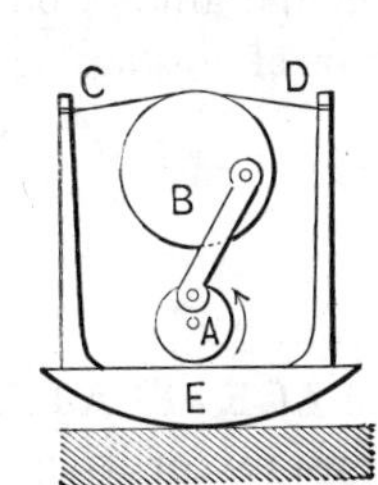

1076. ROCKING MOTION, from a continuous rotary motion of the crank shaft A.

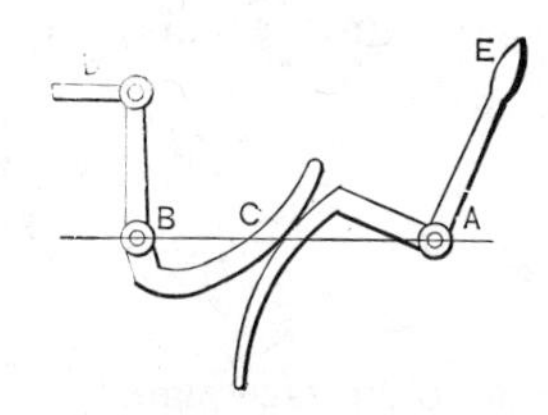

1076a. PAIR OF TOE LEVERS.—Bell-crank order. A and B, fulcrums of the levers; E, handle; C, curved toes. This principle is used as a valve gear.

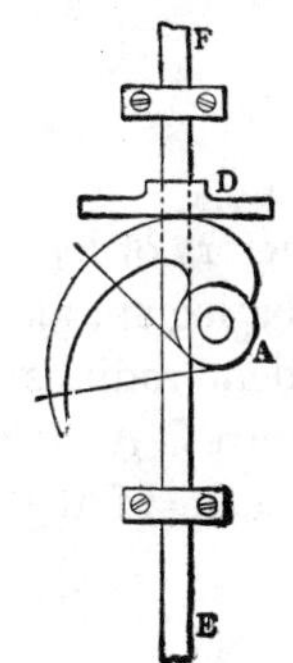

1077. WIPER CAM for stamp mills. A, the wiper; D, flanged chock, allowing the hammer spindle to revolve. Also in use on sewing-machines for throwing the needle bar

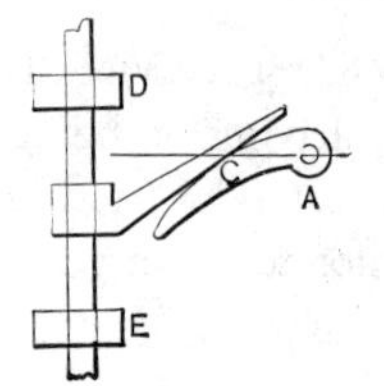

1078. ANGULAR WIPERS, for operating the valves of beam engines. A, the rock shaft; C. the curved wiper, lifting the angular toe and valve rod.

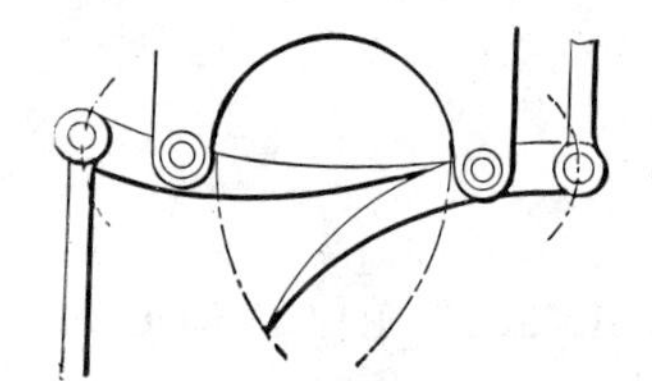

1079. EQUALIZING LEVERS OF TOES, for variable rod movement.

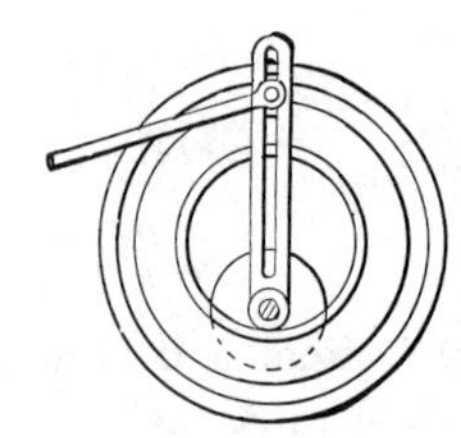

1080. VARIABLE CRANK MOTION.—An eccentric slot in a stationary face plate guides a slide block and wrist pin in a slotted crank. Connecting rod drives the cutter bar of a shaping-machine.

1081. SPIRAL-GROOVED FACE PLATE, for feed motion. Obsolete; but useful for irregular motion, in which the spiral grooves may be wavy or zigzag.

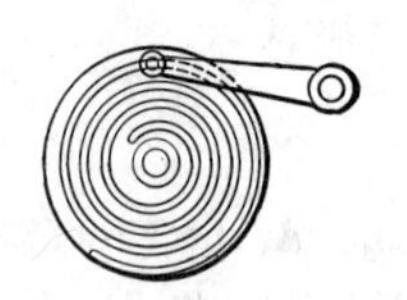

1082. LEVER, guided by a volute face plate.

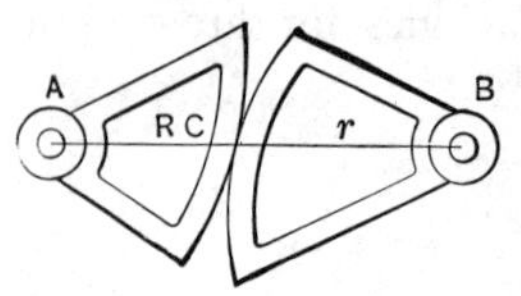

1083. CAM SECTORS, or sectors of log-spiral wheels. When laid out as a log spiral, the sum of each pair of coincident radii is equal to the distance of the centres, A, B. As a pair of pressure cams, the sum of the radii varies to meet the required throw of the cams.

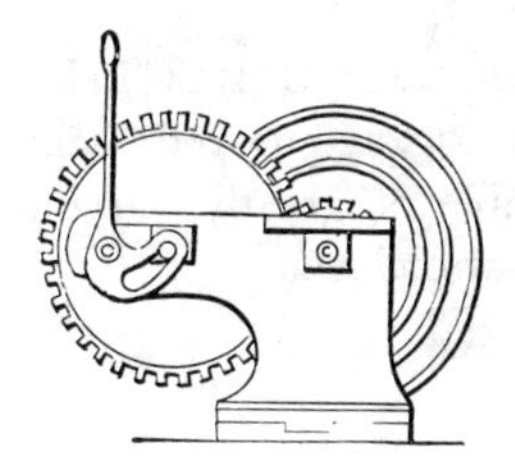

1084. GEAR-DISENGAGING CAM LEVER. — The eccentric slot in the lever throws the slow driving gear out of lock by throwing the lever back.

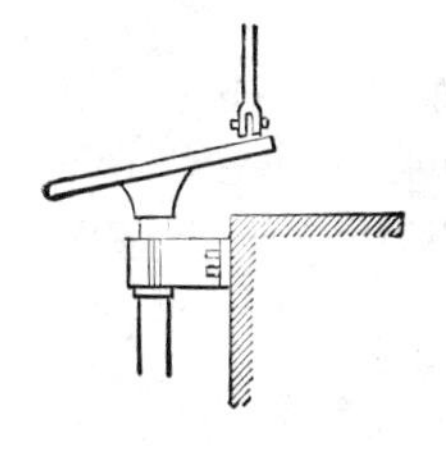

1085. OBLIQUE DISC MOTION.—A disc fixed at an angle upon the end of a shaft gives a variable rectilinear motion to a rod and roller by varying its distance from the centre.

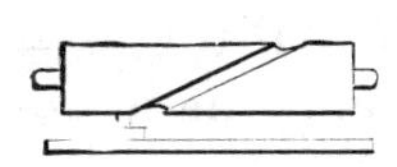

1086. GROOVED CYLINDER CAM.—Used to convert reciprocating into rotary motion.

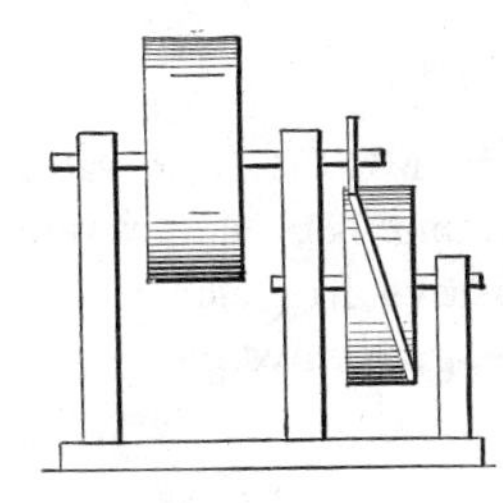

1087. TRAVERSE MOTION of a shaft by a rolling cam. The disc, rolling in the groove of the drum, gives an ever-varying traverse motion to the disc shaft, according to the proportions of the size of disc and cam drum.

1088. FOUR-MOTION FEED of the "Wheeler & Wilson," and other sewing-machines. The traverse bar A is forked and encloses the push bar B, pivoted to it, and is held back by the spring at D. The revolving cam C has its periphery cam-shaped, to lift the push bar, and its face, also cam-shaped, to push the bar forward, when the teeth are in contact with the goods.

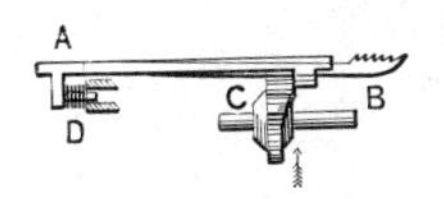

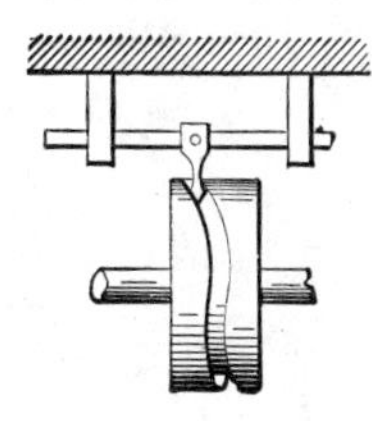

1089. RECIPROCATING RECTILINEAR MOTION, from the circular motion of grooved cams; may be made uniform or intermittent, by the direction of the groove on the cam.

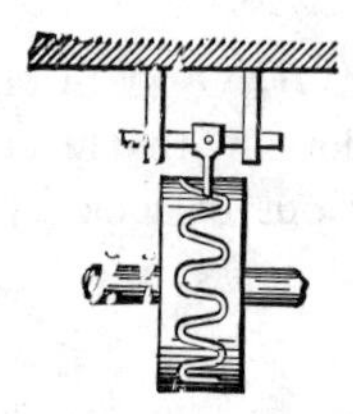

1090. QUICK RECIPROCATING RECTILINEAR MOTION, from a zigzag-grooved cam. Form of cam groove is capable of greatly varying the rectilinear motions of a bar or lever.

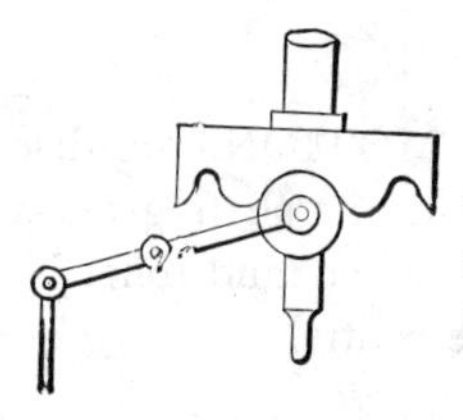

1091. CYLINDRICAL CAM, giving any required special motions through a lever, roller, and connecting rod, according to the curves given to the cam.

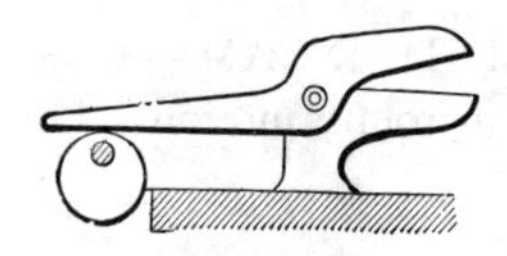

1092. CAM-OPERATED SHEARS.—Many modifications of this device are in use.

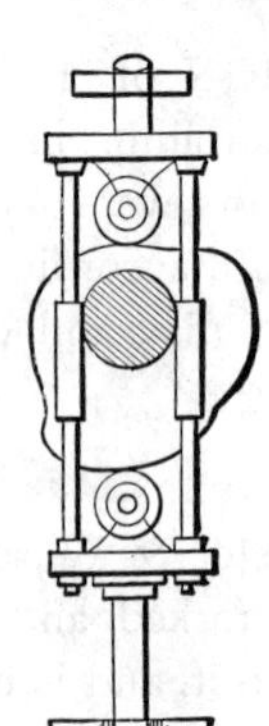

1093. IRREGULAR CAM MOTION to valve rods. An irregular cam, acting between friction rollers in a yoke frame. Positive irregular rectilinear motion. An old steam-engine valve gear.

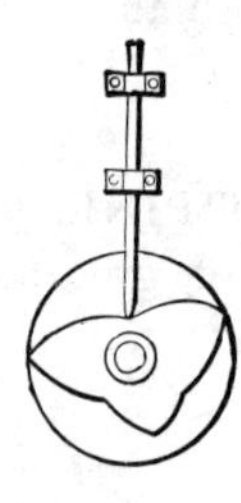

1094. VIBRATING RECTILINEAR MOTION, from a revolving trefoil cam.

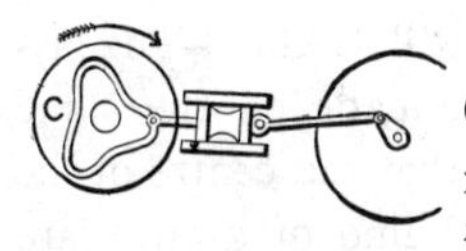

1095. IRREGULAR VIBRATING CIRCULAR MOTION, from continuous circular motion of a cam slot. Any form of cam slot in a face plate may be made to produce a vibratory motion on a crank pin, which may be transmitted to circular or rectilinear motion.

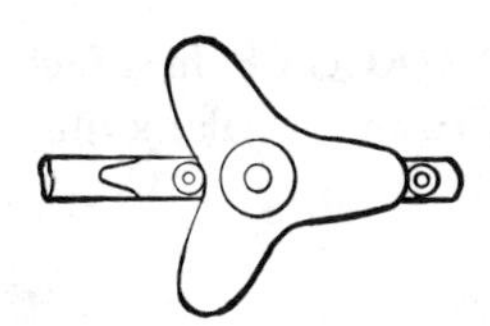

1096. CLOVER-LEAF CAM, for rectilinear motion by follower rollers on a bar. The cam is so designed that the rollers have a bearing in all its positions.

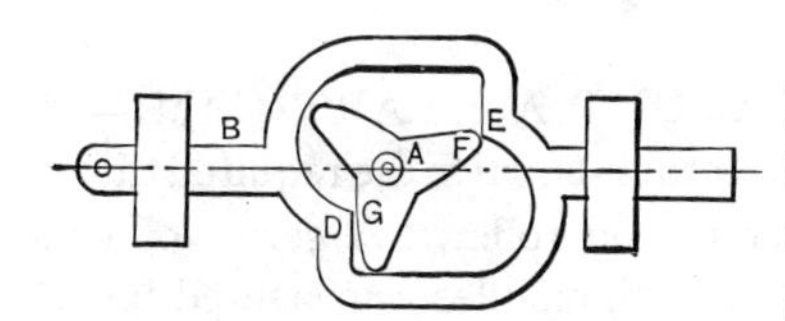

1097. POWER ESCAPEMENT for heavy machines. The traverse bar may be vibrated by the positive motion of the cam arms.

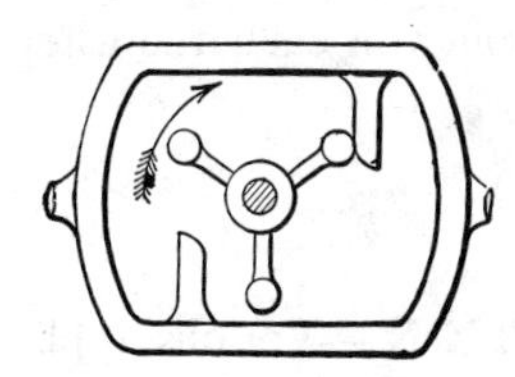

1098. ROTARY MOTION of a three-arm wiper produces a reciprocating rectilinear motion of the toothed frame, and *vice versa.*

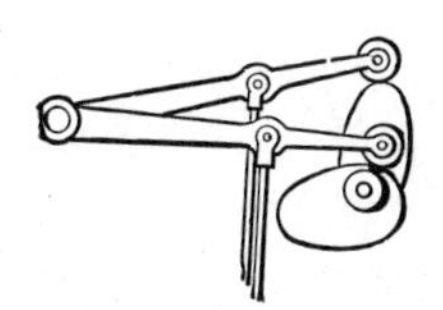

1099. IRREGULAR RECIPROCATING MOTION of connecting rods and levers, moved by alternating oval cams.

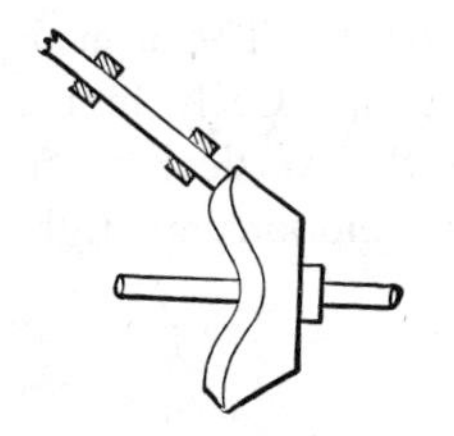

1100. BEVELED DISC CAM, tor variable reciprocating motion of a bar at an angle with the shaft.

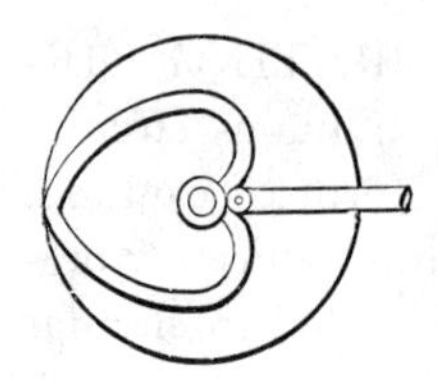

1101. GROOVED HEART CAM.—The layout of a grooved cam may be made on the same principles as No. 1103, only that the centre of the roller or pin and the central line of groove are the measurements for the amount of motion.

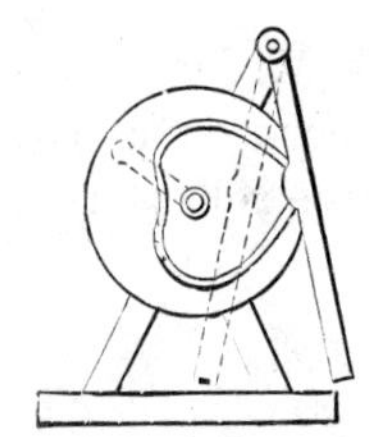

1102. HEART-SHAPED GROOVE in a face plate, vibrating a lever, produces an irregular swinging motion of the lever.

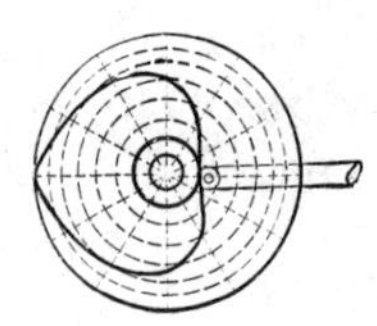

1103. LAYING OUT A HEART CAM.—A circle is drawn on a radius equal to the required throw, plus the diameter of the roller. A series of concentric circles and radii enables a measured layout of the cam, which must be as much larger than the required motion as is equal to the radii of the roller on each radius of the plan.

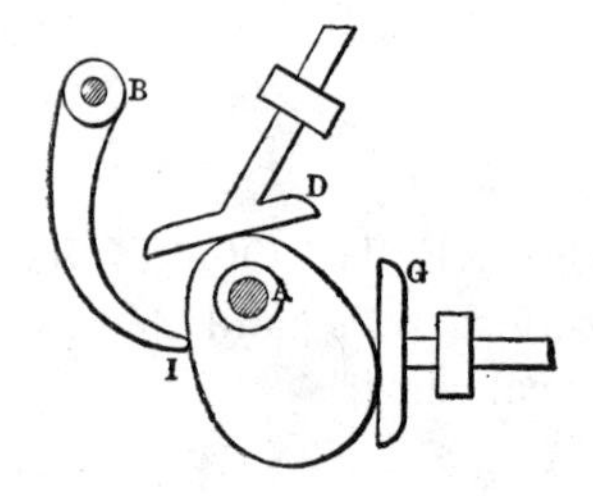

1104. CAM MOTION.—Various applications of cam followers, with direct and oscillating motion.

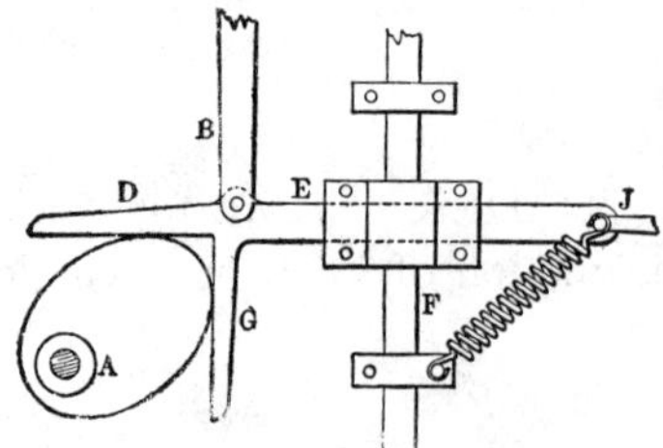

1105. DOUBLE-CAM MOTION, from a sliding follower. The arm E of the follower, slides freely in the box, clamped to the vertical shaft, giving two equal motions at right angles.

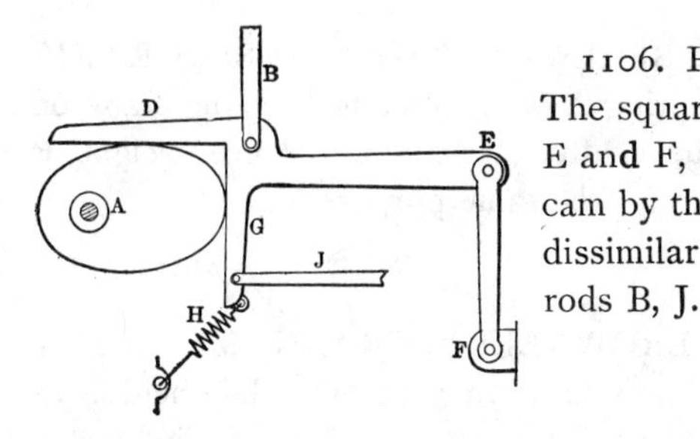

1106. PIVOTED FOLLOWER.—The square-armed follower, pivoted at E and F, is kept in contact with the cam by the spring H, and so produce dissimilar motions in the connecting rods B, J.

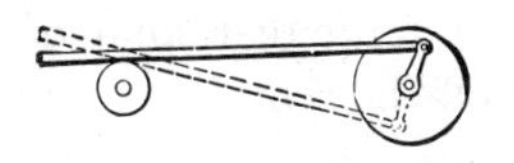

1107. RECIPROCATING MOTION, from two cranks on opposite ends of a shaft.

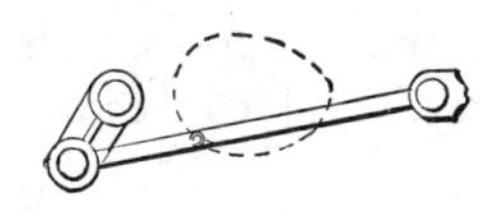

1108. OVOID CURVE is made by any point between the pivots of a single-crank connecting rod, the other end of which is guided by a rectilinear slide.

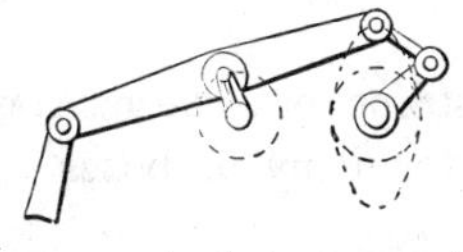

1109. VARIABLE POWER TRANSMITTED from a crank linked to a lever-beam, driving a second crank. In this case there is no pressure on the driven crank when both cranks are vertical, but greatest pressure when the cranks are horizontal.

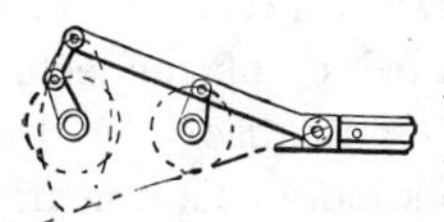

1110. ELLIPTICAL CRANK.—The arm moves in a slot. The inner crank pin, making a revolution, marks an ellipse by a pencil at the outer end of the arm, while the outer crank pin, linked to the arm, makes a circle.

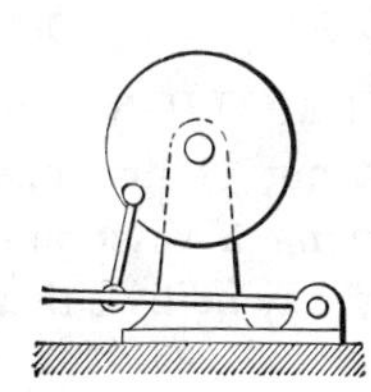

1111. CURVILINEAR MOTION of a treadle gives circular motion to a crank or disc. The foot-lathe motion.

1112. SPRING LATHE-WHEEL CRANK.—The spring A is intended to keep the crank off the dead centre. A counterbalance weight is also used for the same purpose.

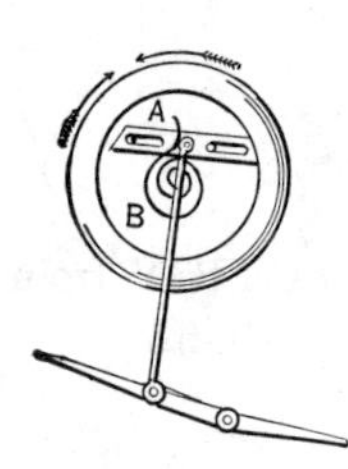

1113. "BROWNELL" CRANK MOTION.—The wrist pin is fixed on a tangent slide held in its forward position by a volute spring attached to the face plate. The slide is retained by pins in traverse slots. Can be arranged for either kind of treadle, to keep the crank pin off the center.

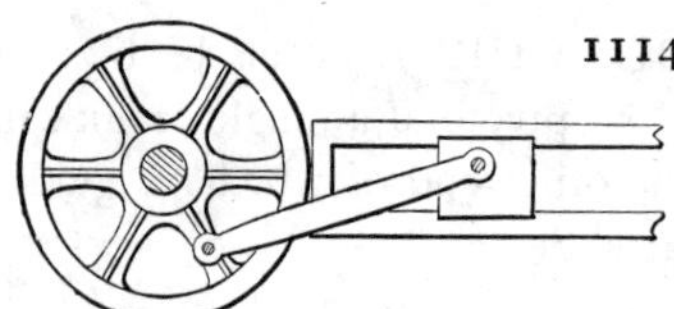

1114. ORDINARY CRANK MOTION for engines or other purposes, with cross head, slides, and connecting rod.

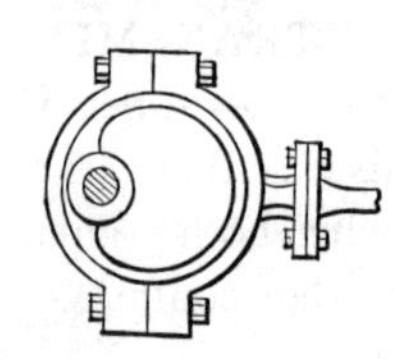

1115. ECCENTRIC and straps for valve motion, also used in place of a crank for many purposes.

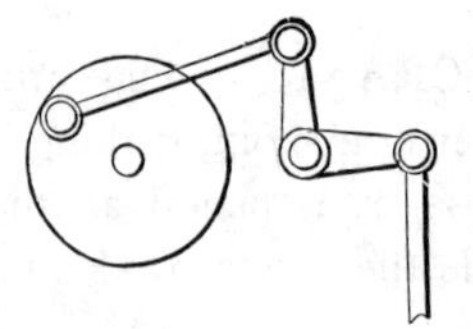

1116. RECIPROCATING MOTION of a connecting rod through a bell crank connected directly with a wrist on crank disc. In this case the forward and back motions are nearly alike depending upon the proportional length of the driving arm of the bell crank and crank motion, as well also to the length of the connecting rod between the wrist pin and bell crank.

1117. VARIABLE CIRCULAR MOTION from two cranks on shafts parallel, but out of line, one crank being slotted, the other carrying a wrist pin, passing through the slot. Driving may be by either crank.

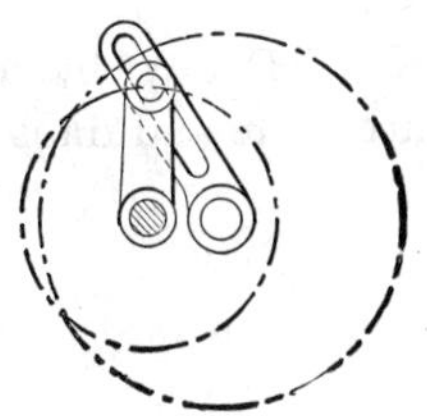

1118. IRREGULAR MOTION of one crank from the regular motion of another crank. A quick-and-slow alternate motion of the slotted crank is made by the regular motion of the smaller crank.

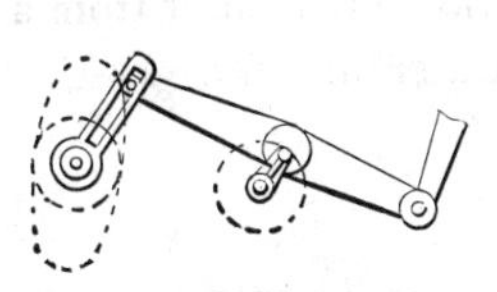

1119. VARIABLE POWER transmitted from a slotted crank driver to a fixed driven crank pin through a lever beam, the opposite end of which is held by a swinging connecting rod. The pressure on the driven crank is continuous, but greatest on and near the central line of the two shafts.

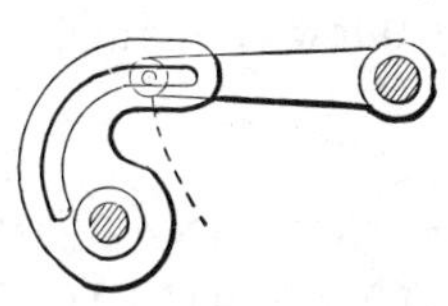

1120. VIBRATING MOVEMENT from a slotted curved arm, gives a variable vibrating movement to straight arm.

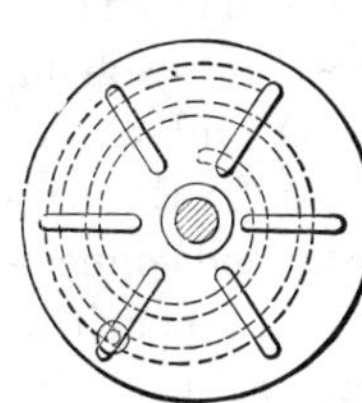

1121. VARIABLE CRANK PIN.—A slotted face plate backed by a spiral slotted plate by which the revolution of one plate upon the other moves a crank pin to or from the centre. The same principle is used in the universal lathe chuck in which each slot carries a grip jaw.

1122. VARIABLE RECTILINEAR MOTION of a shaft from a vibrating, curved, slotted arm.

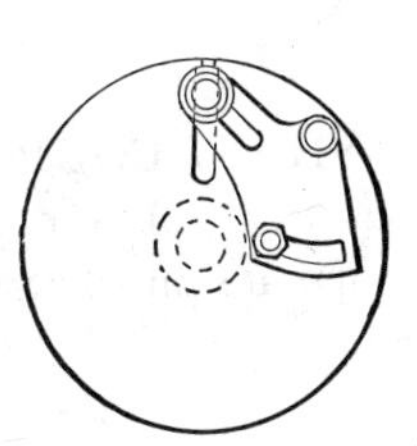

1123. VARIABLE CRANK THROW by a slotted sector on a face plate.

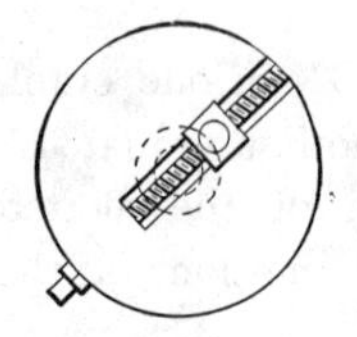

1124. VARIABLE CRANK THROW by a movable pin block in a slotted face plate and transverse screw.

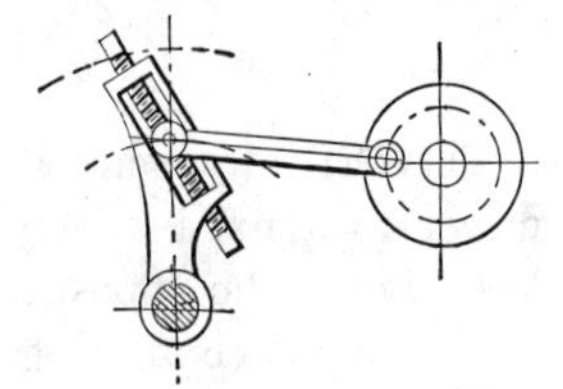

1125. VARIABLE RADIUS LEVER for reciprocating motion of a shaft from a continuous motion of a crank pin.

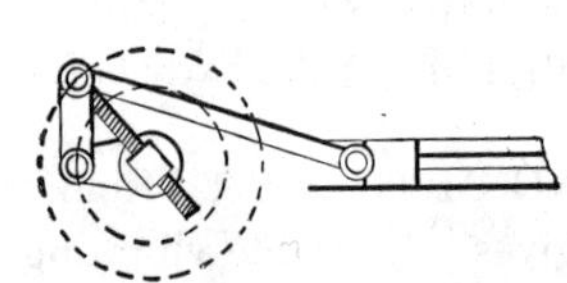

1126. VARIABLE CRANK THROW.—The jointed crank and radial screw give a large variation to the throw of a crank.

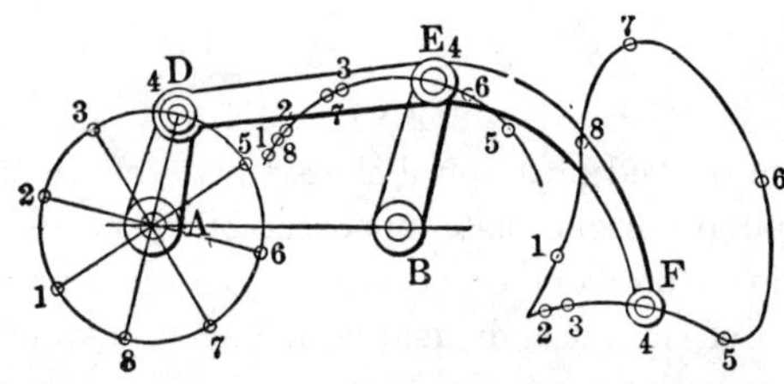

1127. COMBINATION CRANK-MOTION CURVES.—A revolving crank A, D and the vibrating link B, E carrying an extended connecting arm with a pencil at the end, F. A great variety of figures and curves may be made by different proportions of all the parts. The figures on the crank pin circle D correspond with the figured diagram.

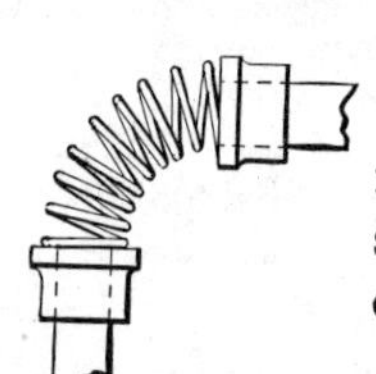

1128. FLEXIBLE ANGULAR COUPLING, for light work. May be a helical spring, round or square, wire or a tube, sawed or a spiral. Used on driving handles for telescopes and other instruments.

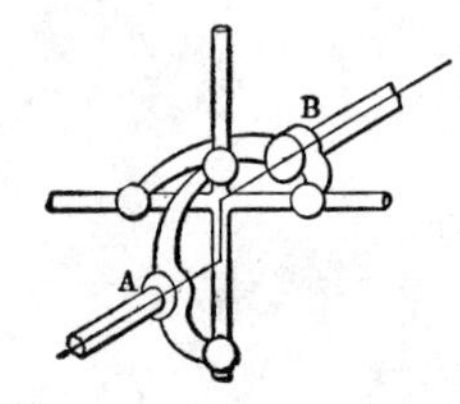

1129. SLIDING CONTACT-SHAFT COUPLING.—A cross bar sliding in two yokes on shafts in offset lines. Will also operate on shafts somewhat out of line or at an angle.

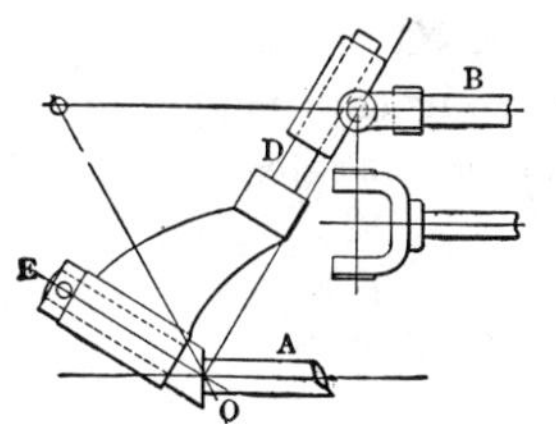

1130. RECTILINEAR MOTION from the rotation of an angular crank pin. A, rotating shaft carrying crank pin E; D, arm with sleeve jointed to yoke and sliding rod B.

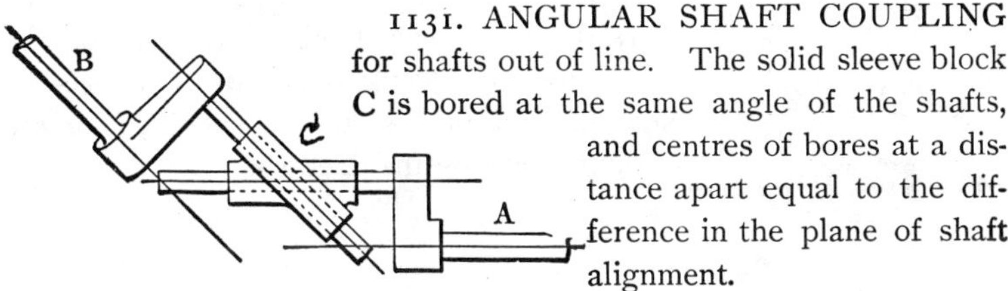

1131. ANGULAR SHAFT COUPLING for shafts out of line. The solid sleeve block C is bored at the same angle of the shafts, and centres of bores at a distance apart equal to the difference in the plane of shaft alignment.

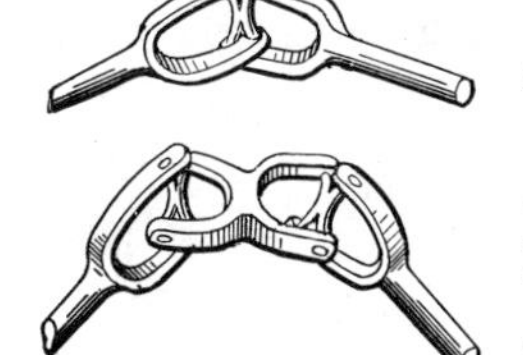

1132. UNIVERSAL JOINT, with a single cross link. Good for angles of 45° and under.

1133. DOUBLE LINK UNIVERSAL JOINT, good for larger angles than above. The connecting link may be made short and guarded, with a sleeve to prevent kinking.

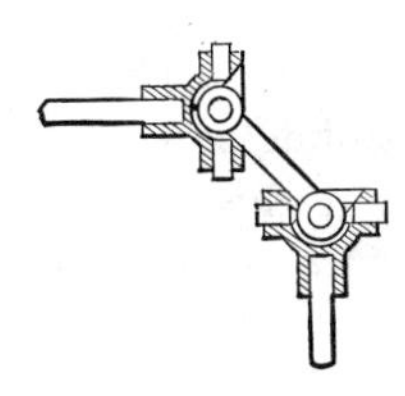

1134. UNIVERSAL ANGLE COUPLING, "Hooke's" principle. Each shell carries a double trunnion ring, the connecting link being pivoted at each end to the rings.

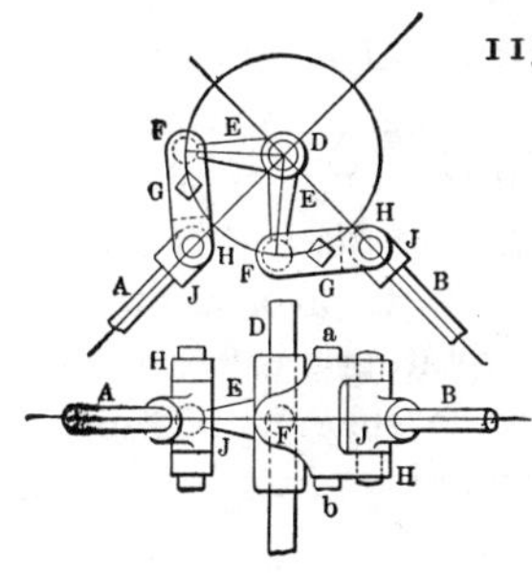

1135. "ALMOND" ANGULAR SHAFT COUPLING.—The yoke links G, G are pivoted to the sockets on the ends of the shaft, and to the right-angled arms on the sleeve which slides freely on the fixed shaft D. The sockets at F, F are ball joints. Angle of shafts may vary within limits.

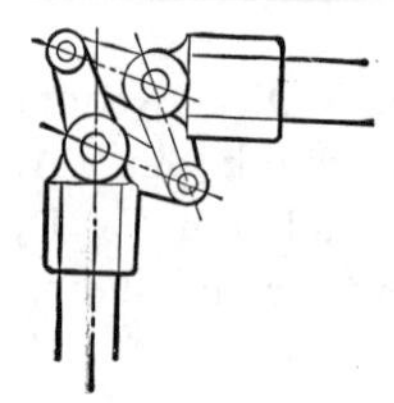

1136. "HOOKE'S" ANGULAR SHAFT COUPLING, the knuckle universal joint. Shaft joints are double-pivoted at right angles.

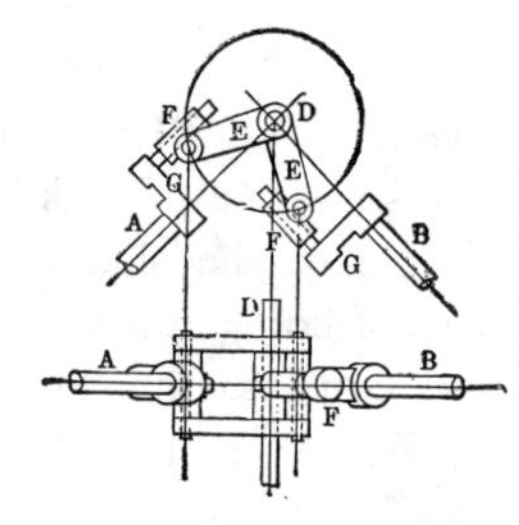

1137. ANGULAR SHAFT COUPLING.—In this arrangement the shafts have cranks and elongated crank pins, on which sleeves slide that are pivoted to the arms E, E of the sliding sleeve on the fixed shaft D.

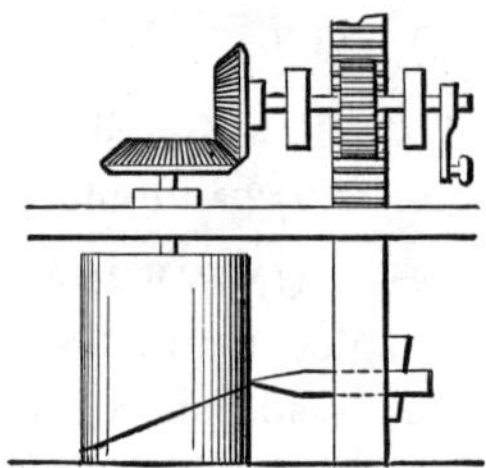

1138. RACK AND PINION MOVEMENT for tracing spiral grooves on a cylinder.

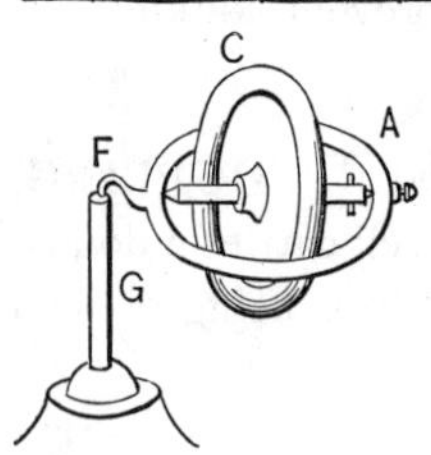

1139. GYROSCOPE.—The heavy disc C, rotating at great speed in the ring A, is suspended by the point F, resting on bearing. The rotation of the disc keeps it from falling and slowly revolves the holding ring A around the point F. An illustration of the tendency of rotating bodies to preserve their plane of rotation.

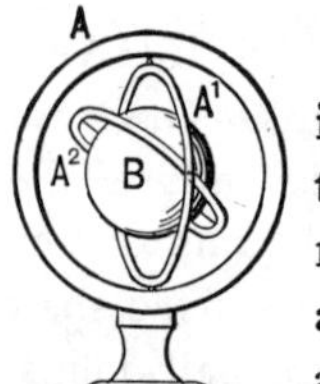

1140. GLOBE GYROSCOPE.—The outer ring A is fixed to a stand. The second ring A^1 is pivoted vertically to the outer ring; the inner ring is pivoted at right angles in the second ring, and the ball is pivoted at right angles in the inner ring to its pivot in the secand ring. This gives the ball, rotating on its own axis, a direction free to move to every point in the sphere. When the heavy ball is made to rotate rapidly in any direction of its axis, much pressure must be made to change its direction.

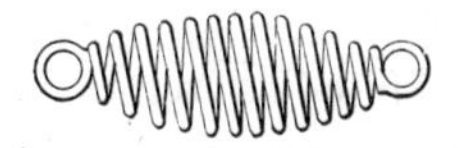

1141. TENSION HELICO-VOLUTE SPRING.

1142. DOUBLE HELICO-VOLUTE SPRING, for compression.

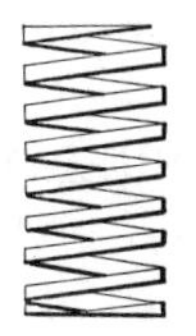

1143. COMPRESSION HELICAL SPRING, square rod.

1144. SINGLE VOLUTE HELIX SPRING.

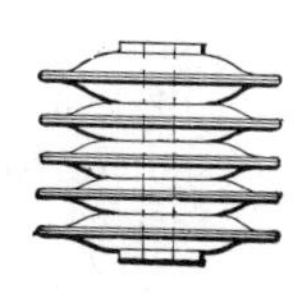

1145. COMPOUND DISC SPRING.—The discs are dished and perforated for a guide pin.

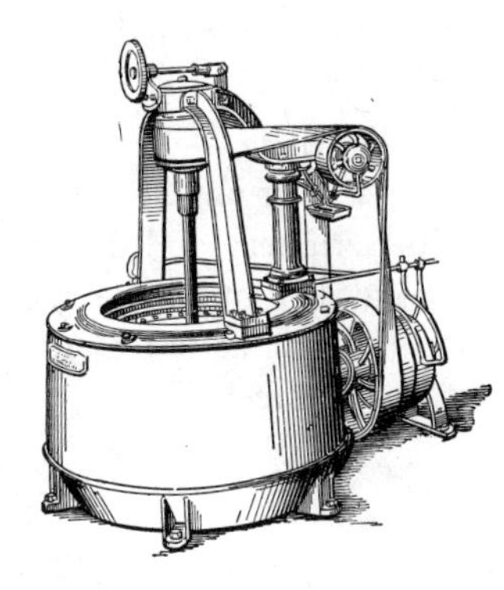

1145*a*. HYDRO - EXTRACTOR, showing method of belting with adjustable idler and cone pulleys.

Type for laundry work. At a speed of from 1,000 to 2,000 revolutions per minute the water flies off by centrifugal force and the material is left nearly dry.

1145*b*. REVERSING PULLEY.—A conical disk fixed to the shaft; a pulley loose on the shaft with a clutch; a disk loose on the shaft, fixed as to motion by an arm, carries a set of conical rollers, which are pushed into a bearing by the shipper as shown in the right-hand figure for reversing at increased speed as the ratio of the diameter of the two conical surfaces. The clutch is operated by the shipper bar.

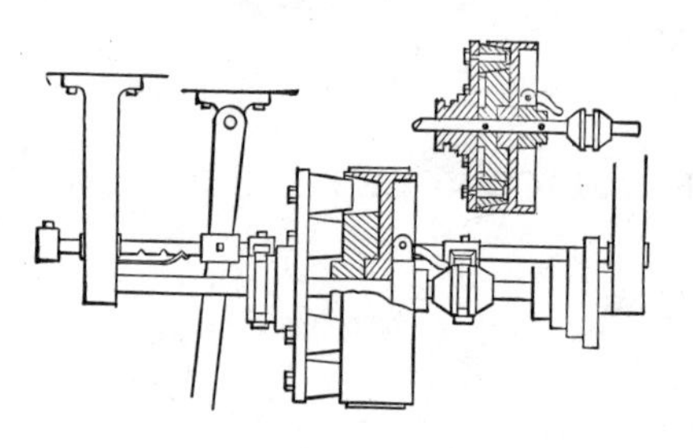

1145*c*. FOUR SPEED CHANGE GEAR. — A hollow spindle with change gears running loose upon it. A rack spindle B carries a hinged pawl or key A, held out by a spring. A lever C carries a sector meshing in the rack, which by its movement draws the key A to catch the keyway in any of the speed gears.

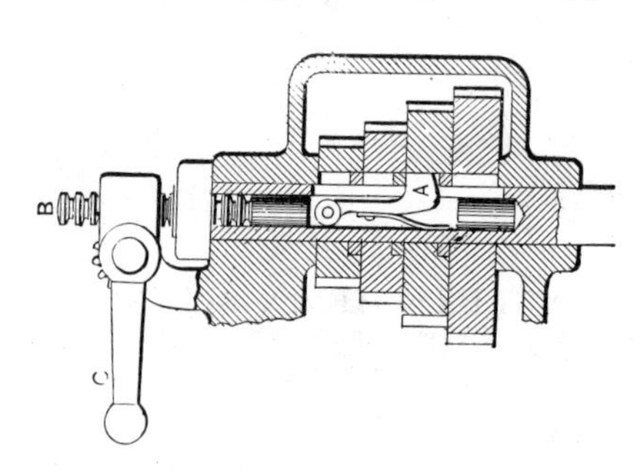

1145*d*. HEDDLE CAM, used in weaving.—The twilling cam K is attached to the grooved hub L, which slides freely on the feathered spindle and moves three times one way and returns by a sliding switch over which the grooves traverse.

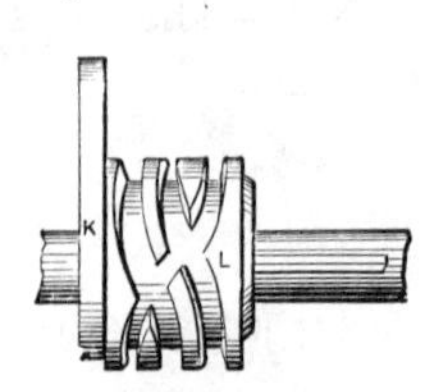

1145*e*. FERRIS WHEEL.—A steel wheel 250 feet diameter carrying a series of balanced cars on its periphery and driven by steam power. Total height above the ground 265 feet. Remarkable as one of the great modern structures of steel.

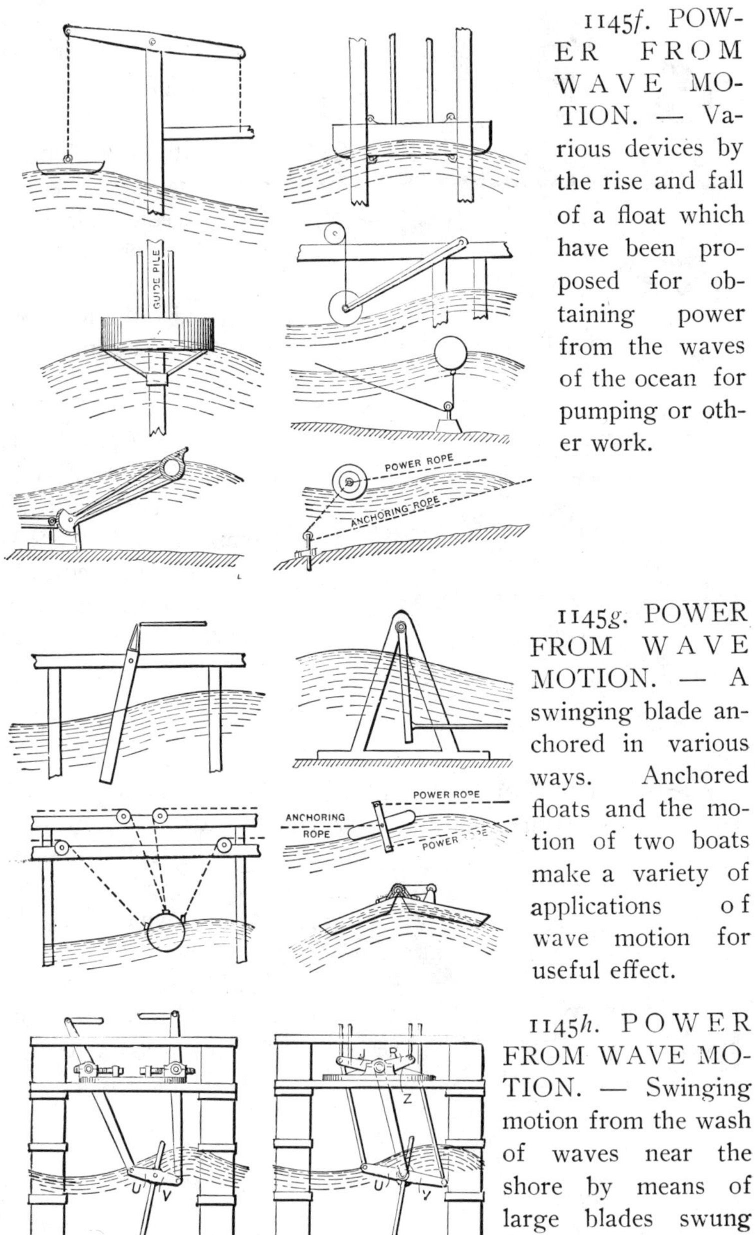

1145*f*. POWER FROM WAVE MOTION. — Various devices by the rise and fall of a float which have been proposed for obtaining power from the waves of the ocean for pumping or other work.

1145*g*. POWER FROM WAVE MOTION. — A swinging blade anchored in various ways. Anchored floats and the motion of two boats make a variety of applications of wave motion for useful effect.

1145*h*. POWER FROM WAVE MOTION. — Swinging motion from the wash of waves near the shore by means of large blades swung from a pier. The two cuts represent a single and double acting transmission.

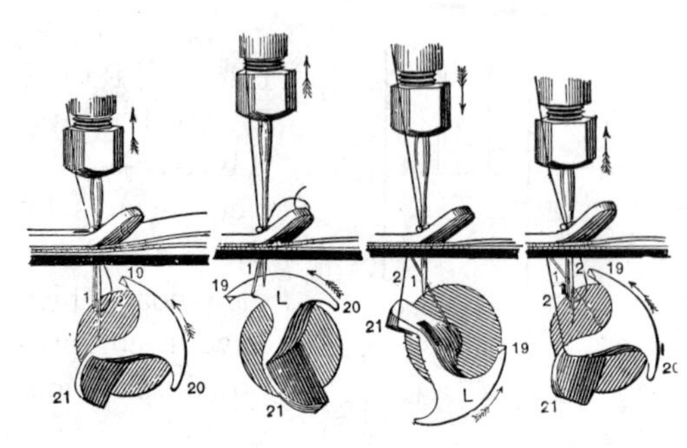

1145*i*. ACTION OF THE HOOK in the Willcox & Gibbs sewing machine.—1st, the loop formed by the up stroke of the needle; 2d, hook catches the loop; 3d, loop reversed and spread; 4th, next loop caught by the hook and carried through the preceding one.

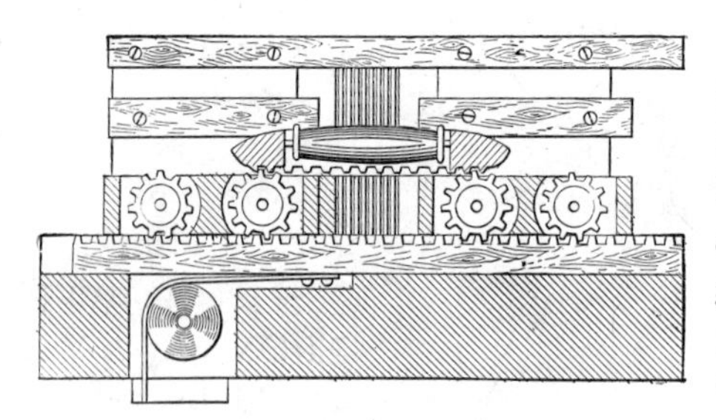

1145*j*. POSITIVE SHUTTLE MOTION for a narrow fabric loom. The shuttle has a narrow recessed rack geared through a set of pinions to the reciprocating rack.

1145*k*. A CURIOUS PADLOCK.—The key is like a cork screw. The circular recess in front contains a rotating cylinder with a spiral keyway and graduated face plate, which must be set to a number that will allow the key to enter the internal spiral passage to push back the bolt.

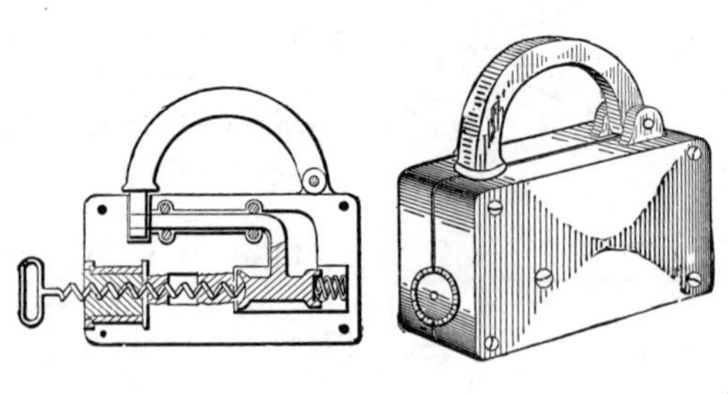

Section XIII.

HOROLOGICAL.

Clock and Watch Movements and Devices.

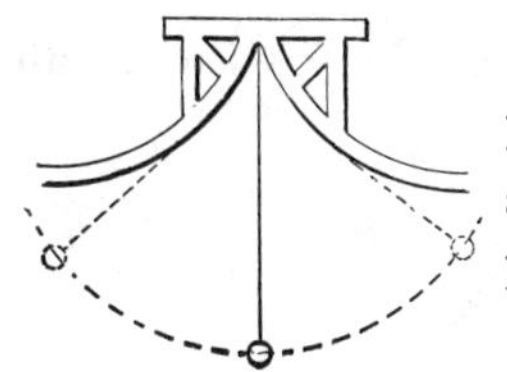

1146. CYCLOIDAL PENDULUM MOVEMENT.—A curved frame, acting as a stop to a flexible pendulum, gives the bob a cycloidal path.

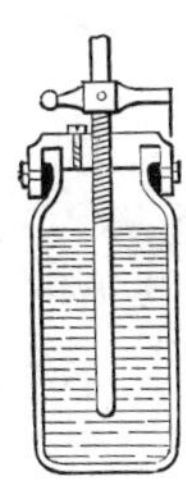

1147. COMPENSATING PENDULUM BOB or weight. A glass jar of mercury is used for the weight, and is adjusted for length of pendulum by turning on the screw and locking in place by the cross-piece and catch. The expansion of the pendulum downward is balanced by the expansion of the mercury in the fixed bottle upward, and *vice versa.*

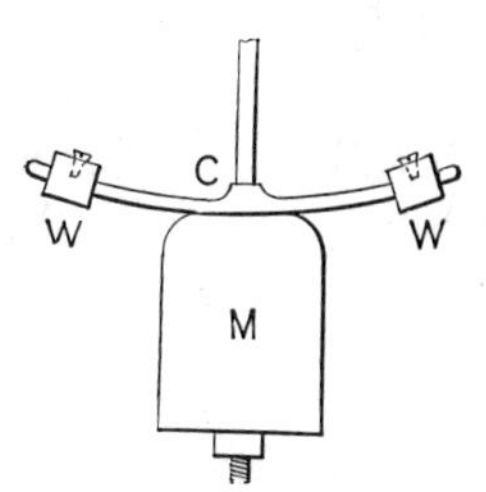

1148. COMPOUND COMPENSATION PENDULUM.—The arms of the pendulum carrying the weights W, W are composed of two metals; steel, which has the least change of length by change in temperature, for the top section, and brass, which has a longer range of length, for the lower section. Heat, by differential expansion of the parts, raises the weights to compensate for lengthening of the pendulum rod, and *vice versa.*

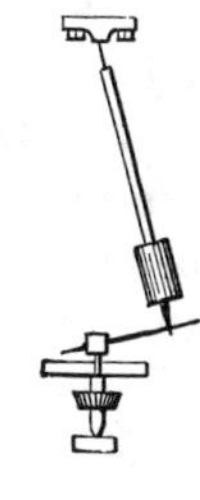

1149. CENTRIFUGAL PENDULUM.—The weight or ball is hung by a thread or very fine wire from an eye, and is driven in a circle by an arm attached to a vertical spindle, rotated by the clock movement. Adjustment is made for time of beat by the vertical movement of the suspension eye of the pendulum.

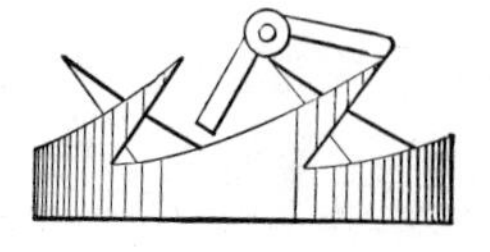

1150. ANTIQUE CLOCK ESCAPEMENT.— The oscillation of the pendulum arbor and attached pallet stops and releases the teeth of the crown wheel.

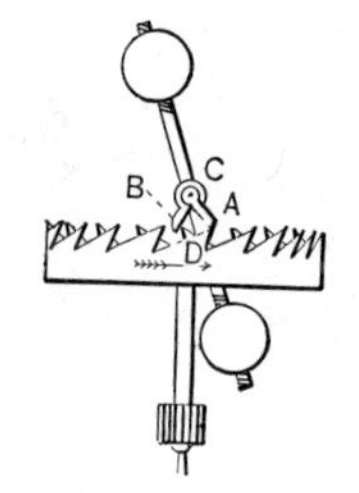

1151. CROWN TOOTH ESCAPEMENT, with ball balance.

B, the stop pallet.

A, the impulse pallet.

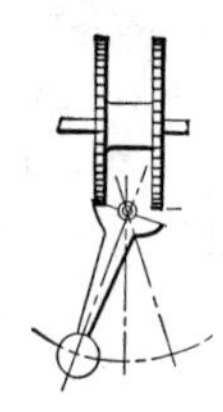

1152. DOUBLE RATCHET-WHEEL ESCAPEMENT and pendulum. The teeth in the escapement wheels alternate with the pallets of the pendulum.

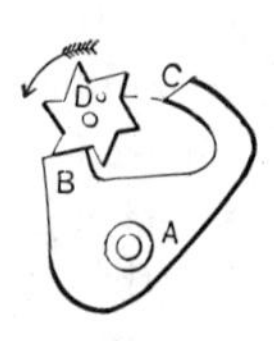

1153. STAR-WHEEL ESCAPEMENT.—B, C, the pallets of the escapement vibrating on its centre at A; D, star wheel.

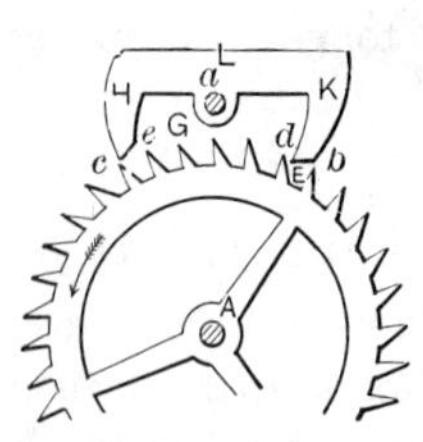

1154. ANCHOR ESCAPEMENT for clocks. The anchor pallet H, L, K oscillates on its axis *a*, by the swing of the pendulum. The teeth of the escapement A are radial on their forward face, and strike the curved faces of the pallet K or H, which are concentric with their axis *a*. By this form of teeth and pallets the escapement is anchored or in repose during the extreme parts of the pendulum stroke, and gives an impulse to the pendulum while the teeth are in contact with the planes of the pallets *c*, *e* and *b*, *d*.

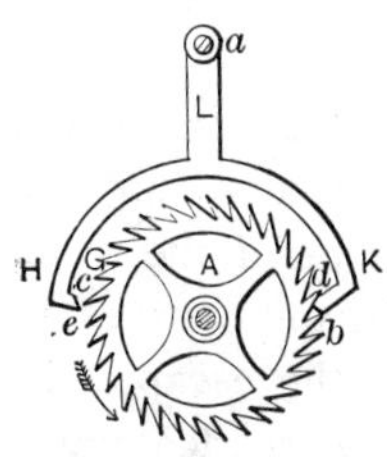

1155. RECOIL ESCAPEMENT. — In this form the forward face of the teeth of the escapement A leans forward from the radial lines. The front face of each pallet is in line with the front face of the teeth, so that the extreme part of the pendulum stroke gives a recoil movement to the escapement wheel. The points of the escapement teeth, acting upon the planes of the pallets *c*, *e* and *b*, *d*, give the impulse to the pendulum.

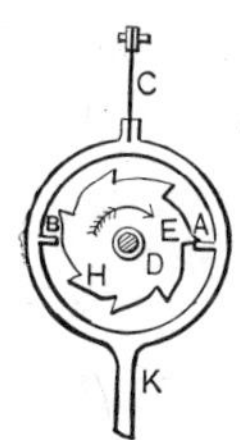

1156. PENDULUM ESCAPEMENT.—In this form the upper part of the pendulum terminates in a ring around the escapement wheel, with pallets A, B projecting inward and with a forward pitch to their face, to give the proper impulse to the pendulum.

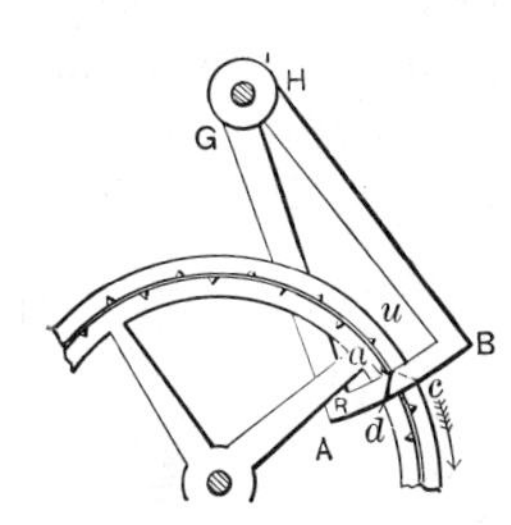

1157. STUD ESCAPEMENT, used in large clocks. Alternate studs are set on front and back of the escapement wheel. The pendulum swings on the axis of the pallet at F. The concentric curve of the stop-faces of the pallet, with its axis at F, gives the escapement a dead-beat action, the incline planes of the pallets giving the alternate impulse.

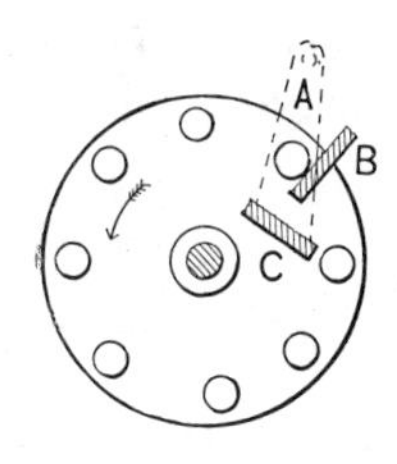

1158. LANTERN-WHEEL ESCAPEMENT. —The pallet arm A is attached directly to the pendulum, swinging upon the axis A, and receives its impulse from the inclined faces of the pallets C, B. Used for large clocks.

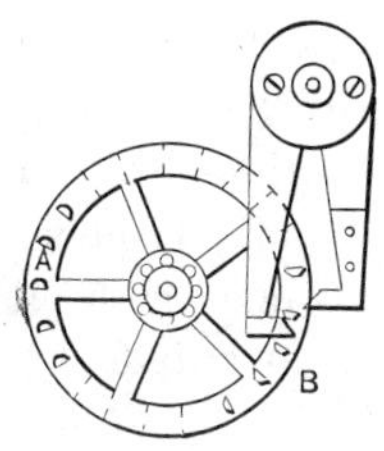

1159. PIN-WHEEL ESCAPEMENT, with a dead-beat stop motion. For short-beat pendulum clocks.

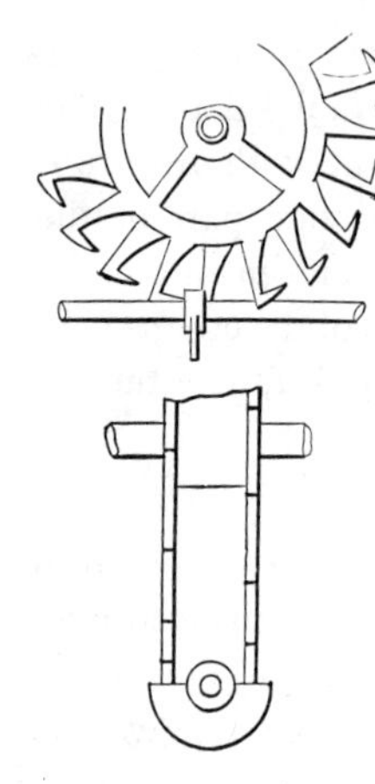

1160-1161. HOOK-TOOTH ESCAPEMENT. —The teeth are arranged alternately on two escapement wheels. The oscillation of the semi-circular pallet alternately releases and receives an impulse from the hook teeth of the escapement wheel. The curved outer face of the teeth acts upon the edge of the straight edge of the disc.

1162. SINGLE-PIN PENDULUM ESCAPEMENT. —The pin is set in a small face plate close to the arbor, which makes a half-rotation at each stroke of the pendulum. The impulse is given on the vertical faces of the quarter sections in the pendulum.

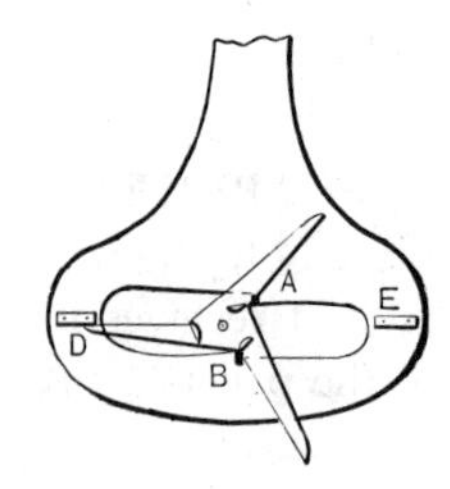

1163. THREE-TOOTHED ESCAPEMENT with long teeth and stops on the pendulum frame. A, B, pallets; E, D, stops. A nearly dead-beat movement.

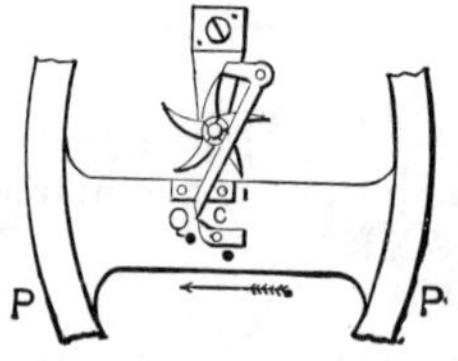

1164. DETACHED PENDULUM ESCAPEMENT.—In this movement the pendulum is detached from the escapement, except at the moment of receiving the impulse from the single pallet I. The bell-crank lever unlocks the escapement tooth by contact with the balanced click C as the pendulum nears the middle of its stroke.

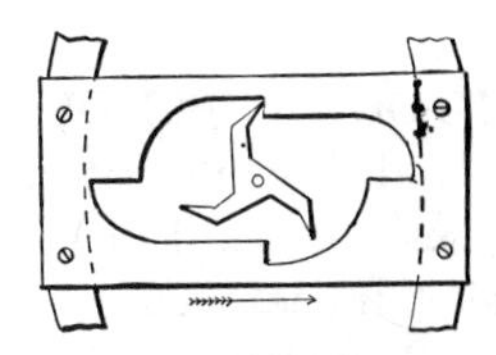

1165. THREE-TOOTHED ESCAPEMENT for a pendulum. The pallets are made in a plate attached to a pendulum. The escapement makes one rotation to every three beats of the pendulum.

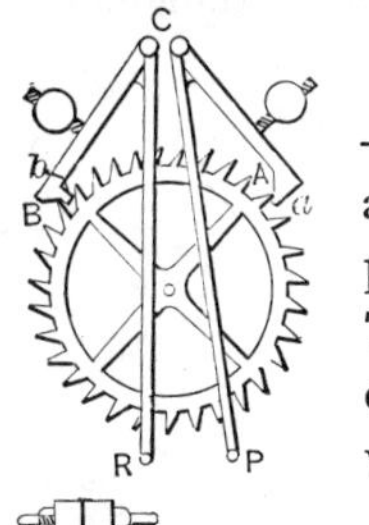

1166. MUDGE GRAVITY ESCAPEMENT.—The pallets A, B are on separate arbors, with arms extending down to the pendulum contact pins R, P, between which the pendulum swings. The pallets are loaded with weights. The pendulum lifts the pallet over the tooth, and the weight gives the impulse.

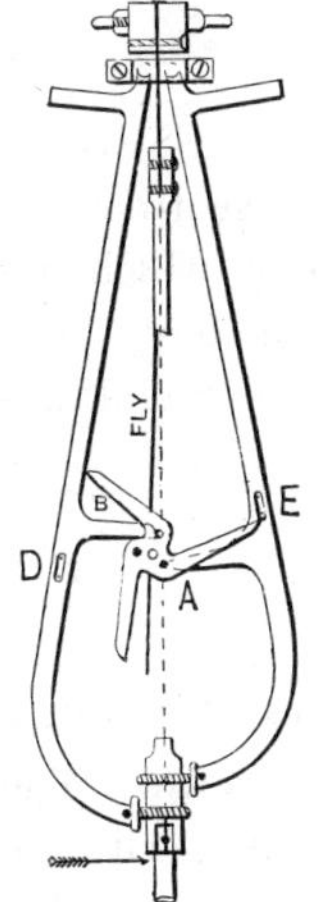

1167. TRI-TOOTH PENDULUM ESCAPEMENT.—Impulse is given to the pendulum by contact of the pins against the pallets A and B alternately. The stops D and E hold the escapement during the extreme part of the pendulum stroke. The escapement makes one rotation every third stroke of the pendulum. The fly softens the strike of the pins upon the pallets.

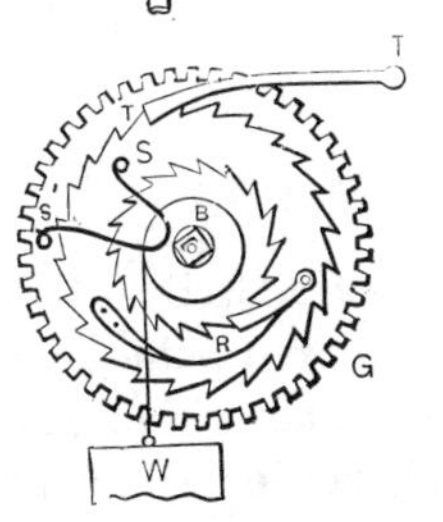

1168. "HARRISON" WINDING DEVICE for clocks, and which may also be adapted to a spring barrel. G is the driving spur gear. The larger ratchet has a fixed check pawl, T; is loose on the arbor, but attached to the gear wheel by a curved spring, S, S′. The smaller ratchet is fixed to the winding barrel and arbor. The spring and pawl R are pivoted to the larger ratchet, and stop the barrel against the weight W. The curved spring S is compressed and drives the gear wheel, and by its elasticity continues, while winding, by the check pawl T falling into the teeth of the large ratchet.

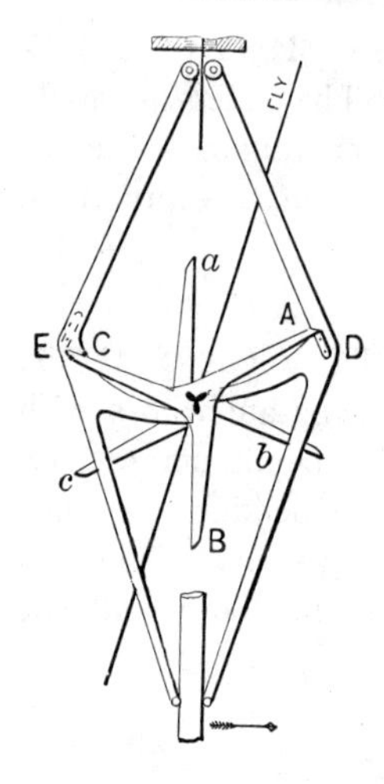

1169. DOUBLE TRI-TOOTH PENDULUM ESCAPEMENT with fly regulator. The alternate teeth of the escapement lock on opposite sides of the pallet frame. The impulse is given by the small triangular arbor striking the curved pallets.

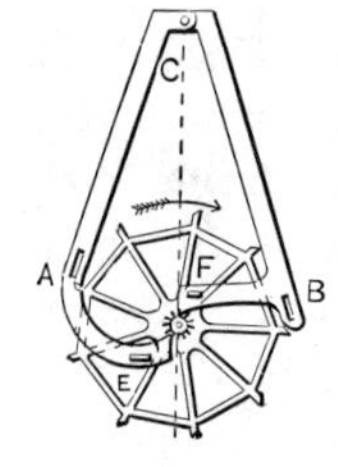

1170. "BLOXAM'S" GRAVITY ESCAPEMENT.—The pallets receive an impulse from the small toothed wheel, the long arms of which are stopped by the studs A and B alternately. The studs at F and E are the fork pins which embrace the pendulum bar.

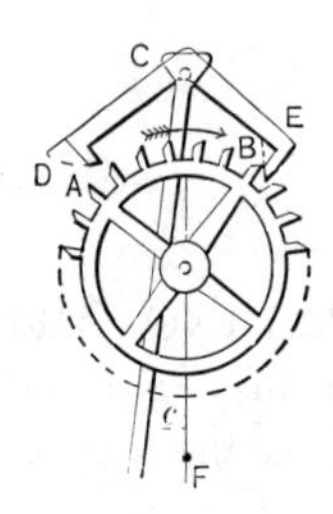

1171. DEAD-BEAT CLOCK ESCAPEMENT.—The face of teeth is slightly pitched forward. The stop-faces of the pallets A, B are concentric with the axis, which gives the dead-beat stop.

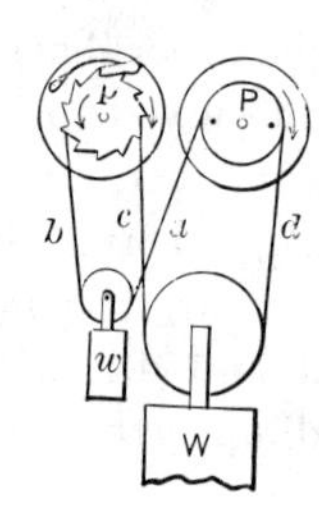

1172. ENDLESS CORD-WINDING DEVICE for clocks. The cord runs over grooved pullies. P is the driving wheel, and *p* the ratchet winding arbor, the turning of which by crank, key, or by pulling the cord *b* raises the driving weight W, and lowers the balance weight *w*. By this device the movement of the escapement is not suspended while winding the clock.

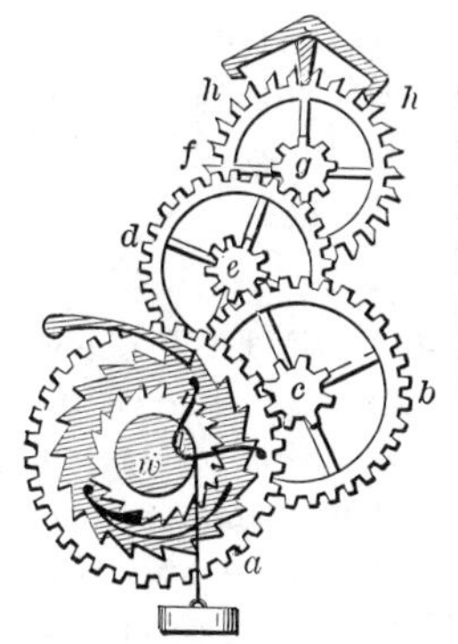

1173. CLOCK TRAIN, showing the method of sustaining the movement of the train during the time of winding. The bent spring keeps a tension on the large gear by the locking of the large ratchet to which the bent spring is attached, when the winding of the barrel can be made without a back-set in the train.

See No. 1168.

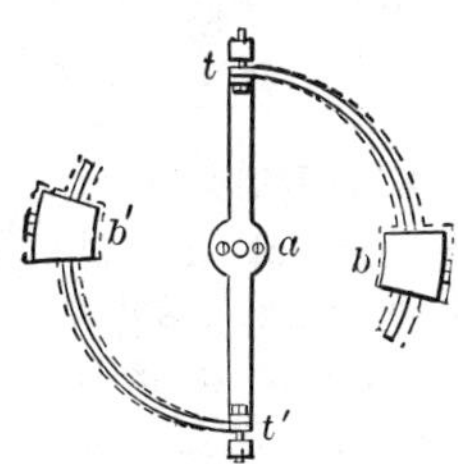

1174. COMPENSATION WATCH BALANCE.—At the ends of the balance bar are attached compound sector bars, the inner section of which is of steel, and the outer section of brass. The weights *b*, *b* regulate the momentum of the balance wheel, while the change in length of the arms is compensated by a reverse distance of the weights. Adjustment is made by moving the weights along the compensating sector.

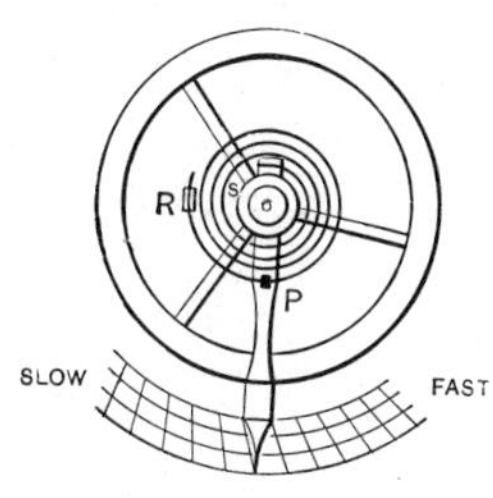

1175. WATCH REGULATOR.—The outer end of the balance spring is fixed to a stud at R, and the inner end to the balance wheel arbor. The index hand carries two curb pins at P, between which the spring vibrates, forming a neutral point in its length which limits the arc of movement of the balance wheel, and by its change of position (by moving the index hand) adjusts the time beat of the balance wheel.

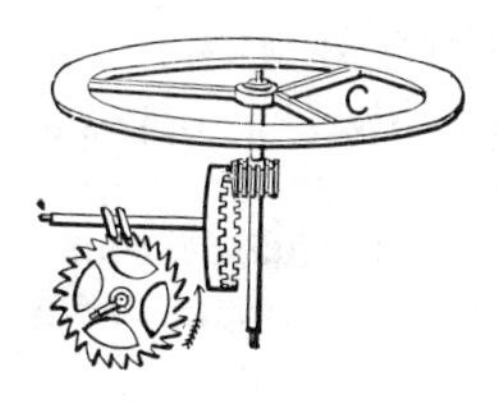

1176. ANTIQUE WATCH ESCAPEMENT.—A pinion on the balance-wheel arbor meshes in a crown gear, on the shaft of which a mutilated screw of large pitch releases the teeth of the escapement and gives an impulse by the incline of the screw.

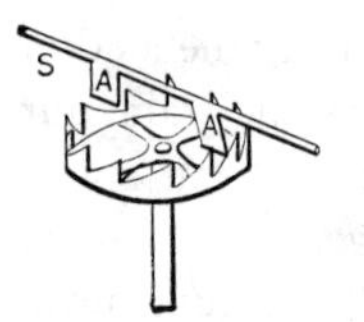

1177. VERGE ESCAPEMENT.—The arms of the escapement are set at an angle with each other, and its oscillation allows a tooth of the crown wheel to pass with each oscillation.

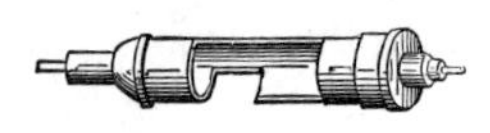

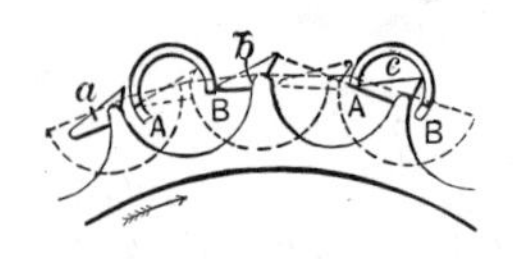

1178. CYLINDER ESCAPEMENT, shows the form of the cylinder, and 1179 shows the method of action. The oscillation of the cylinder allows the teeth of the escapement wheel to pass under the open hollow side and stop against its outside. The impulse from the escapement teeth is given to the edge of the cylindrical section.

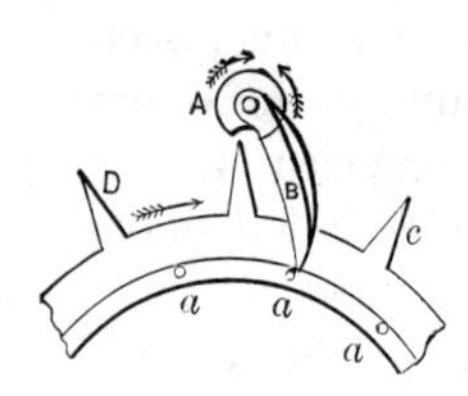

1180. DUPLEX ESCAPEMENT.—A, the balance-wheel stop; B, the oscillating pallet fixed to the balance-wheel shaft and adjusted to receive a strong impulse from the studs *a*, *a*, *a* at the moment the escapement tooth falls into the notch in the stop A.

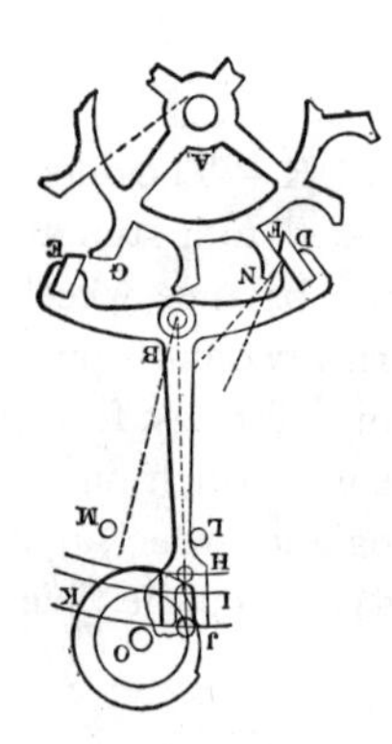

1181. JEWELLED DETACHED LEVER ESCAPEMENT.—D, E, jewel pallets; J, roll jewel in the arbor disc; L, M, lever stops; H, balance-wheel stop.

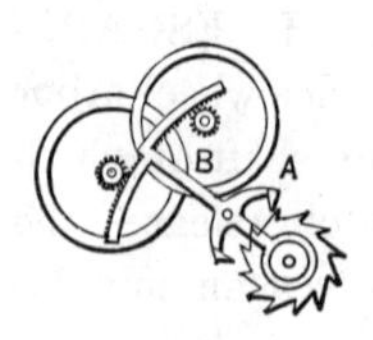

1182. "GUERNSEY" ESCAPEMENT, consisting of two balance wheels driven in opposite directions by an inside and outside sector gear on the pallet lever, with the ring guard around the escapement axle. To prevent stopping of a watch by a jar.

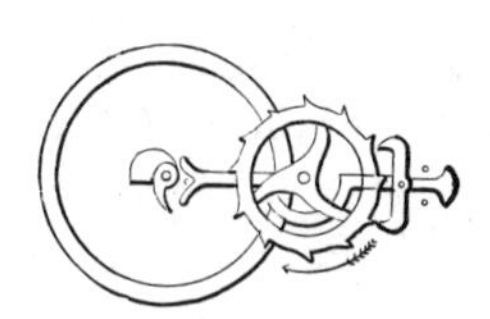

1183. ANCHOR AND LEVER ESCAPEMENT for watches. "Reed's" patent.

1184. LEVER ESCAPEMENT.—The anchor pallet B is attached to the lever C E, at the end E of which is a notch to receive the pin in the balance-wheel disc D. The impulse is given to the balance wheel at the middle of its oscillation by the escape of the teeth from the stop surface to the impulse planes of the pallets.

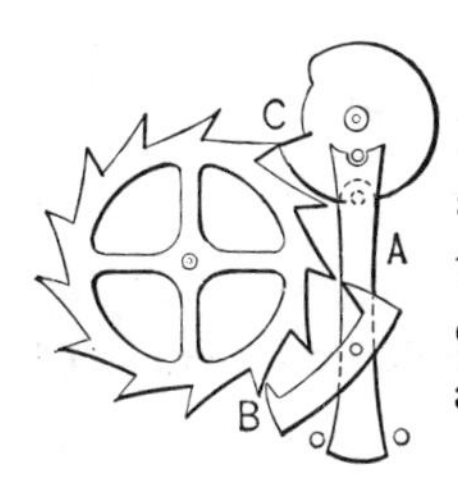

1185. LEVER CHRONOMETER ESCAPEMENT, single-pallet impulse. The lever pallets alternately lock the escapement by the throw of the lever; the oscillating pin on the pallet disc drops into the fork of the lever, throwing it against the stop pins at its other end.

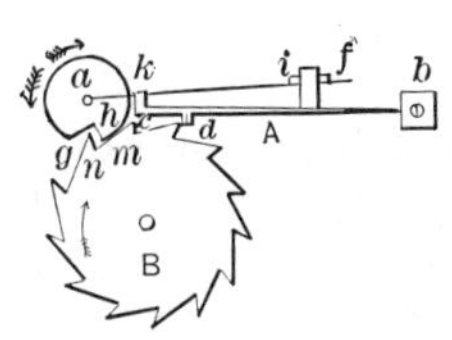

1186. "ARNOLD" CHRONOMETER ESCAPEMENT.—The spindle of the oscillating pallet *a* carries a small stud that vibrates the light spring *i*, in the hook *k*, of the stop spring A. The stop *a* catches and holds a tooth of the escapement while a reverse oscillation of the pallet *a* is made, when the stop *d* is lifted by the action of the stud at *a*, and an impulse given to the balance wheel by the tooth *n*, striking the face of the notch at *h* in the pallet.

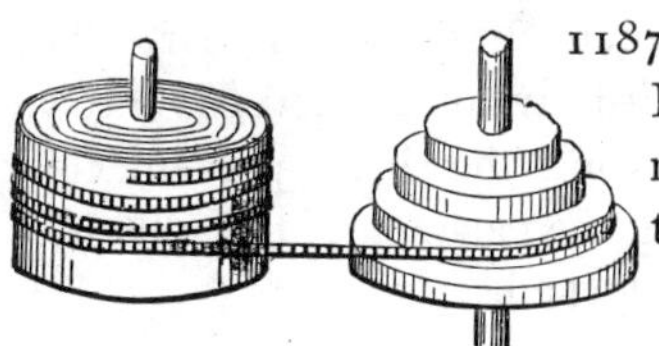

1187. FUSEE CHAIN AND SPRING DRUM, used in watch and clock movements. This device compensates for the variation in the force of the spring.

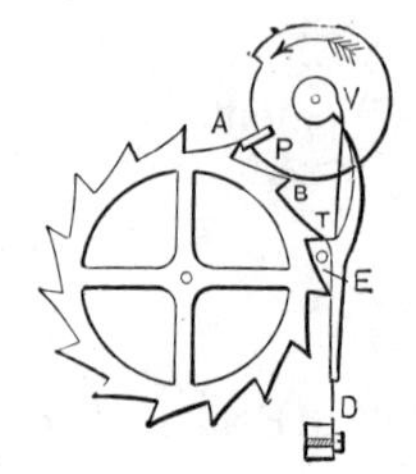

1188. CHRONOMETER ESCAPEMENT.—P, the impulse pallet on the arbor disc of the balance wheel; V, a release tooth on the arbor which strikes the end of the stop lever and releases the escapement at the moment that the tooth A falls in mesh with the pallet P. At the return oscillation of the balance wheel the tooth V on the arbor carries the spring forward, holding the lever and catch in lock against the pin E.

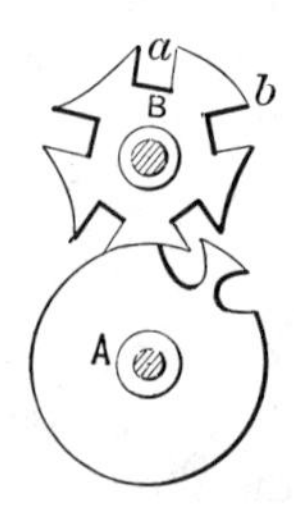

1189. "GENEVA STOP."—A winding-up stop used on watches. Winds as many turns of the wheel A as there are notches in wheel B, less one. The curve *a b* is the stop.

1190. GEARED WATCH STOP.—Contact of the two arms makes the stop.

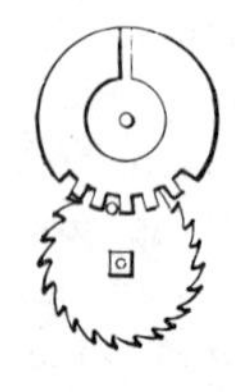

1191. WATCH STOP.—The number of turns of the ratchet pinion is limited by the number of teeth in the stop. The pin moves one tooth for each turn.

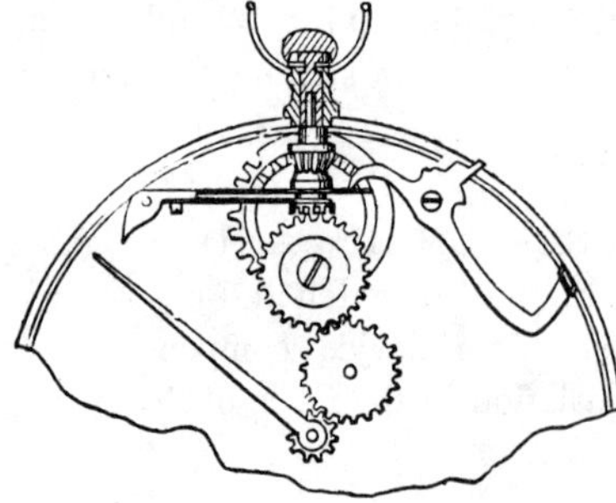

1192. STEM-WINDING MOVEMENT of a watch. The movement of the lever with an arm outside of the rim locks a clutch on the hand gear. The third arm of the lever is thrown beyond the rim to prevent closing the case until the clutch is unlocked.

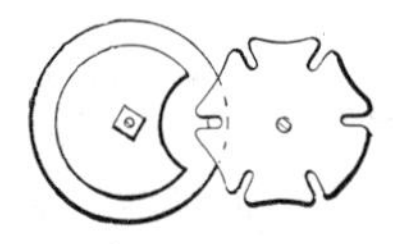

1193. PIN-GEARED WATCH STOP.—The winding stops at the convex tooth of the stop.

1194. WATCH TRAIN.
a, key stem.
b, barrel and spring.
c, *e*, *h*, *i*, pinions.
d, *h*, spur wheels.
l, *l*, pallets and escapement.
k, lever and balance wheel.

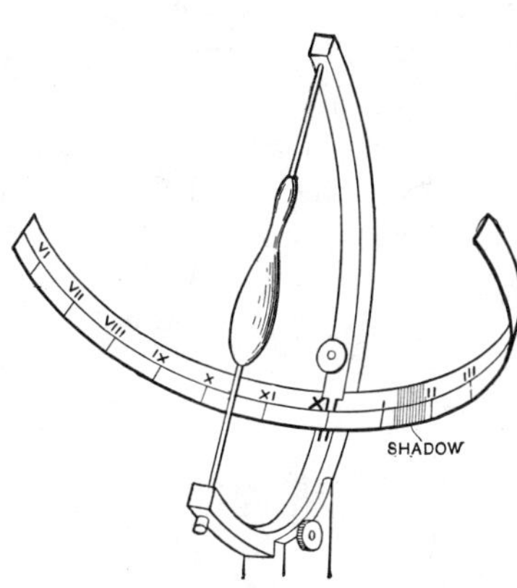

1194*a*. EQUATED SUN DIAL.—The curved bulbs on the shadow stile are made to conform to the equation of solar time. The end of the upper bulb represents the sun's declination at the summer solstice, the lower end of the large bulb the winter solstice of the shadow on the gnomon. The following edge of the shadow is the correct time when the sun is fast, the middle of April to the middle of June, and from September 1 to December 24. The forward edge of the shadow is the correct time from the middle of June to Septembe 1, and from December 24 to the middle of April.

1194*a*. CLOCK-SETTING DEVICE.—This invention shows the use of a simple clutch on the shaft which carries the escapement device, throwing the train of wheels out of connection with the shaft which carries the hands. Thus the latter could be rigidly mounted on their shaft and sleeve, instead of being revolved by frictional contact.

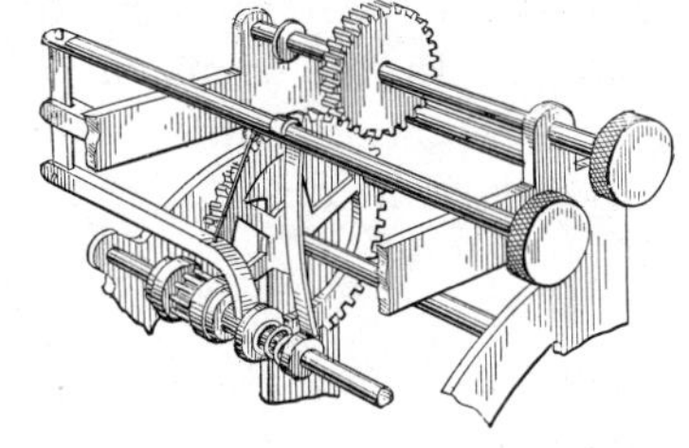

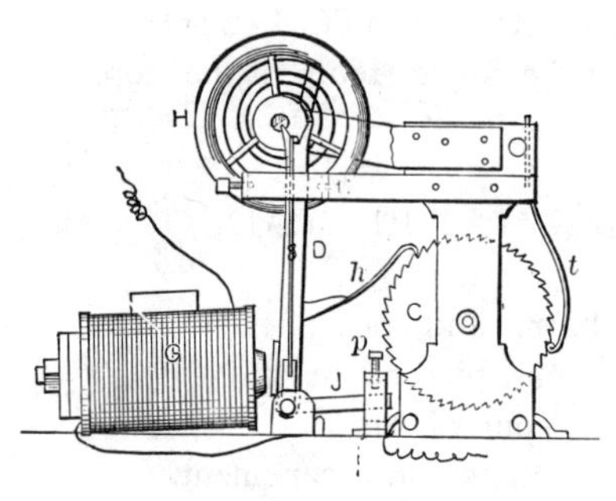

1194*b*. ELECTRIC BALANCE WHEEL CLOCK.—G represents the magnet, H the balance wheel, D armature, *m* a pin on the balance wheel, C ratchet wheel, *h* and *t* are pawls, P adjusting screw. When the circuit is closed the magnet draws the armature D forward, whereby the ratchet wheel C is turned the distance of one of its teeth. At the same time the crotched end of the lever, by means of the pin, gives an impulse to the balance wheel in one direction. The pin is suddenly released from the spring S, which in its recoil, aided by the weight of the arm J, breaks the circuit. The return movement of the balance wheel, caused by the recoil of the hair spring, moves, by means of the pin, the lever D away from the magnet, so as to set the pawl *h* on the next tooth of the wheel C, and the pin will again pass by the upper end of the spring S.

1194*c*. COMPENSATING PENDULUM.—The heavy black lines represent steel rods; the open lines the brass rods. The relative expansion of steel is 2, brass 3. The center rod is fixed to the lower cross head at the top and slides freely through the cross heads at the bottom. In the combination shown the length of the compensating frame should be one-third the length of the pendulum.

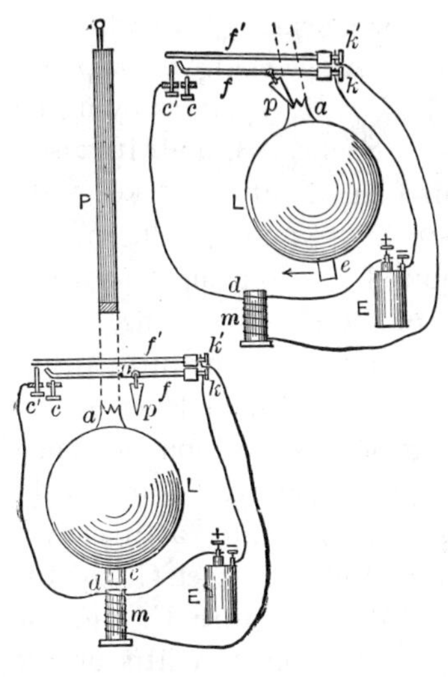

1194*d*. ELECTRO-MAGNETIC CLOCK PENDULUM.—P is an ordinary pendulum with a notched piece *a* and an iron piece *e* attached, *m* is an electro-magnet. E is a battery; *f f'* are the springs which act as contact pieces attached to the battery. *p* is a steel piece called the pallet. As long as the pendulum is at its full swing the pallet will pass over the notched piece *a*, but should the arc of oscillation be lessened the pallet will catch in the notch, raise the spring *f*, complete the circuit, and the pendulum will receive an impulse from the magnet.

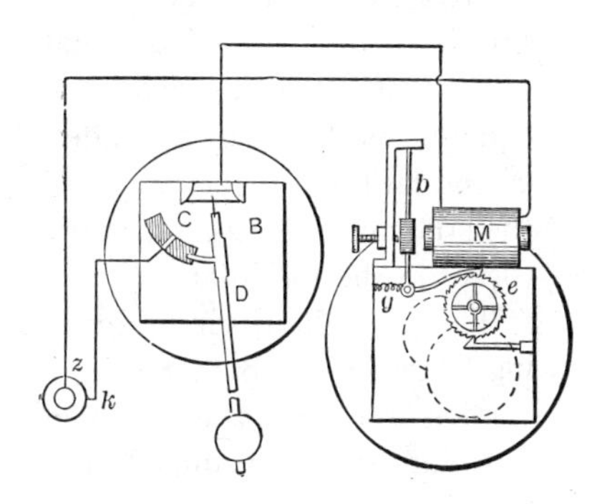

1194*e*. ELECTRIC TIME CLOCK TRANSMISSION. —Simultaneous beat of two clocks. B is the primary clock. M *b g e*, the secondary. ZK is the battery. The copper contact D is fastened to a pendulum of the primary clock. Every second this copper piece makes contact with the plate C, completing the circuit and energizing the magnet M of a secondary clock. This attracts its armature *b*, operates the pawl which moves forward and catches one of the teeth of the wheel *e*. As soon as the contact is broken at *c* a spring acting upon the armature *b* draws it away from the magnet, and at the same time the pawl moves the wheel one tooth forward. The wheel *e* may be connected direct to the second hand of the electric dial. If so, this hand will move in unison with the pendulum of the primary clock; that is, once in every second.

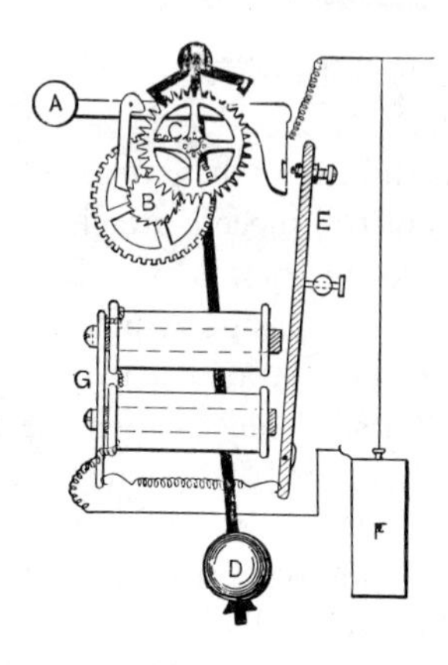

1194*f*. ELECTRIC WINDING DEVICE for clocks.—A weighted L-shaped lever A, working on a pin at the corner of the L, operates by means of a pawl the wheel B, and, if raised, falls of its own weight, and keeps the pendulum D swinging. The motive force required to effect this is provided by means of the electro-magnet G, the battery F, and the armature switch E. When the weighted lever A has fallen to its lowest position, it makes contact with the screwed point at the end of the armature E which rocks about a center at its lower end. This completes the circuit of the electro-magnet G and the battery F. The magnet then attracts the armature E, and the screw pressing the short arm of the cranked lever A lifts up the weighted end, so that the pawl rests on the next tooth of the ratchet B.

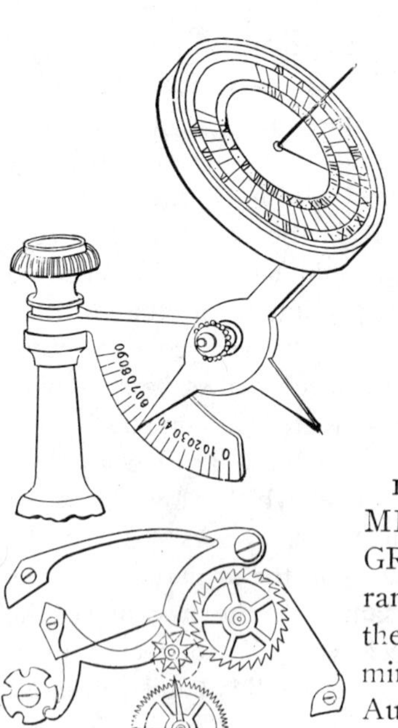

1194*g*. UNIVERSAL SUN DIAL.—The face of the dial to be placed parallel with the equator, as shown by the index of the latitude. The range of the stile and the 12-hour mark to be on the meridian.

The inner hour circle figuring is reversed so that by inverting the dial the summer morning and evening time may be observed.

1194*h*. NEW MOTION FOR THE MINUTE HANDS IN CHRONOGRAPHS.—A new mechanical arrangement by which in chronographs the instantaneous movement of the minute hand is effected. Invented by August Baud, of Geneva, Switzerland.

Section XIV.

MINING.

QUARRYING, VENTILATION, HOISTING, CONVEYING, PULVERIZING, SEPARATING, ROASTING, EXCAVATING, AND DREDGING.

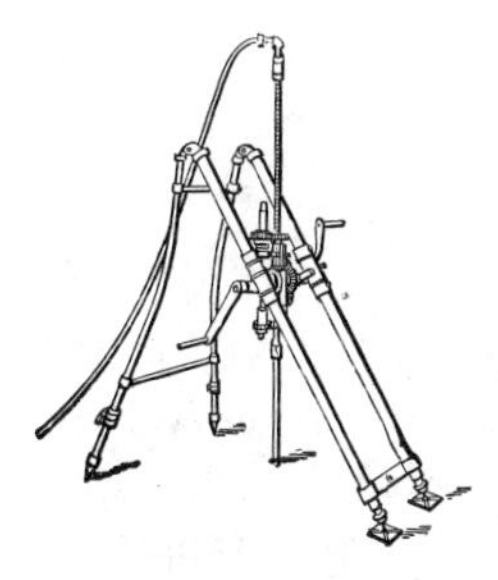

1195. DIAMOND PROSPECTING DRILL, operated by hand. The drill rod is hollow, with a hose connection at the top, through which water is forced to the bottom and up outside of the drill to wash out the borings. The drill point is set with bort or black diamonds, and is revolved quickly by the cranks and bevel gear.

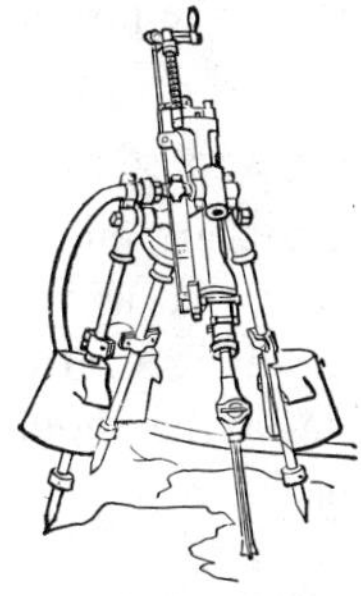

1196. ROCK DRILL, "Ingersoll" model. The loaded tripod gives stability to the reciprocating action of the drill.

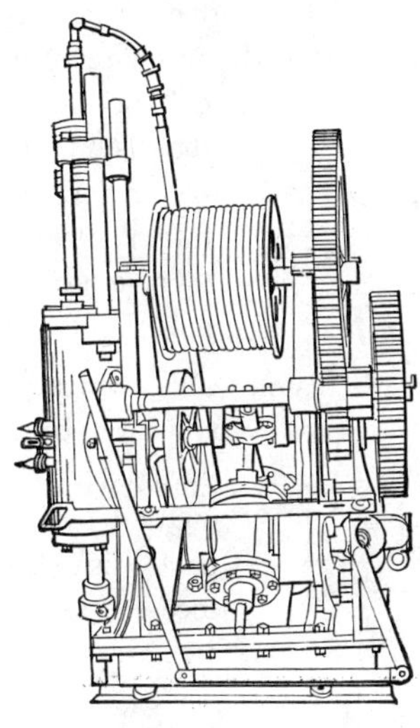

1197. DIAMOND WELL-BORING MACHINE.—A small oscillating engine and gear train drives the hollow boring auger at great speed, and also serves to hoist the drill rods by the drum and a rope over the block in the top of the derrick frame. Water is fed through the hollow drill rod by a pump.

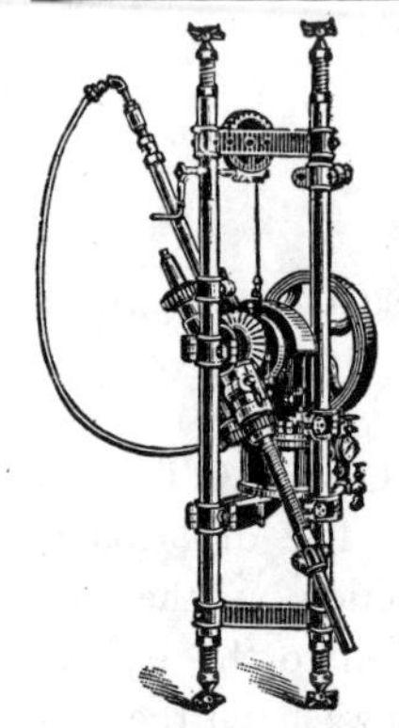

1198. PORTABLE DIAMOND DRILL, for tunnel work or mine drifting. A swivelled hose connection for feeding water to the drill. Screw-jacks in the frame for clamping. Hand-driven by crank and speed gear.

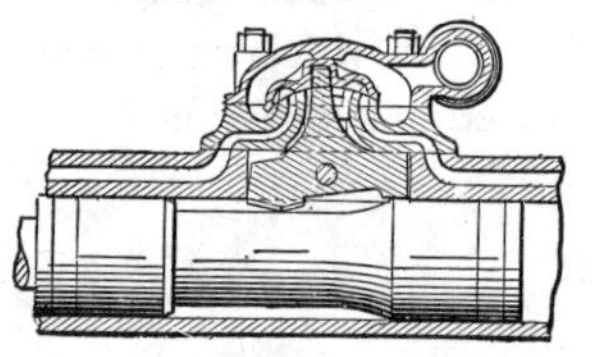

1199. ARC TAPPET VALVE MOTION, for a rock drill. The valve is moved on a circle radial with the tappet centre, and is thrown by the tappet-arm contact with the shoulders on the piston. "Sergeant" model.

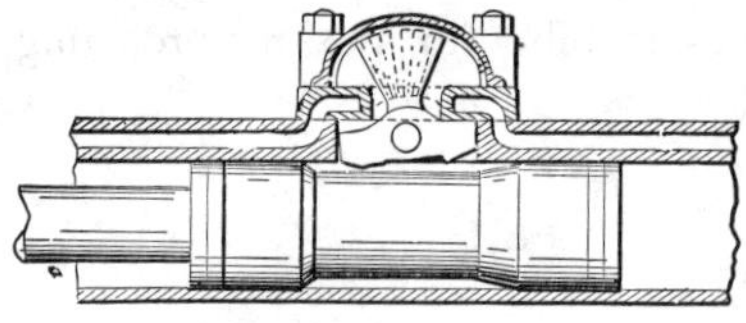

1200. TAPPET VALVE, for a rock drill. The ports are radial, and are opened and closed by the swing of the valve on its centre. The valve is thrown by the shoulders on the piston, striking the valve arms. "Sergeant" model.

1201. ROCK DRILL, with balanced piston valve, which is thrown by compressed air inlet through ports opened by the reciprocal motion of the piston. B, piston; M, rotation device. "Ingersoll" model.

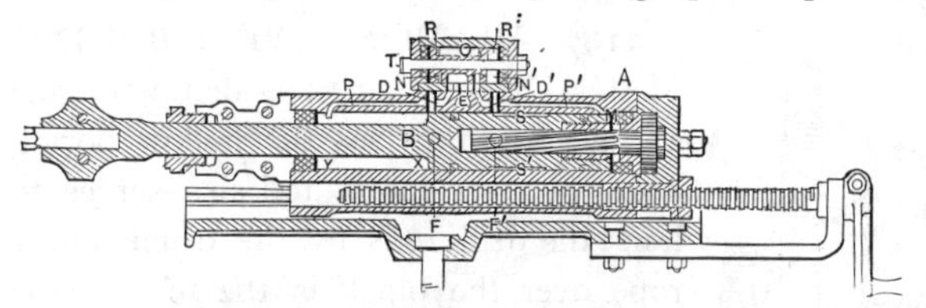

1202. ROCK DRILL, with balanced piston valve, which is thrown by a ported sector, moved by impact with the recessed shoulders on the piston. "Sergeant" model.

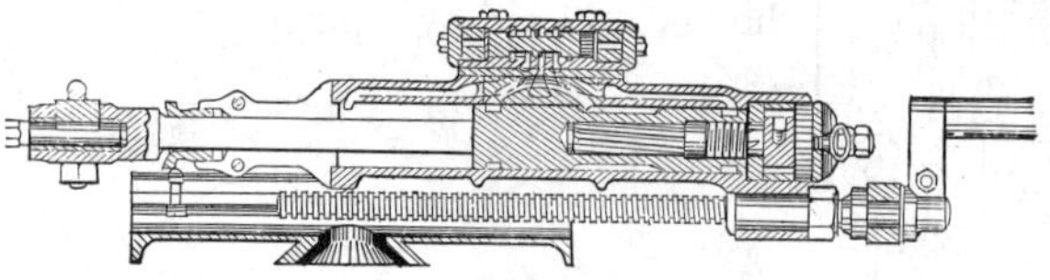

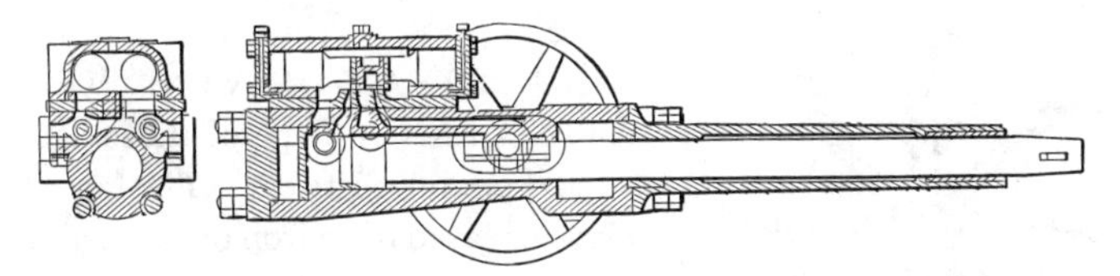

1203. COAL-CUTTING MACHINE, "Ingersoll-Sergeant" model. The piston and drill rod are automatically operated by the alternating motion of two piston valves. Operated by compressed air, and only has to be held against the coal wall to under-cut, when the face can be broken down.

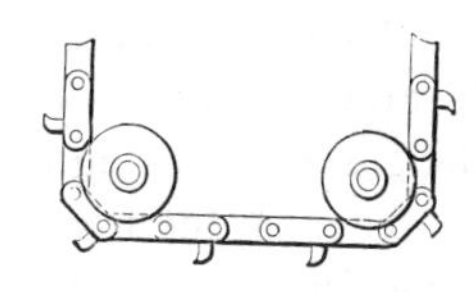

1204. LINK CHAIN CUTTER, used in coal-cutting machines.

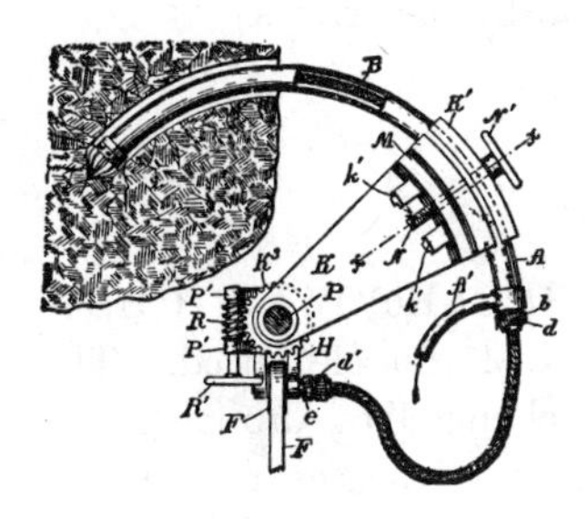

1205. DRILL FOR CURVED HOLES, used in coal mining. The drill is on the end of a curved tube, and is driven by a flexible shaft. The tube is fed forward by a pivoted arm and worm gear.

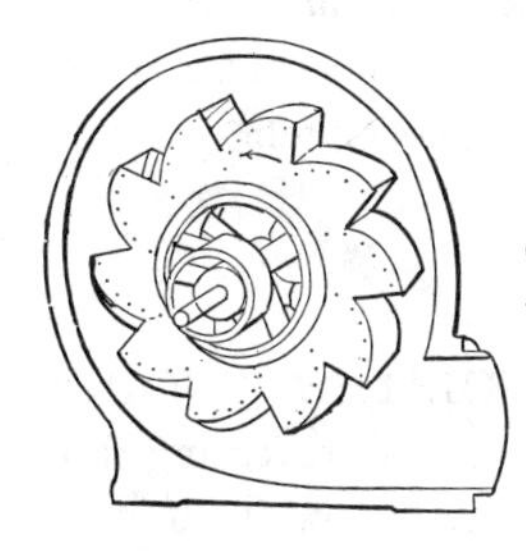

1206. BOX-WING BLOWER.—The discharge openings of the disc are rectangular, with the sides enclosed. Made of sheet metal.

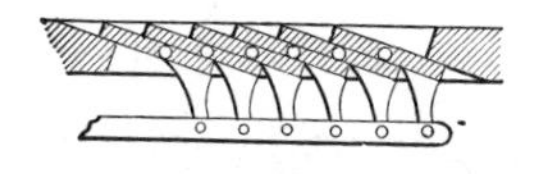

1207. MULTIPLEX BUTTERFLY VALVE, for ventilating shafts.

1208. STEAM-DRIVEN VENTILATING FAN.—Type of those used in the coal-mining districts. The fan wheel may be encased in an iron or wooden shell.

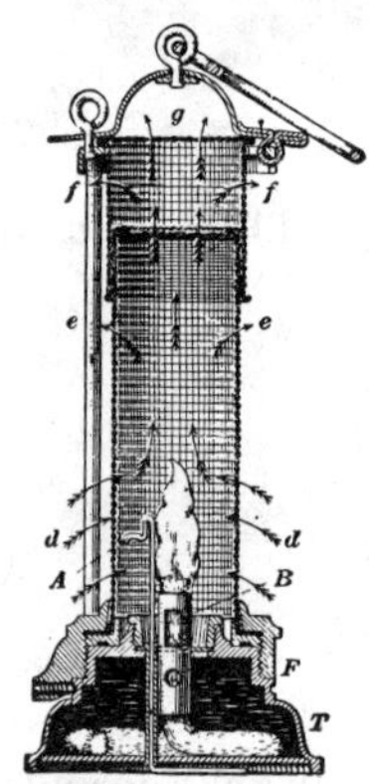

1209. MINER'S SAFETY LAMP.—The flame is surrounded with wire gauze and a double wire gauze cap. In explosive mine gases, the firing of the incoming air and gas takes place on the inside of the wire gauze. The flame does not pass through fine wire gauze. The course of air for the lamp burner is shown by the arrows.

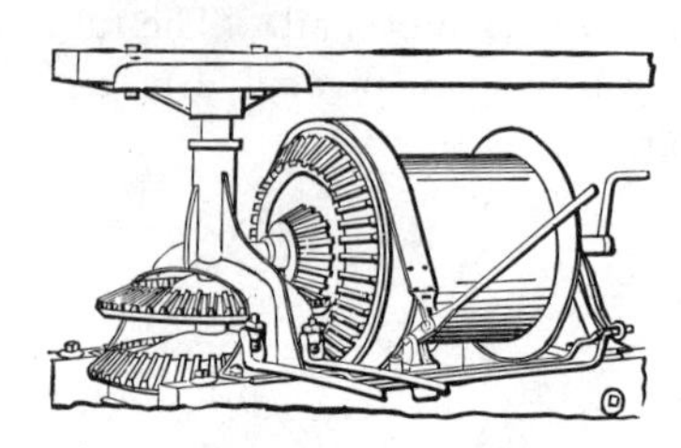

1210. HORSE-POWER HOISTING DRUM, double speed. The speed is changed by dropping one or the other driving gear by the levers. A release for running back is made by turning the crank which disengages the gear clutch.

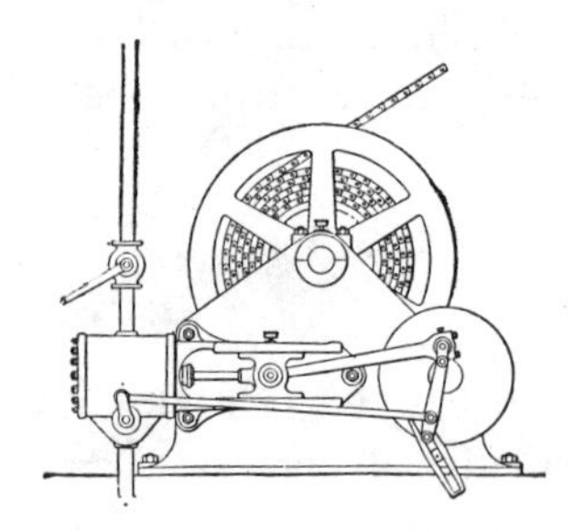

1211. STEAM HOISTING ENGINE, with flat chain drum and reversing link. The flat chain winds upon itself on a narrow drum.

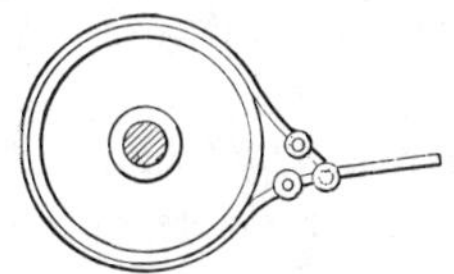

1212. STRAP BRAKE, used on hoisting drums and wheels. The strap is usually made of a steel band with its ends jointed to a lever.

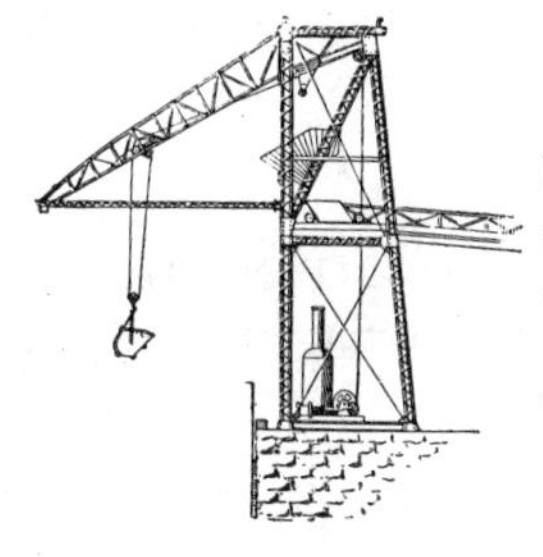

1213. ELEVATOR TOWER with inclined boom. The bucket is lifted to the trolley by the double tackle, drawn up the incline, and the load dumped automatically into a car.

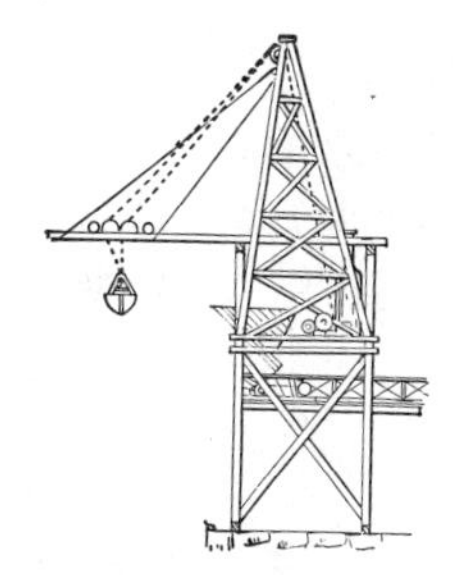

1214. HORIZONTAL BOOM TOWER, with traversing trolley and automatic shovel bucket.

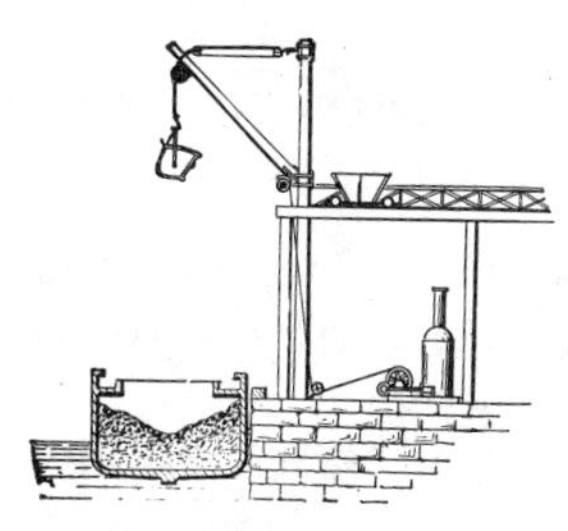

1215. MAST AND GAFF HOIST, for unloading coal barges to an elevated track. A portable boiler and steam hoist or an electric motor hoist, with occasionally a horse pull, are the motive powers.

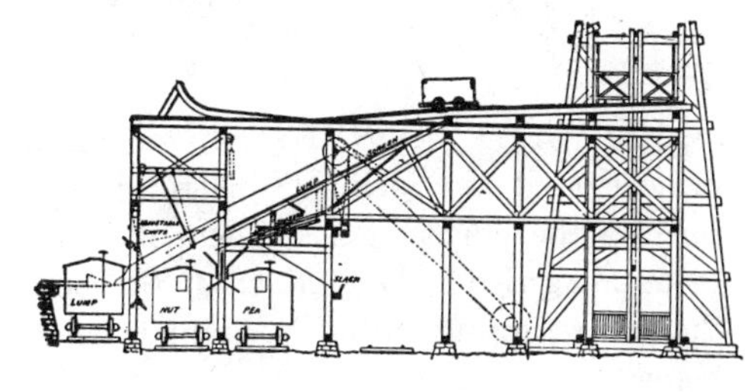

1216. COAL-LOADING TIPPLE and sorting screens for loading cars. The screens are inclined at the sliding angle and drop the slack, pea, nut, and lump into separate cars.

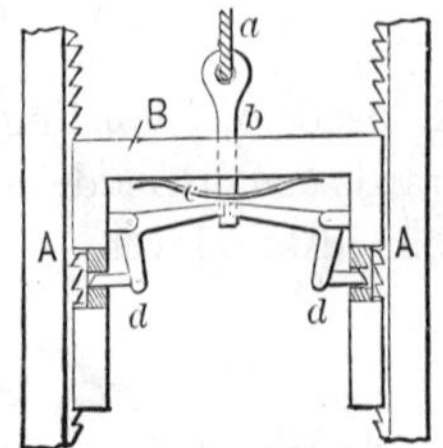

1217. "OTIS STOP" for elevator cars. B, car frame sliding on the ratchet posts A, A; *d*, *d* are the stop-dogs operated by bell-crank levers to thrust the dogs into the ratchets on the release of the eye bar *b*, by a break in the rope or hoisting machine. The spring *c* quickens the operation of throwing out the dogs.

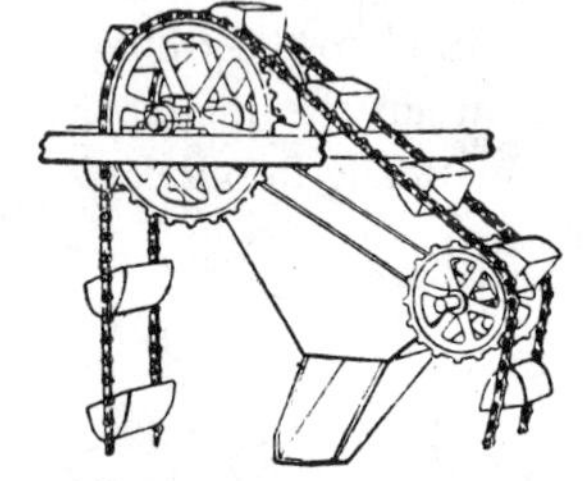

1218. ELEVATOR DUMPING HEAD, showing method of inverting the buckets over a hopper spout.

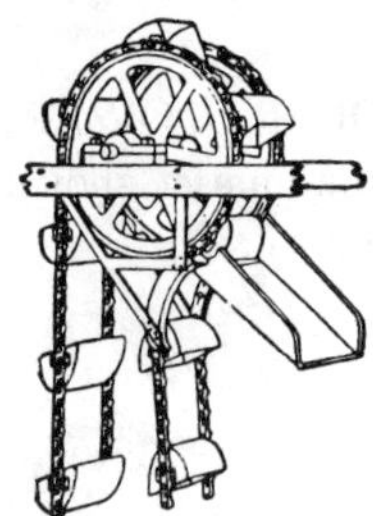

1219. ELEVATOR DUMPING HEAD.—An inverted sector frame guides the bucket chain under the head wheel, which allows the buckets a clean discharge.

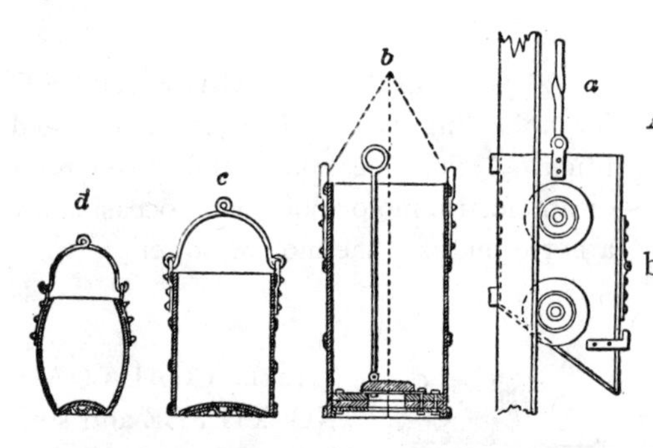

MINING BUCKETS AND SKIP.

1220. *d*, Cornish kibble.

1221. *c*, Hooped straight bucket.

1222. *b*, Water bucket.

1223. *a*, Tram skip.

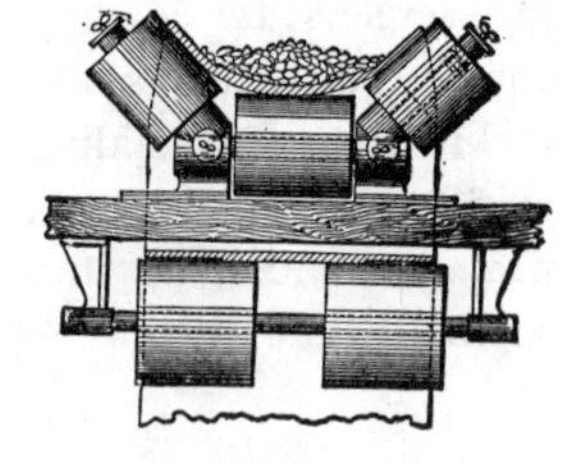

1224. BELT CONVEYOR.—A series of horizontal and inclined rollers serve to turn up the edges of a belt, enabling the material carried to be retained on the belt; the belt returning on the horizontal rollers below.

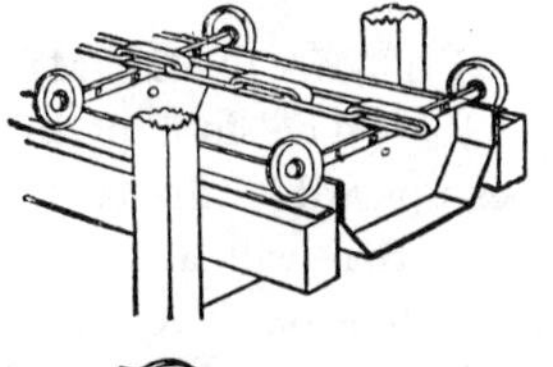

1225. CHAIN SCRAPER CONVEYOR.—A chain supported on rollers and axles to which scrapers are fixed that fit the conveyor trough.

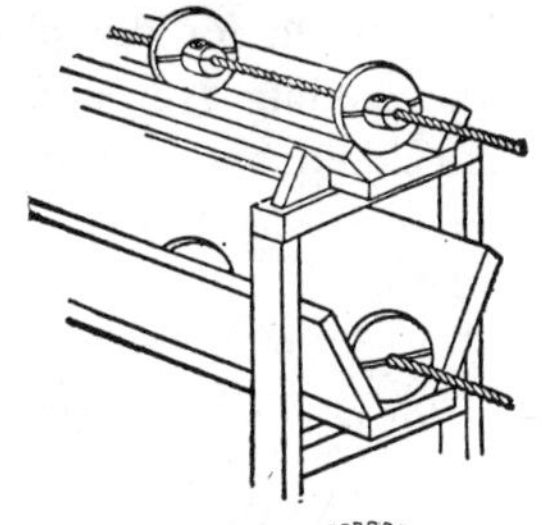

1226. CABLE CONVEYOR.—Discs fixed to a cable running in a trough and returning overhead.

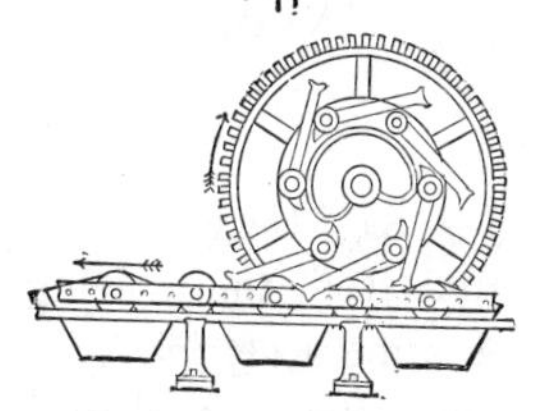

1227. DRIVING MECHANISM for a coal or grain conveyor. "Hunt" model. The heart cam is fixed. The face plate carrying the pawls revolves with the driving gear. The cam guides the pawls to lock with the pins in the chain and lifts them again into position for their next push.

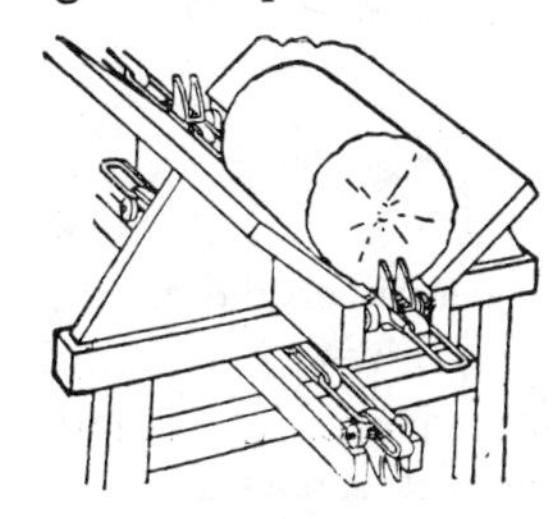

1228. LOG CONVEYOR.—A link chain with hooks running in a trough.

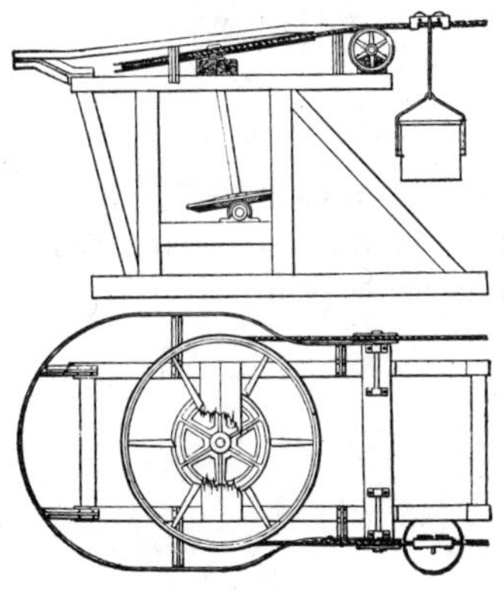

1229. ROPE TRAMWAY, overhead system. Elevation, showing the switch rails for transferring the carrier bucket around the terminal to the return rope. Loading or unloading of the bucket is done at the transfer switch.

1230. Plan showing the crossing of the switch rail over the carrier rope.

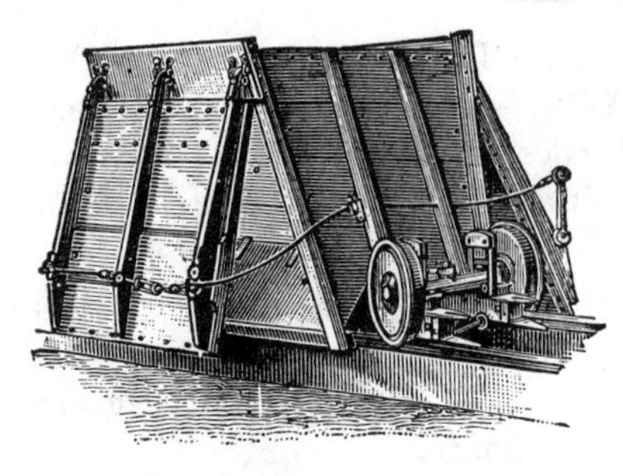

1231. AUTOMATIC DUMPING CAR. — The floor of the car slopes upward to the centre at an angle that will allow the material to slide out. A chock at any point desired for dumping trips the holding-lever and releases both side doors at once.

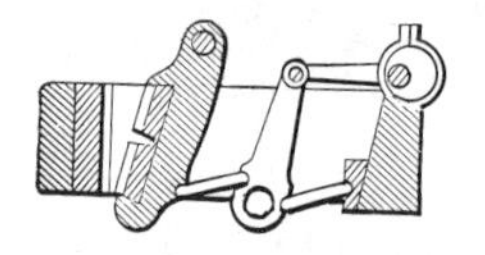

1232. TOGGLE JOINT, for a stone breaker.

1233. STONE CRUSHER. — The power is transmitted from the driving shaft by a cam operating a vertical connecting link and toggle joint. "Blake" pattern.

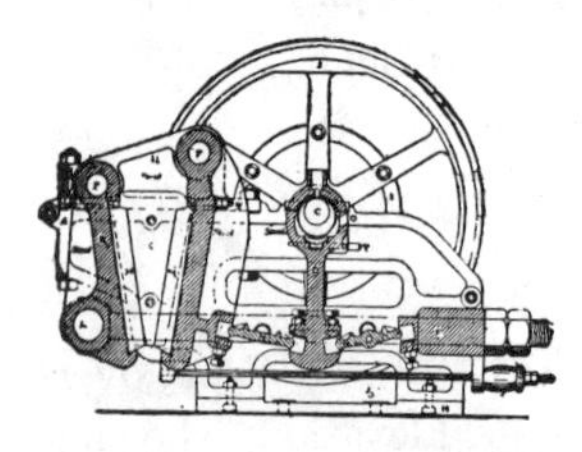

1234. "BUCHANAN" ROCK CRUSHER.—An eccentric on the driving shaft and toggle arm gives a powerful pressure to the crusher jaws. The adjustment is made by the back screws and side rods to set up the outside jaw.

1235. ROLLER COAL CRUSHER. —Driven by a direct-connected steam engine with screw gear.

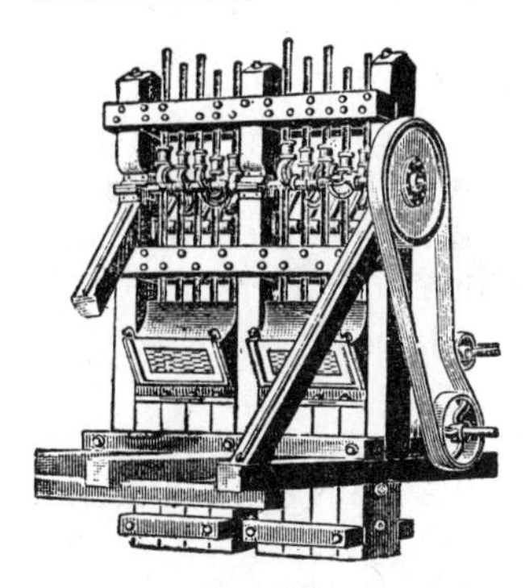

1236. EIGHT-STAMP ORE MILL, for pulverizing gold quartz or other ores. Cams on a power-driven shaft lift the bars successively to equalize the belt tension.

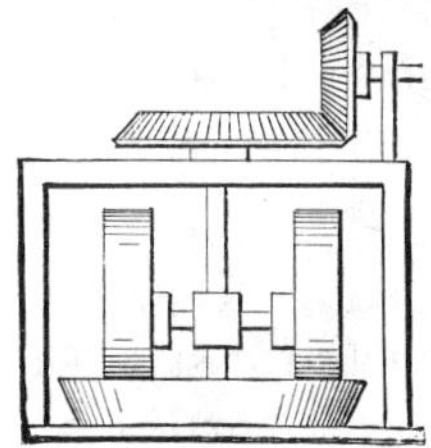

1237. ROLLING CRUSHER.—The "Arastra." Rolling wheels on a cross arm of vertical shaft.

1238. "ARASTRA" ORE MILL.—Two heavy rolls revolving in a circular trough, driven through a central shaft and overhead gear.

1239. "CHILI" MILL.—A three-roller ore mill. Rollers carried around by a shaft and three-armed crab. Ore is fed inside the rollers. The crushed ore washes into the annular trough and is carried to the amalgamators.

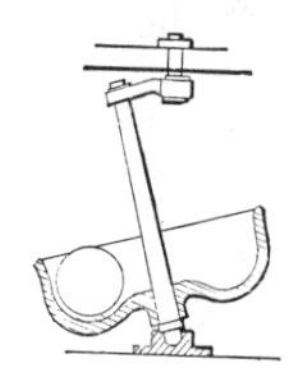

1240. PULVERIZING BALL AND PAN MILL.—The pan is continually tilted by being swung around the vertical centre, rolling the ball down the slope side of the pan.

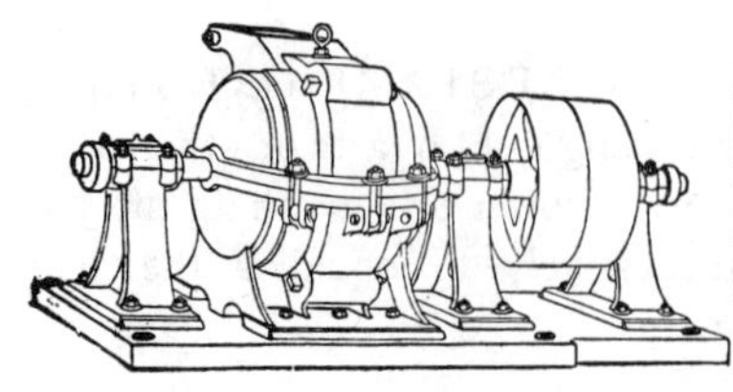

1241. REVOLVING PULVERIZING MILL.—The material is reduced to a fine powder by the high-speed impact of the revolving arms, within an iron casing.
"Frisbe-Loucop" model.

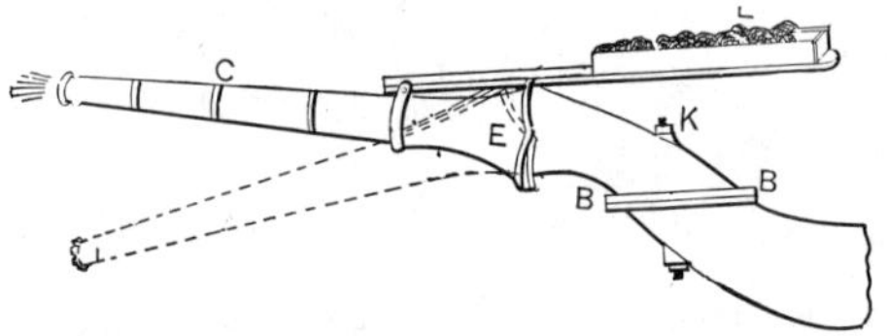

1242. HYDRAULIC BALANCED GIANT NOZZLE.—Used in hydraulic mining for washing away gravel banks. The nozzle turns on a movable joint at B B, and also in the vertical by the socket at E.

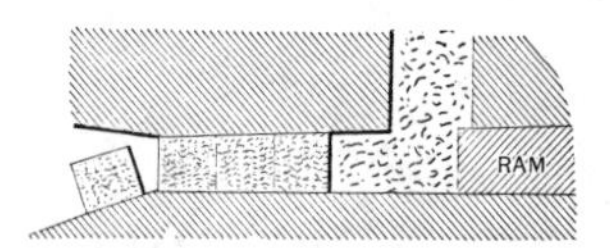

1243. COAL DUST PRESS for bituminous coal. The fine dust is fed down from a hopper. The nozzle has a slight taper, which gives the ram sufficient resistance to produce a solid cake at each stroke.

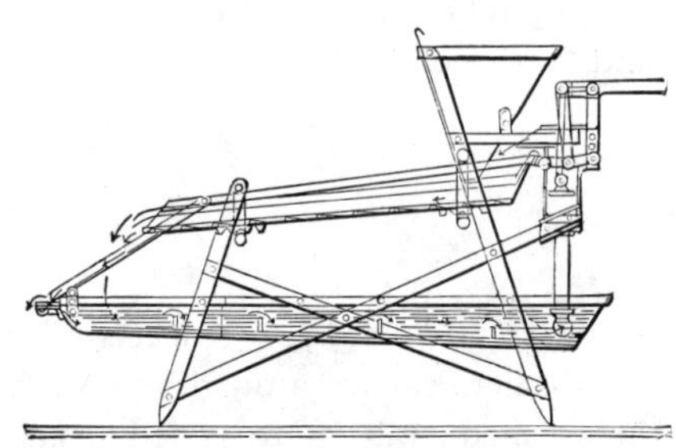

1244. KLONDIKE MINING MACHINE.—The gold-bearing gravel is shovelled into the hopper and is fed to the riffle pan, which is vibrated by the pump handle. The pump supplies water to the riffle pan, from which it falls into the settling pan beneath, and is kept from freezing by a fire underneath. "Lancaster" model.

1245. GOLD SEPARATOR; dry process. A bellows furnishes an air blast, which separates the fine sand and dust from the gold on the riffle screen and blows the dust away.

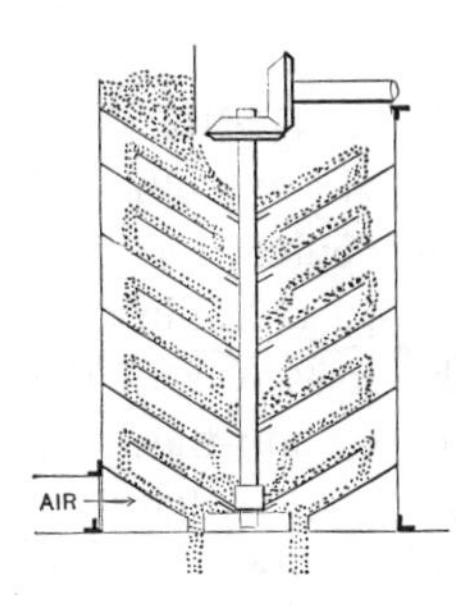

1246. CENTRIFUGAL SEPARATOR.—A central revolving shaft carries a set of conical perforated plates, between which perforated plates are fixed to the shell of the machine. Grain or other material is fed at the top, and an air blast at the bottom. Centrifugal action discharges the material at the periphery of the revolving plates, returning by gravity on the fixed plates.

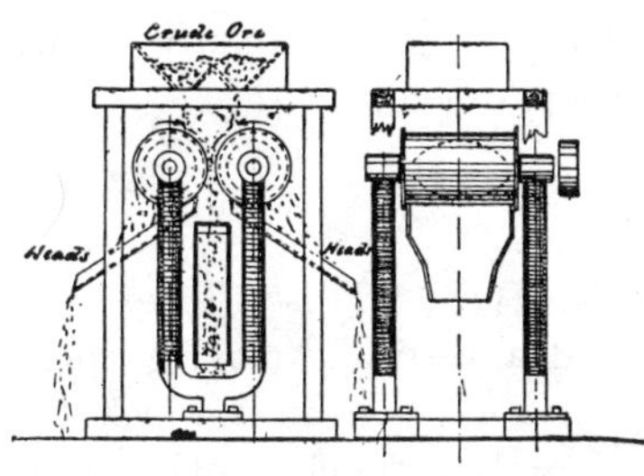

1247. MAGNETIC ORE SEPARATOR, "Buchanan" type. Two cylinders, magnetized by powerful horseshoe electro-magnets, are revolved at considerable speed. The pulverized ore is fed from hoppers on top of the rolls; the iron is held to the rolls and thrown off after passing the chutes. The tailings drop directly into a box.

1247 *a*. Front end view.

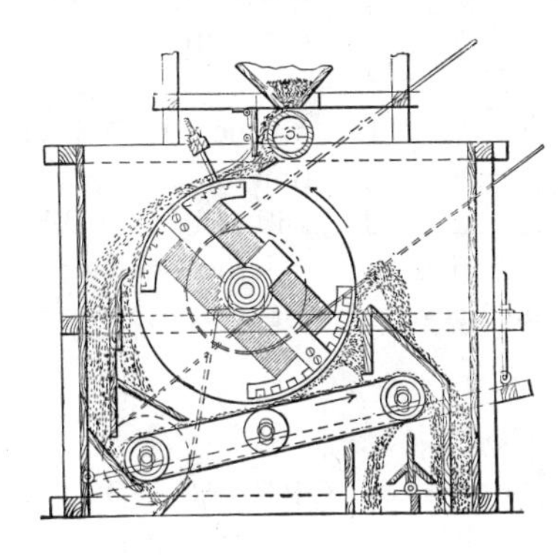

1248. IRON ORE SEPARATOR, "Buchanan" model. The pulverized ore is fed from a hopper to a revolving drum, a section on each side of which is magnetized by a fixed electro-magnet. The magnetic particles are carried around by the drum to a part of the neutral section and discharged. An apron below, travelling over magnetic rollers, further separates the ore.

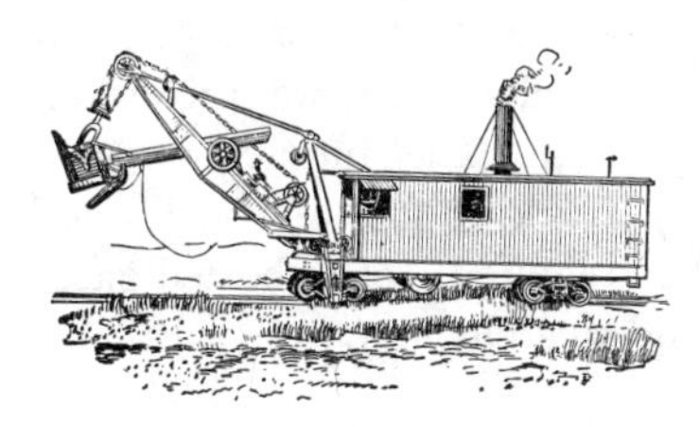

1249. RAILWAY STEAM SHOVEL, the "Bucyrus" model. For railway or other excavating on movable trucks.

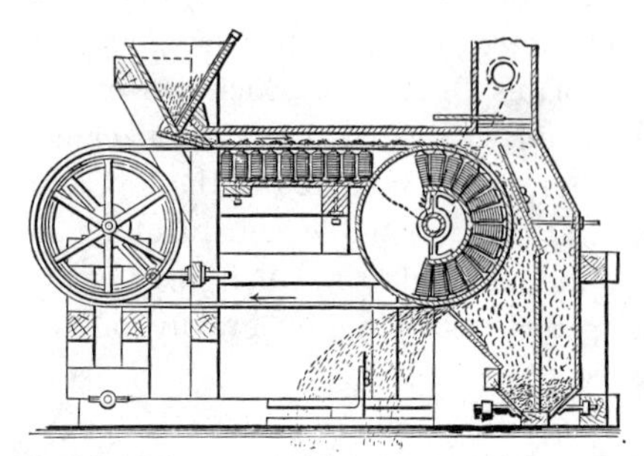

1250. MAGNETIC ORE SEPARATOR, "Hoffman" type. The pulverized iron ore is fed to a travelling apron, which passes over a series of magnets beneath the apron and over a drum where the magnetized iron particles are held to the belt until they pass the bottom side of the drum. The unmagnetized particles are thrown off, and drop into a separate compartment.

1251. MAGNETIC ORE SEPARATOR, "Edison" type. A series of electro-magnets are set behind a vertically moving apron against which the pulverized ore is discharged from a hopper spout. The concentrates move along the line of magnets by the action of the apron, and fall into buckets attached to the apron, and are carried over the top, while the tailings are drawn away from the front by an exhaust blower.

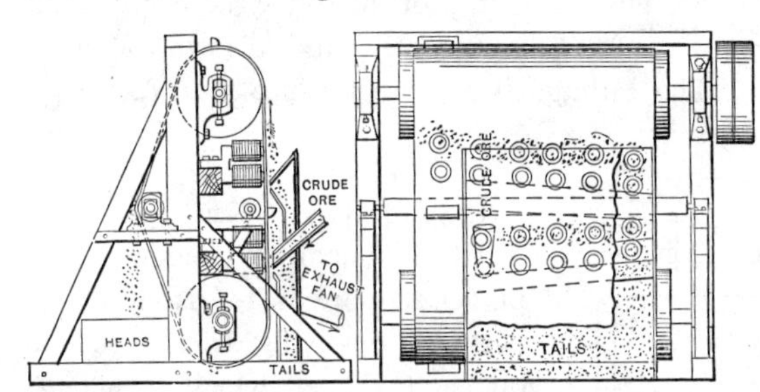

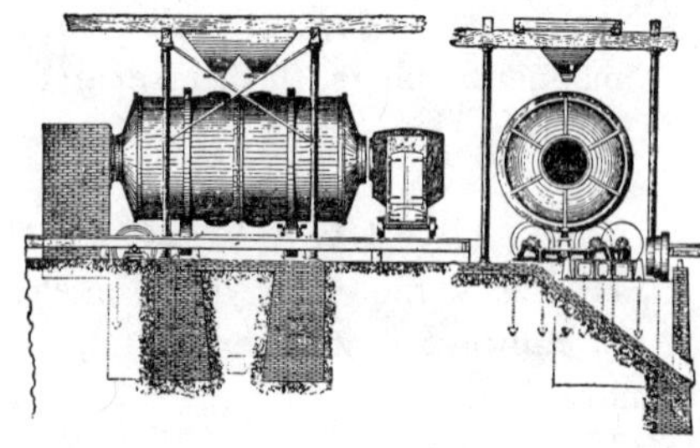

1252. ORE ROASTING FURNACE, revolving type. The large cylinder takes charge by the manholes, and revolves on power-driven rollers. The furnace is on a truck to be removed when required. The heated gases pass through the revolving cylinder and to a chimney.

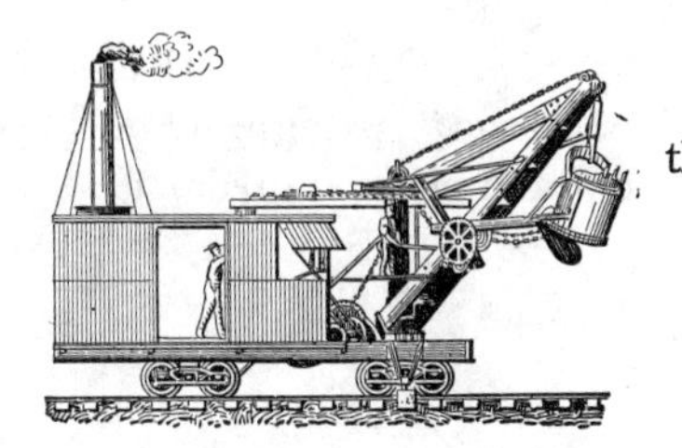

1253. RAILWAY EXCAVATOR, the "Otis" pattern.

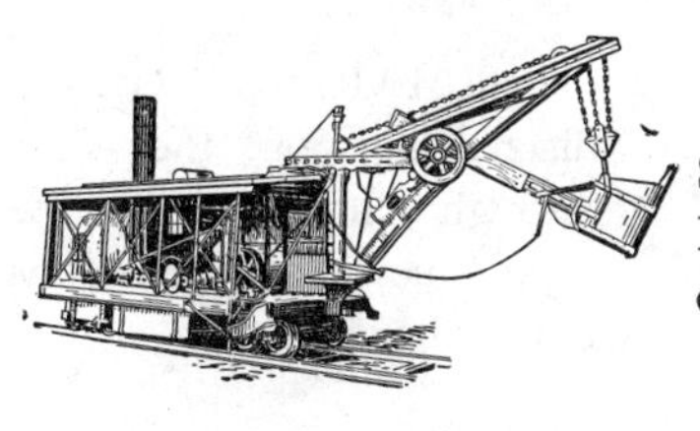

1254. RAILWAY STEAM SHOVEL, the "Victor" model. For excavating railway cuts, or general work on temporary rails.

1255. CONTINUOUS DITCHING DREDGE. — Discharging overhead on the banks by a carrier from under the bucket discharge.

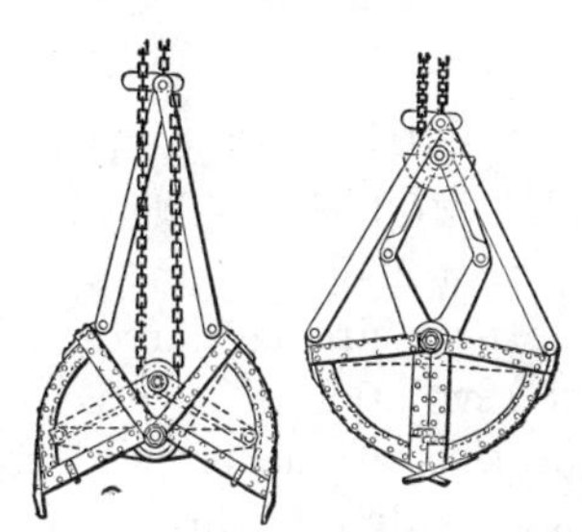

1256. CLAM-SHELL BUCKET, for dredging. Operated by a double chain. One chain is attached to the joint of the long arms, the other chain passes around a sheave in the joint of the lazy tongs that opens the bucket, and is made fast to the first chain. The bucket is suspended by the first-named chain to keep it open, the second chain is then pulled to close the bucket on its load.

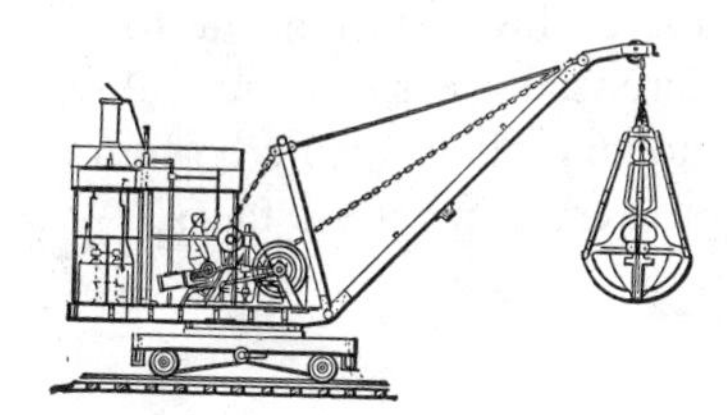

1257. REVOLVING HOISTING DREDGE, balanced on railway truck. "Lancaster" pattern, with clam-shell bucket.

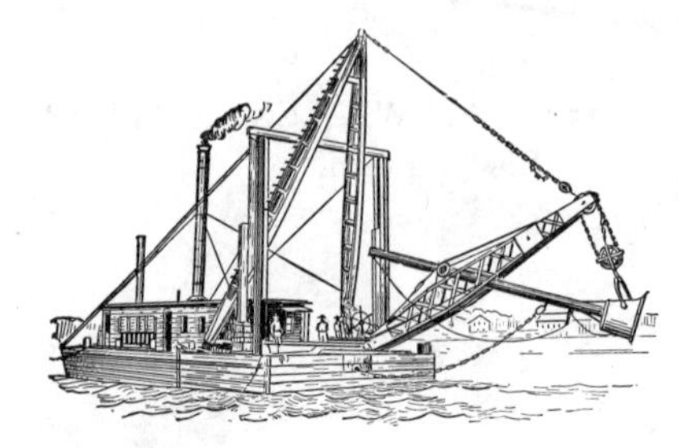

1258. FLOATING DREDGE, "Osgood" pattern. For harbor and channel dredging.

1259. MARINE DREDGE, discharging on the shore through a long floating pipe. Pipe buoyed by pontoons. For harbor work.

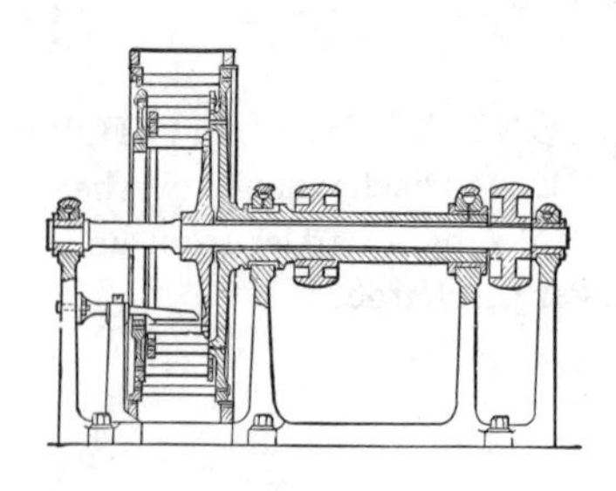

1259*a*. DISINTEGRATOR, for pulverizing ore. Two concentric shafts with disks and beating bars running at high velocity in opposite directions.

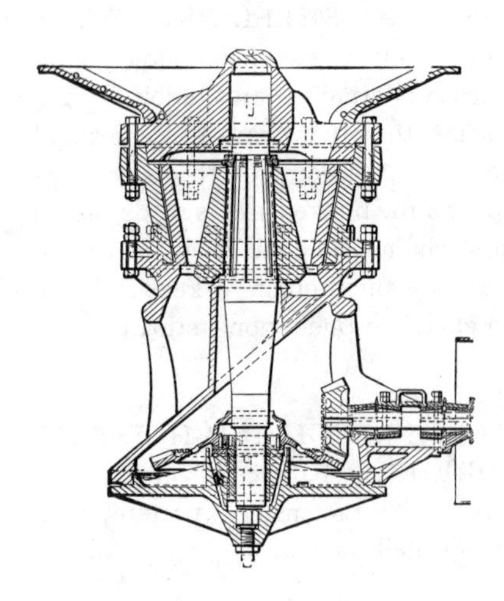

1259*b*. ORE CRUSHER, GATES MODEL.—The cone on the central shaft is made to vibrate in a circular direction by the revolution of an eccentric bearing at the bottom of the shaft, driven by bevel gearing. The crushing cone has a slow rotation due to differential areas of cone and stationary plates at the bottom or nearest contact surfaces.

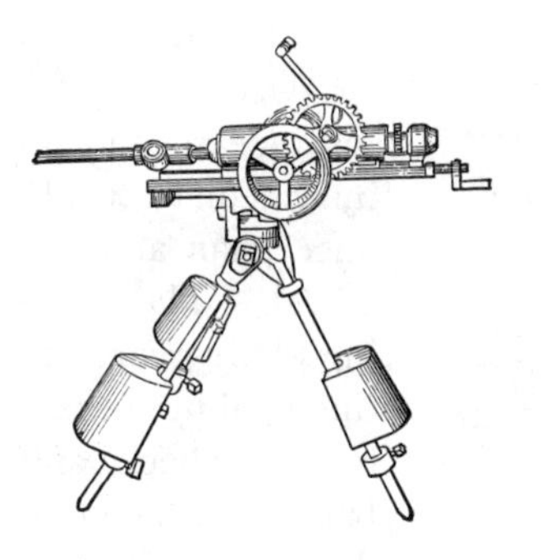

1259*c*. HAND POWER ROCK DRILL, Jackson model.—A powerful helical spring drives the drill forward. The crank operates a cam wiper for drawing back the drill; the motion is regulated by the fly wheel.

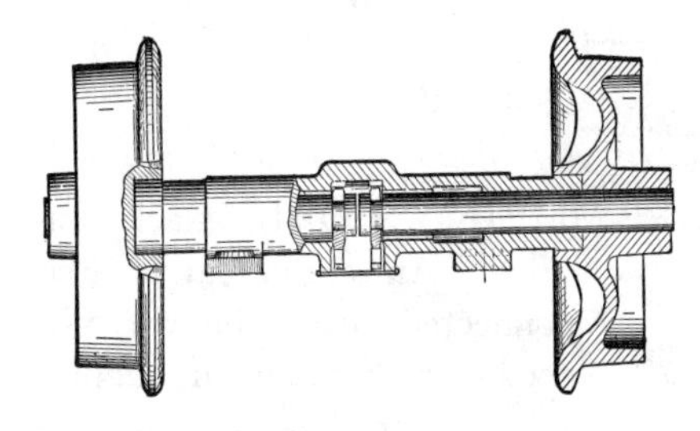

1259*d*. FREE RUNNING AXLES for mining cars. The divided axle held together by grooved bearings, makes a light running car on the small curves in mines.

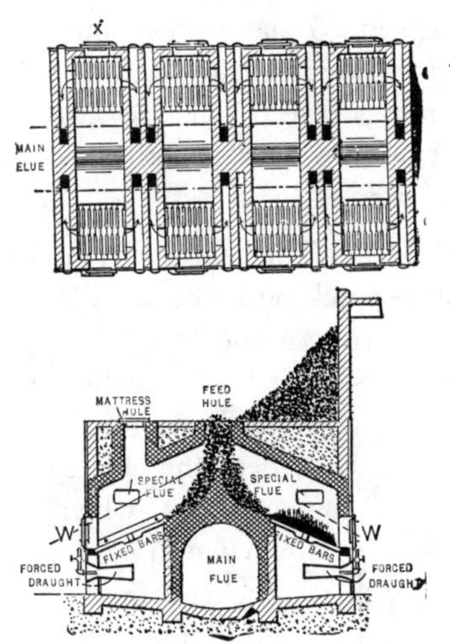

1259*e*. GARBAGE FURNACE.—Plan and elevation. The garbage falls through openings at the top and is divided on a curved parting hearth and dried by the heat of the fire on the grate and then slides on to the grate to be burned. Coal may also be used to facilitate the burning. Plan and vertical section.

1259*f*. ROPE DRIVE FOR MINE HAULAGE.—Two grooved drums with gears and an intermediate balance gear and fly-wheel. The driving shaft geared to the hauling drum; brake wheel and band on the same drum shaft. Shafts are horizontal. Plan and side view.

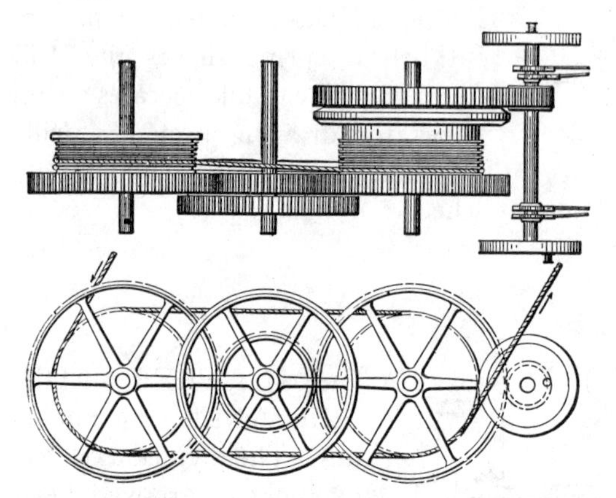

1259*g*. AIR BLAST FOR MOVING COAL, slack and dust. An engine and direct connected Root blower. A feed screw from the hopper to the air pipe adjusted in speed to the proper quantity for the air blast. Can be used for refuse from ore concentration works or other refuse that can be conveyed by compressed air.

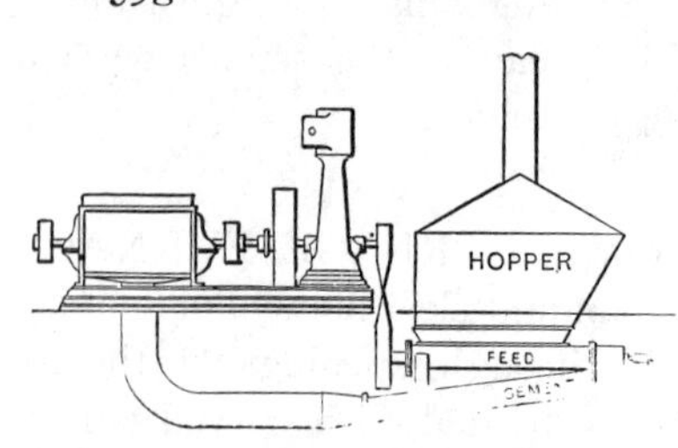

1259*h*. AUTOMATIC AIR DUMP.—The cable from the drum runs over a sheave on a movable truck on inclined rails. A stop at each end of the run limits the run of the truck. A bar across the frame at the middle post holds a Y-slot into which the chain and ball under the bucket catches, when by lowering the bucket tips over and its contents dumped. Again hoisting the bucket is released, and being light runs back over the pit.

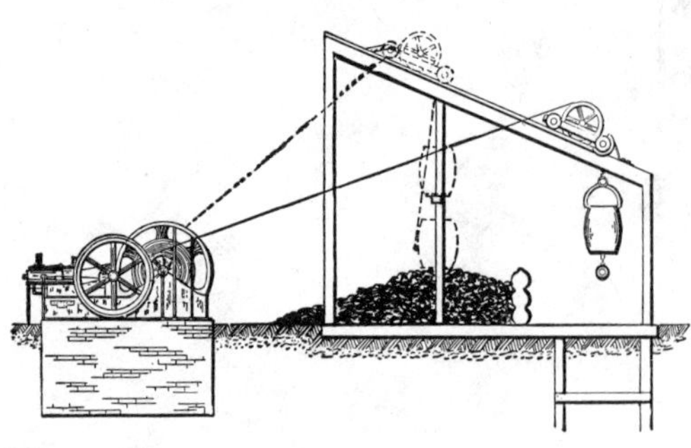

Section XV.

MILL AND FACTORY APPLIANCES.

Hangers, Shaft Bearings, Ball Bearings, Steps, Couplings, Universal and Flexible Couplings, Clutches, Speed Gears, Shop Tools, Screw Threads, Hoists, Machines, Textile Appliances, Etc.

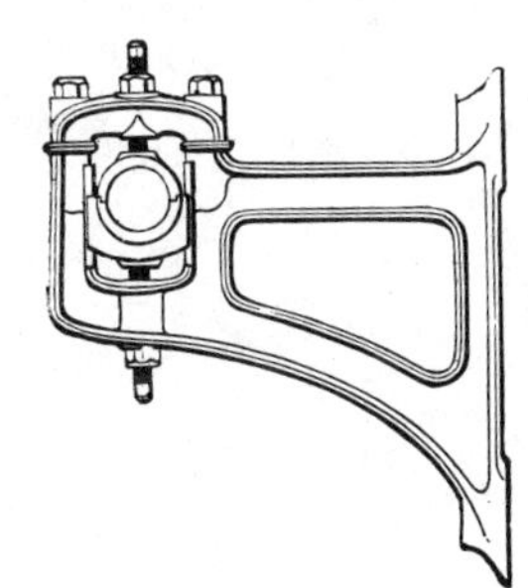

1260. ADJUSTABLE BRACKET HANGER.

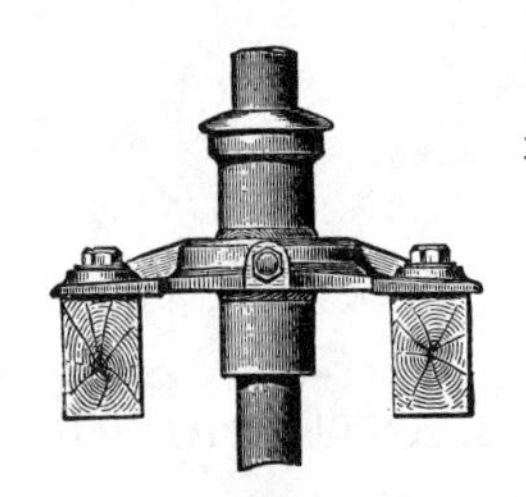

1261. ADJUSTABLE FLOOR BEARING for vertical shaft.

1262. Elevation.

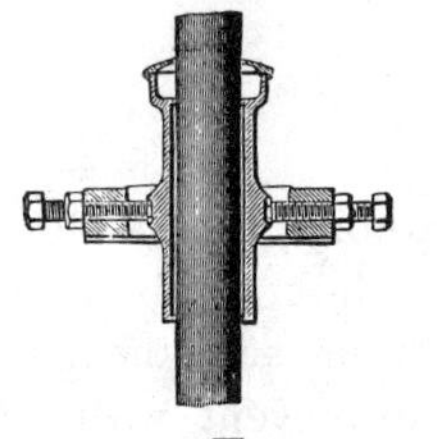

1263. Section.

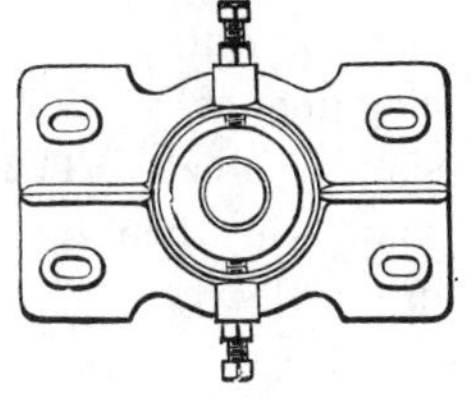

1264. Plan.

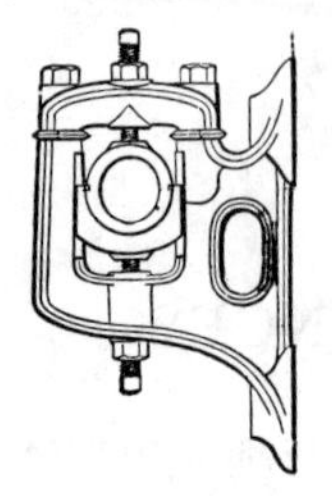

1265. ADJUSTABLE POST HANGER.

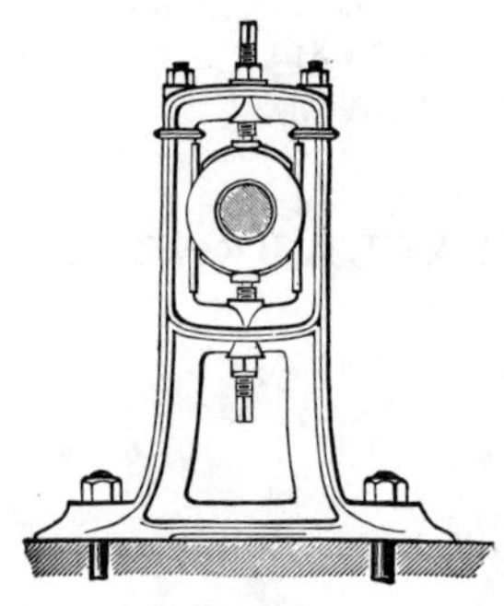

1266. ADJUSTABLE FLOOR STAND, shaft bearing.

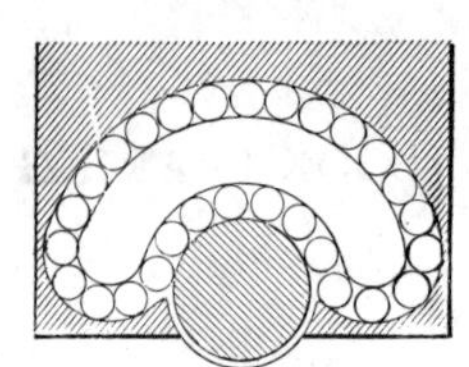

1267. CONTINUOUS TRAVERSING ROLLER or ball bearing for an axle.

1268. ROLLER WHEEL ANTI-FRICTION BEARING.

1269. BALL BEARINGS in an adjustable journal box. A loose sleeve is inserted between the balls and the shaft to prevent wear of shaft, and to prevent clogging if a ball should break. The shaft will then turn in the sleeve.

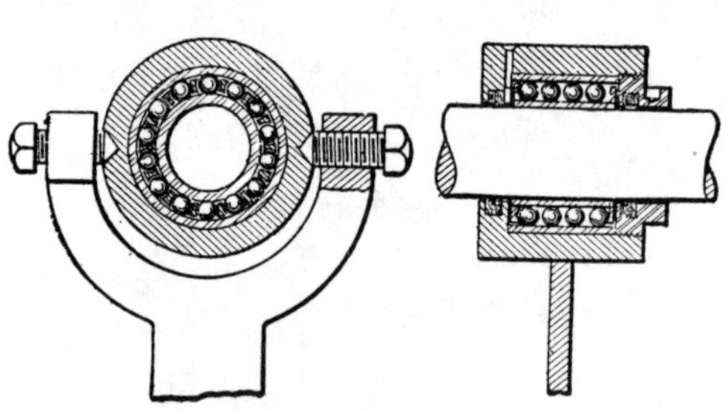

1270. Longitudinal section.

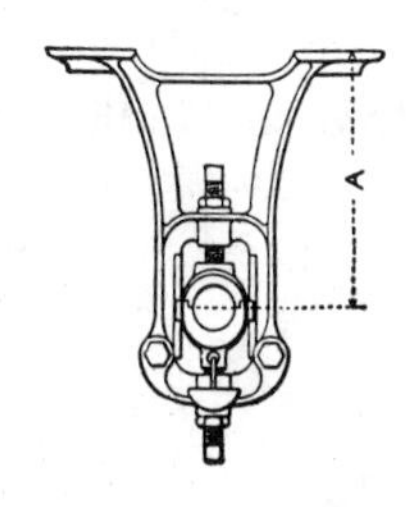

1271. ADJUSTABLE HANGER for shafting. A, drop of the hanger. Jointed cap to allow of removal of shaft.

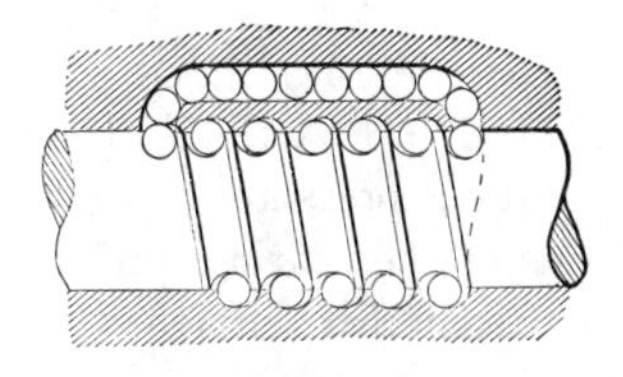

1272. SCREW TRAVERSING BALL BEARING, with balls returning through outside passage. Grooves recessed in shaft.

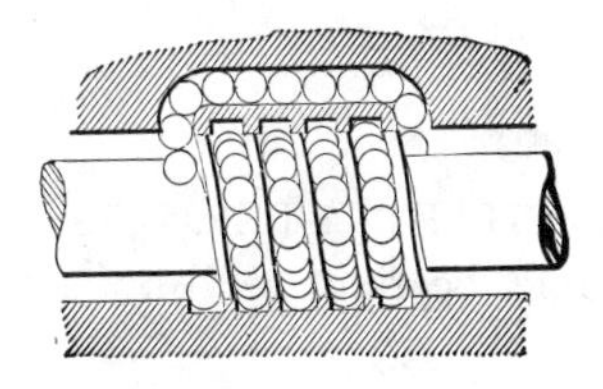

1273. SCREW TRAVERSING BALL BEARING. The balls returning by a side passage. Ball grooves enlarged for full strength of shaft.

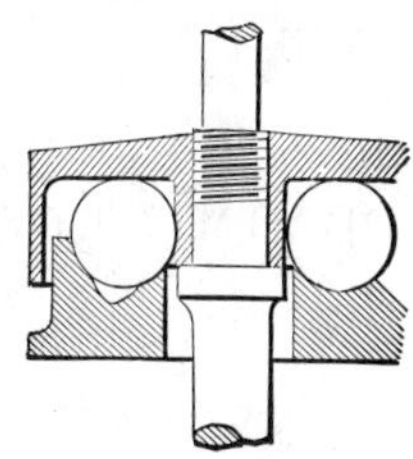

1274. HANGING SHAFT on ball bearings.

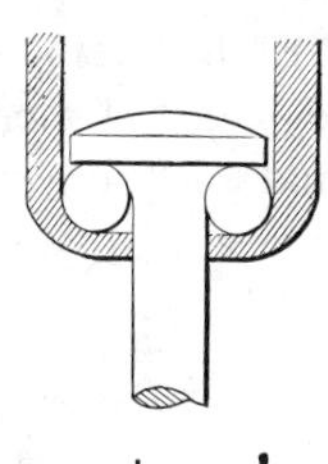

1275. SUSPENDED SHAFT on ball bearings.

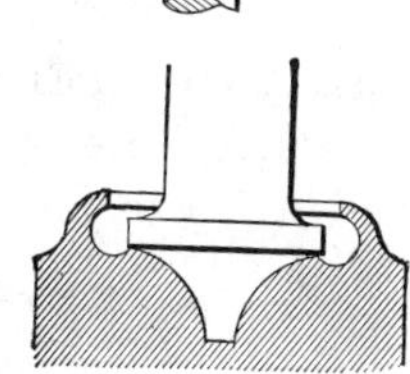

1276. CURVED STEP BEARING, with oil reservoir.

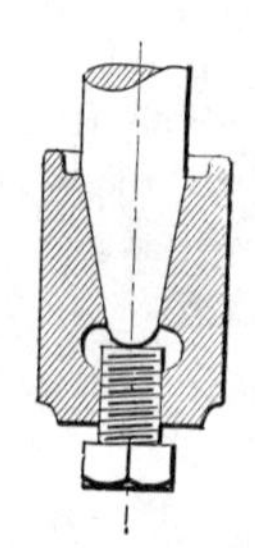

1277. CONICAL PIVOT BEARING and adjusting screw.

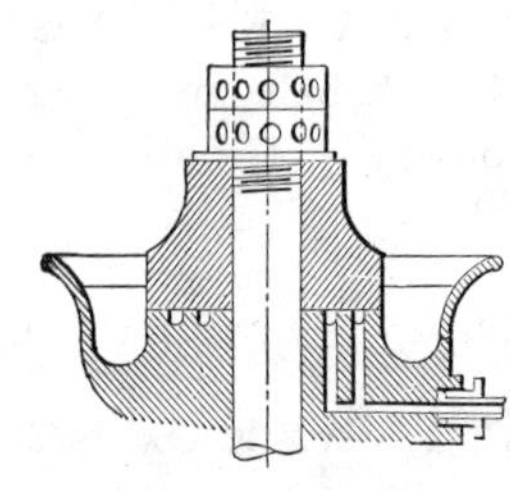

1278. LUBRICATION OF A HANGING BEARING by hydraulic pressure. Oil is forced into the grooves of the bearing through the small holes and discharges into the cup around the outside.

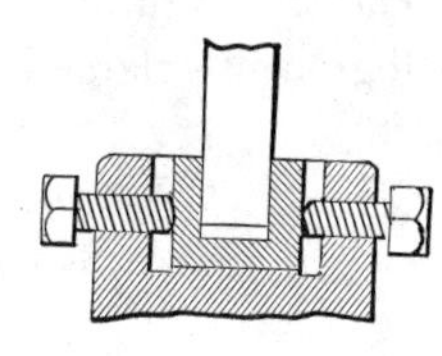

1279. VERTICAL SHAFT STEP.—Made adjustable by a movable bearing held by set screws in the foot block.

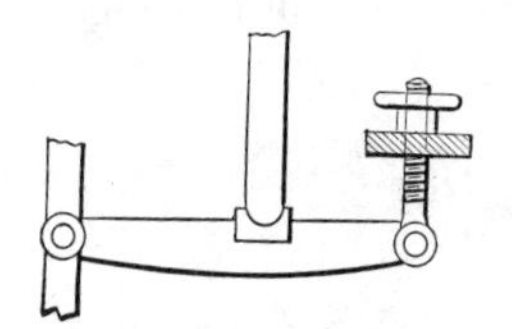

1280. SHAFT STEP ADJUSTMENT for spindles of millstones.

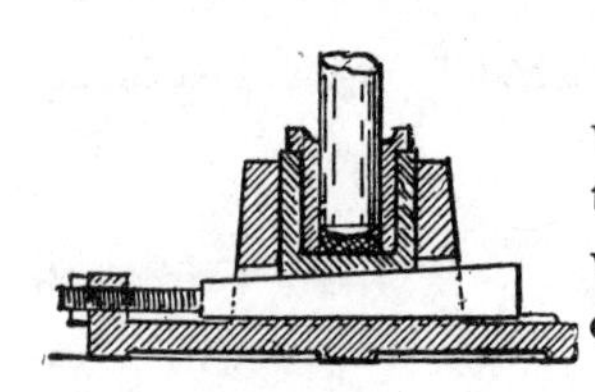

1281. ADJUSTABLE STEP BEARING, with hard bronze bush and step. A mortise through the iron base and a key drawn with a screw extension and nut are for vertical adjustment.

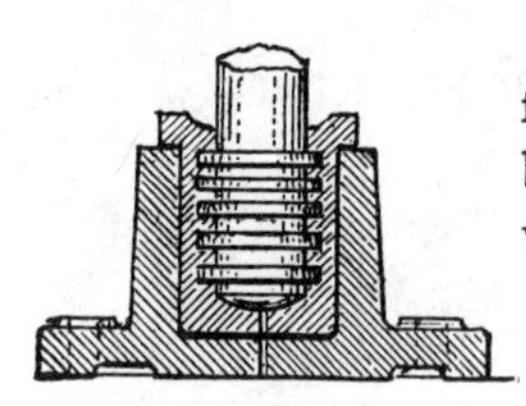

1282. COLLAR BEARING AND STEP for a vertical shaft. The thrust sleeve of bronze is split and should have a key to prevent rotation.

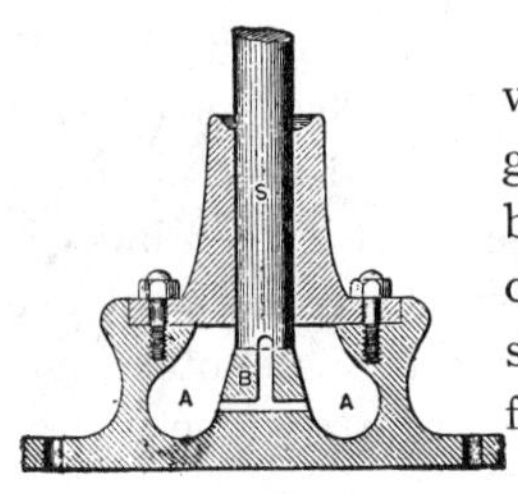

1283. OIL CIRCULATING STEP for a vertical shaft. The foot of the shaft has a groove cut across its centre. The cast-iron bearing has a hole down the centre to meet a cross hole from the oil well. The joint of the sleeve and step is packed oil tight, oil being fed at the upper end of the sleeve.

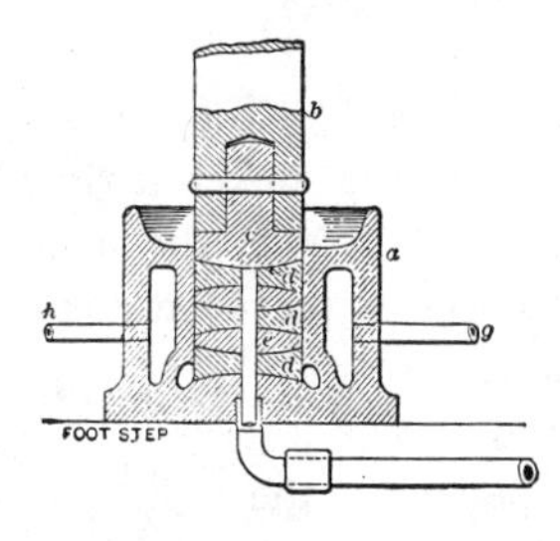

1284. LENTICULAR BEARING for a vertical shaft. Each section is lubricated by the pressure oil feed from beneath, through the central hole. The concave discs are of hard bronze, and the convex discs of steel. The shaft terminates in a steel toe, *c*. The cast-iron step is chambered for water circulation.

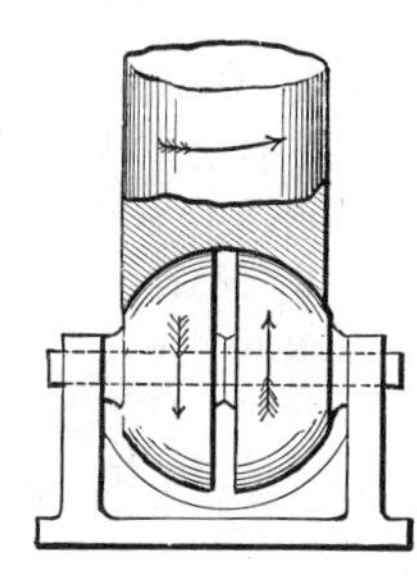

1285. SPHERICAL STEP BEARING.—Two semi-spheres, rolling on a horizontal shaft, support a vertical shaft having a concave spherical end. The semi-spheres roll in opposite directions in oil, and by the cross direction of the bearing surfaces preserve a perfect contact.

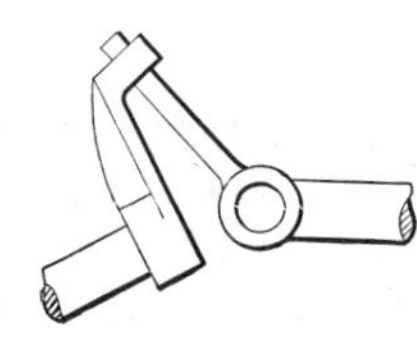

1286. ANGLE COUPLING for shafts. The jointed rod on one shaft slides in the bent crank eye of the other shaft. For small angles and light work.

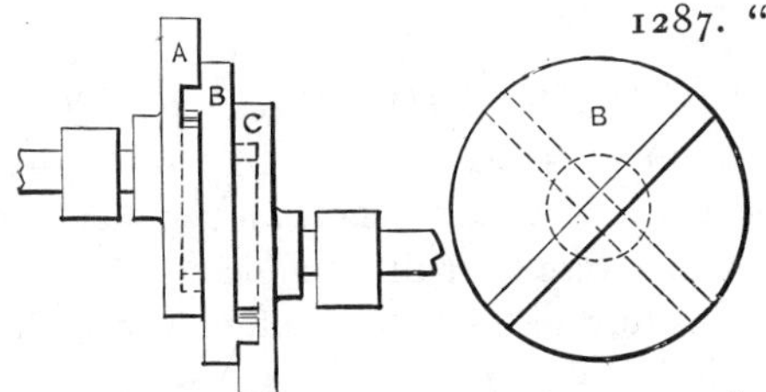

1287. "OLDHAM" COUPLING for shafts slightly eccentric in alignment. The double-splined disc B runs free against the grooved face plates A, C.

1288. Disc showing grooves at right angles, front and back.

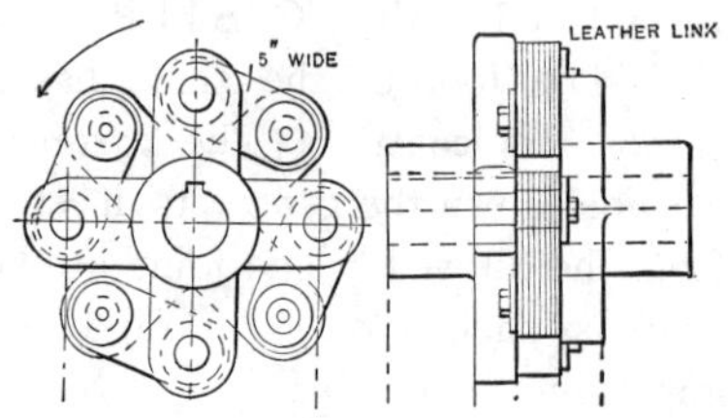

1289. FLEXIBLE LINK COUPLING.—The end of each shaft is fitted with a four-armed hub. A series of leather links is inserted between the arms of one hub and those of the other hub, and secured with stud bolts.

1290. Side view.

1291. FLEXIBLE SHAFT COUPLING.—A ball and socket shaft ends with a slot in the ball and a mortise in the socket at right angles, in which the right-angled cross piece has a free sliding motion.

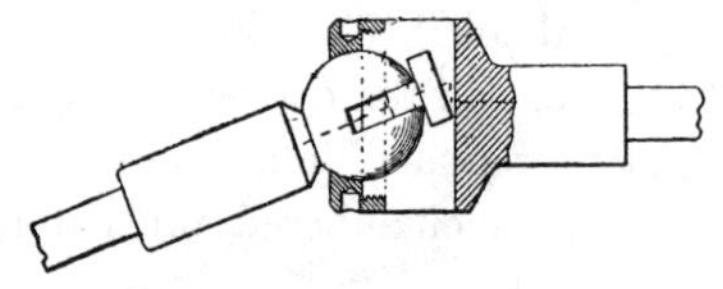

1292. The cross key in perspective at the right.

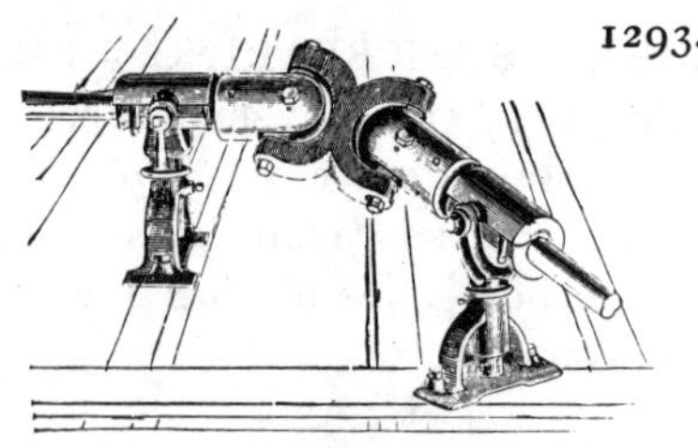

1293. ANGLE SHAFT COUPLING, "Robes" patent. The shaft heads are slotted, in which cross bars are pivoted; the ends of the cross bars are also pivoted to the arms of the double yoke, giving a free motion to the driven shaft at any angle greater than a right angle.

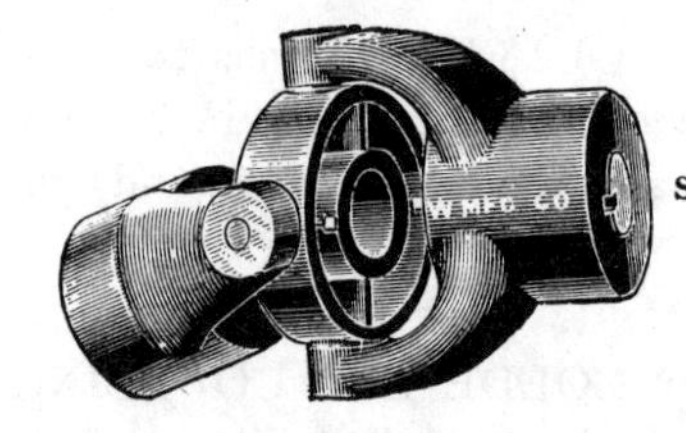

1294. UNIVERSAL JOINT, for shafting. Ring gimbal.

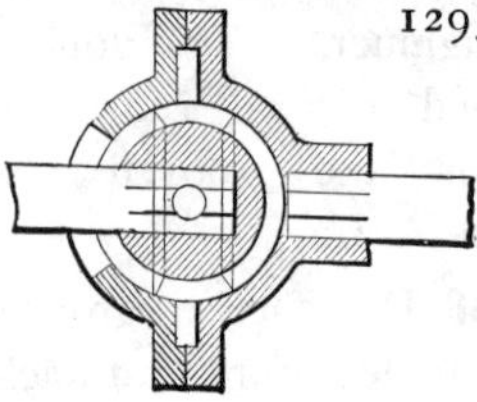

1295. "HOOKE'S" UNIVERSAL JOINT.—One shaft end is keyed into a ball with trunnions, which turn in a ring with trunnions at right angles with the ball trunnions. The ring trunnions turn in the outer shell to which the other shaft is keyed.

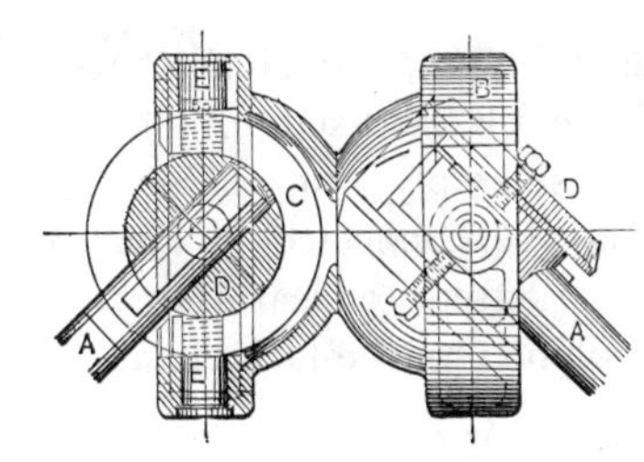

1296. "GOUBET'S" UNIVERSAL SHAFT COUPLING.—A, A, shafts; C, a trunnion ring recessed in a ball, D. Each shell is alike, and in itself a universal joint for 45°. Both together equal to 90°.

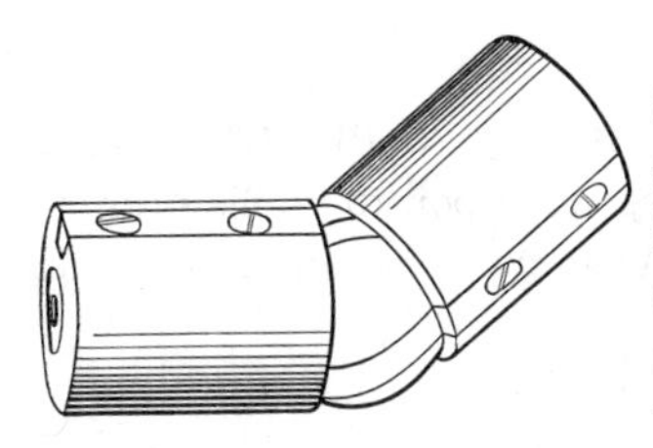

1297. BALL SOCKET UNIVERSAL JOINT.—A ball with grooves around it at right angles and bearing in the spherically recessed ends of the shafts. Straps fitted in the grooves, and screwed in slots in the shaft, hold the ball in position.

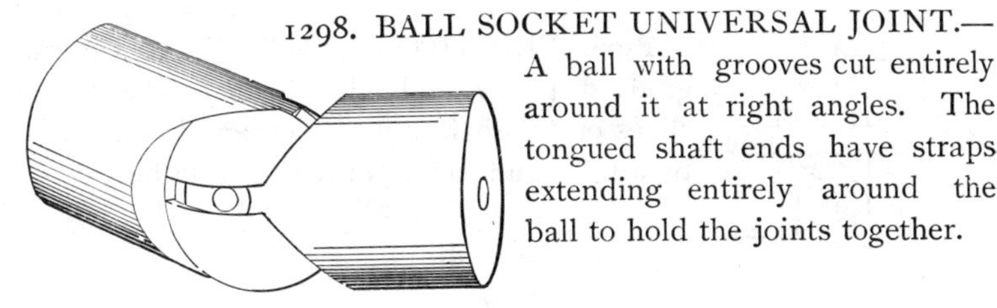

1298. BALL SOCKET UNIVERSAL JOINT.—A ball with grooves cut entirely around it at right angles. The tongued shaft ends have straps extending entirely around the ball to hold the joints together.

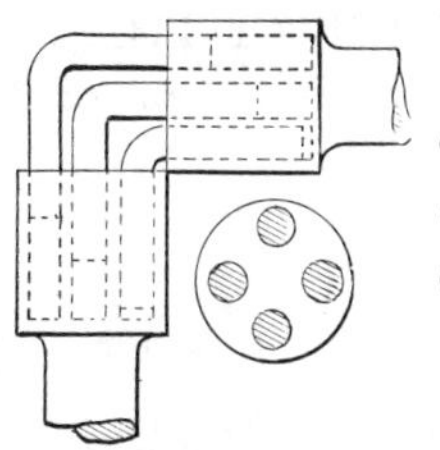

1299. RIGHT-ANGLE SHAFT COUPLING, "Hobson" and other patents. Right-angle crank pins revolve and slide in holes in the shaft couplings.

1300. RIGHT-ANGLE SHAFT COUPLING, "Hobson" patent.—A number of right-angle steel rods move freely in perforated guide flanges on the ends of shafts that run at right angles. The rods draw out and in through the flanges to suit the conditions of revolution of the shafts. A larger angle rod serves as a centre bearing over which the shafts revolve.

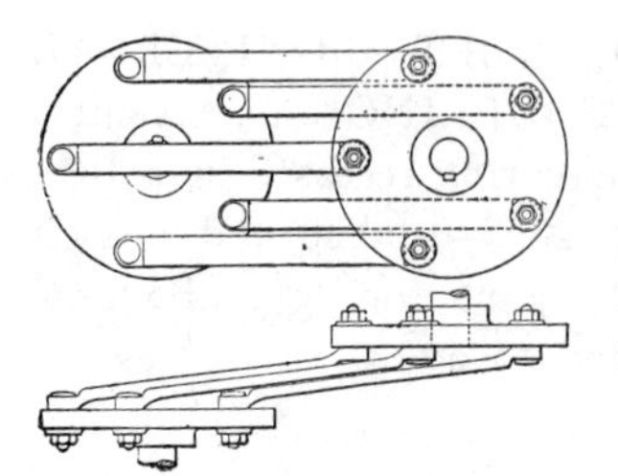

1301. ECCENTRIC LINE COUPLING.—Face plates, fixed to ends of shafting considerably out of line but parallel, may be connected by four or five bars with offsets to clear each other in their revolution on the face plates.

1302. Side view of offset links.

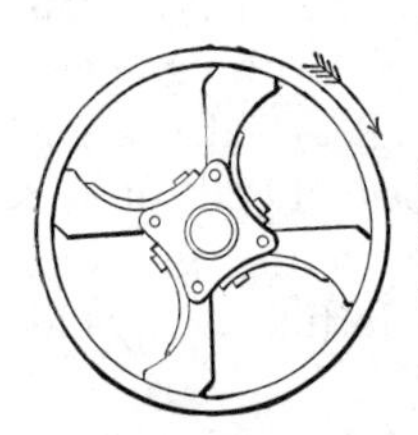

1303. SIMPLE FRICTION PULLEY.—The self-acting clutch arms act upon the pulley rim in one direction only. When shaft motion is reversed, the pulley is free.

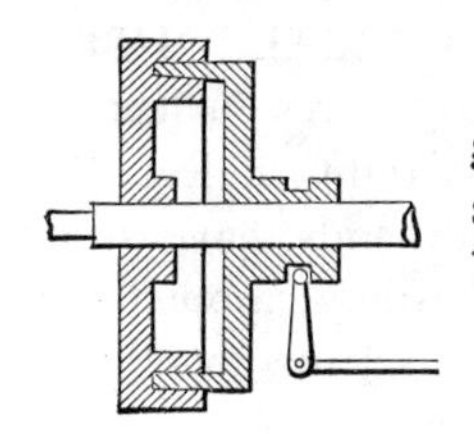

1304. FRICTION CLUTCH.—A conical-grooved pulley and clutch rim. The clutch slides on the shaft and feather, and is controlled by a lever and carrier in the grooved hub.

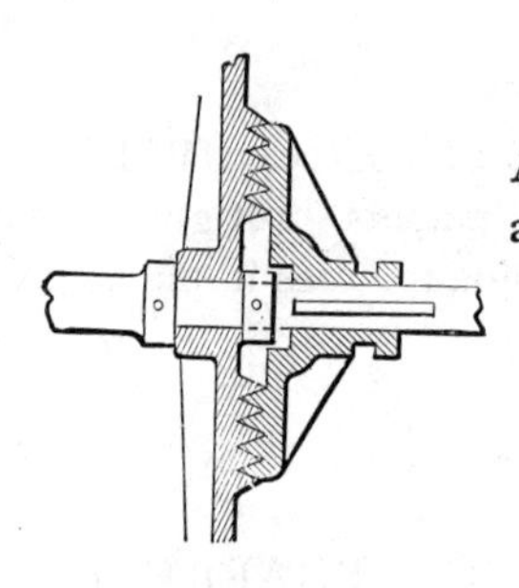

1305. V-GROOVED FACE CLUTCH.—A very effective clutch with teeth of small angle.

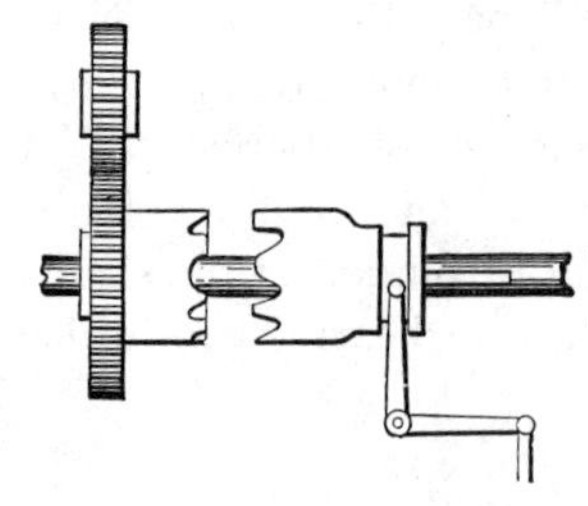

1306. CLUTCH AND GEAR.—The clutch slides on the feathered shaft, and throws the gear into motion by the operation of the bell-crank lever and runner.

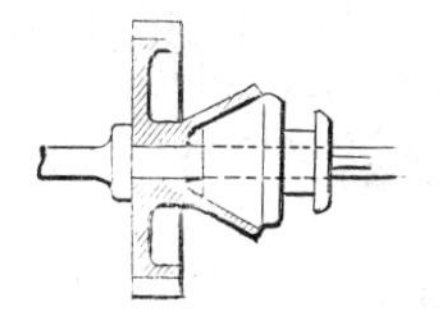

1307. CONE CLUTCH.—Can be made at any angle greater than will cause the clutch to stick.

1308. MULTIPLE PLATE FRICTION CLUTCH.—Several plates of iron or steel are fitted loosely on a three-feather shaft, between which plates of wood or other hard material, sometimes steel, are placed and keyed in an iron housing or coupling to move loosely on the keys. The coupling is keyed to the next shaft in line. A follower sleeve and springs compress the plates, giving a very large frictional surface, which is relieved by drawing the sleeve back by a yoke lever.

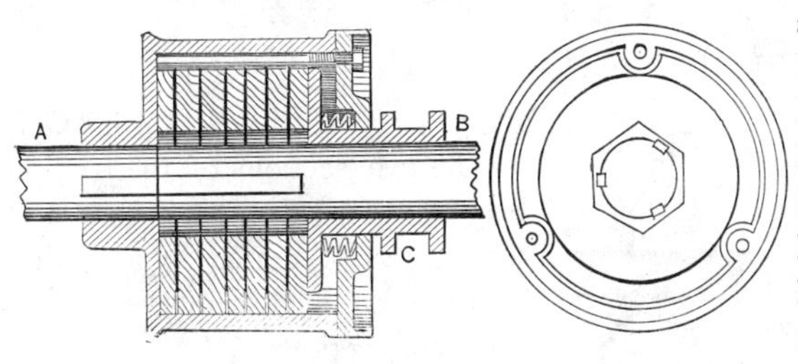

1309. Section showing stops in outer case and keys on shaft.

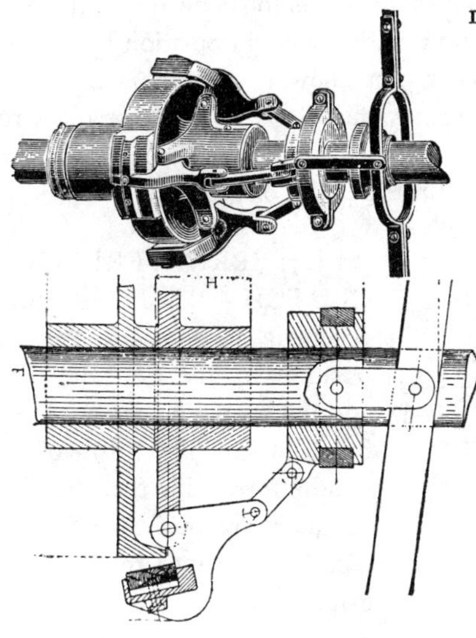

1310. FRICTION CLUTCH, outside view, with toggle-joint thrust, sleeve, and yoke lever.

1311. Section of outside bearing, clutch, toggle joint, and sleeve.

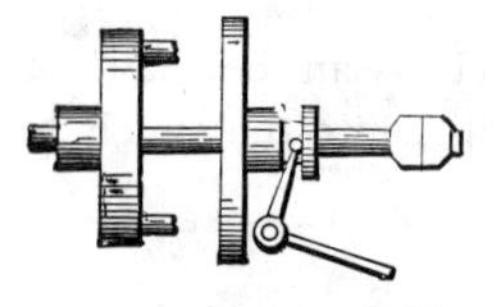

1312. PIN CLUTCH.—The pin plate is fast on the shaft. The hole plate slides on a feather, and is operated by a bell-crank Y-lever in a hub slot.

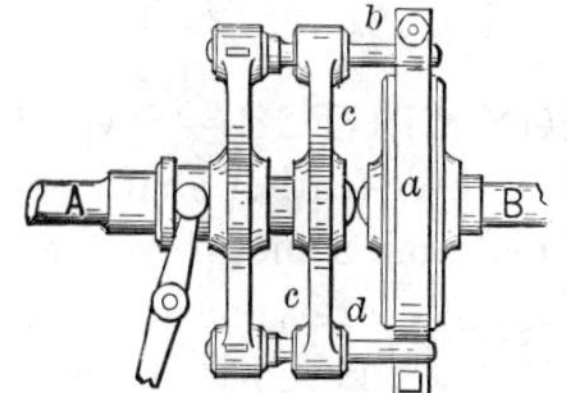

1313. FRICTION PIN CLUTCH.—A or B may be the driving shaft; *a* is a friction band that slips to prevent shock when the pins are thrown into contact with it.

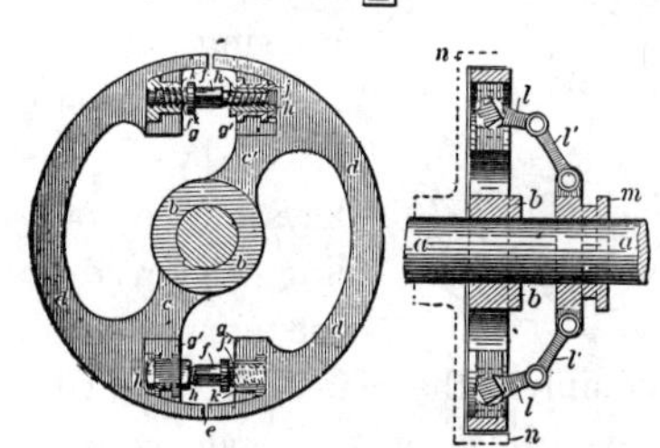

1314. FRICTION CLUTCH.—The two sections of the friction ring are pressed out by right and left screws, operated by a sliding spool on the shaft and the toggle-joint connections, *l*, *l'*.

1315. Longitudinal section.

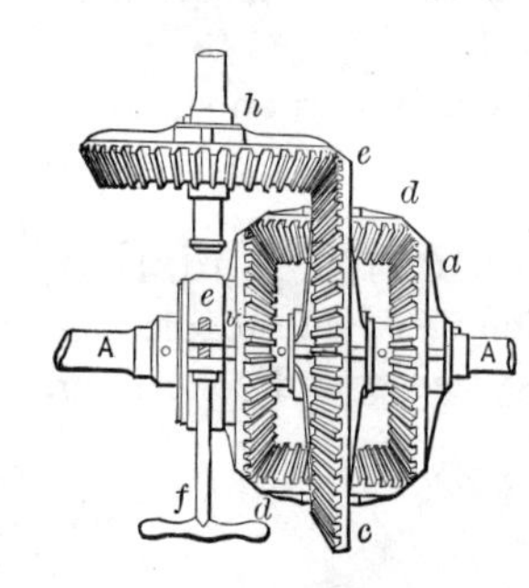

1316. FRICTION CLUTCH BEVEL GEAR.—A A is a driving shaft extended through the gear hubs; gear *a* is fast on the shaft; gear *b* is loose on the shaft, with a friction clutch fixed in position by a lever extension not shown. Clutch is tightened by the screw handle *f*, when the gear *e c* rotates to drive gear *h*. The pinions are pivoted in the plane of gear *e c*.

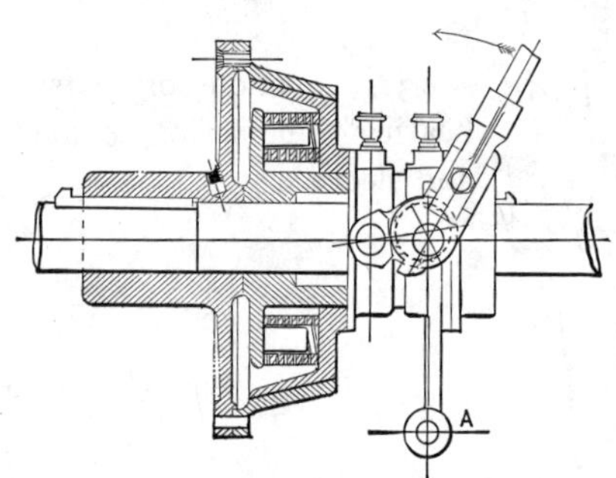

1317. SPRING FRICTION CLUTCH.—The lever handle, eccentric, and link are held in position by the arm A. The springs keep the cones closed for driving. The throw of the handle forward in the direction of the arrow pushes the inner cone back and releases the grip.

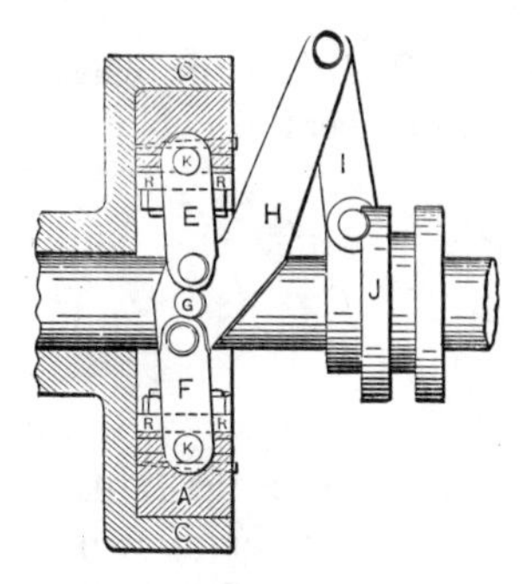

1318. DOUBLE TOGGLE-JOINT FRICTION CLUTCH. — The movement of the grooved sleeve J opens or closes the grip A, upon the rim wheel C. The lever H throws the toggle links E, F into line for the grip.

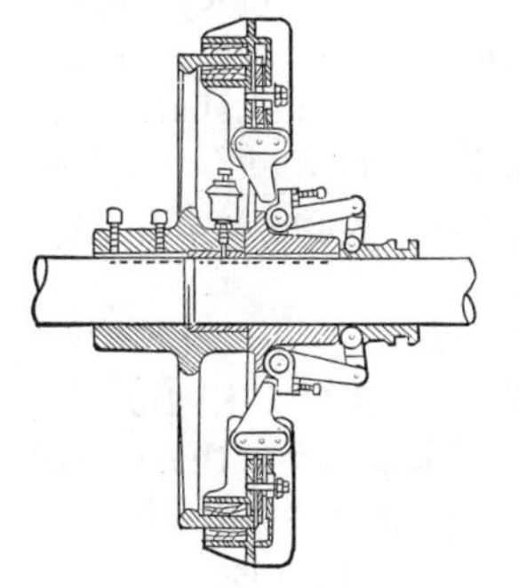

1319. ADJUSTABLE FRICTION CLUTCH, with double-grip bearings. Adjustment tightness is made by locked set screws in the arm of the bell-crank levers. The jaws are held open by a ring spring running around the clutch.

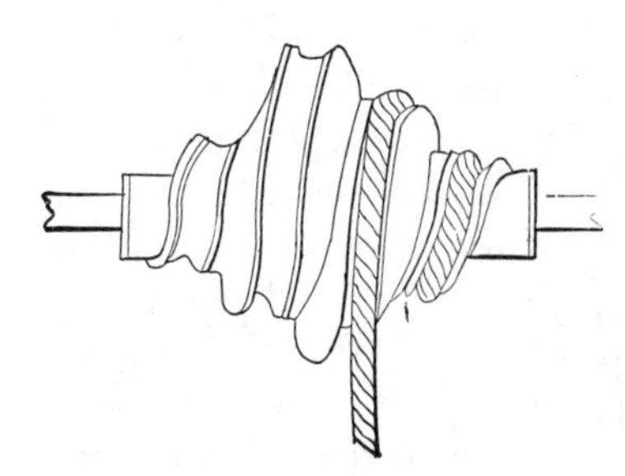

1320. DOUBLE-CONIC ROPE DRUM.—Used on some forms of winding engines, and as a fusee in a spinning mule.

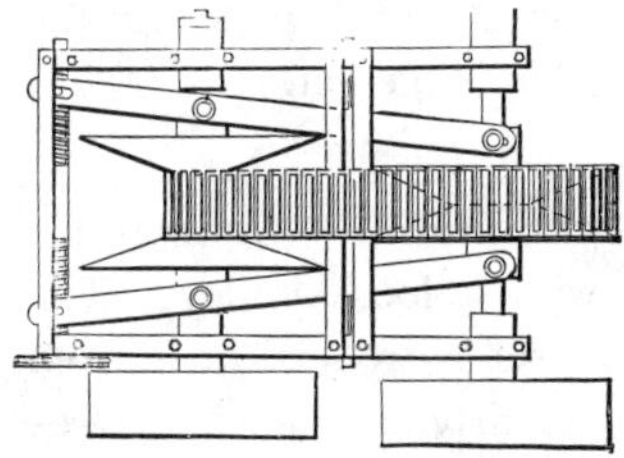

1321. VARIABLE SPEED DEVICE.—Transmission is made by a stiff belt running over two coned spools, which have their inside cone bearings simultaneously changed to meet requirement for equal belt tension, by two levers pivoted to nuts on a right- and left-hand screw, with a fulcrum central between the shafts. Both expanding spools slide on feathered shaft keys.

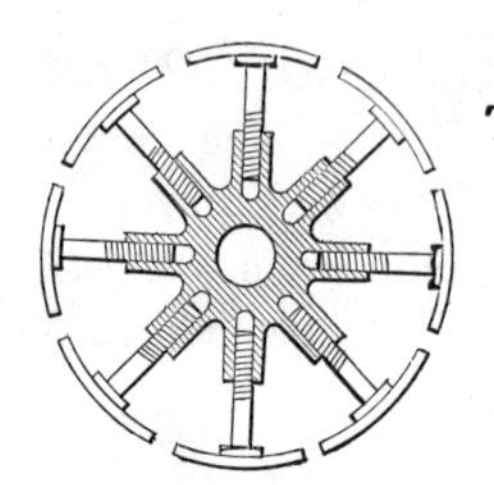

1322. EXPANDING PULLEY or wheel. The rim sections screw into a central hub.

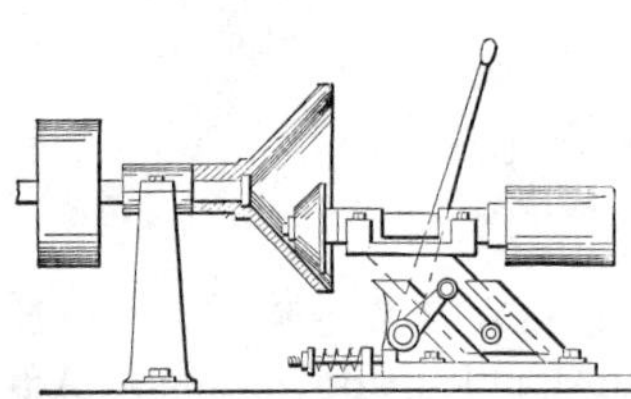

1323. VARIABLE SPEED DEVICE.—An internal driving-cone pulley, with a smaller cone pulley rolling on its internal surface on a shaft parallel with the driving shaft, but drawn eccentric to it for higher speed by an inclined slide operated by a lever, rock shaft, and crank connection.

1324. VARIABLE SPEED TRANSMITTING DEVICE.—A thin disc is fast on the counter shaft. Two discs drive the speed shaft, between which and the driving disc are two rollers pivoted to transverse spindles. The rollers are kept to their slow-speed position between the discs by springs. A connecting rod draws the rollers toward the high-speed position. Friction pressure on the rollers is made by a spring pressing the discs together.

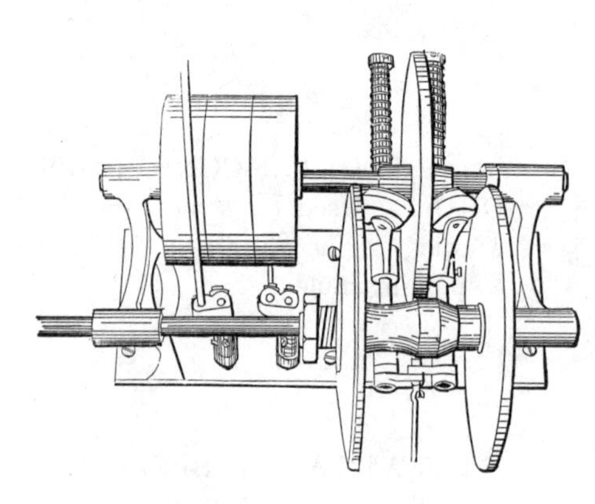

1325. BELT HOLDER, "Wellington" model. Does away with a loose pulley. The belt is guided on to a set of rollers in a fixed frame at the side of the driving pulley. Saves time and avoids danger in putting on belts.

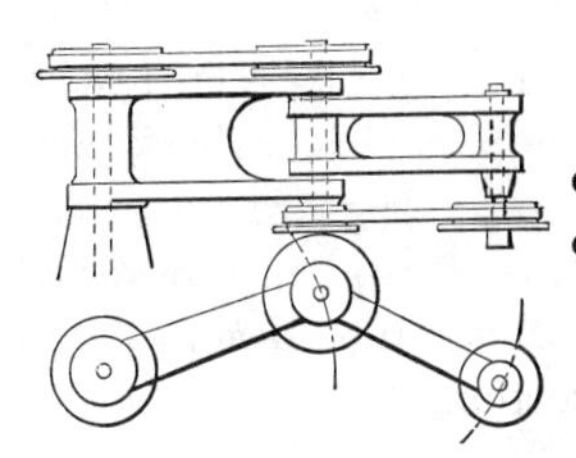

1326. JOINTED RADIAL ARM, for drilling machines, marble polishing, and other similar machines. Elevation.

1327. Plan, showing joints and action.

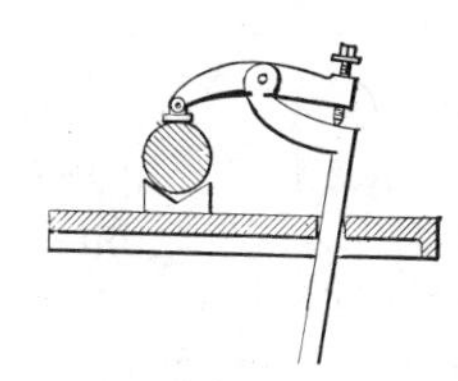

1328. DRILLING MACHINE CLAMP.—A handy tool about a drill press. The shank is pushed loosely through a hole in the drill-press table until the lever bears on the work, when a turn on the set-screw makes a tight grip.

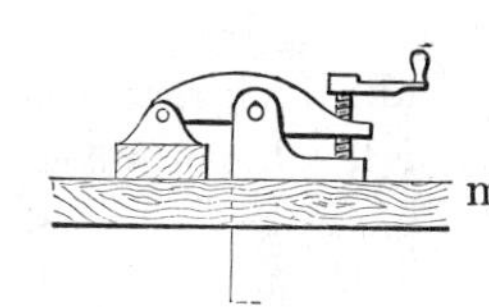

1329. SCREW BENCH CLAMP, for cabinet-makers.

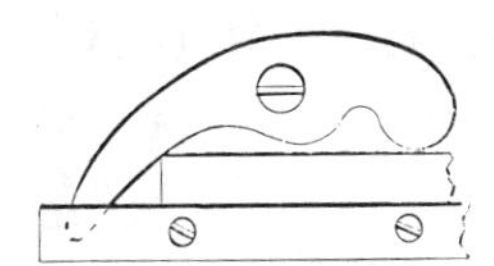

1330. AUTOMATIC BENCH CLAMP, for carpenters and cabinet-makers. Used for holding work on the flat.

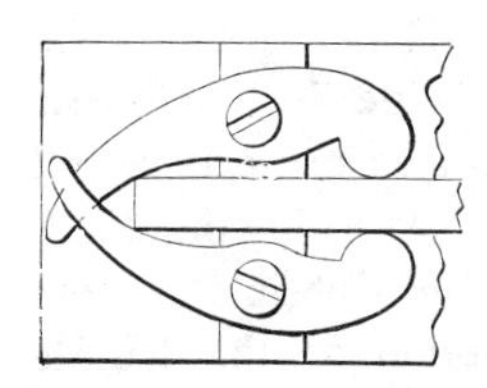

1331. AUTOMATIC BENCH CLAMP used by carpenters and cabinet-makers for holding work on edges for planing.

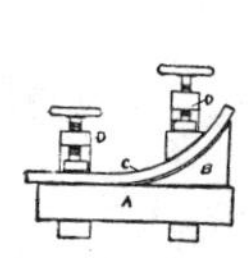

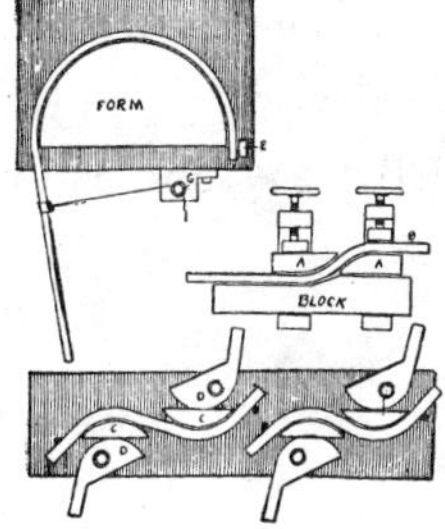

1332. WOOD-BENDING CLAMPS AND FORMERS.—Strips of wood are thoroughly steamed and bent while hot over the formers and clamped.

1332A. Offset clamp.
1333. Thill clamp.
1334. Bend clamp

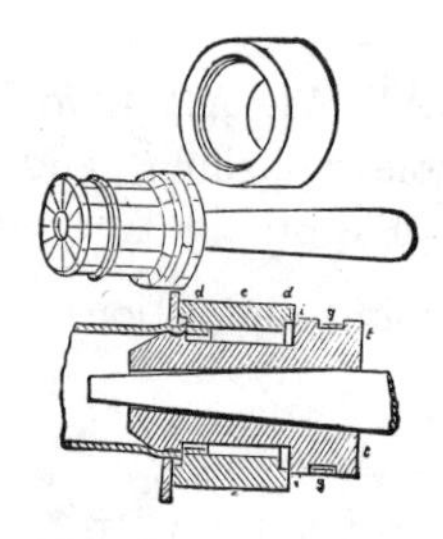

1335. BOILER TUBE EXPANDER.—A series of sets surrounding a conical driving pin. "Prosser" percussion type. A guard ring fixes the proper position of the expanding grooves of the sectional sets to match the tube head.

1336. Longitudinal section.

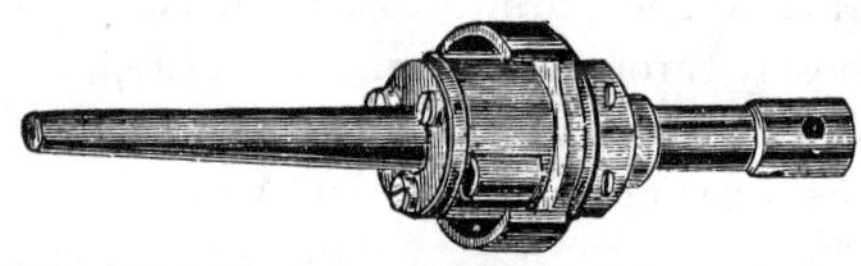

1337. ROLLER TUBE EXPANDER.—The rollers are loosely fitted in a case to hold them in position. The slightly tapered mandril is pushed or driven within and bearing on the rollers and revolved by a bar in the mandril head, which revolves the rollers, rolling them over the interior surface of the boiler tube. "Dudgeon" model.

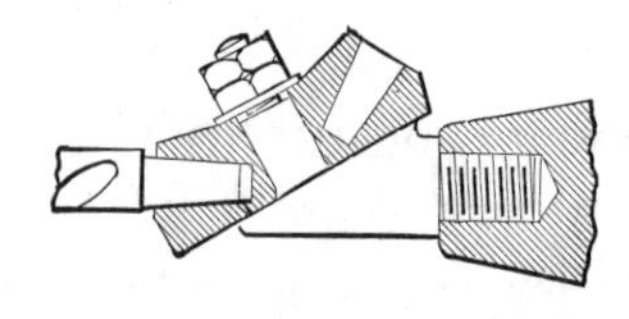

1338. REVOLVING TOOL HEAD for a Monitor lathe.

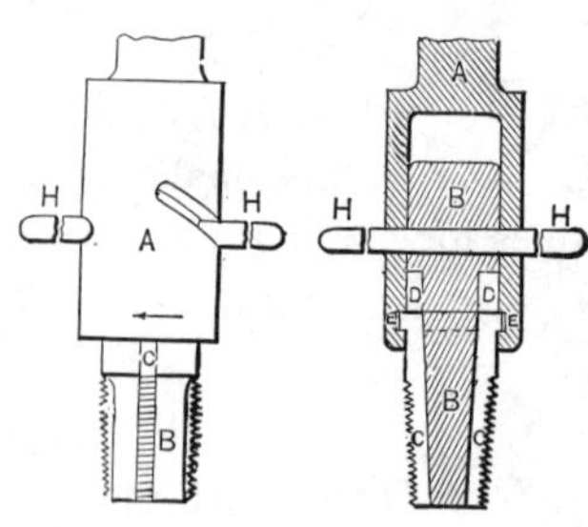

1339. COLLAPSING TAP.—The hook cutters C, C, slide in the taper shank B, and are drawn up to their full diameter for cutting by turning the shank handle in the inclined slot in the shell. and the reverse motion of the handle for collapsing the tap.

1340. Longitudinal section.

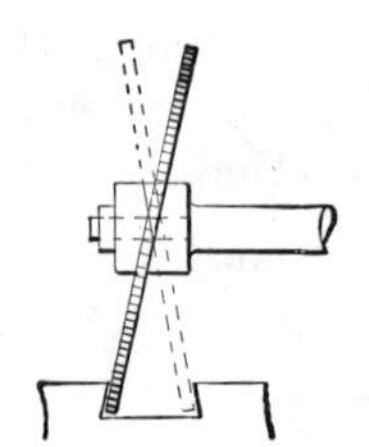

1341. WABBLE SAW, for cutting dovetail and rabbet grooves.

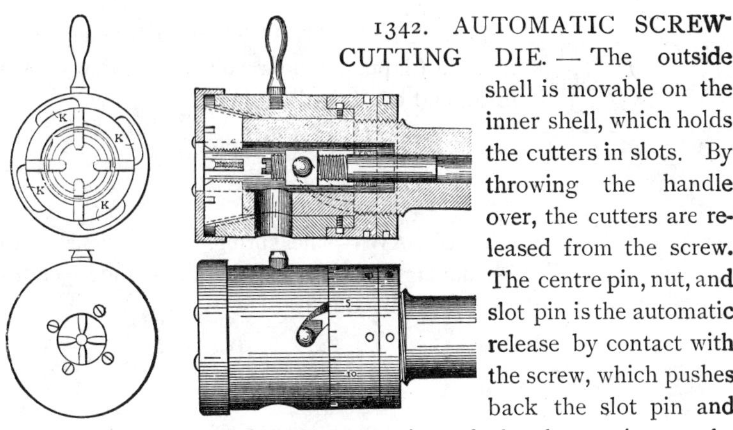

1342. AUTOMATIC SCREW-CUTTING DIE. — The outside shell is movable on the inner shell, which holds the cutters in slots. By throwing the handle over, the cutters are released from the screw. The centre pin, nut, and slot pin is the automatic release by contact with the screw, which pushes back the slot pin and revolves the outer shell. Adjustment is made by the set rings at the back of the die. A circular spring throws out the cutters. Cross section.

1343. Front view.
1344. Longitudinal section.
1345. Outside view.

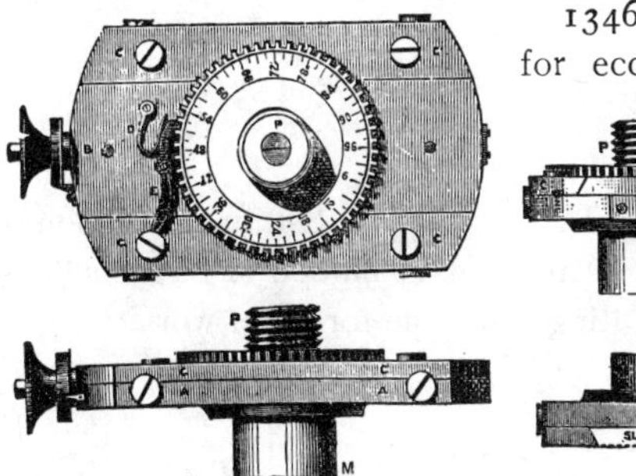

1346. UNIVERSAL CHUCK, for eccentric turning. The divided gear plate and chucking screw are revolved and held at any division by the spring pawl. The slide is given its eccentric position by a screw with an index. A great variety of designs may be made with this simple chuck. Front view.

1347. Side view.
1348. End view.
1349. Nut and screw.

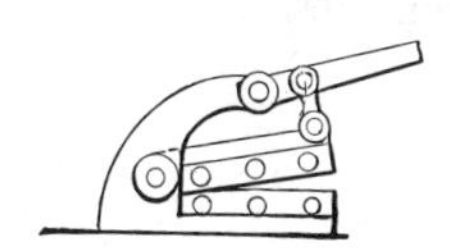

1350. COMPOUND LEVER SHEARS.

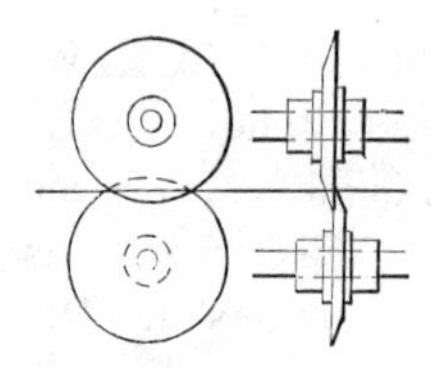

1351. DISC SHEARS.—Two bevelled edge discs just lapping, and revolving. Largely used in tin and cardboard cutting.

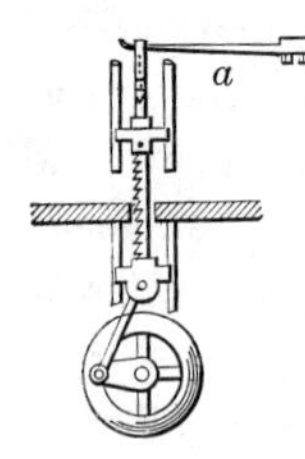

1352. GIG SAW.—The spring *a* gives tension to the saw running between guide frames, and operates by crank and connecting rod.

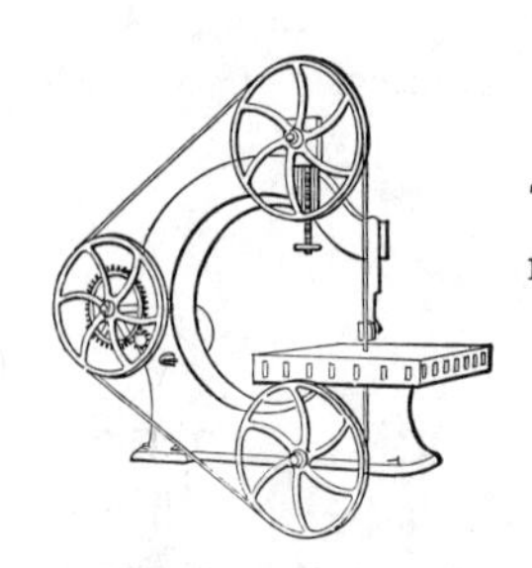

1353. BAND SAW, for sawing metals. The frame and third wheel are set back to give room for large plates.

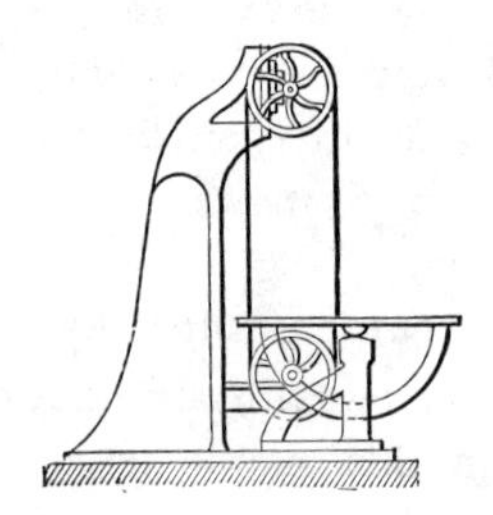

1354. BAND SAW.—Rectilinear motion of saw blade from rotary motion of band pulleys, with a tilting saw-table for bevel work.

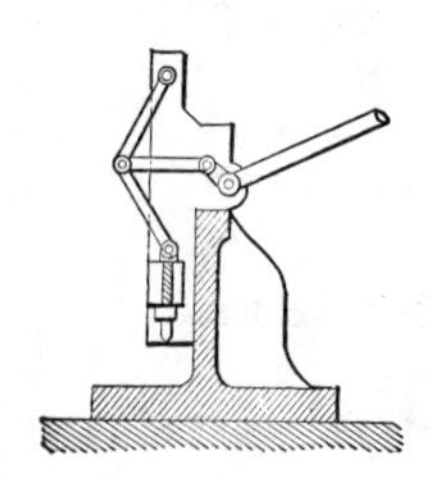

1355. TOGGLE-JOINT LEVER PRESS or punch. A type of toggle-joint used in the old form of printing and stamping presses.

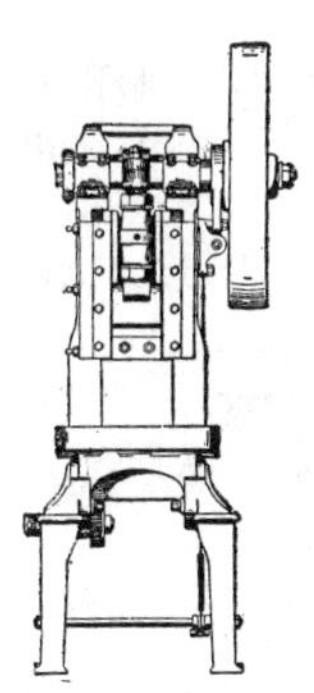

1356. POWER STAMPING PRESS. — Driven from a pulley with crank or cam shaft. A miss impression is made by a stop-clutch operated by a foot treadle.

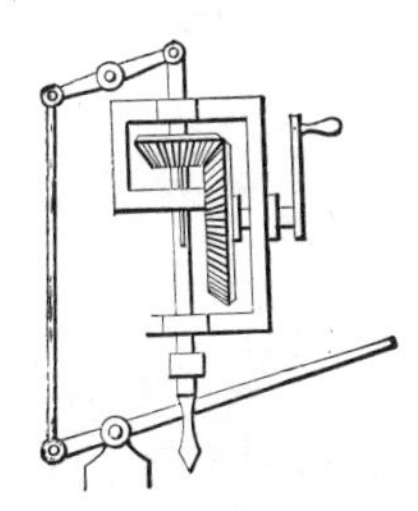

1357. HAND DRILLING MACHINE, with lever feed.

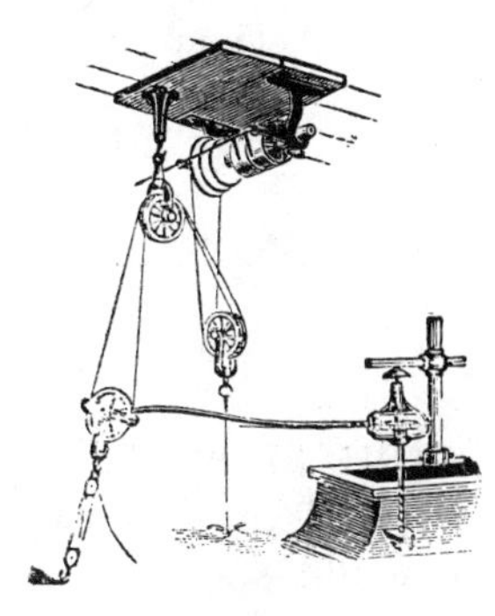

1358. PORTABLE DRILL, rope transmission and flexible shaft. One continuous rope over driving pulley, two double sheaves anchored, and flexible shaft pulley; allowing the driving sheave of the flexible shaft to be anchored in any position, and for tightening the driving rope.

1359. MULTIPLE DRILLING MACHINE, for close drilling or perforating plates. Drills are operated close together by converging spindles.

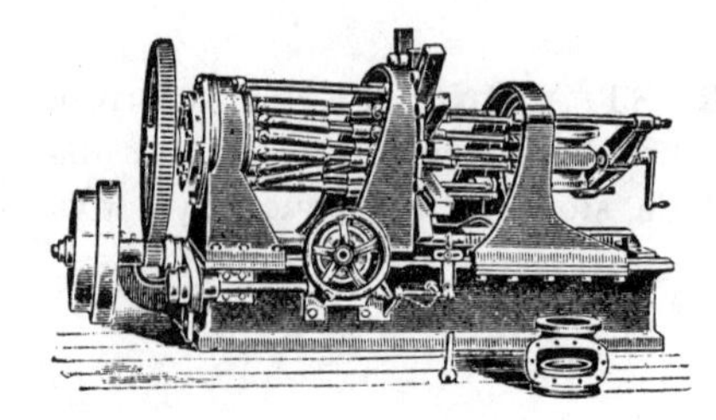

1360. MULTIPLE DRILLING MACHINE.—For drilling a number of holes in flanges at one time. The drill chucks are adjusted in a spider for any size circle and connected to the driving head with jointed rods.

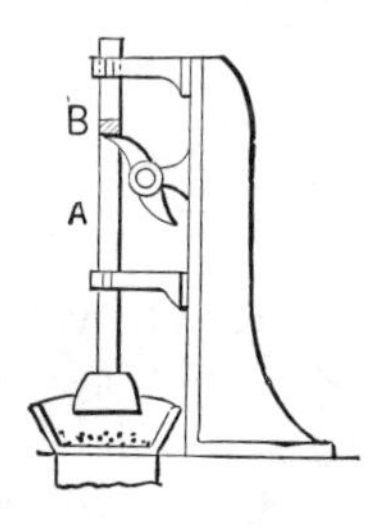

1361. STAMP MILL CAM MOTION.—The revolution of two or more cam wipers lifts the stamp hammers to drop by gravity.

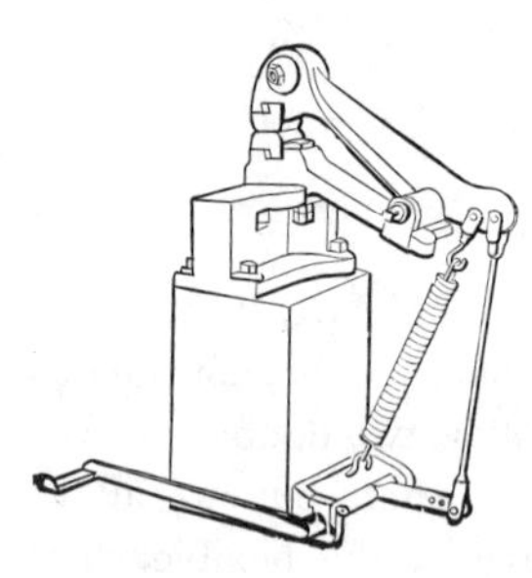

1362. BLACKSMITH'S HELPER, or foot helve hammer. Operated by the foot on the treadle. Hammer held up by the spring.

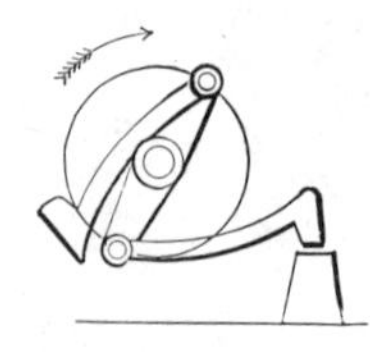

1363. REVOLVING RAPID-BLOW HAMMER.—The centrifugal action of the revolving arms throws the hammers outward.

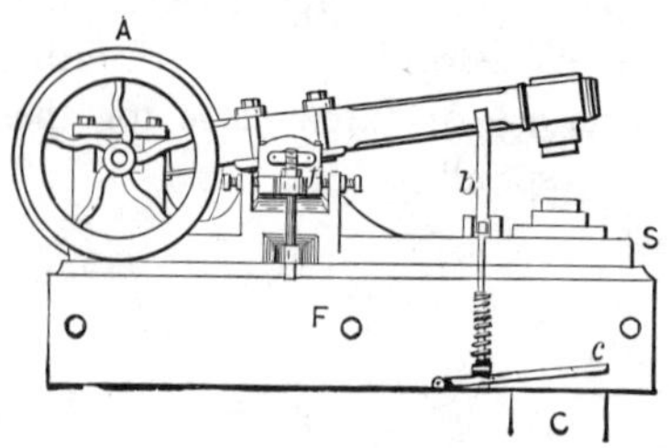

1364. HELVE TRIP HAMMER.—An ancient device yet in use. The treadle stops the action of the hammer by disengaging the bell-crank catch *b*. Used for small work.

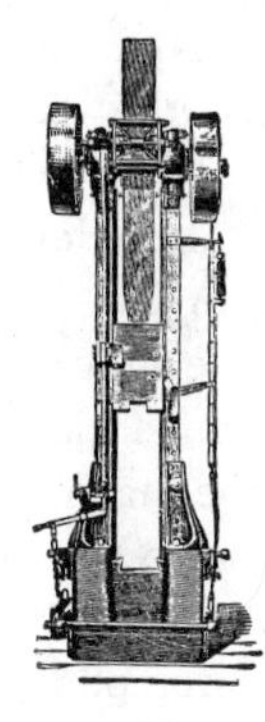

1365. FRICTION DROP HAMMER.—The hammer head is attached to a hardwood board running between friction rolls. One of the rolls has an eccentric sleeve shaft with a lever and lanyard to throw the roll out of contact with the board at the proper time for long or short drop. The other roll and shaft carry the driving pulley and are in constant motion.

1366. BEAM TRIP HAMMER.—The beam is vibrated by an eccentric on the driving shaft. The cushions intensify and regulate the blow of the hammer. The treadle operates the brake and controls the blow of the hammer. "Bradley" pattern.

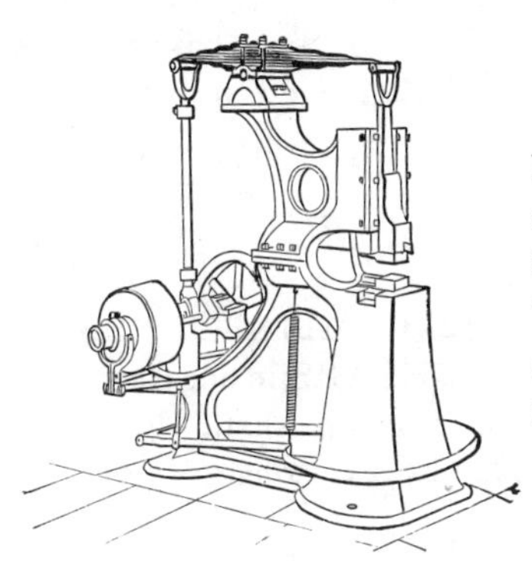

1367. SPRING HAMMER.—The height of the hammer, to suit the size of the forging, is adjusted by changing the length of the connecting rod. The treadle controls the stroke by operating a friction gear on the driving pulley.

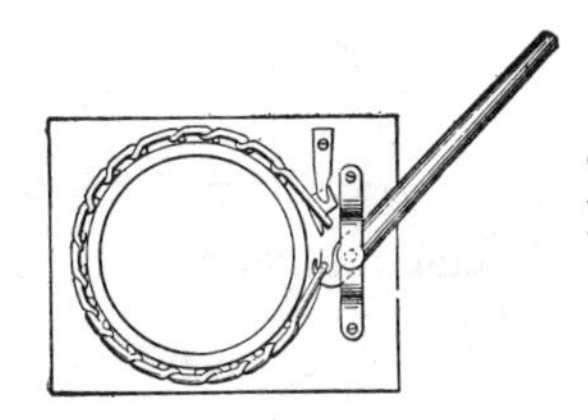

1368. TIRE SHRINKER.—A link chain around the tire terminates in a fixed hook, and the hook on a powerful lever.

1369. COMBINED TIRE UPSETTING AND PUNCHING MACHINE.—The tire is made fast by the cam jaws, and the movable cam is set forward by the sector cam lever and pinion. A punch is attached to the movable jaw with a punch die in the horn of the machine, so that the same operation of upsetting a tire may be used for punching iron.

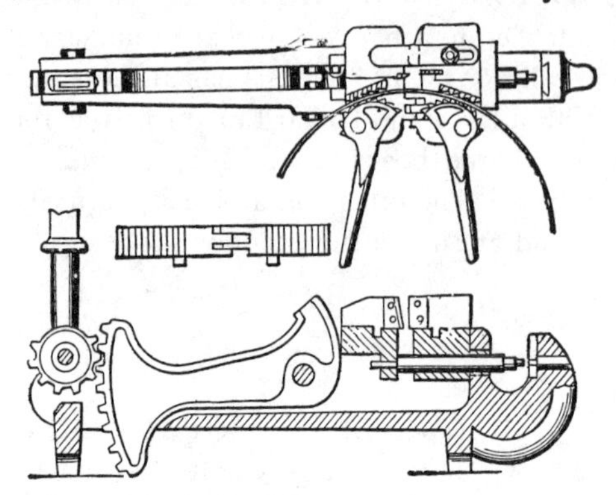

1370. Vertical section.

1371. PLATE SAWING MACHINE.—A slow-running steel saw blade lubricated by dipping in an oil box. The saw is automatically fed to the plate by a worm gear, but has a quick return by the hand wheel.

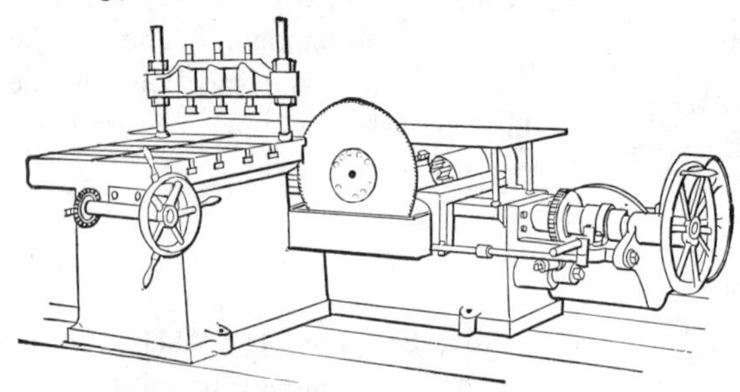

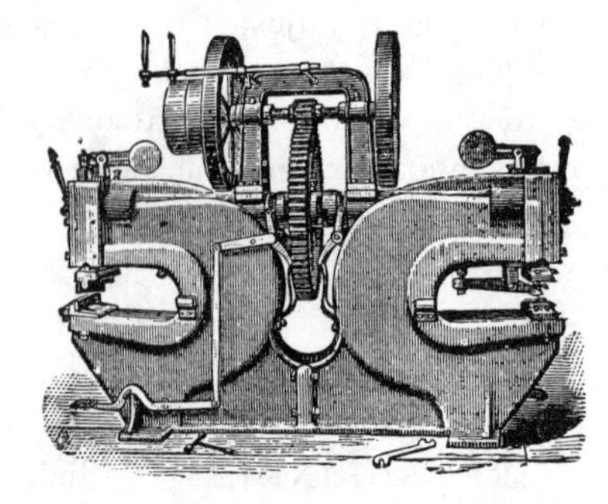

1372. COMBINED PUNCH AND SHEARS in one frame and driven from one shaft. Each controlled by a treadle.

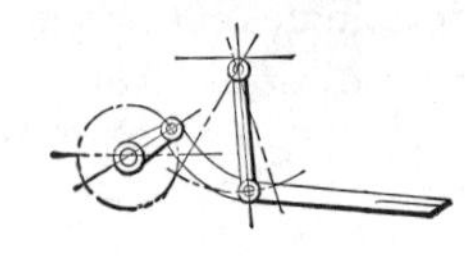

1373. SUSPENDED SWING TREADLE.—The foot takes a circular motion, no dead centre.

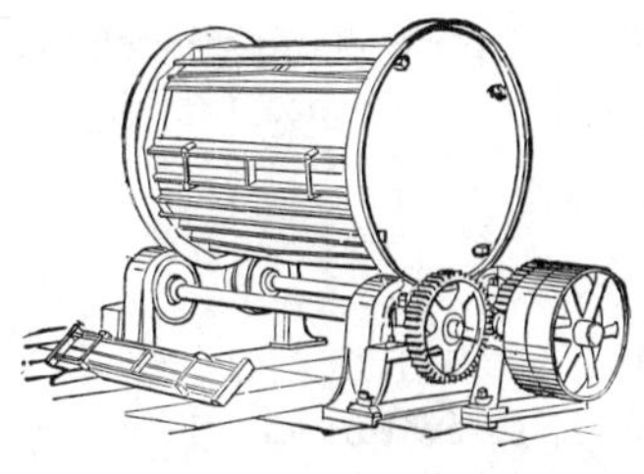

1374. POWER RUMBLING MILL, for cleaning sand from castings, polishing metal articles by tumbling with sand, charcoal, leather scrap, or any polishing powder.

1375. CENTRIFUGAL SEPARATOR, for removing oil from iron chips and turnings. The iron pan A is fixed to the spindle and pulley. The unequal loading of the pan is balanced by the elastic swivelled box B, held in a central position by springs. A cover with felted edge closes the top of the pan. The friction stop C acts as a brake to stop the motion of the pan.

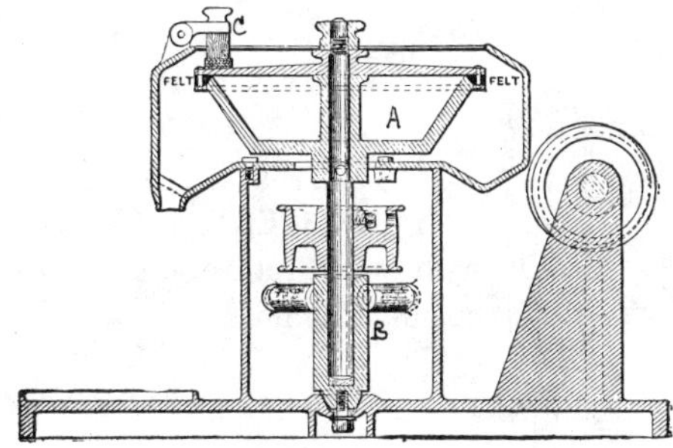

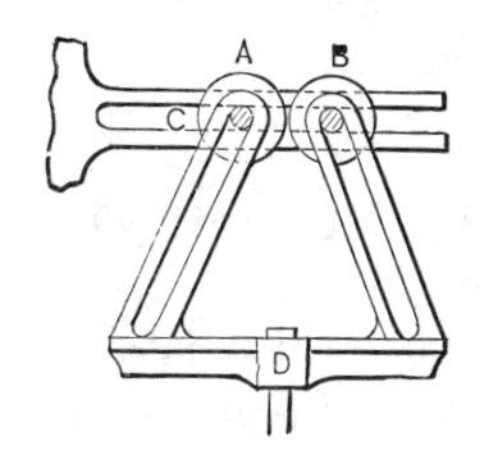

1376. CLOSURE OF ROLLERS by traversing the angular slots guiding the roller bearings. The slot guide C is fixed. The piston-rod head D carries the angular slots that move the rollers forward and backward.

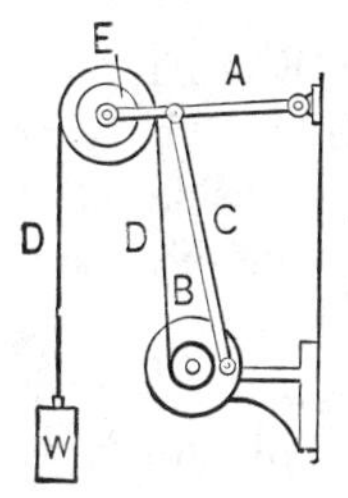

1377. VIBRATING LIFT.—The revolving drum B lifts the weight W, while the crank-pin connecting rod C gives the arm A and sheave E a vibrating vertical movement. With certain proportions between the size of the drum B, the distance of the crank pin and connecting arm at A, a variety of motions to the cord D may be made.

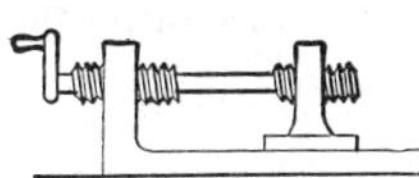

1378. DIFFERENTIAL PITCH MOVEMENT.—The motion of a traversing stud by the revolution of a differential screw allows of measurement of minute motions and distances. A micrometer device.

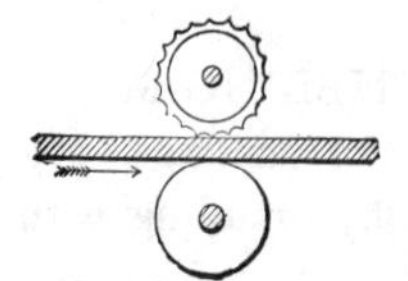

1379. FEED WHEEL for a planing machine. The corrugated upper wheel pushes the lumber to the cutter.

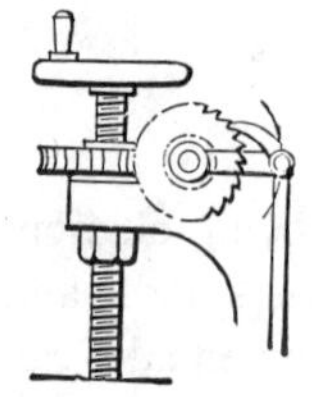

1380. COMBINED RATCHET AND HAND FEED GEAR.—The hand screw turns in the worm-gear nut, and may be used for quick adjustment.

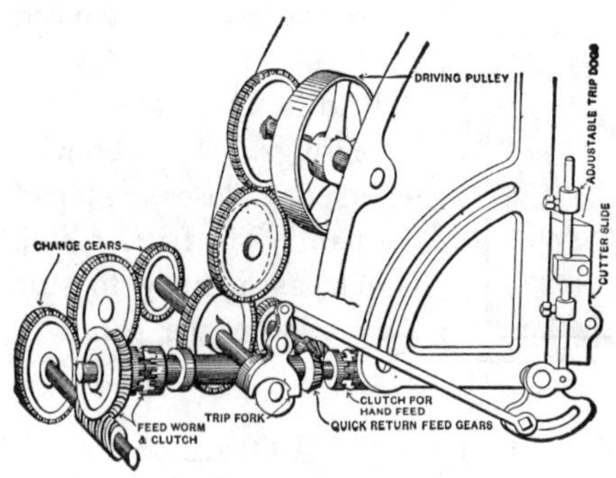

1381. GEAR TRAIN, with quick return, for a gear-cutting machine.

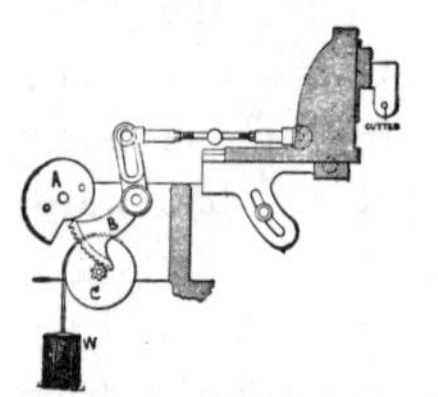

1382. QUICK RETURN MOVEMENT for a cutter head. A constant rotation of the cam operates the bell-crank sector, which is quickly drawn back by the weight W and pinion C.

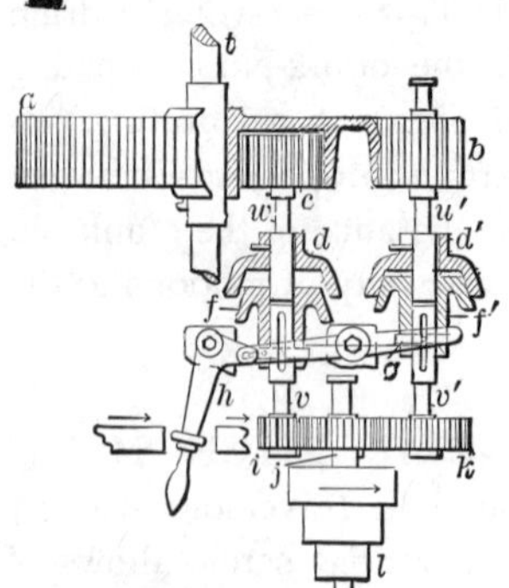

1383. REVERSING GEAR, from a single belt and cone pulley. The gear wheel *a* has an outside and inside set of teeth with the pinions *b*, *c* meshing and running in opposite directions.

The friction clutches operated by a lever reverse the motion of the large gear by alternately putting in motion the inside or outside pinion.

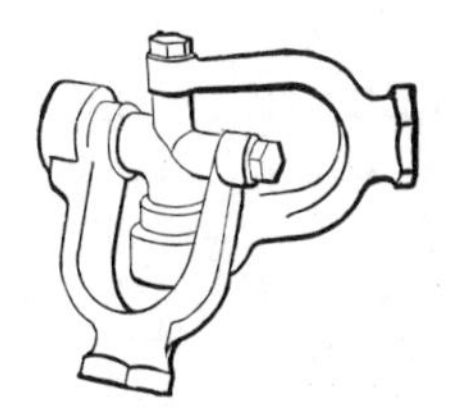

1384. FLEXIBLE UNIVERSAL STEAM JOINT.—"Hampson" model. The steam flows through the thick arms of the Y's, which have ground joints.

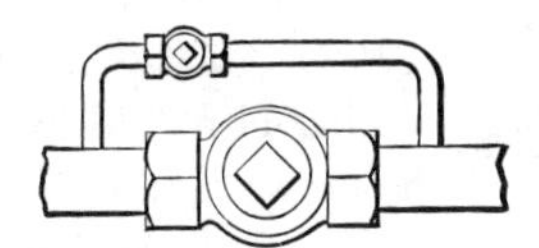

1385. BYE PASS COCK OR VALVE. —To allow of a small delivery when the large valve is closed, or for relief of pressure against a large valve.

1386. SIGHT-FEED LUBRICATOR.—The amount of feed is seen by the frequency of drops at the sight hole. Adjusted by a needle-point valve with milled head and screw.

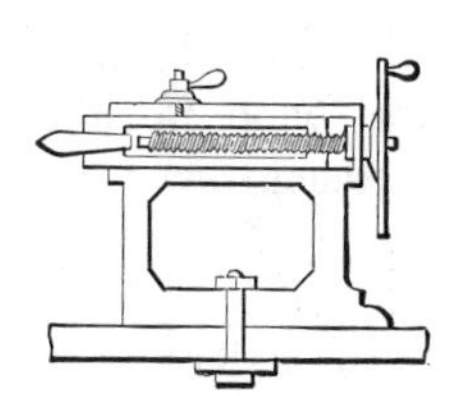

1387. SCREW MOVEMENT, for the tail stock of a lathe. The spindle moves in a key slot to prevent turning. The screw has a collar and is shouldered on the outside by the wheel hub. The back end of spindle has a thread acting as a nut on the driving screw.

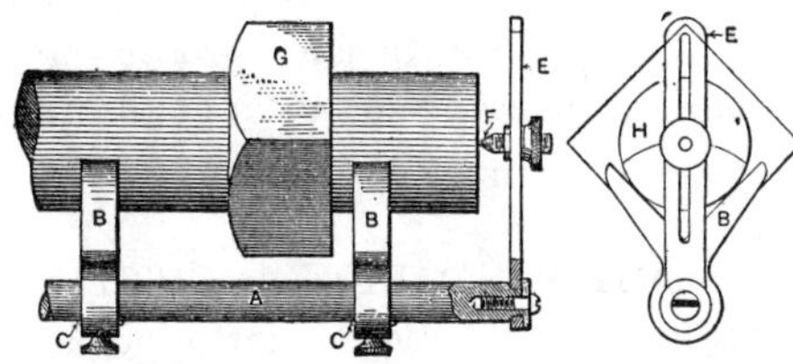

1388. CENTERING TOOL. Used for scratching the centre on round shafting or rods. The slotted arm E swings on the spindle A, as it traverses around the shaft to be centered.

1389. End view.

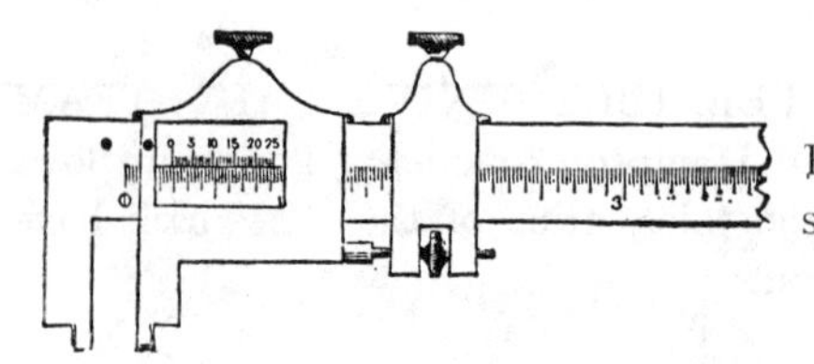

1390. VERNIER CALIPER, with slow-motion stop screw.

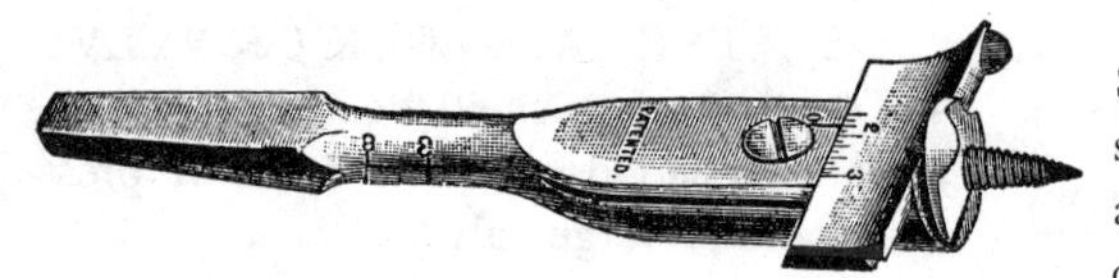

1391. EXPANSION BIT.—The spring clip held by a screw clamps the cutter in position to bore any size hole within its limits of expansion.

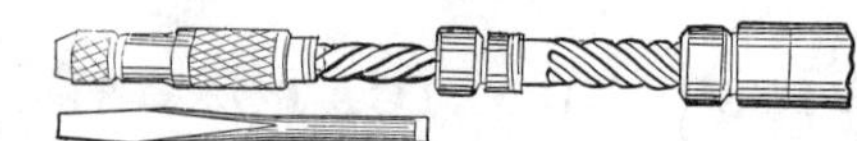

1392. DOUBLE-ACTING SCREWDRIVER. — The inside spindle has a left-hand screw, the outside hollow spindle a right-hand screw; and both with nuts that can lock either spindle by screwing to the thread on the lower end of each or either spindle.

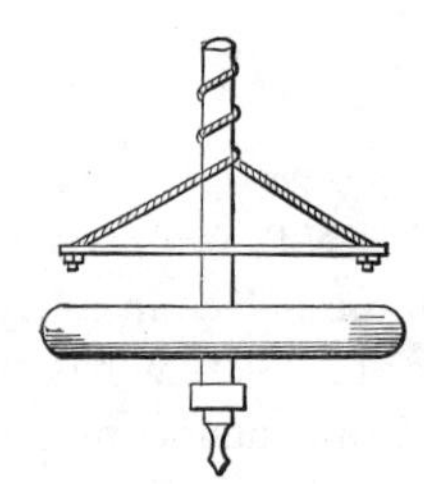

1393. PUMP DRILL STOCK.—A very ancient device, yet largely in use at this date in the jewelry and other light manufacturing establishments. The heavy revolving disc keeps up the momentum to rewind the band upon the spindle in contrary direction for each downstroke of the bar.

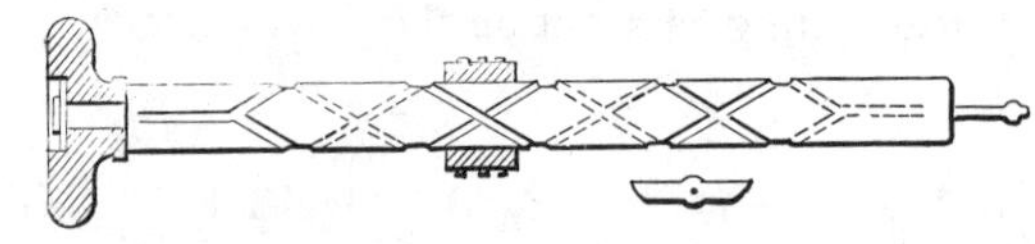

1394. RECIPROCATING DRILL STOCK.—By the double groove and follower, the drill turns the same way at each movement of the ring and follower.

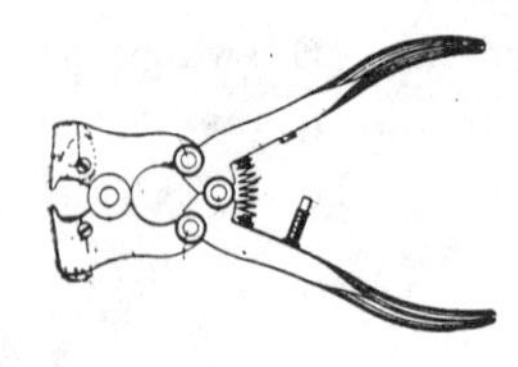

1395. COMPOUND LEVER CUTTING PLIERS, in which the toggle-joint principle is used to give the greatest power at the closure of the jaws.

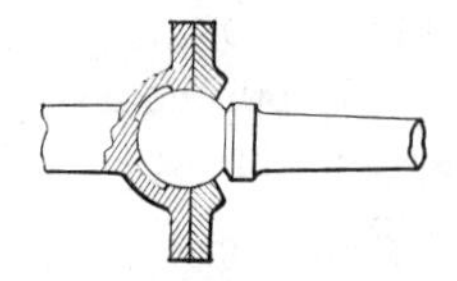

1396. BALL SOCKET, used on surveyor's compasses. The gland is tightened with countersunk screws.

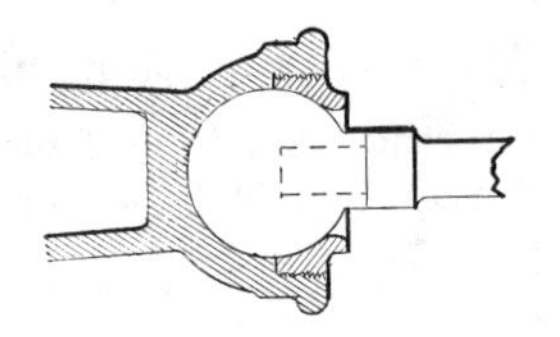

1397. BALL SOCKET, with a screw gland.

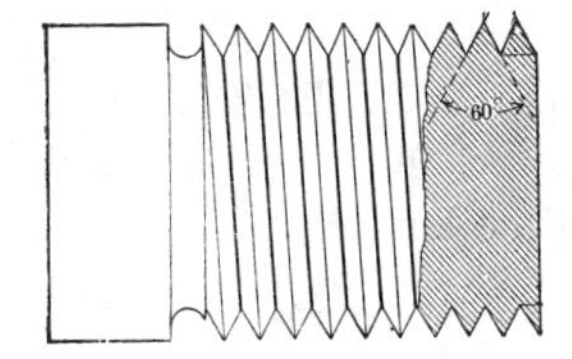

1398. SCREW THREADS.—Standard V thread, sharp at top and bottom. Depth equals 0.85 of the pitch. Angle 60°.

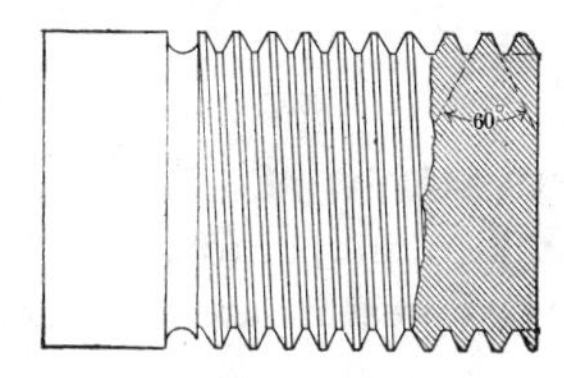

1399. SCREW THREADS.—United States Standard Thread. Flat top and bottom. Depth equals 0.65 of the pitch. Angle 60°.

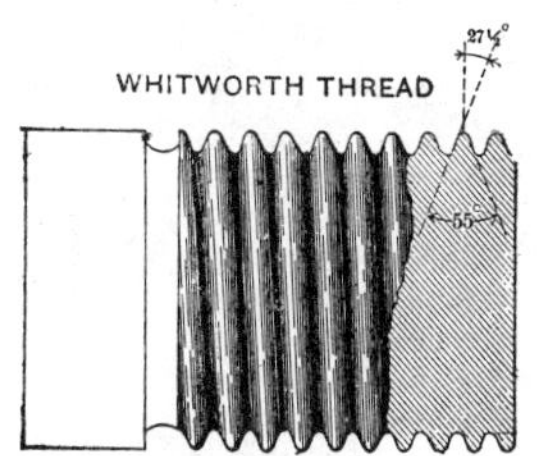

1400. SCREW THREADS, "Whitworth" thread. Rounded top and bottom. Depth equals 0.75 of the pitch. Angle 55°.

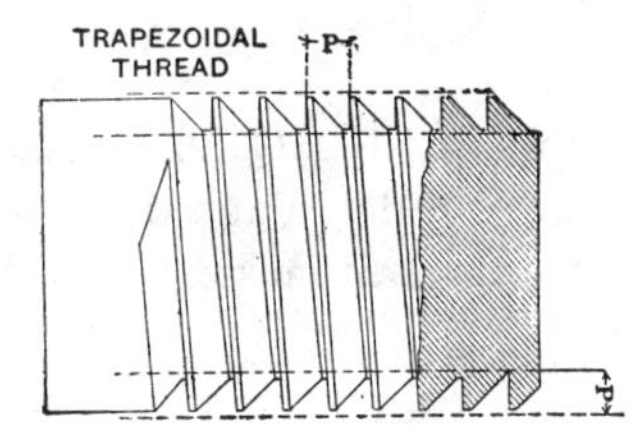

1401. SCREW THREADS, Trapezoidal thread. Angle 90° face, 45° back. Depth equals 0.75 of the pitch.

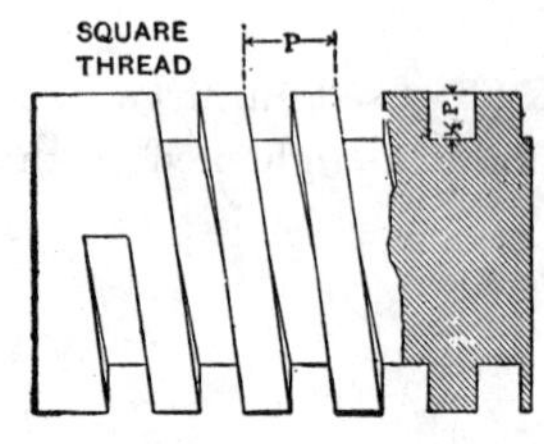

1402. SCREW THREADS, square thread. Angle square. Depth equals + half pitch. Width between threads equal + half pitch, for clearance.

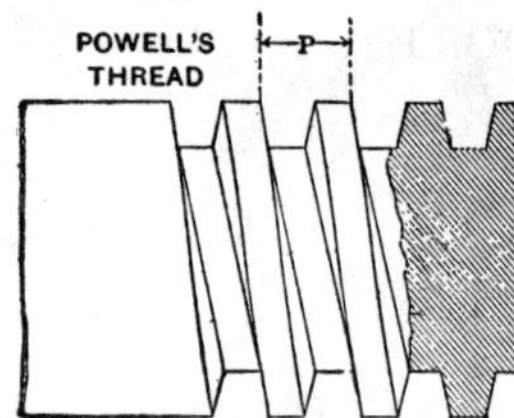

1403. SCREW THREADS, "Powell's" thread. Depth of thread equals + half pitch. Width of top of thread, 0.37 – of pitch. Width of bottom, 0.37 + of pitch. Angle of side, 11 1/4 °.

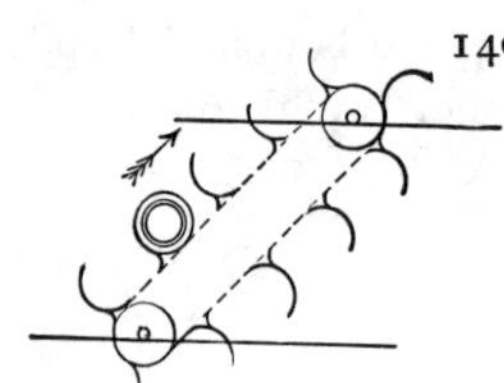

1404. CONTINUAL BARREL ELEVATOR. —Sprocket wheels and link chains with curved arms to hold the barrels.

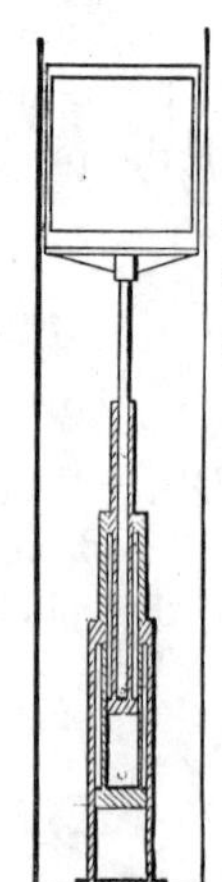

1405. TELESCOPIC HYDRAULIC ELEVATOR. —The several piston cylinders take a proportional lift by their differential areas and balanced pressure areas in each compartment.

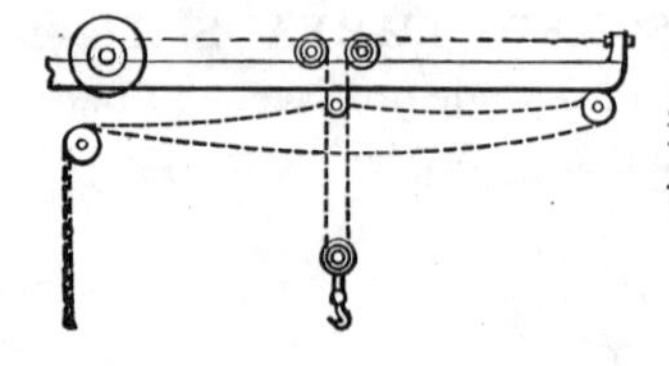

1406. TRAVELLER HOIST, showing the principles of the balanced counter pull and the traverse tackle.

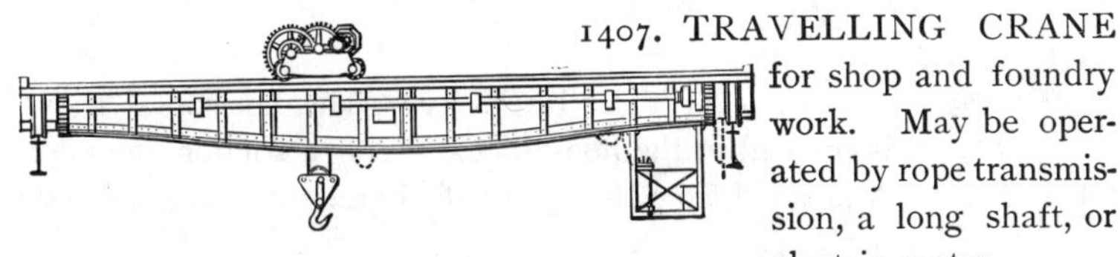

1407. TRAVELLING CRANE for shop and foundry work. May be operated by rope transmission, a long shaft, or electric motor.

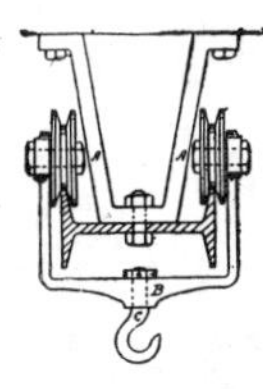

1408. I-BAR TRAVELLING TRAMWAY, an easily made shop device. The I bar lies sidewise, bolted to brackets from the ceiling. The double trolley cannot run off.

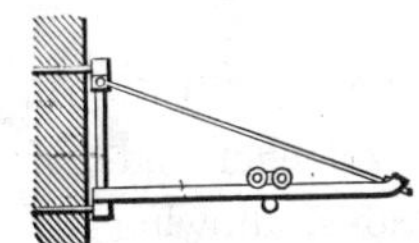

1409. SWING BRACKET CRANE, with trolley.

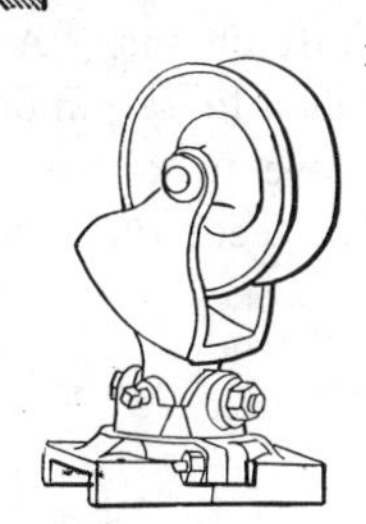

1410. ADJUSTABLE UNIVERSAL SHEAVE. It can be set in any desired direction and canted by the double-swivel foot.

1411. "HARRINGTON" CHAIN HOIST.—A worm gear operates a double-chain sprocket, with chains yoked at hook.

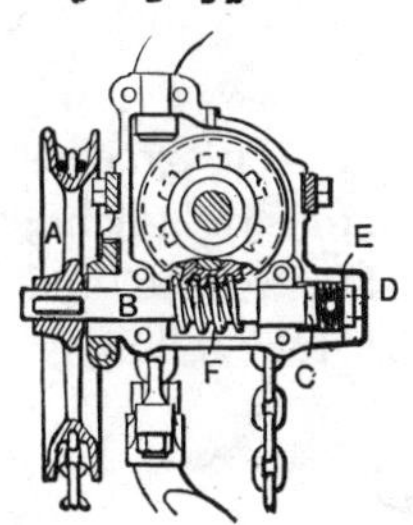

1412. "YALE" DUPLEX HOIST.—A worm F meshed in a gear on the same shaft with the hoisting-chain sprocket. A, Hand-chain sprocket on worm shaft B; C is a friction plug which holds the worm from running back. For self-running down, the plug may be reversed, presenting a smaller friction surface to the worm shaft. A pin holds the plug from turning.

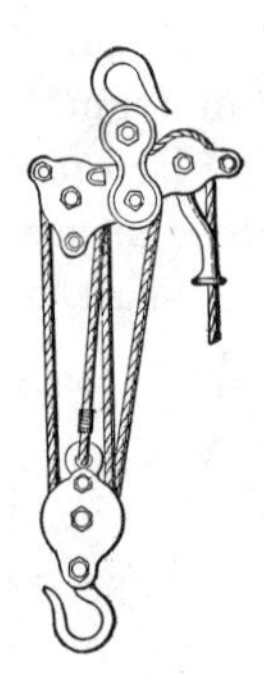

1413. SAFETY TACKLE.—The horizontal frame is pivoted in the hook block having a friction shoulder. A lanyard from the eye of the horizontal frame releases the grip.

1414. DIFFERENTIAL CHAIN-PULLEY BLOCK.—The chain sprockets, one on each side of the gear drum, run in different directions, allowing the surplus chain to hang between the draft chains. An eccentric on the hand-wheel shaft rolls a loose pinion around the discs, causing them to move in opposite directions by the differential number of teeth on each side of the pinion.

1415. DOUBLE SCREW-GEAR HOIST.—A right-and-left screw turns the chain sprockets in mesh with the lifting chain. "Box & Co." model.

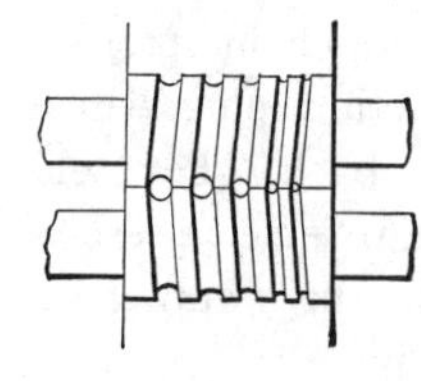

1416. TAPER TUBE ROLLS.—The grooves are turned as a taper screw. One rolls right-hand, the other left-hand to match. Much care and management are required in taper tube-rolling.

1417. "YALE-WESTON" DIFFERENTIAL GEAR HOIST.

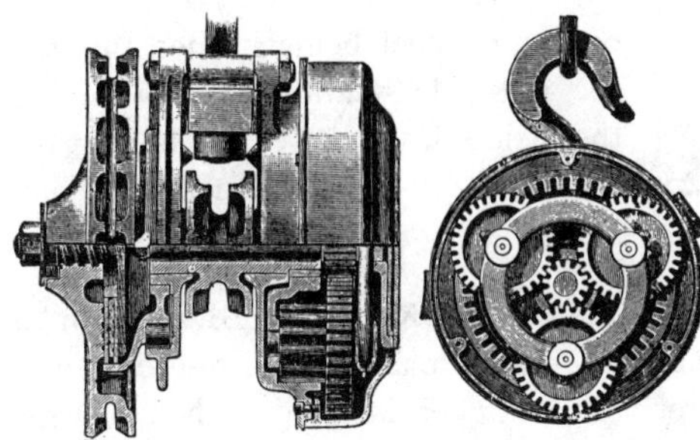

—The hand-chain sprocket shaft runs loose in a sleeve which carries the hoist-chain sprocket. A small pinion on the right-hand end of the central shaft drives three spur gears pinioned on a circular movable frame attached to the chain sprockets. To each of the three spur gears are fixed a pinion, which meshes in an internal tooth gear fixed in the case.

1418. Section, showing gear.

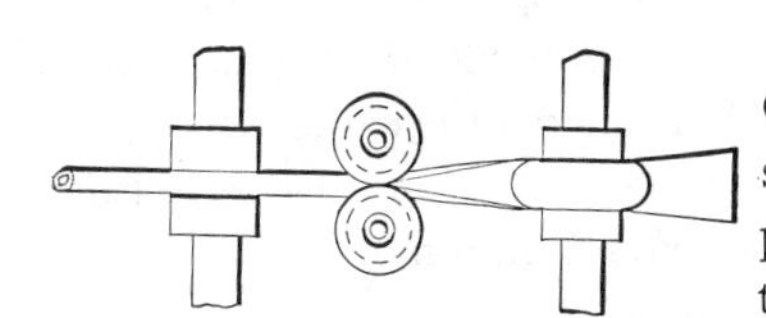

1419. TUBE-ROLLING MACHINE.—The first roller turns the strip of metal to a half-circle. The pair of vertical rolls close up the tube.

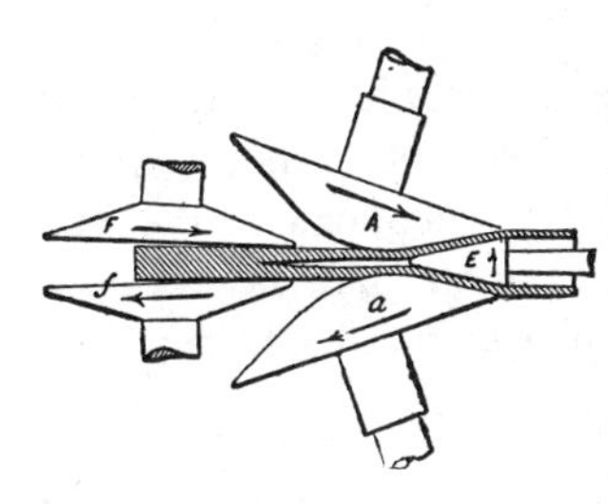

1420. SEAMLESS TUBE MAKING. — Rolling a solid bar between a pair of angular-axled disc rollers opens a cavity within the bar which is further expanded by a second pair of disc rollers. The rolling of the tube between the discs pushes the tubular bar over a revolving conical mandrill.

1421. WIRE-BENDING MACHINE.—A marvel of complex motions. Hooks and eyes, and any special shapes of wire-work can be made on these machines.

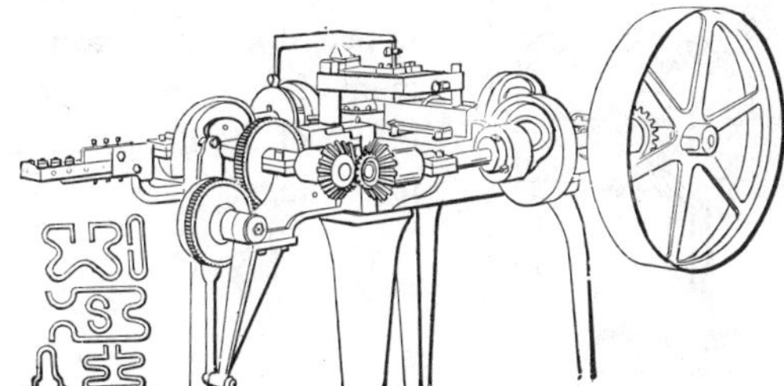

1422. Samples of wire bending.

1423. SEAMLESS TUBE MAKING.—The "Mannesmann" process. A, *a*, conical corrugated rolls; B, guide tube; B″, hot bar of iron or steel being pushed through the rolls; D, mandrill for widening the inside of the tube, the hollow being started by the action of the outside rolls.

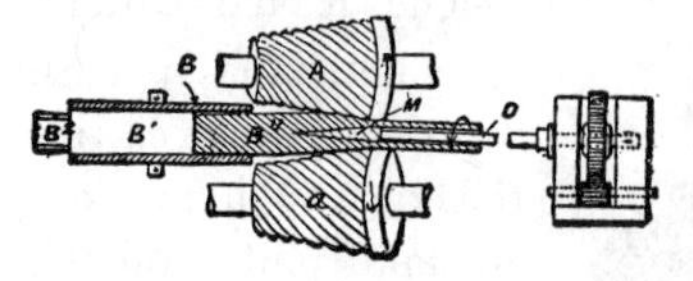

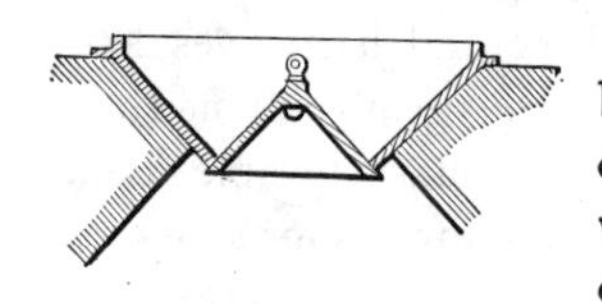

1424. HOPPER AND BELL, for a blast or other furnace, for feeding coal and ore. The hopper is filled with a charge, when the bell is quickly lowered and the charge drops into the furnace.

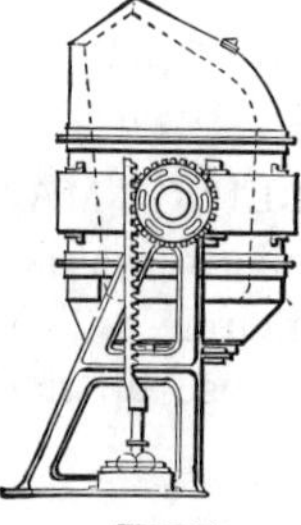

1425. "BESSEMER" STEEL CONVERTER. A large crucible on trunnions, through which air is blown to passages in the bottom of the shell and through the cast iron, burning out the excess of carbon, when the crucible is turned over and the cast iron, converted into steel, is poured into moulds.

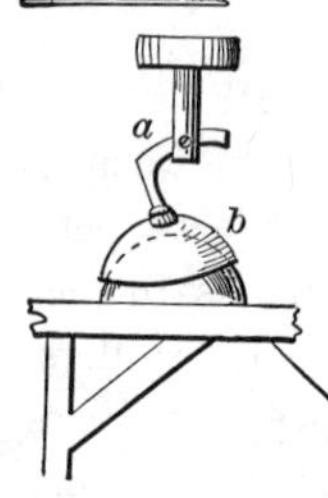

1426. LENS-GRINDING MACHINE. — The bell-crank arm *a* is made adjustable in the vertical shaft, and is pivoted for a free motion in the grinding cup *b*, to give a variety of motions to the cup over the lens; or the operation may be reversed and the lens given a circular motion in the cup.

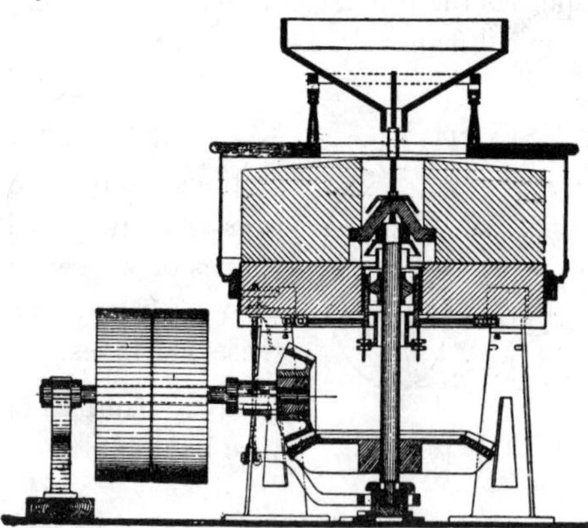

1427. GRINDING MILL in section, showing the balancing of the upper stone and adjustment of step, and the centering of the hopper and feed gauge.

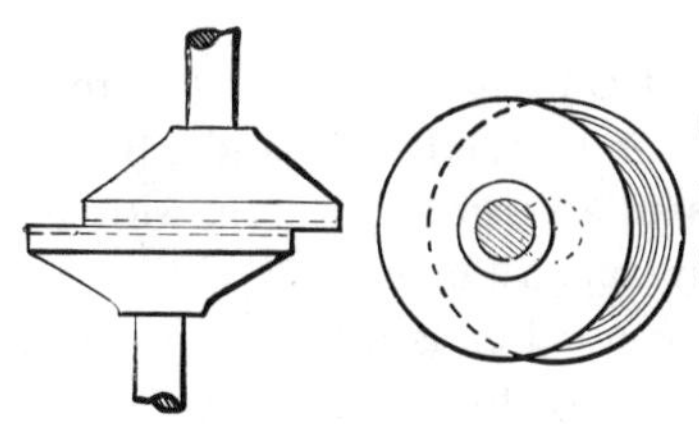

1428. "BOGARDUS" MILL.—Grooved steel discs running eccentric to each other. Largely used for grinding paints and drugs.

1429. Plan showing grooves.

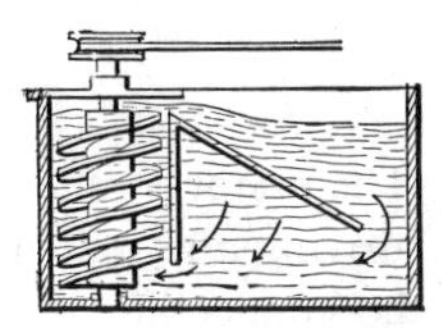

1430. CIRCULATING SCREW PROPELLER AND MIXING TANK.—Is used in various forms in laundries, soap crutching, and oil refining.

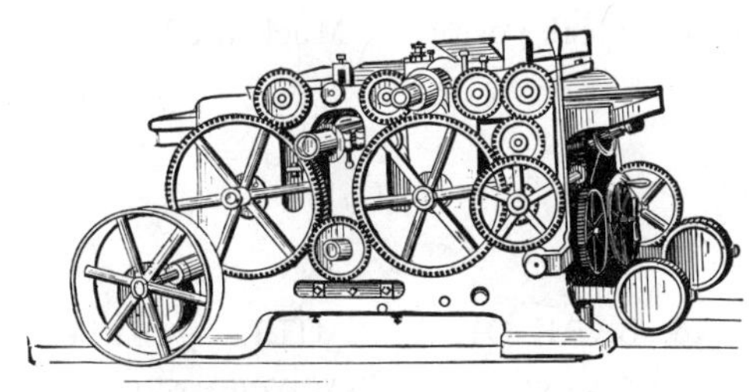

1431. DOUBLE CYLINDER PLANER, for lumber. Takes a rough and finishing cut by once passing the lumber through the mill.

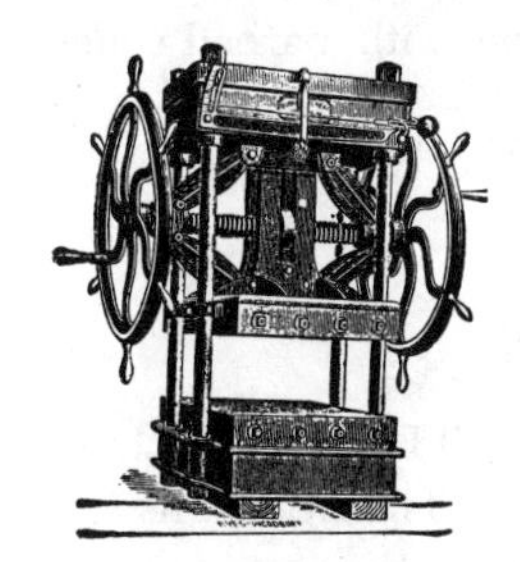

1432. DOUBLE TOGGLE-JOINT SCREW PRESS with steam-heated platens for vulcanizing rubber or embossing by heat and pressure

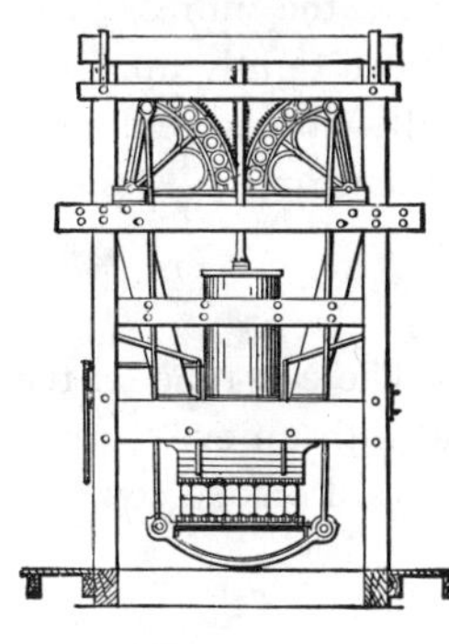

1433. STEAM COTTON PRESS, for repressing and condensing baled cotton. The geared sectors, driven by the double-rack piston rod and piston, increase the pressure immensely at the latter part of the stroke by the toggle-joint action of the connecting rods as they approach the radial bearing of the sectors.

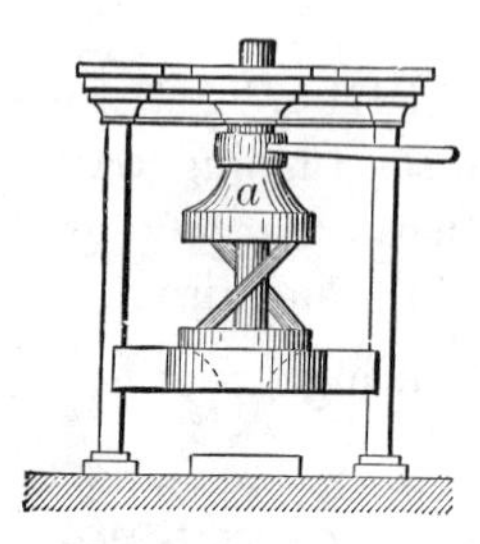

1434. TOGGLE-BAR PRESS.—The rotation of the disc *a* by the lever handle brings the toggle bars to a vertical position, with increasing pressure upon the platen. The toggle bars have spherical ends fitted to spherical cups in the top and bottom discs.

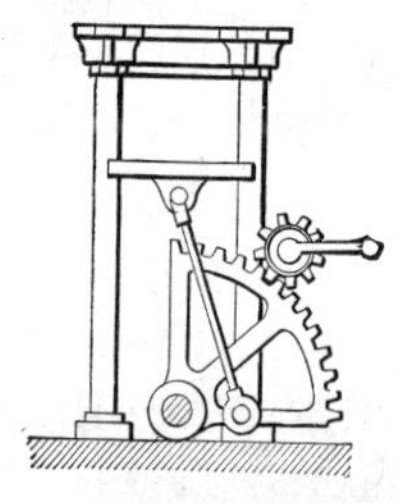

1435. SECTOR PRESS.—The sector is rolled up by the crank and pinion, driving the platen up with increased force until the connecting rod reaches its vertical position. Much used on cotton presses.

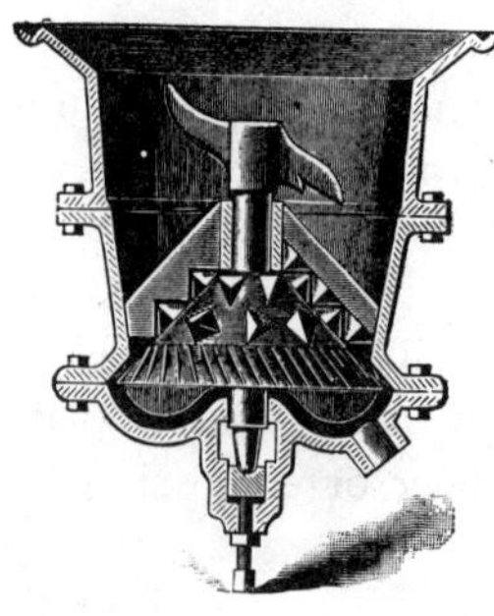

1436. BARK OR COB MILL.—A barbed and corrugated cone revolving within a spider and counter cone, with barbed cones and corrugations.

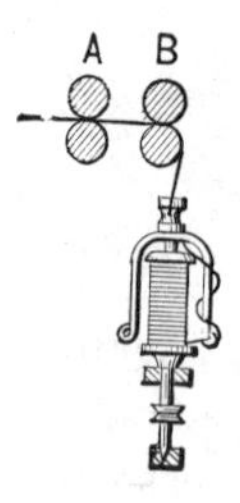

1437. DRAWING AND THROSTLE TWISTING ROLLS AND BOBBIN WINDER.—The front rolls run faster than the feed rolls, and draw the fibre. The throstle twists the thread which is drawn tightly upon the spool that runs loose on the spindle, and is held by a friction spring to give it the winding tension.

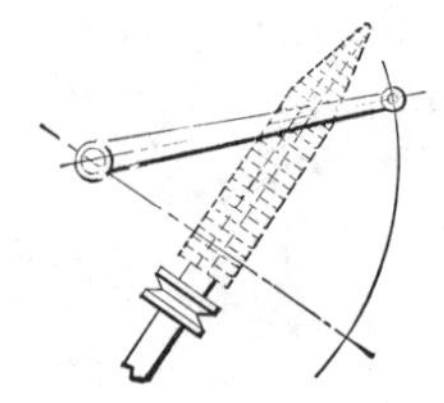

1438. COP WINDER.—The cop tube on the spindle revolves. The arm with an eye, carries the thread forward and backward on the cop.

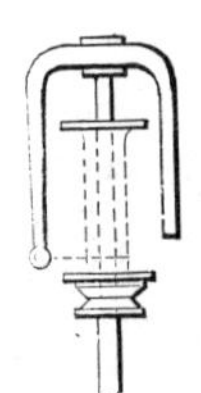

1439. BOBBIN WINDER.—The flyer revolves, while the bobbin is moved up and down the spindle for even winding. Thread passes through the hollow spindle down the arm and through the eye of the flyer arm.

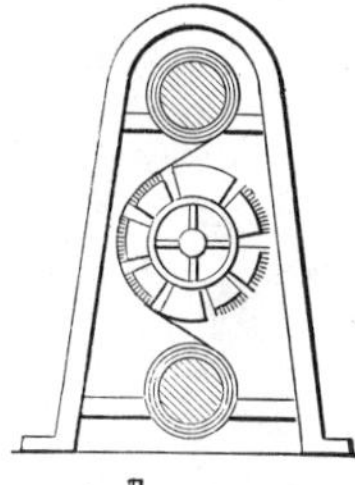

1440. CLOTH DRESSER.—The central wheel is the teazel drum. The cloth is guided by the rollers above and below.

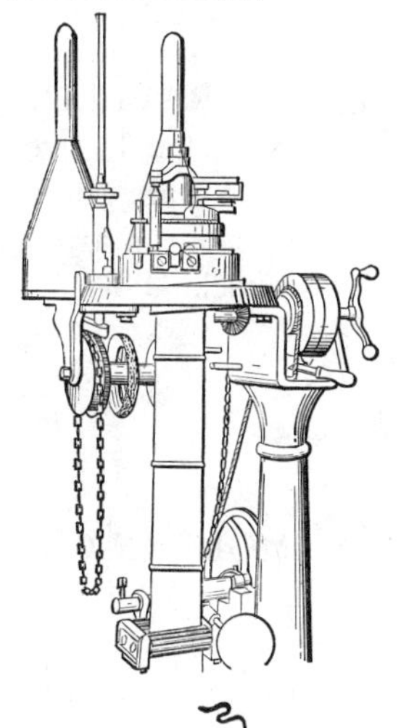

1441 KNITTING MACHINE, automatic rib knitter, "Heginbotham" model. Vertical needles and two bobbins.

1442. KNITTING MACHINE, seamless knitter, "Bellis & Weinanmayer" model. Vertical needles.

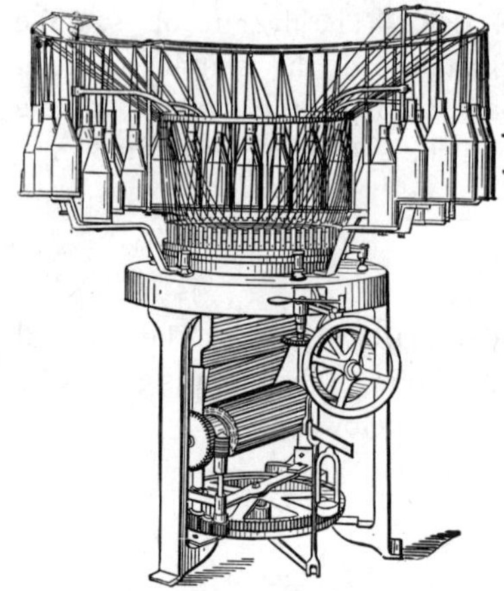

1443. KNITTING MACHINE.—Multiple thread knitter, "Hepworth" model, for web goods.

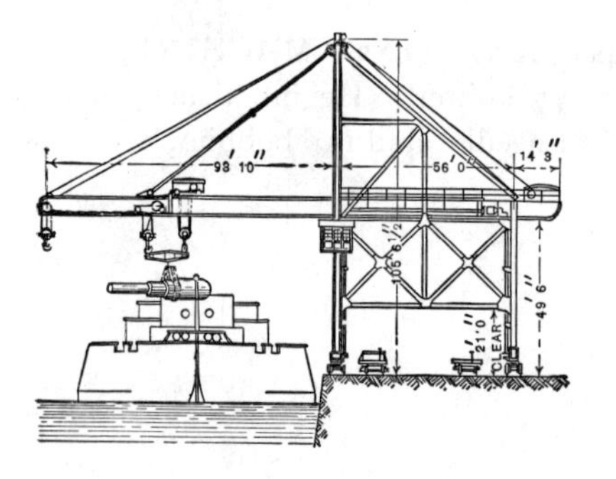

1443*a*. TRAVELING DERRICK.—Double trolleys and lever beam putting a 75-ton gun on an armor-clad war vessel.

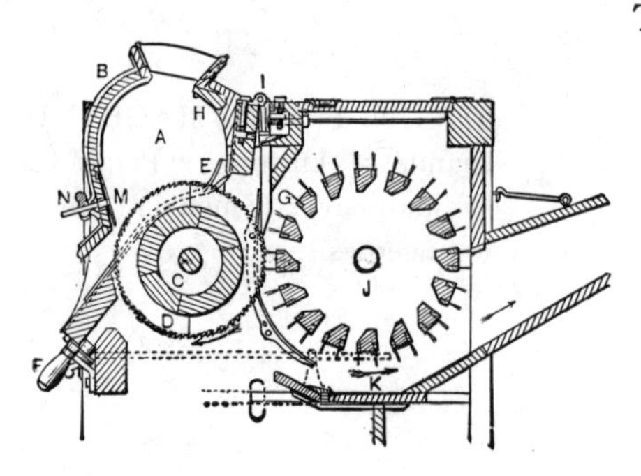

1443*b*. MODERN COTTON GIN.—

D, nest of saws.

E, saw grate between each saw to hold back the seed.

A, feeder trough and hopper.

J, cylinder brush stripping the cotton fibre from the saw.

F, adjusting lever.

K, sliding mote board.

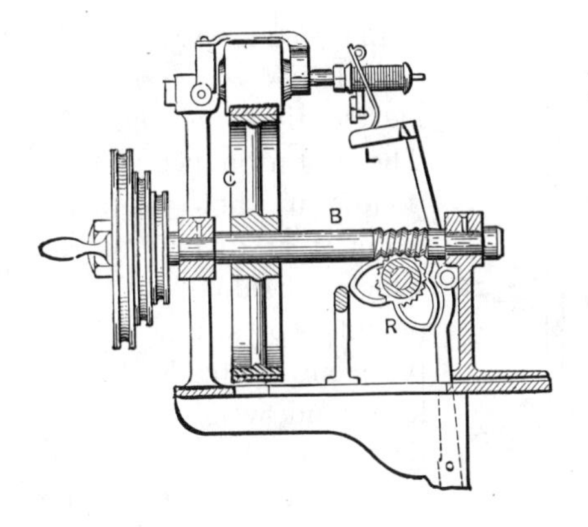

1443*c*. SPOOL WINDING MACHINE.—A worm screw B and gear drives a set of cams R on a cross shaft and oscillates a lever and thread guide to and fro. The spool spindle driven by friction gear from the shaft B.

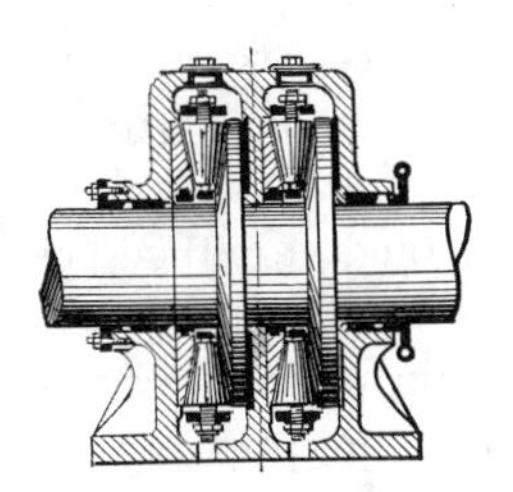

1443*d*. CONICAL ROLLER THRUST BEARING.—The conical rollers are held in ring travellers, inside and outside, which are connected together between the rollers.

The conical lines meet at the center of the shaft.

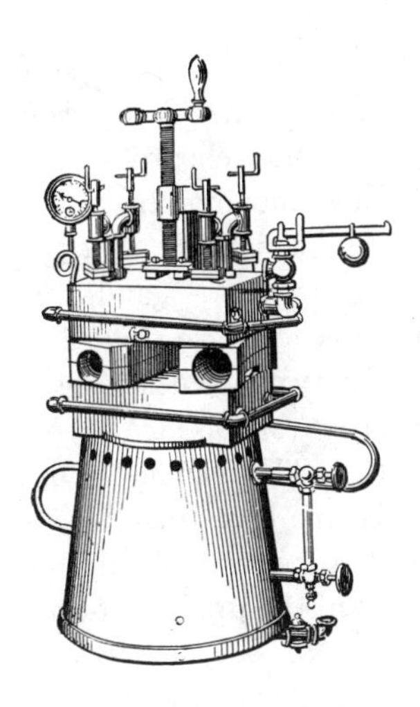

1443*e*. STEAM VULCANIZER for repairing bicycle tires.—Two steam slabs and tightening screw set on a small boiler heated by a lamp or torch; steam pressure should be 75 lbs. per square inch. Small clamps and screws are attached to the top steam box for vulcanizing other articles. The lower section is the boiler; the lamp or burner is set beneath it and not shown in cut.

1443*f*. STEAM VULCANIZER for repairing bicycle tires and bands.—The cast iron bed piece is hollow to hold sufficient water for making steam, which should be at 75 pounds pressure. Compression is made by the lever and weight for both tires and bands. A gas or gasoline torch for heating.

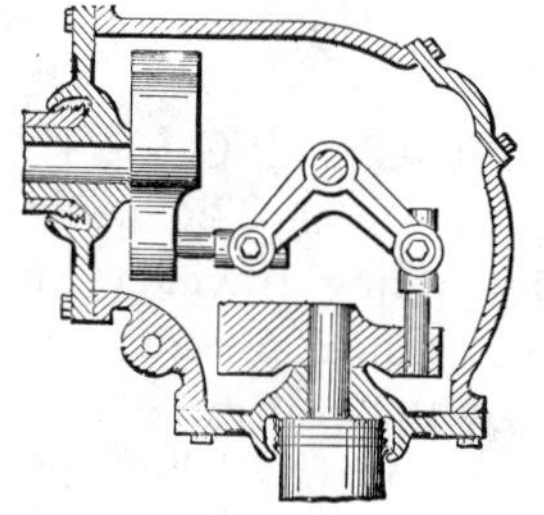

1443*g*. RIGHT ANGLE SHAFT TRANSMISSION.—A bell crank sliding on a cross bar is pivoted at its ends to the crank pins of shafts at right angles. The crank pins have sliding sleeves pivoted to the bell crank arms. The movement is enclosed in an oil-tight case to which the sliding bar is fixed. Horizontal section and plan.

1443*h*. TAKE-UP AND LET-OFF MOTION FOR LOOMS.—The detailed parts are: Take-up roller I, let-off roller K, a pawl U pivoted to the vibrating lever V operated by the shuttleboard frame G, H, and moving the ratchet T, bevel gear S, S′, shaft P, worm gear R, R, and gear connection to move the rollers I, K. E, warp with constant feed from spools C on the creel B. The inclined worm gear is made adjustable by the sliding boxes O, O′, shown in detail in the upper left-hand corner.

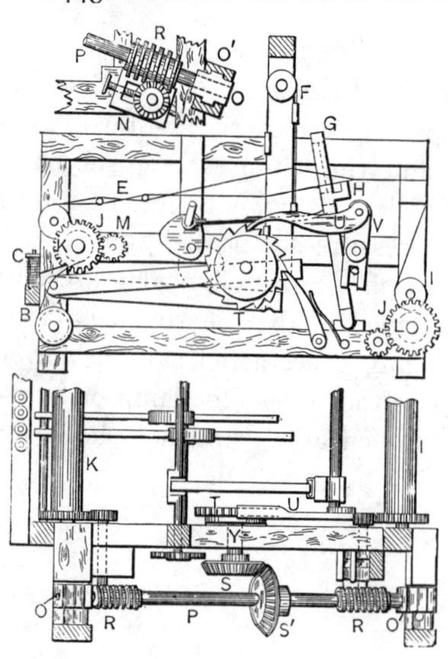

Section XVI.

CONSTRUCTION AND DEVICES.

MIXING, TESTING, STUMP AND PILE PULLING, TACKLE HOOKS, PILE DRIVING, DUMPING CARS, STONE GRIPS, DERRICKS, CONVEYER, TIMBER SPLICING, ROOF AND BRIDGE TRUSSES, SUSPENSION BRIDGES.

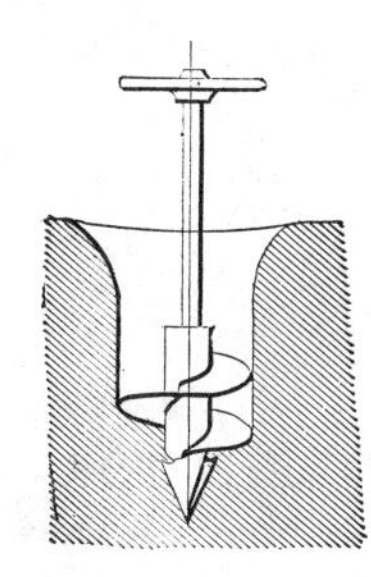

1444. POST AUGER.—Often made with a single turn to the blade. Used also for prospecting for foundations.

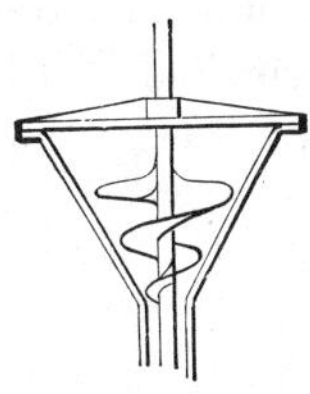

1445. PUG MILL, with spiral worm in a conical shell, for mixing mortar, concrete, or other material.

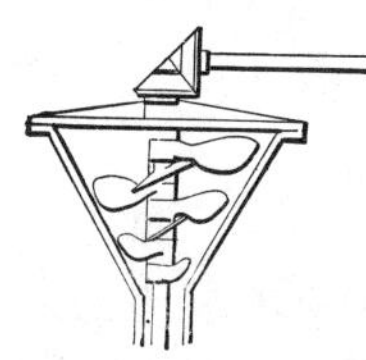

1446. CONICAL PUG MILL for mixing clays, mortar, concrete, and other material.

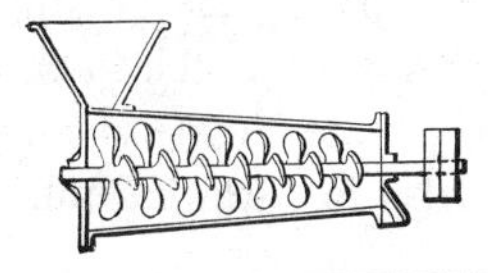

1447. CONICAL MIXING BARREL for mortar, concrete, or other material.

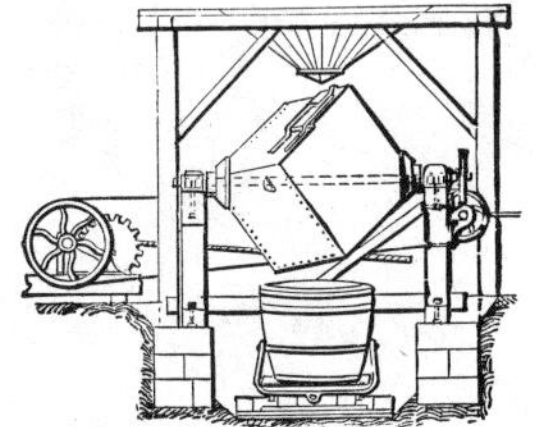

1448. CONCRETE MIXER.—A rectangular box of iron revolves on trunnions at opposite corners. A hopper for charging and a dumping car to receive the mixed charge.

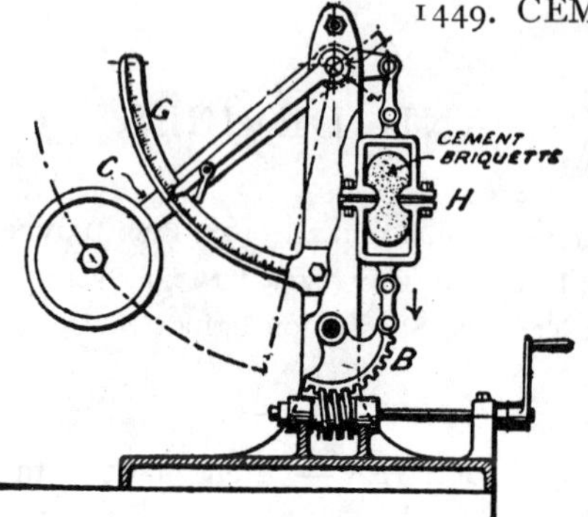

1449. CEMENT-TESTING MACHINE. —The cement sample is placed in the jaws at H. The sector B is turned by the worm screw until the weight on the arm C is raised to the limit of the breaking strain, where the index hand on the graduated arc is caught by the pawl, when the weight falls.

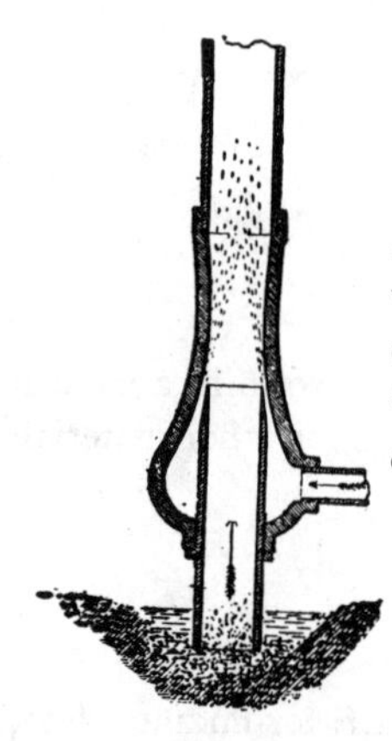

1450. HYDRAULIC SAND EJECTOR.—A thin annular jet of water, under high pressure, will eject sand and water from a sump and discharge at an elevation. The principle of the "Eads" ejector dischargings and from the caissons of the St. Louis Bridge.

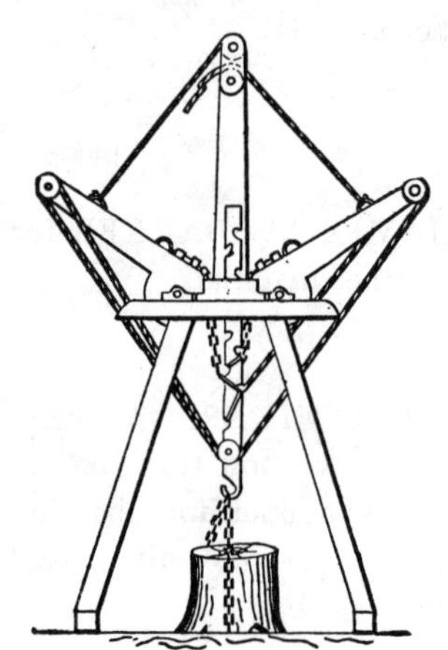

1451. TOGGLE STUMP PULLER. —By pulling up the two toggle levers, the chain and links slip down a notch in the draw bar when the double tackle draws the levers down. Also for drawing piles and sheet piling.

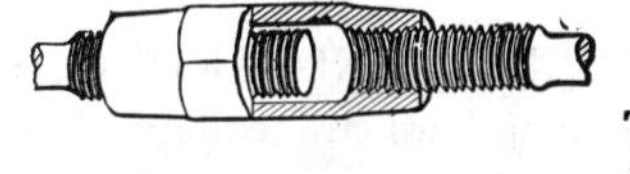

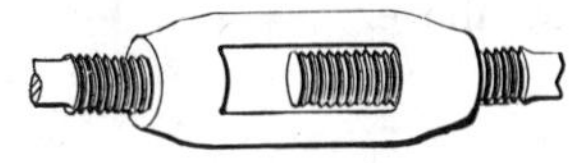

1452. RIGHT- AND LEFT-HAND TURNBUCKLE, sleeve and yoke pattern.

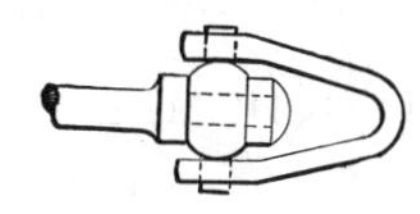

1453. SWIVEL SHACKLE.

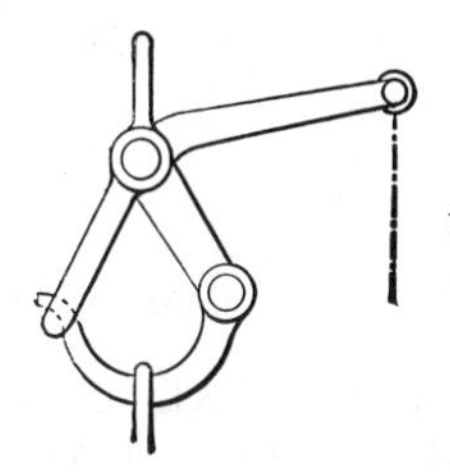

1454. SLIP HOOK, for drop weights and temporary pile hammer.

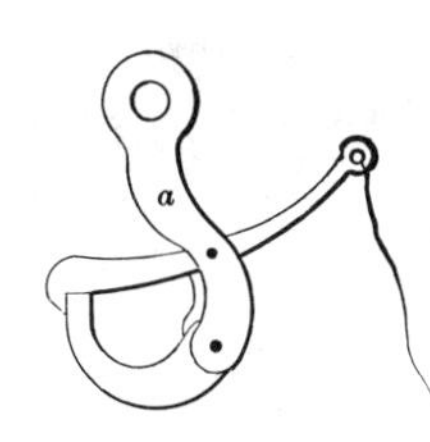

1455. TRIP HOOK.— A split shank with tongue and catch pivoted between the sides of the shank as shown.

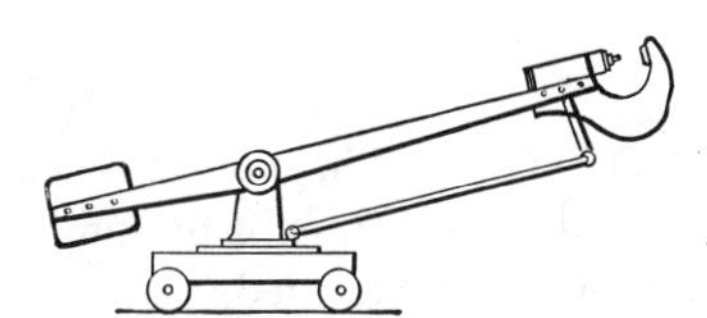

1456. BALANCED RIVETING MACHINE on a truck. For yard service, and iron and steel structural work.

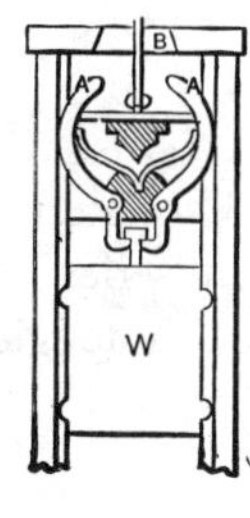

1457. RELEASING GRIP of a pile-driving machine. The bow ends of the grip are compressed when they reach the slot B in the frame and cast off the ram W. The springs between the bowed handles of the grip close the jaws to pick up the ram.

1458. AUTOMATIC DISENGAGING GRIP for a pile driver. The arms of the grip jaws are collapsed by contact with the inclined chocks above.

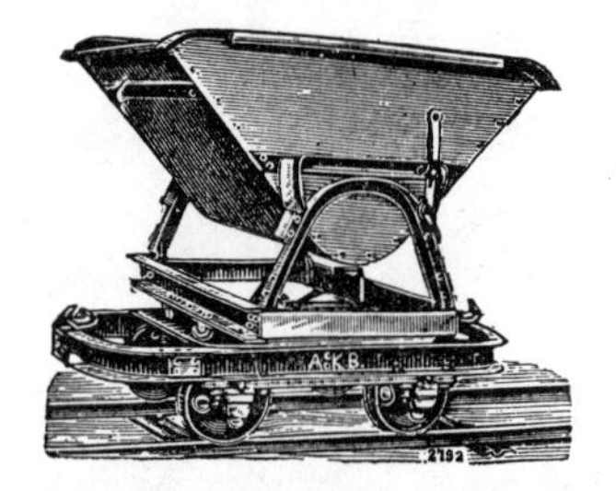

1459. SWIVELLING DUMPING CAR.— By turning the box and its frame, which is pivoted on the truck, the load can be dumped in any direction.

1460. SQUARE BOX SIDE-DUMPING CAR. —The side boards are hinged and locked by a snap lever.

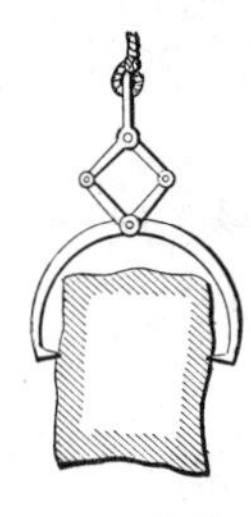

1461. LEVER GRIP-TONGS.—The pull on the shackle connecting the links and upper arms of the tongs causes a strong grip on any object to be lifted.

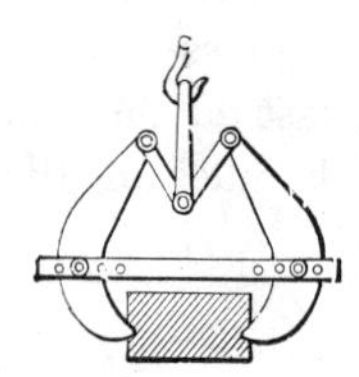

1462. ADJUSTABLE GRIP TONGS, for stones and heavy boxes. The link bars have a series of holes to vary the opening of the jaws. A toggle grip.

1463. PNEUMATIC DUMPING CAR.—A small compressor, operated from the axle, pumps air into a receiver under the platform. An oscillating cylinder, with direct connection with the bottom of the car, lifts it to the proper angle for dumping and returns it to the horizontal position by the mere movement of a valve.

1464. LEWIS WEDGE, for lifting stone. A central taper wedge, with eye and ring at the small end. A taper wedge is inserted in a reverse position on each side of the double-taper wedge, so that the outside of the combination is parallel in the hole in the stone. A pull on the centre wedge pushes the outer wedges against the side of the hole with force sufficient to lift the stone by the friction of their contact surfaces.

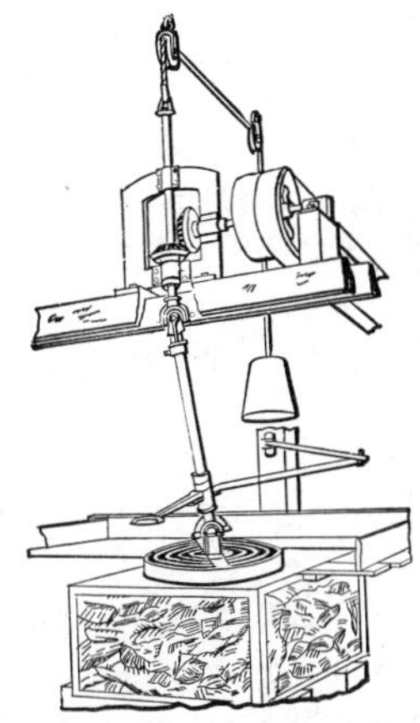

1465. STONE GRINDING AND POLISHING MACHINE. The lap for grinding is of cast iron in a concentric series of rings, through which sand and water is fed. The rod connecting the lap with the driving shaft has a universal joint at each end and a swivel handle for guiding the lap. The upper shaft is balanced, feathered, and moves freely through the gear hub.

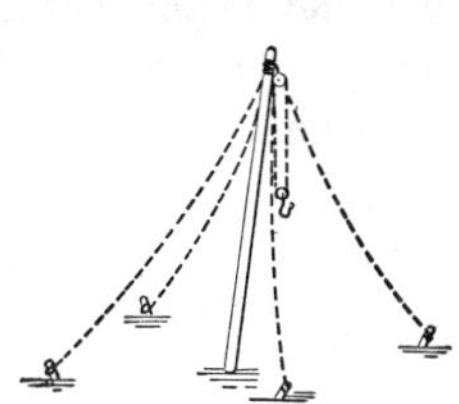

1466. FOUR-GUY MAST DERRICK pole or gin.

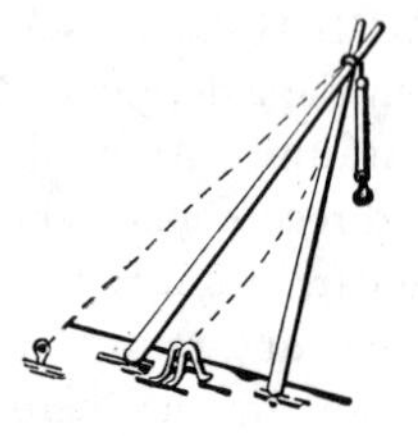

1467. SHEARS WITH WINCH or tackle blocks.

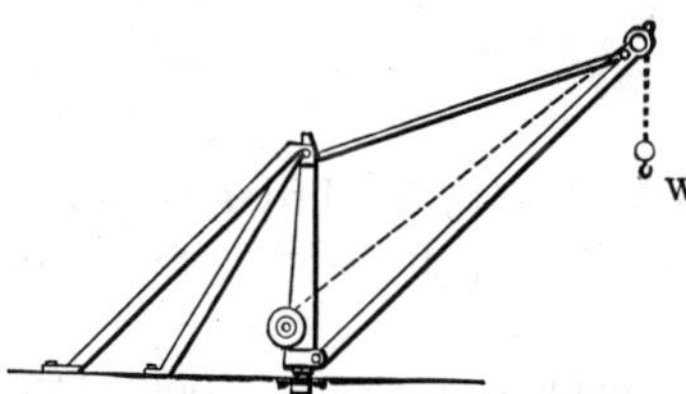

1468. SWING-DERRICK CRANE, with fixed guys and hand gear.

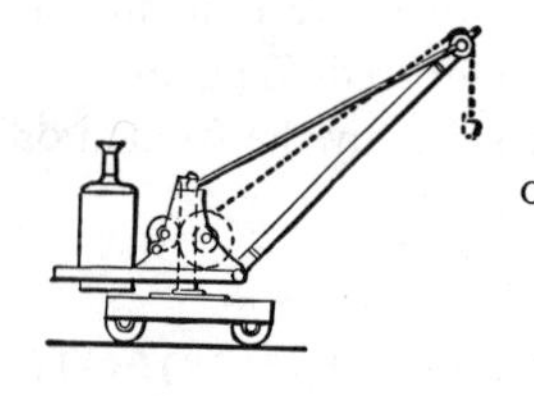

1469. PORTABLE STEAM DERRICK, on swivelled platform, balanced by boiler.

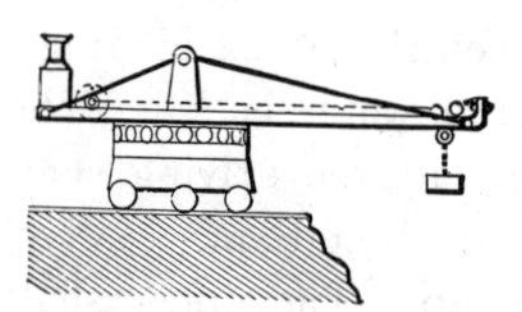

1470. SWING-BOOM CRANE, with a travelling truck and trolley lift. Boom revolving on radial rollers.

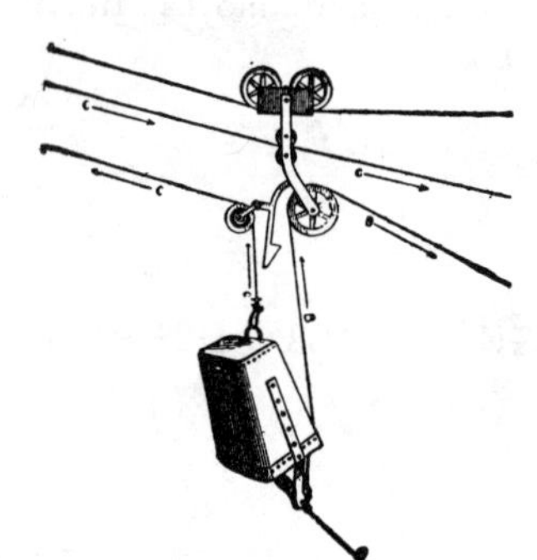

1471. CABLE HOIST AND CONVEYER, for excavating canals and trenches. The upper line is the cable, middle line the traveller, and lower lines operate the dumping device.

1472. CANTILEVER HOISTING AND CONVEYING MACHINE, "Lancaster" system. The trussed booms and standing frame revolve on rollers on the truck. The truck moves on rails. The buckets swing with the truss booms for loading and discharging.

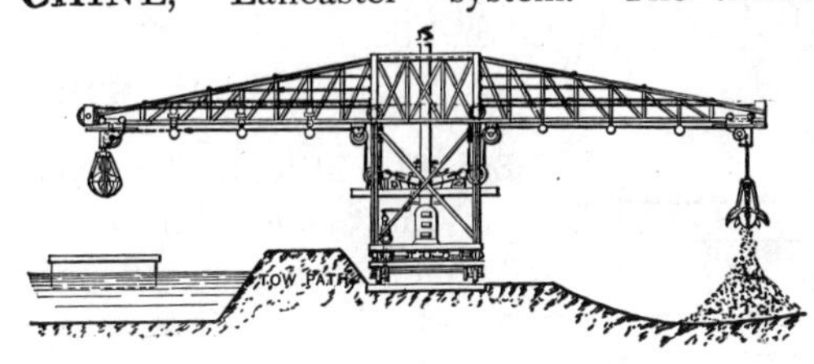

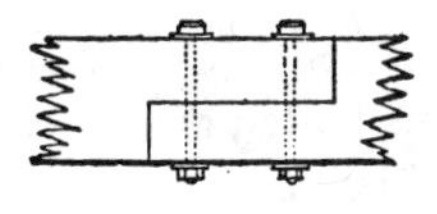

1473. TIMBER SPLICING.—The straight splice bolted.

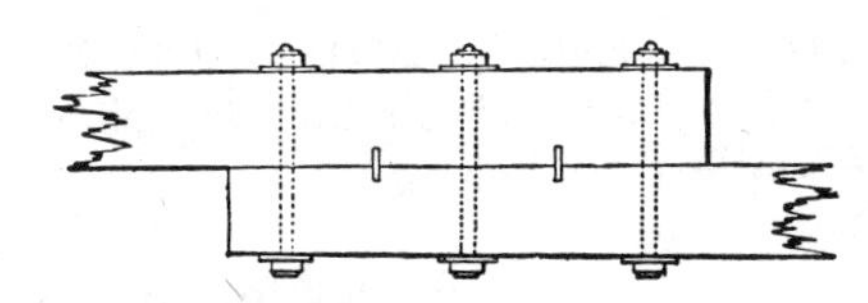

1474. TIMBER SPLICING.—The lap splice with iron keys and bolts.

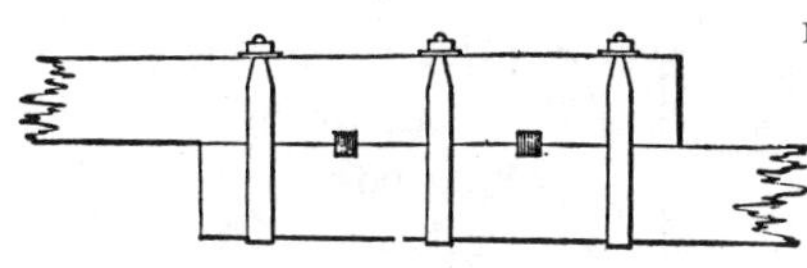

1475. TIMBER SPLICING.—The lap splice with oak keys and yoke straps.

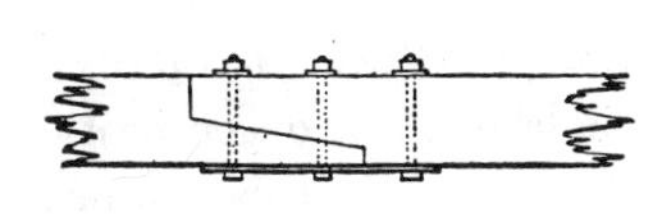

1476. TIMBER SPLICING.—A scarf and butt joint with one fish plate, bolted.

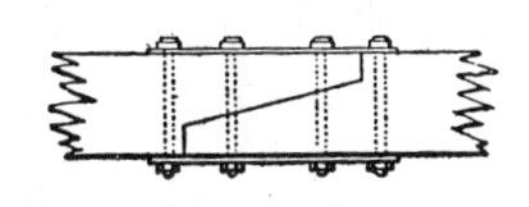

1477. TIMBER SPLICING.—The scarf and butt splice with iron fish plates, bolted.

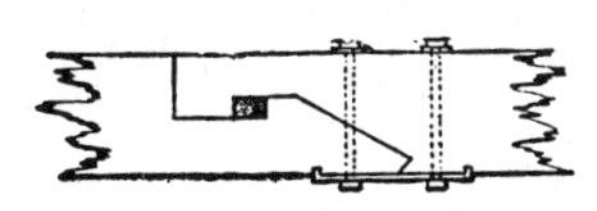

1478. TIMBER SPLICING.—A lap and scarf butt joint, keyed with oak and locked with anchor fish plate and bolts.

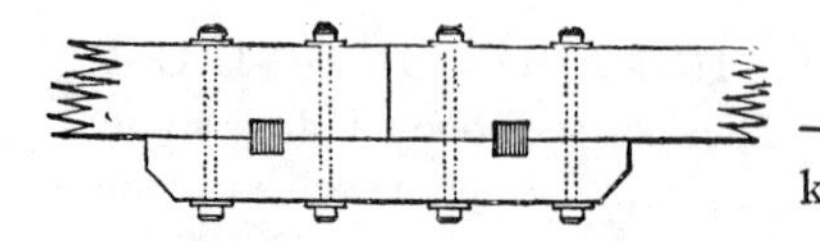

1479. TIMBER SPLICING.—Butt joint with timber fish plate, keyed and bolted.

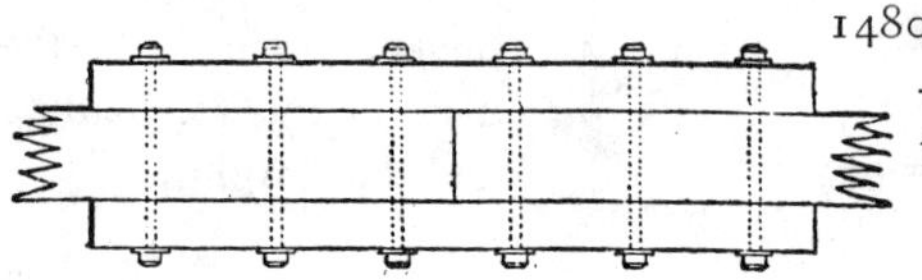

1480. TIMBER SPLICING.—Butt joint with double timber fish plates, bolted.

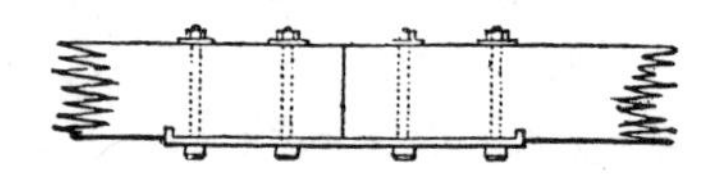

1481. TIMBER SPLICING.—Compression beams butted and held by a fish plate and bolts.

1482. TIMBER CHORDS AND ARCHES.—Splicing by breaking joints and bolting.

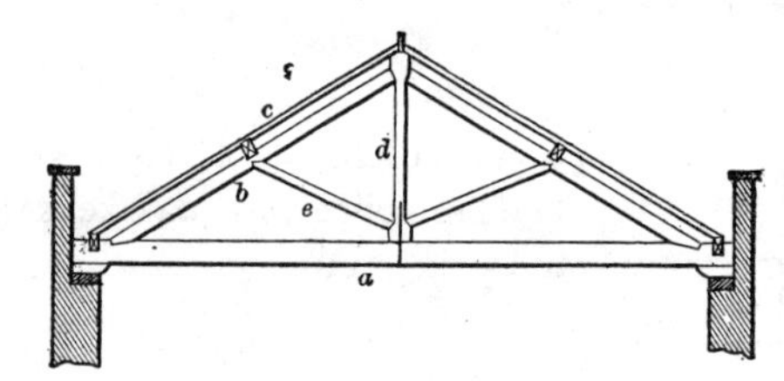

1483. TRUSS ROOF.

a, tie beam.
b, principal rafter.
c, common rafter.
d, king post.
e, strut.

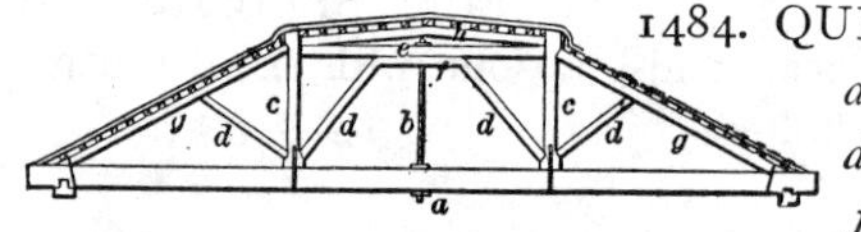

1484. QUEEN POST ROOF TRUSS. *a*, tie beam; *c*, *c*, queen posts; *d*, *d*, braces; *e*, truss beam; *f*, straining piece; *g*, *g*, principal rafters; *h*, cambered beam; *b*, iron string bolt to support tie beam.

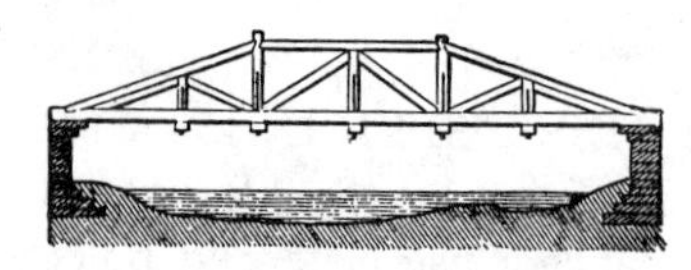

1485. WOODEN ROAD BRIDGE TRUSS.

DECK BRIDGE TRUSSES.

1486. Single strut deck truss for short spans, 30 to 40 feet.

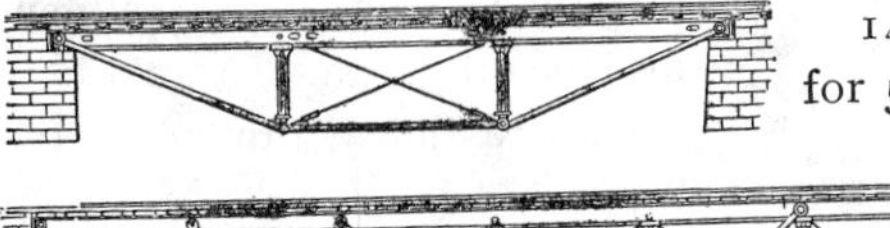

1487. Double strut deck truss for 50 to 70 feet span.

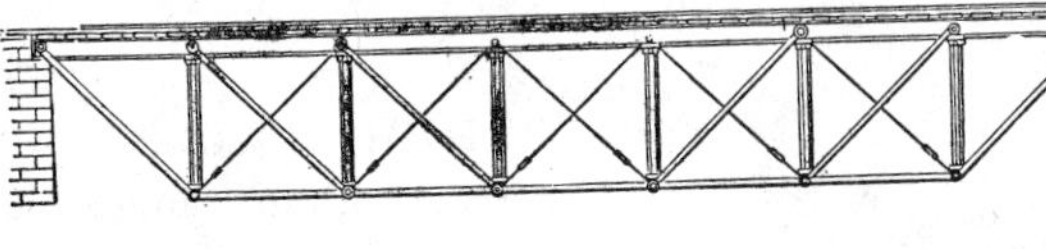

1488. Multiple strut deck truss for 100 feet span.

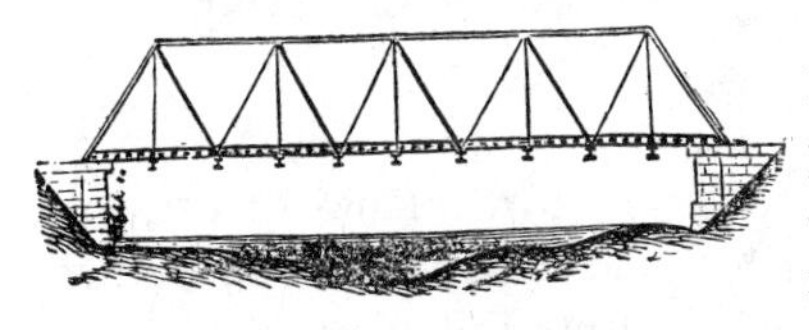

1489. BRIDGE TRUSS.—Inclined strut and tie rod for each panel, with stiff compression upper chord. Vertical members are tie rods.

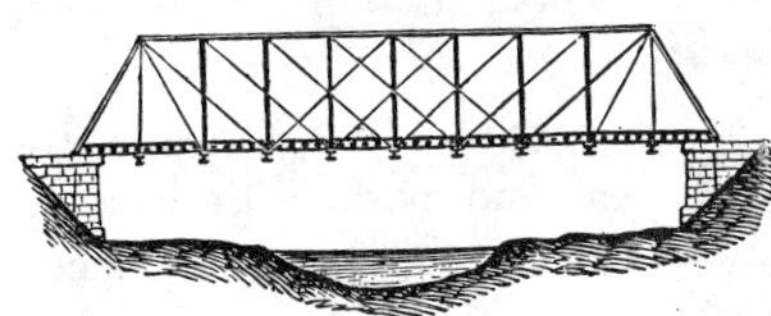

1490. BRIDGE TRUSS.—Vertical struts except in end panels, which have vertical tie rods. Inclined end struts and diagonal tie rods.

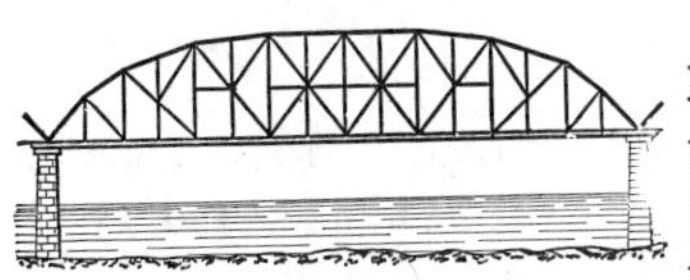

1491. ARCHED DECK TRUSS BRIDGE.—The arch takes the pressure and gives tension to the chord. Struts and tie rods give stability to the structure.

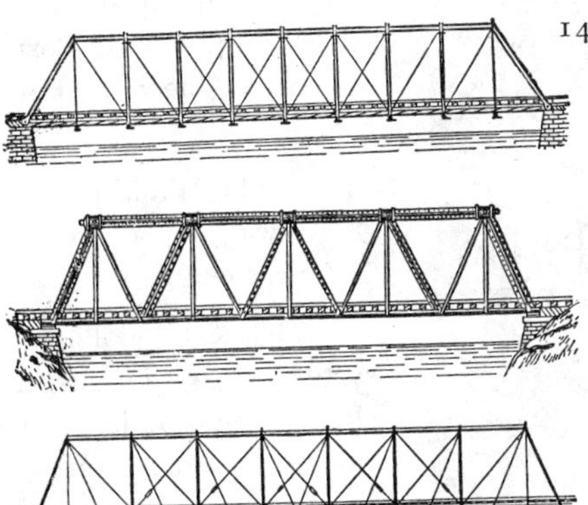

1492. BRIDGE TRUSSES.—The "Whipple" truss. Vertical and end posts are struts; vertical tie rods from end posts; diagonal tie rods in panels.

1493. Inclined posts and vertical tie rods. Baltimore model.

1494. "Whipple" truss, with interpanel tie rods.

1495. ARCH TRUSS BRIDGE.—The entire load is not supported by the wood or iron arch alone. The truss bracing is made to equalize the load by stiffening the arch and so to throw a compression strain upon the chord, which is thickened in the middle.

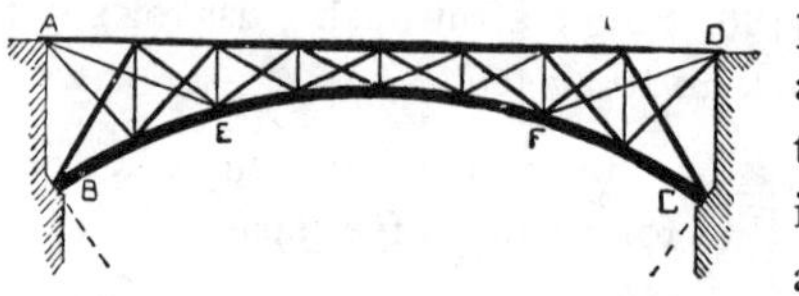

1496. BRIDGE TRUSSES.—The "McCallum" inflexible arched truss. A wooden bridge.

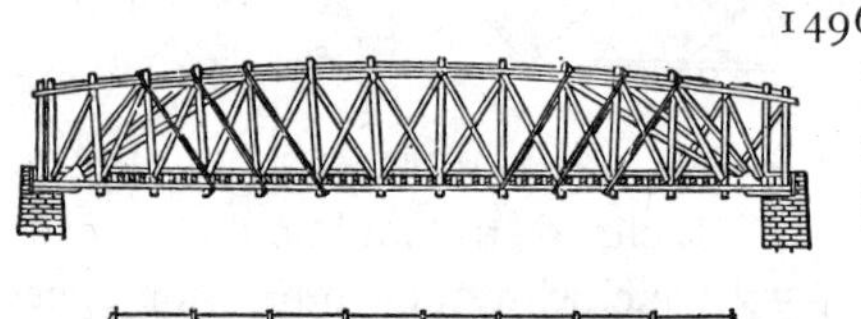

1497. "Howe" truss, with inclined end posts, vertical struts and bi-panel tie rods.

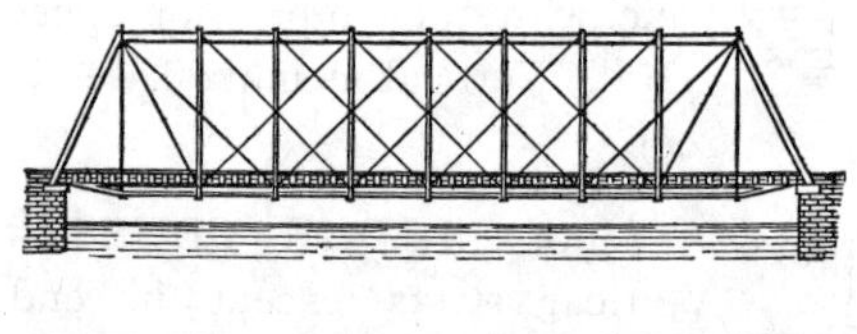

1498. "Post" truss, vertical end posts with inclined struts from each end meeting at the centre.

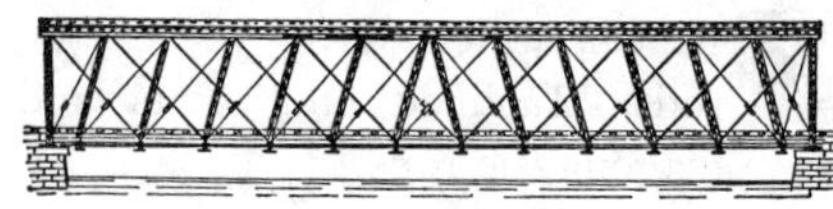

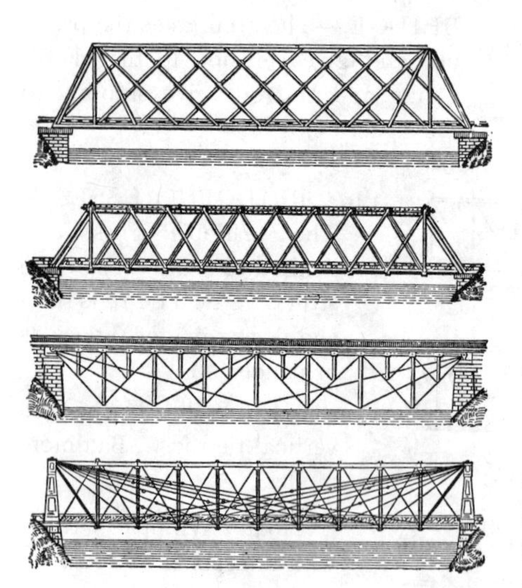

1499. Modification of the "Whipple" and "Warren" systems.

1500. Modification of the "Whipple" and "Post" systems. The "Warren" bridge.

1501. The "Fink" system. A railway deck bridge. No lower chord.

1502. The "Bollman" system. A girder suspension. The top girder carries the compression load due to suspension.

1503. SWING BRIDGE, "Whipple" system.

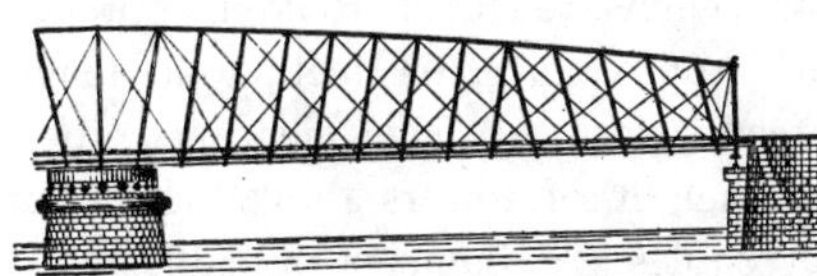

1504. SWING BRIDGE. "Post" system.

1505. CANTILEVER BRIDGE.—The ends, being anchored, balance all other parts on the piers. This cut shows the principle of Cantilever construction.

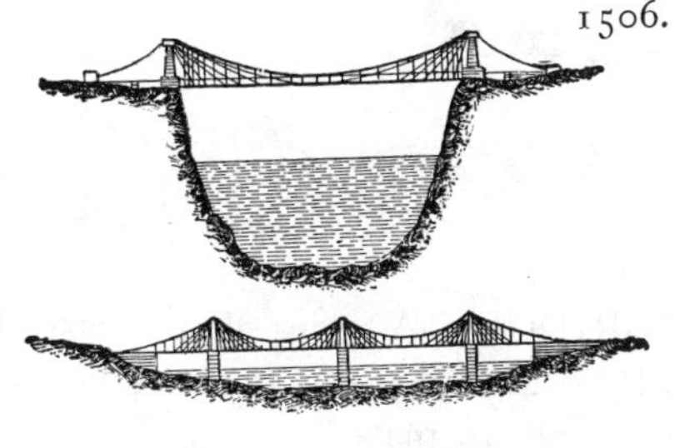

1506. SUSPENSION BRIDGES.—The old railway bridge at. Niagara. Eight hundred and twenty-one feet span.

1507. A four-span suspension bridge. Allegheny River, at Pittsburgh.

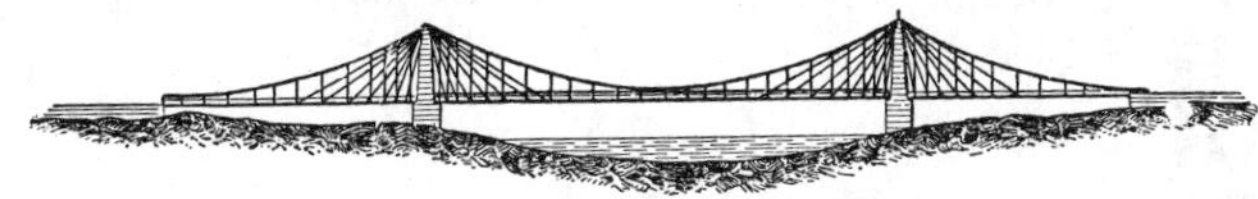

1508. SUSPENSION BRIDGE.—The Cincinnati bridge, "Roebling" system. Ten hundred and fifty-seven feet between piers.

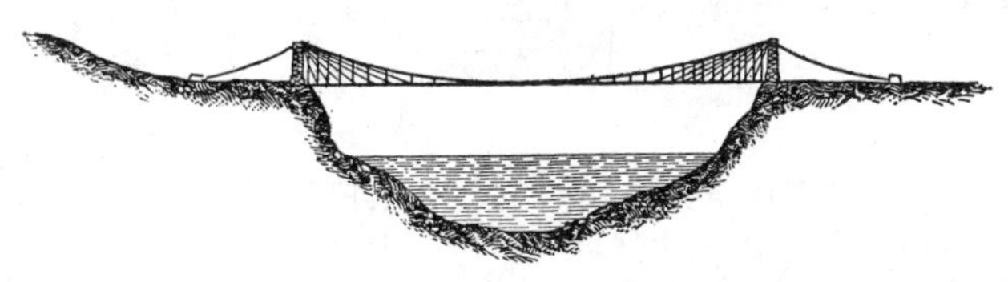

1509. SUSPENSION BRIDGE.—Niagara upper bridge, "Roebling" system. Twelve hundred and fifty feet between piers.

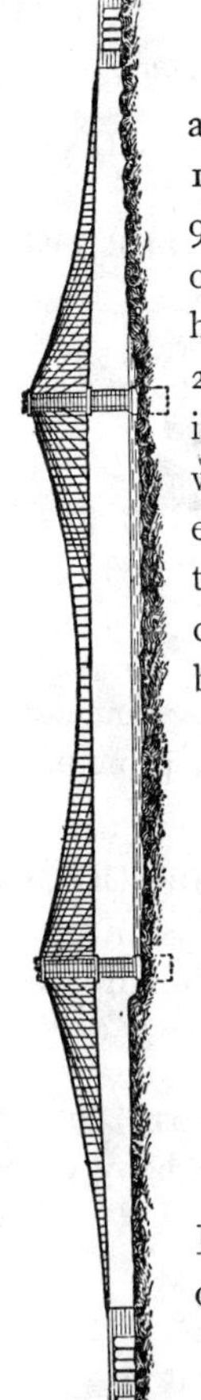

1510. SUSPENSION BRIDGE.—The New York and Brooklyn bridge, "Roebling" system. Centre span, 1,600 feet; land spans, each 920 feet; Brooklyn approach, 998 feet; New York approach, 1,562 feet; total length curb to curb, 6,016 feet; width, 85 feet; clearance above high water, 135 feet; height of towers above high water, 272 feet; number of cables, 4; diameter of cables, 15¾ inches; length of single wires, 3,579 feet; total length of wires in four cables, 14,361 miles; number of wires in each cable, 5,296; strength of each cable, 12,200 net tons; cost of bridge, exclusive of land, $9,000,000; total cost, $15,552,878. Commenced 1870; thirteen years in building.

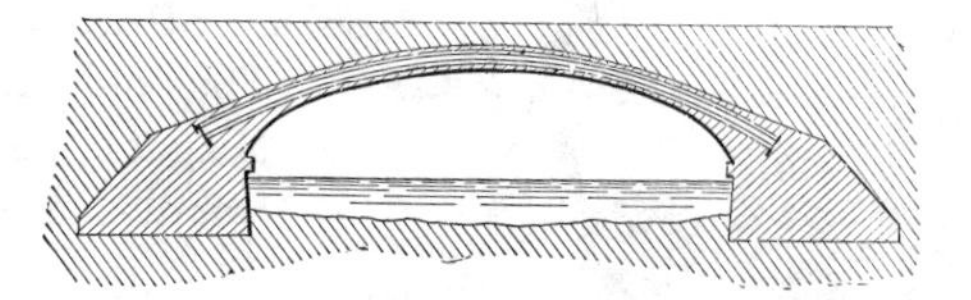

1510*e*. MELAN BRIDGE.—A series of arched steel I beams, filled in with Portland cement concrete. Ends of beams resting against thrust plates.

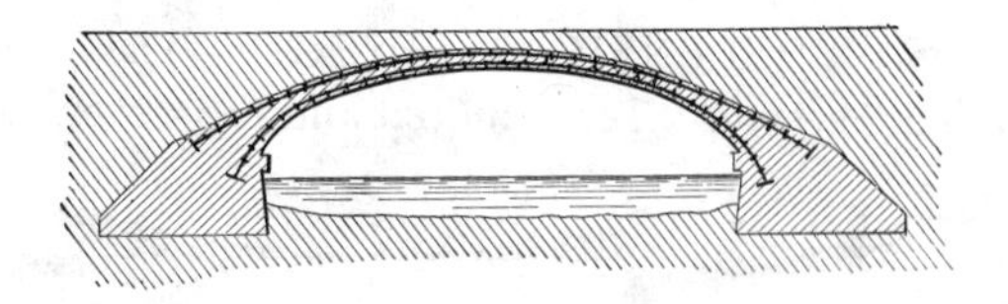

1510*f*. STEEL ARCHED CONCRETE BRIDGE, Thatcher type.—The concrete rib of the arch is reinforced by steel bars on the inner and outer edge of the concrete rib.

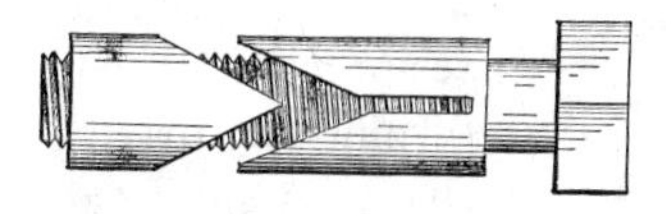

1510*a*. EXPANSION OR ANCHOR BOLT.—A wedge nut and split sleeve pattern.

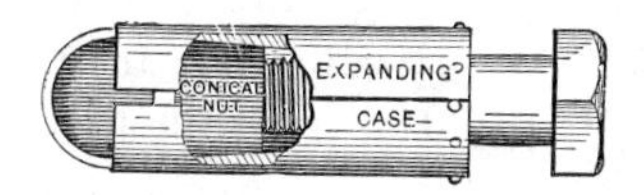

1510*b*. EXPANSION OR ANCHOR BOLT.—A conical nut and split sleeve.

1510*c*. ROLLING LIFT BRIDGE.—Are nearly balanced in operation with many advantages in utilizing the entire width of channel and dock space.

Chicago Type.

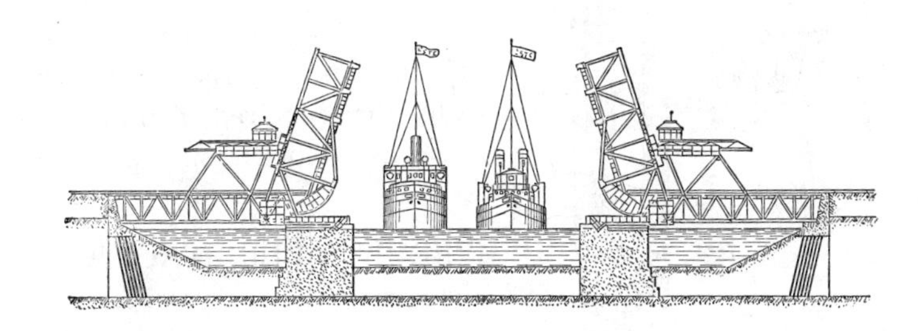

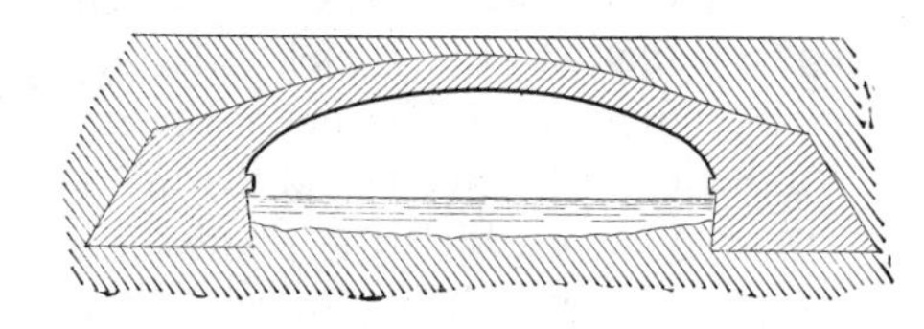

1510*d*. CONCRETE BRIDGE. — For small bridges, a concrete arch of Portland cement, sand and broken stone are reliable and lasting structures.

1510*g*. LOG SAWING MACHINE.—The lever being properly adjusted, and the screw and dogs being placed in position, the saw is rapidly reciprocated by turning the driving shaft. A spring attached to the butt of the saw and center of the pitman presses the teeth down upon the bottom of the kerf; this process can be easily adjusted. The blade is raised after having cut through the log, and is held in the guiding slot by a screw, so as not to interfere with shifting the machine.

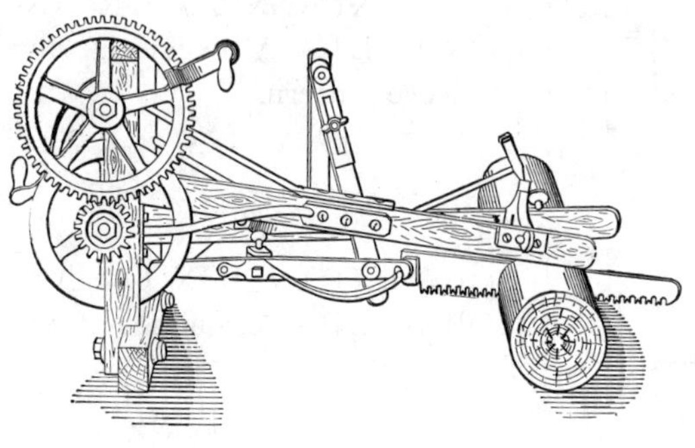

1510*h*. HYDRAULIC PILE DRIVER.—A pile with a groove on its side in which a pipe is laid to the bottom of the pile, loosely clipped in place to enable its withdrawal after the pile is set. A strong stream of water from a pump excavates a passage for the pile to the required depth. No hammer is needed; only a steady pressure.

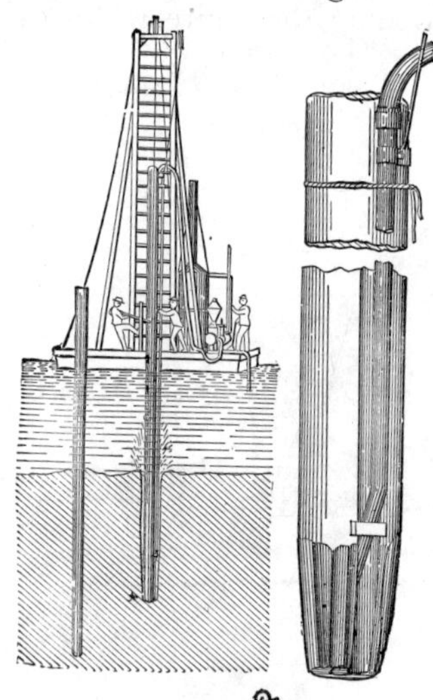

1510*i*. LUMBER STACKER.—A machine has been invented consisting of a conveyor belt which lifts the planks one by one. The machine comprises a skid formed of a pair of arms, one of which bears against the stack already piled, while the other lies adjacent to the conveyor belt and forms a guide against which the planks are supported while being lifted. The stacker is adjustable to any desired angle.

Section XVII.

DRAUGHTING DEVICES.

Parallel Rules, Curve Delineators, Trammels, Ellipsographs, Pantographs, etc.

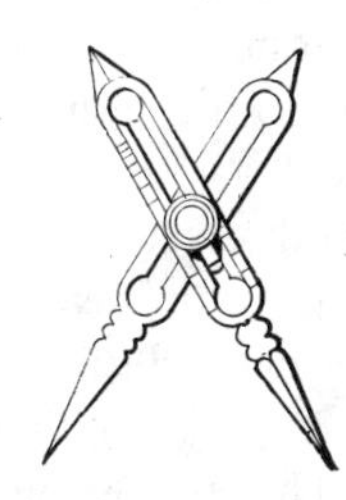

1511. PROPORTIONAL COMPASSES for reducing the scale of drawings.

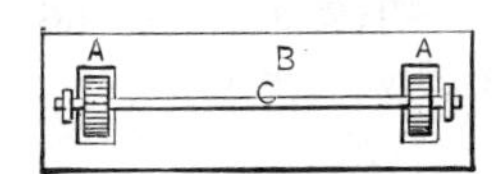

1512. ROLLER PARALLEL RULER. —The two fluted rollers of exactly equal size, on an arbor, project slightly below the under surface of the ruler.

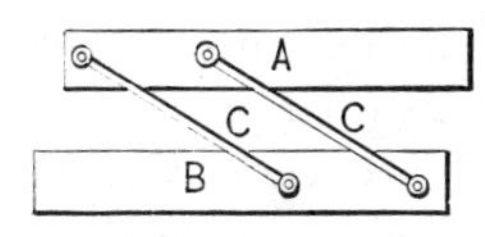

1513. PARALLEL RULER, formed of two bars pivoted to two pieces of metal of exactly equal lengths between pivot centres and at equal distances on the bars.

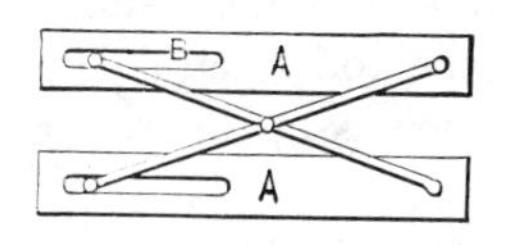

1514. SLOTTED PARALLEL RULER that traverses in line. A, cross bars movable on a central pivot; each bar being pivoted at one end to the ruler bars, the other ends sliding in slots in the bars.

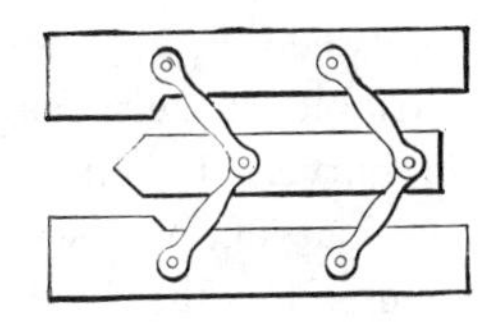

1515. THREE PART PARALLEL RULER.—All connecting arms of equal length. Pivots are at equal distances on each of the blades.

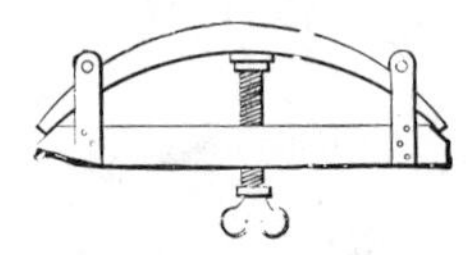

1516. SPRING CYCLOGRAPH. — A spring of elastic material is made thicker in its central part so that in bending its outer edge will take the form of a circular arc. By clamping the ends of the spring to the bar, the screw will bend the spring to the desired curve.

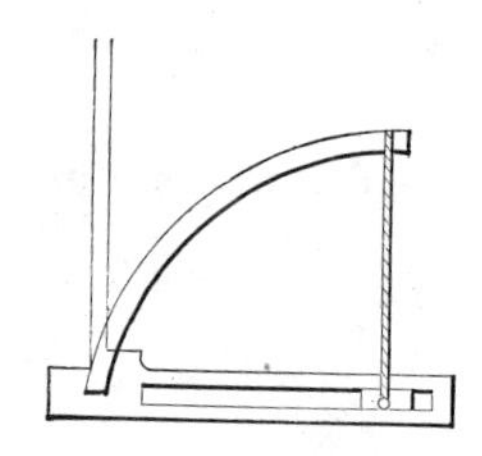

1517. FLEXIBLE CURVE SCRIBER.— A spring of any suitable material may be fixed in a ruler and drawn by a string to the desired curve. There are many forms of this device, such as the string fastened to both ends of the spring and flexible rubber strips with heavy weights to hold it to any form of curve desired.

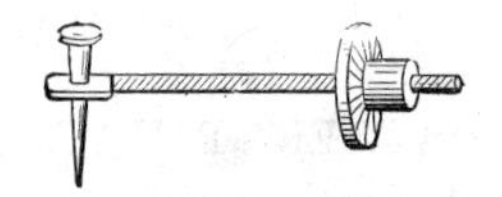

1518. HELICOGRAPH.—The traversing of the disc by moving the screw arm around a fixed centre describes a helical curve.

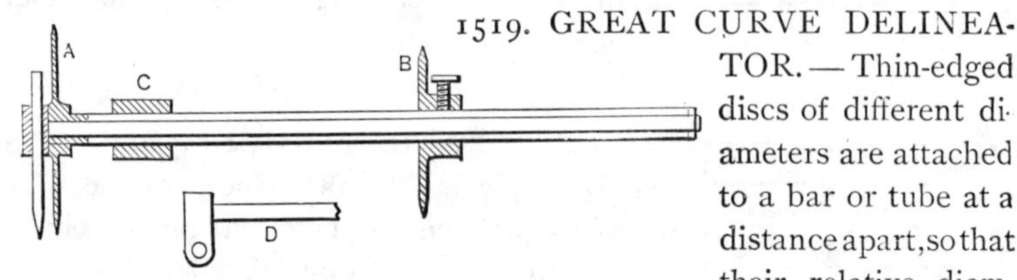

1519. GREAT CURVE DELINEATOR. — Thin-edged discs of different diameters are attached to a bar or tube at a distance apart, so that their relative diameters and distance will correspond to the required radius, which may be computed by the difference in diameter multiplied by the distance of the wheels apart. C is a loose sleeve to roll the rod freely; D is an offset from the inside rod to allow the pencil to press on the paper.

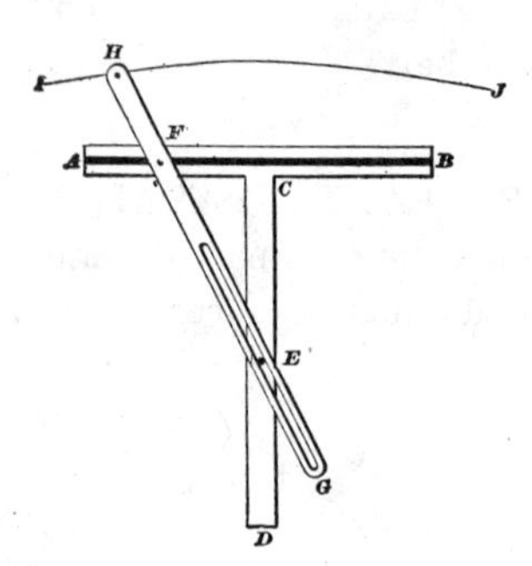

1520. CONCHOID DELINEATOR, of Nicomedes. A slotted head T-piece, A, B; a slotted arm, G H, with traverse pin at F. Distance between F, H, and pin at E may be variable to suit the required condition of curve. Pencil at H delineates a conchoidal curve, used in architectural drawings for the lines of columns.

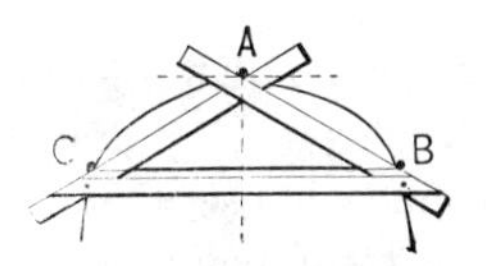

1521. CYCLOGRAPH, for drawing circular arcs with an inaccessible centre. Three straight rules clamped together so that when the outer edge of the rules are against the pins B, C, representing the chord of the arc, the pencil at A will be at the vertex of the versed sine of the arc, when by moving the rules against the pins the pencil will describe a circular arc.

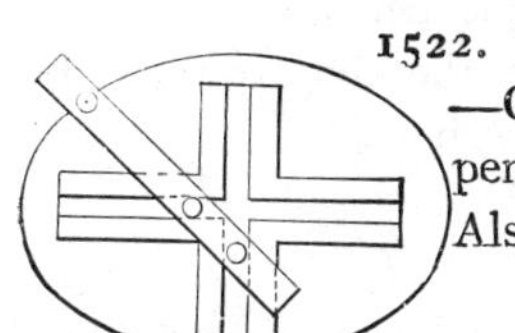

1522. TRAMMEL FOR DRAWING ELLIPSES. —Grooves at right angles direct two studs on a pencil bar for the elliptical motion of the pencil. Also called an ellipsograph.

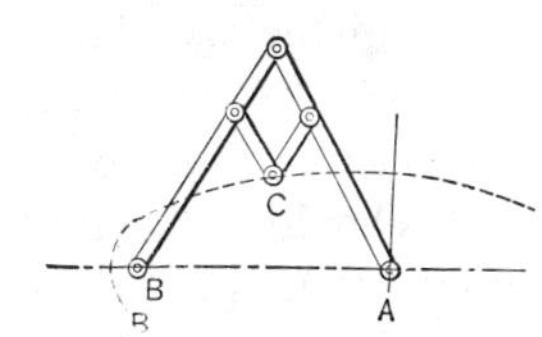

1523. ELLIPSOGRAPH. — A is a fixed centre; B, traversed in a straight line, will make the pencil at C trace an elliptical curve.

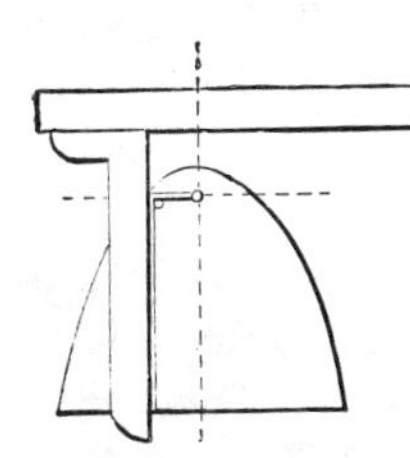

1524. PARABOLA SCRIBER.—The longitudinal focal distance from the apex being fixed with a pin. A straight-edge may be fixed just beyond the apex and traversed by a square. A looped string on the pin with the other end fastened to the longer leg of the square with sufficient sag to allow a pencil point to rest in the bight of the string at the apex of the parabola, when the square is on the axial line, will describe an arc of a parabola by moving the pencil against the square.

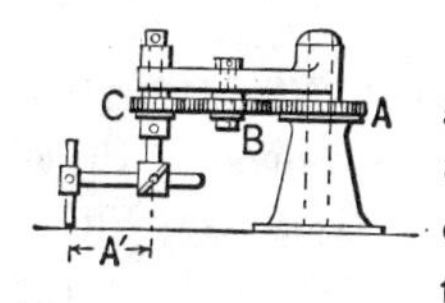

1525. GEARED ELLIPSOGRAPH.—A spur gear A is fixed to the pedestal. An arm carrying an idle gear, B and a gear C, one-half the diameter of the fixed gear. The pencil arm makes two revolutions to one revolution of the arm. The distance A′ equals the difference between the major and minor axes of the ellipse.

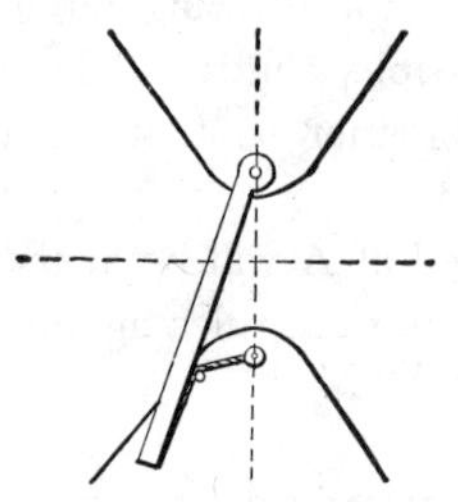

1526. HYPERBOLA SCRIBER.—The foci of the opposite hyperbolas may be drawn on their longitudinal axis and pins set therein. A straight edge moving on focal point of the opposite hyperbola, and a looped string on the pin of the required arc, with the other end attached at the end of the straight edge, with enough sag to allow the pencil to touch the apex of the curve, will, on moving the pencil in the bight of the string and close to the rule, describe an arc of a hyperbola.

1527. GEARED ELLIPSOGRAPH.—The arm and horizontal shaft slide through the frame and second bevel gear. The bevel gear A is fixed to the standard. The proportion of the gears should be such that the pencil spindle should make two revolutions to one revolution of the arm. Then the distance A^1 equals the difference between the major and minor axes of the ellipse.

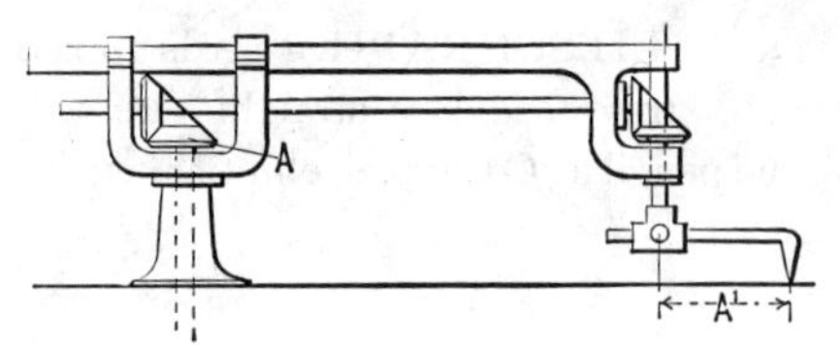

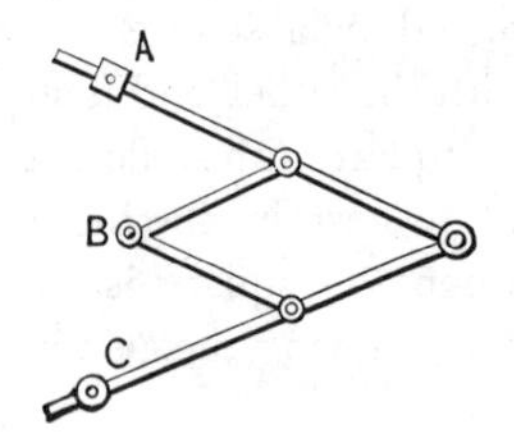

1528. PANTOGRAPH.—For perfect proportions the points A, B, C must always be in line. With the point B fixed, the pencil at A will produce an exact copy of tracing from point C. By changing places for the fixed point a double or half-size tracing may be made.

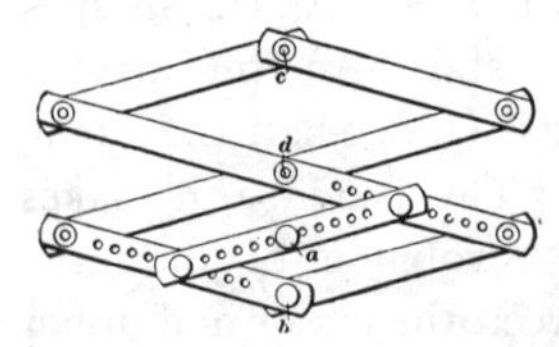

1529. LAZY-TONGS PANTOGRAPH, for reducing or enlarging copies of drawings. *c* or *d* may be the fixed points. Either one being fixed, the other should be the tracer. The pencil at *a* should be exactly in line with *c*, *d*, for accurate delineation.

1530. PERSPECTIVE CENTROLINEAD.—The edges of legs on the sides *a*, *b*, and *c* must be in line with their common axis, with clamp screws to hold the movable legs in their set position. The directing pins *b* and *c* should be set on the radial lines of the back point of the perspective, when the long leg will be radial from that point in all directions.

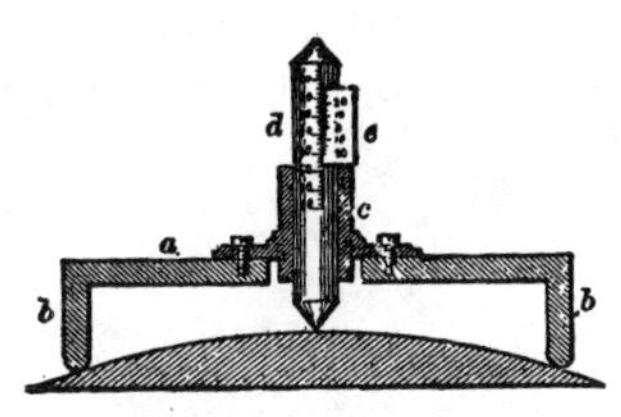

1531. SPHEROMETER.—For measuring the curves of spherical surfaces or of templates of lenses by means of a graduated follower at the centre between two bearings. The scale and nonius are computed for the versed sine of a fixed chord length.

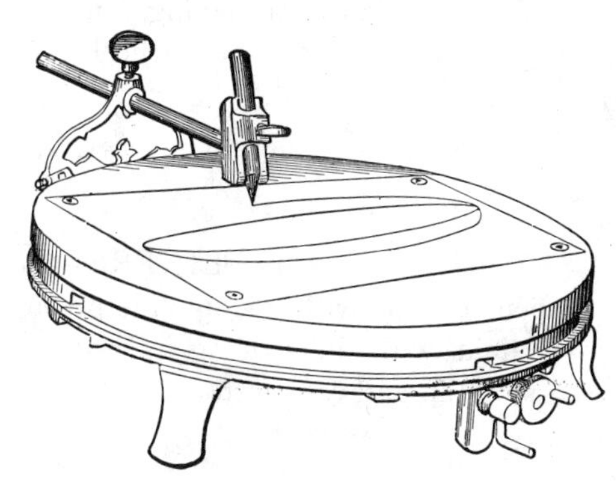

1531*a*. ELLIPSOGRAPHIC TURN-TABLE. — The table sits upon a trammel frame, which moves a bar with gimbal yoke and sliding rod to carry the pen or pencil.

Makes an ellipse from a circle to a straight line.

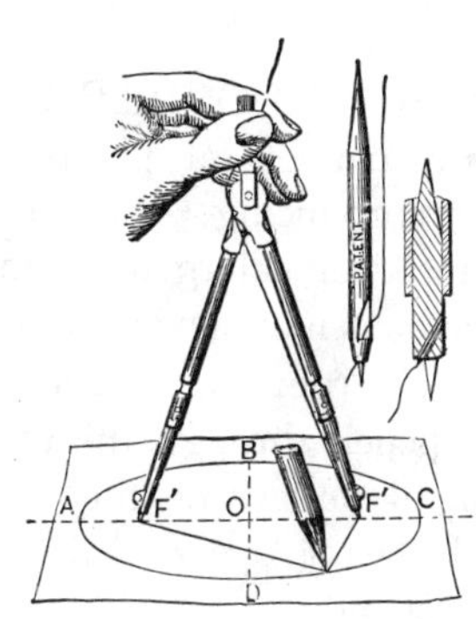

1531*b*. ELLIPSOGRAPH.—A pair of dividers with points perforated like the eye of a needle allows of adjusting the length of the thread to any size curve and holding the end under the thumb on the head of the dividers. The points with their eyes are shown at the right-hand side.

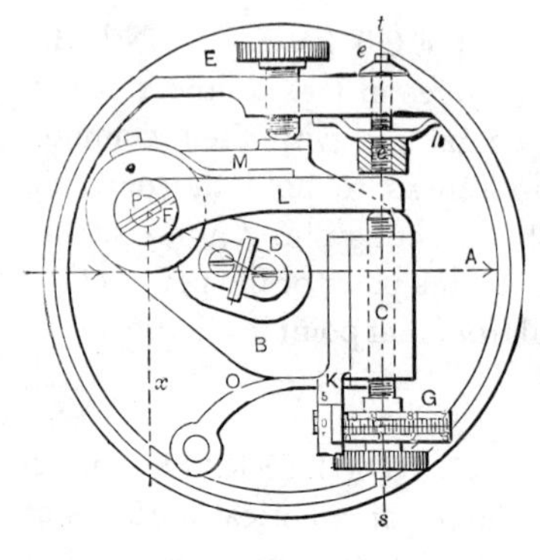

1531*a*. OMNI-TELEMETER.—For obtaining the distance of an object by two observations on a line at right angles. A, direct line of sight through the half silvered mirror D; X, reflected line of sight at 90° or variable by the mirrors D, F. The lever L varies the angle of the mirror, F, by means of the micrometer screw C, and graduated disk G. E, adjusting screw for the mirror D.

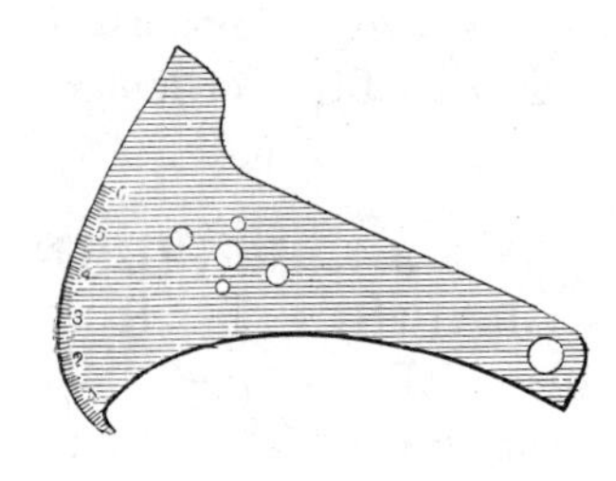

1531*b*. ODONTOGRAPH. — A scribe template for laying out the curves of the teeth of gear wheels.

1531*c*. SECTION LINERS.—An open triangle with a stop slide pivoted to links with the spacing adjusted by a stop pin and cam sector.

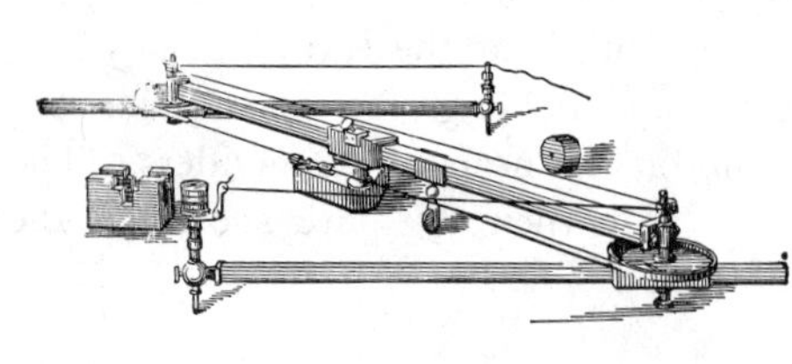

1531*d*. EIDOGRAPH.—An instrument of precision in reproducing drawings, reduced or enlarged. Simultaneous motions are transmitted from the tracer to the pencil through the motion of pulleys on the central beam operated by a steel band.

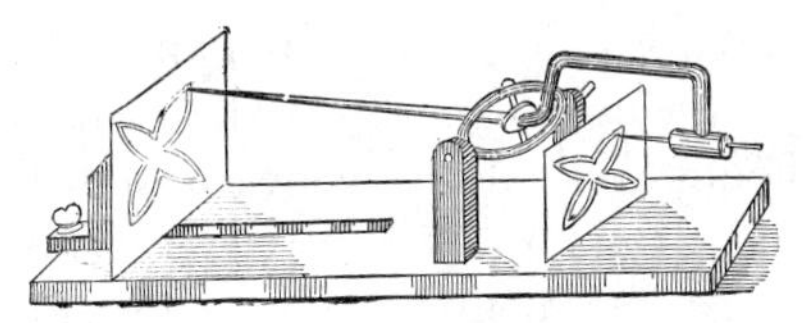

1531*e*. TRACING BAR.—The yoke-shaped needle bar has tracing points in line with the gimbal ring bearing. The points may be extended for any proportion.

1531*f*. REFLECTING DRAWING BOARD.—A vertical plate of glass in a frame on a drawing board. The picture on one side and a plain sheet of paper on the other side of the glass. On looking into the glass on one side the picture is plainly projected on the other, and can be readily traced with a pencil.

1531*g*. SELF REGISTERING BAROMETER.—Four aneroid disks are connected in series and in contact with the lever R and linked to the index lever L, its end holding a pencil that marks the variation of pressure on the graduated paper on a drum driven by a clock.

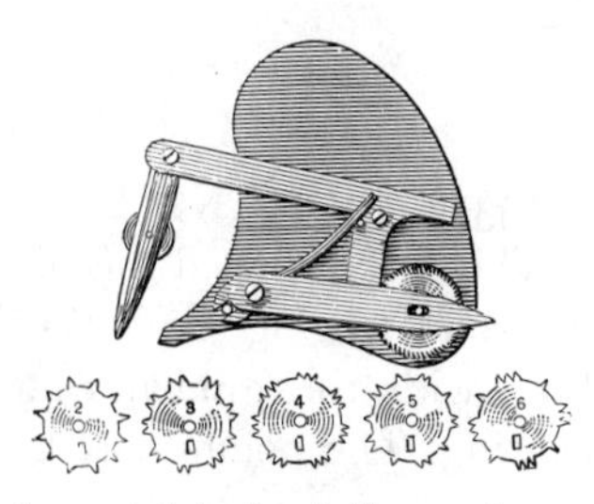

1531*h*. DOTTING INSTRUMENT.—One of the toothed wheels for the kind of dotting required is placed on the pin and held by the spring clip. The wheel should roll on the edge of the T square and the frame against its side. The motion of the bell crank lever and pin follows the spacing of the teeth on the wheels.

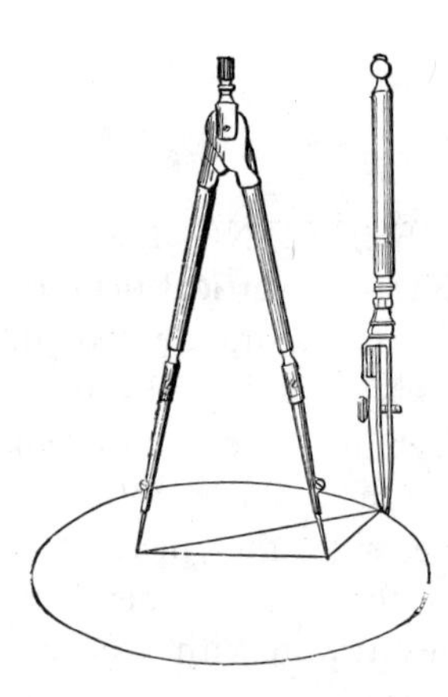

1531*i*. ELLIPSOGRAPH with a pen and dividers. A small friction drum is mounted inside of the pen blades, or may be mounted on the handle upon which a thread is wound for adjusting the size of the ellipse, and is passed through an eye or hole near the end of the pen blade and around the legs of a dividers, set on the foci of the ellipse.

1531*j*. SPIROGRAPH.—An instrument for drawing spiral curves. A pair of compasses with a lengthening leg to allow of a vertical position of the stationary leg *c*. *b* is a sleeve and knurled button loose on the vertical leg by which the pencil or pen is moved along the curve, controlled by the thread winding around the vertical leg, which may have points of varying sizes to suit the spaces in the curves. The vertical leg is held by the milled head *g*.

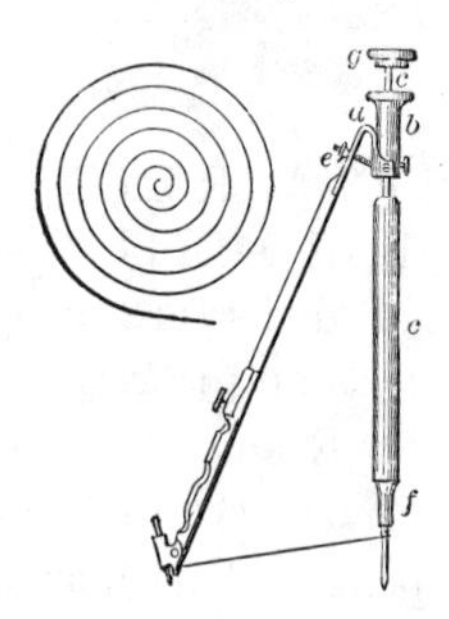

Section XVIII.

MISCELLANEOUS DEVICES.

ANIMAL POWER, SHEEP SHEARS, MOVEMENTS AND DEVICES, ELEVATORS, CRANES, SEWING, TYPE-WRITING, AND PRINTING MACHINES, RAILWAY DEVICES, TRUCKS, BRAKES, TURNTABLES, LOCOMOTIVES, GAS, GAS FURNACES, ACETYLENE GENERATORS, GASOLINE MANTLE LAMP, FIREARMS, ETC.

1532. HUMAN TREADMILL.—Still used in Eastern countries for raising water.

1533. HORSE-POWER TREAD-WHEEL.—One of the many designs for stationary animal power.

1534. HORSE-POWER MACHINE.—An endless chain and rollers, with a slatted platform, roll over a sprocket-wheel driving shaft. The walking platform is elevated to an angle of about 15°.

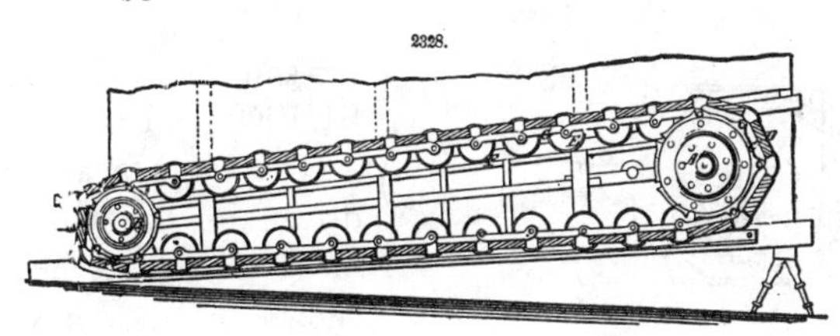

1535. DOG-POWER MACHINE. —The plane of the track wheel is set at an angle of about 20°, with its under edge bearing upon a friction pulley. Shaft and fly-wheel, with crank for operating churn.

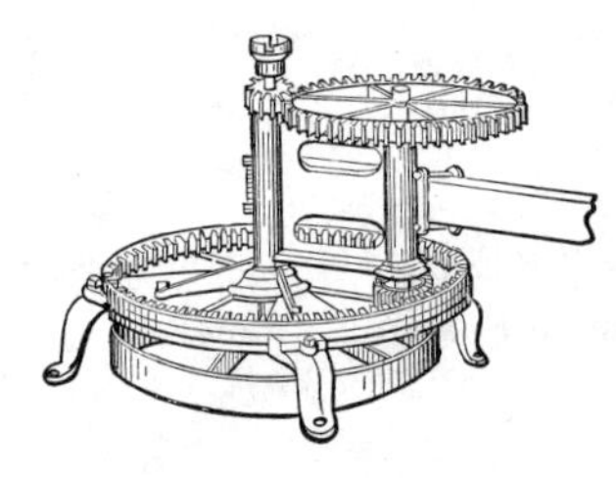

1536. GEARED HORSE-POWER. —The sweep carries the pinion and spur gear on the second shaft around the stationary spur gear, rotating the central shaft and pulley at high speed.

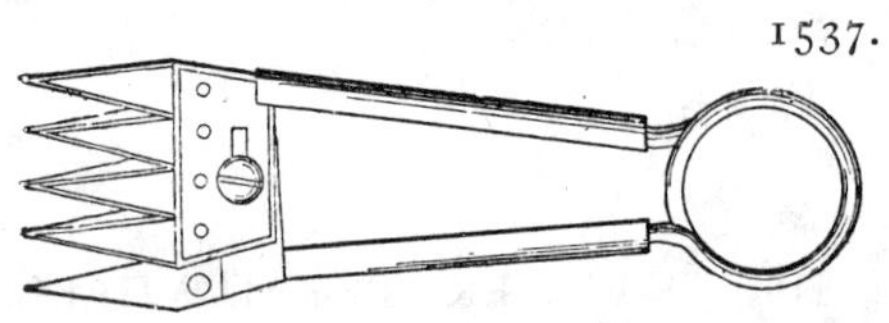

1537. MULTIPLE BLADED SHEEP SHEARS.—Opened by a spring handle. and closed by hand grip.

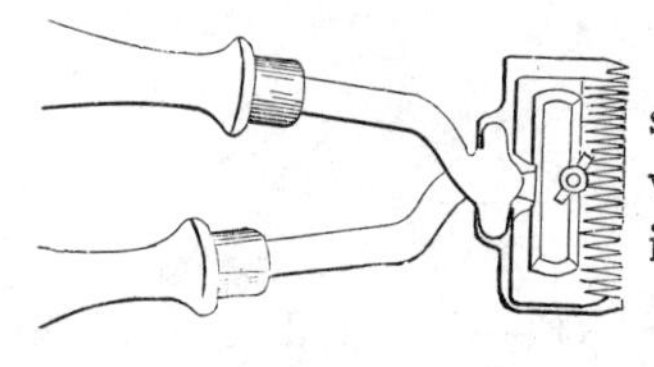

1538. HORSE CLIPPER. — A sharp comb-tooth cutter is made to vibrate across a fixed cutter by vibrating the handles.

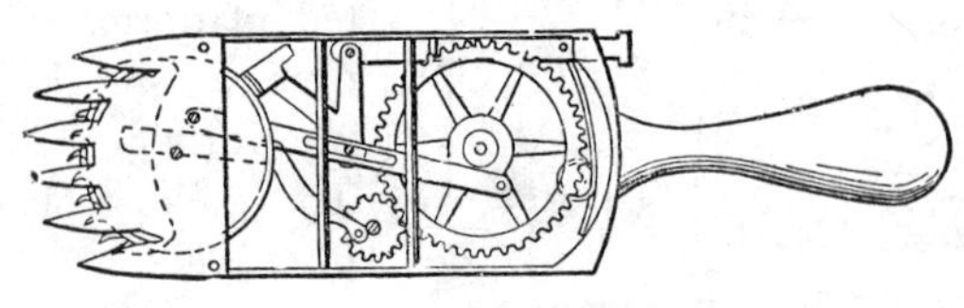

1539. MACHINE SHEEP SHEARS. —The large gear is driven by the hand on a crank, not shown, or by a flexible shaft from another source of power. A revolving serrated blade with guard finger plates.

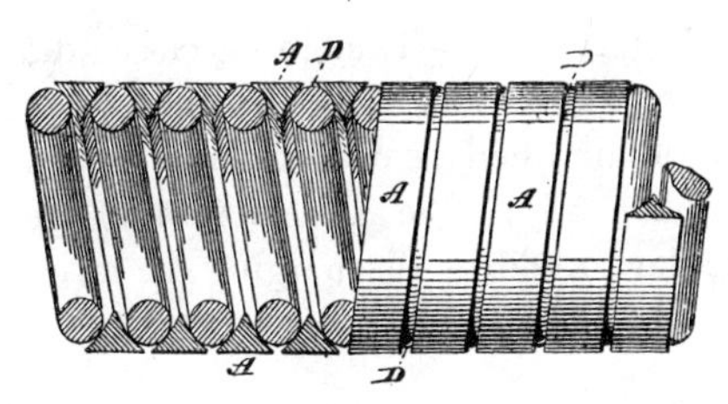

1540. "ALMOND'S" FLEXIBLE METALLIC TUBE.—A coil of round wire, open wound, with a coil of triangular wire wound tightly over it. Bending of the coil tube allows the triangular sections to draw in on the outside of the bend and to push out on the inside, keeping the points of contact tight.

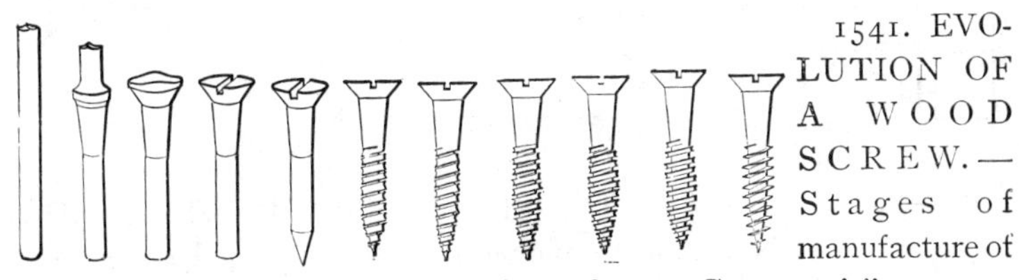

1541. EVOLUTION OF A WOOD SCREW.—Stages of manufacture of the modern wood screw. "American Screw Company's" process. The thread is made by the roller process.

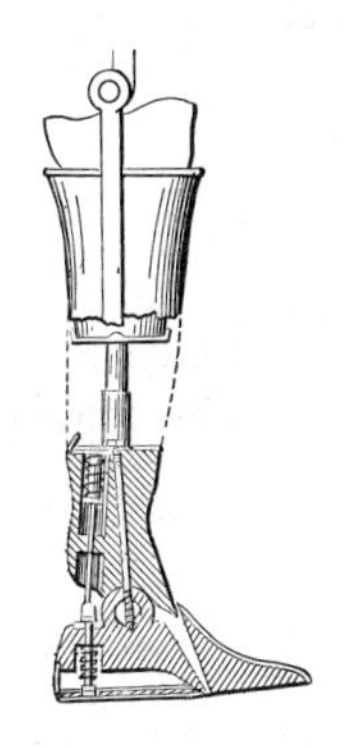

1542. ARTIFICIAL LEG AND FOOT.—Most ingenious combinations of movements are made in producing artificial limbs, not easily explained without a model.

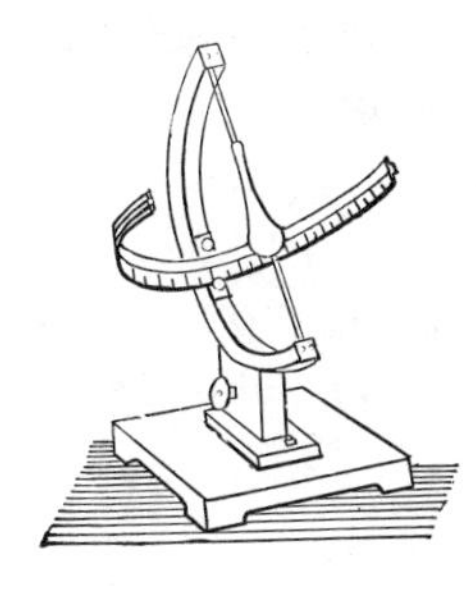

1543. MEAN TIME SUNDIAL.—The length of the stile is made to just cover the entire range of the sun's altitude at the distance of the scale on the hour circle. Its shape and size to be proportionate to the sun's equation of time as marked on the scale. When the sun is fast the reading should be on the left-hand side of the shadow, and when slow on the right side.

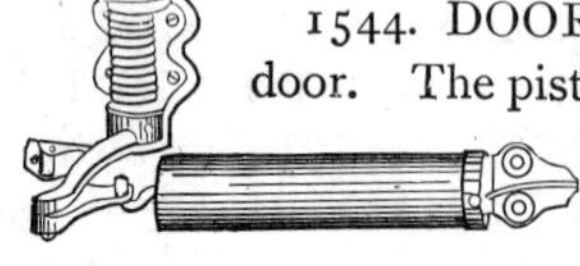

1544. DOOR PUSH CHECK.—The spring closes the door. The piston in the cylinder has a valve to allow quick inlet of air when opening a door, and a small hole adjustable at the bottom of the cylinder for slowly discharging the air.

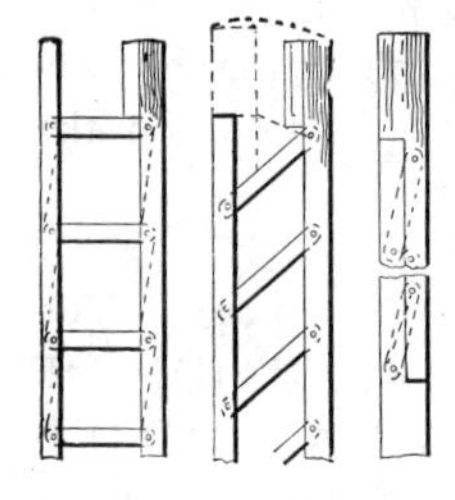

1545. FOLDING LADDER.—The rounds are pivoted to the side pieces, which are recessed to enclose the rounds when the ladder is shut.

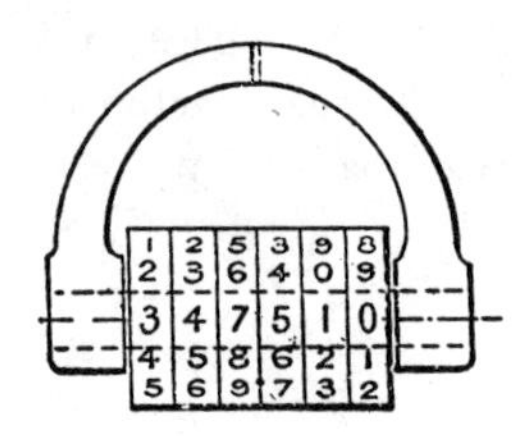

1546. SIMPLE COMBINATION LOCK.—A number of discs arranged on a spindle having a feather key. The discs are notched to match the notches in the key so that they readily turn to be set to the register number to release the spindle.

1547. TRIPOD.—The legs are pivoted on a triangular prism, which allows the legs to be folded into a cylindrical staff.

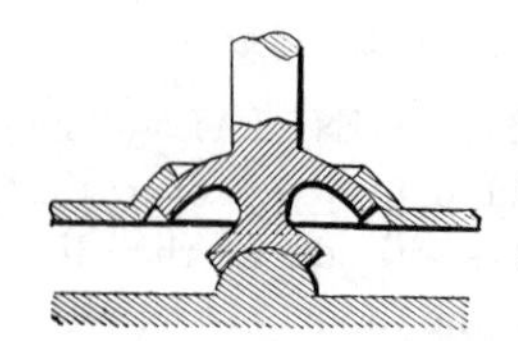

1548. DOUBLE SPHERICAL SOCKET, used mostly on surveying instruments. The socket is clamped by drawing the plates together with thumb-screw.

1549. DISC SLICER, with hopper, for cutting roots, etc. Each slot in the disc has a knife slightly projecting.

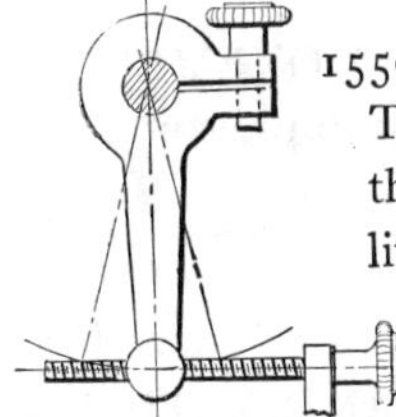

1550. MICROMETER SCREW ADJUSTMENT.—The tangent arm is made fast or loose on the shaft by the spring clip and screw. Used mostly on theodolites and transit instruments.

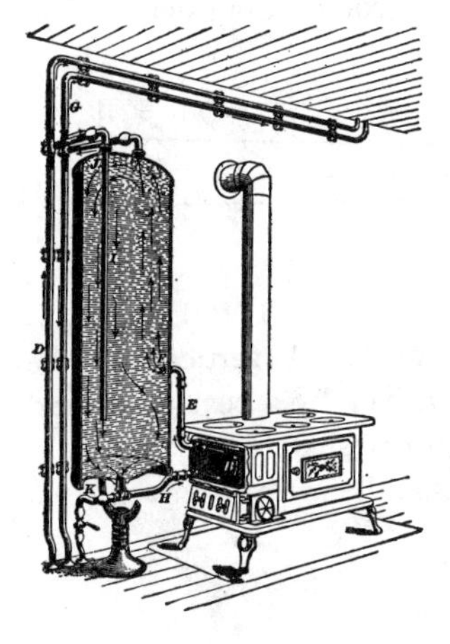

1551. CORRECT PRINCIPLE in setting a hot-water house boiler.

H, E, circulating pipes.

B, water-back or coil.

K, draw-off.

D, cold-water supply, extending down on inside of boiler.

G, hot-water supply taken from top of boiler.

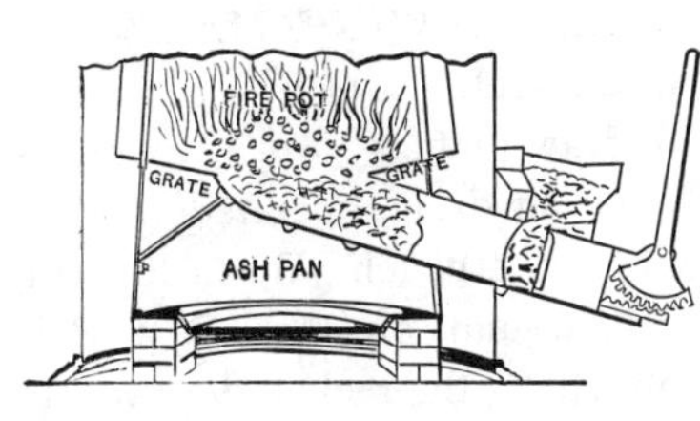

1552. UNDER-FEED HEATING FURNACE, "Colton-Smead" model. A smokeless furnace for house heating with bituminous coal. A plunger is operated by a lever sector and rack to push the coal beneath the fire.

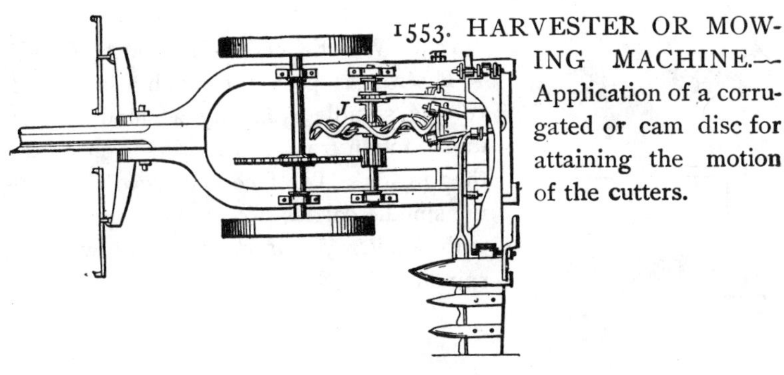

1553. HARVESTER OR MOWING MACHINE.—Application of a corrugated or cam disc for attaining the motion of the cutters.

1554. BELL CLAPPER MOVEMENT.—The outside stroke is the best to prevent cracking in large bells.

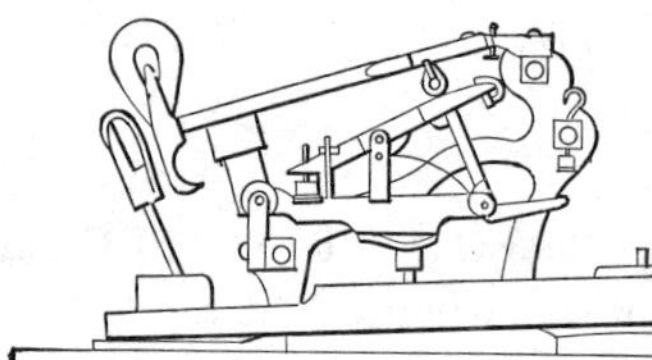

1555. PIANO KEY AND ACTION.—A study of complex movement.

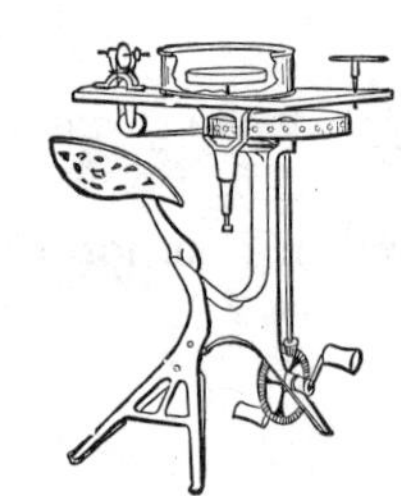

1556. LAPIDARY OR LITHOLOGICAL LATHE for amateur work. A vertical spindle with disc lap of lead, driven by a bevel gear and cranks, through a vertical shaft pulley and belt. A splitting disc and spindle are also driven from the main pulley.

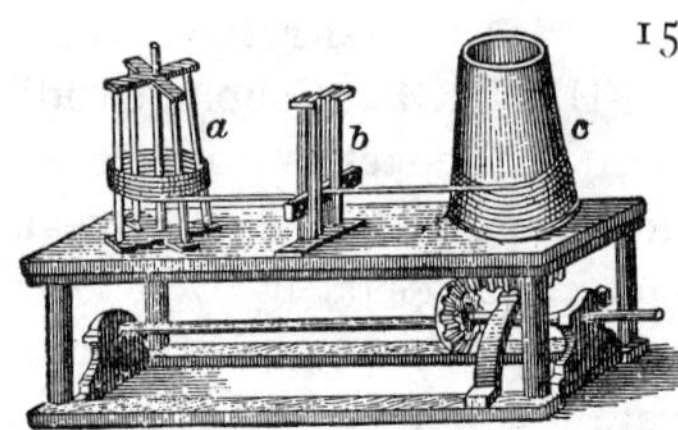

1557. WIRE-DRAWING MACHINE.

a, the reel.

b, draw plate.

c, power drum, operated by gear beneath the bench. When the wire is all wound on the drum it is changed to the reel and drawn in a reverse direction.

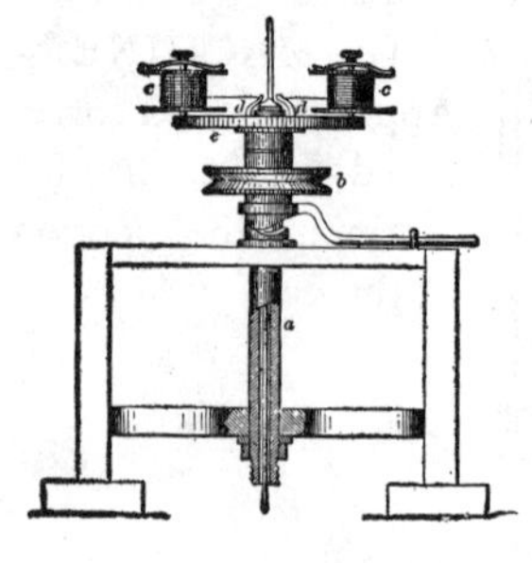

1558. WIRE-COVERING MACHINE.—The wire is passed through a hollow revolving spindle *a*, having a small longitudinal motion from a vibrating cam to lap the threads. The face plate revolving with the spindle carries two or more spools, *c*, *c*, with guide eyes, *d*, *d*, vibrating with the spindle.

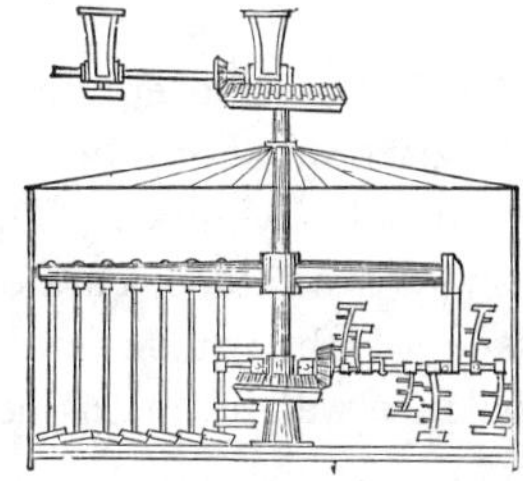

1559. STIRRING MACHINE, for grain mash or other material in water. One arm carries a vertical set of arms with bottom scrapers. The other arm, a revolving shaft and arms for vertical stirring.

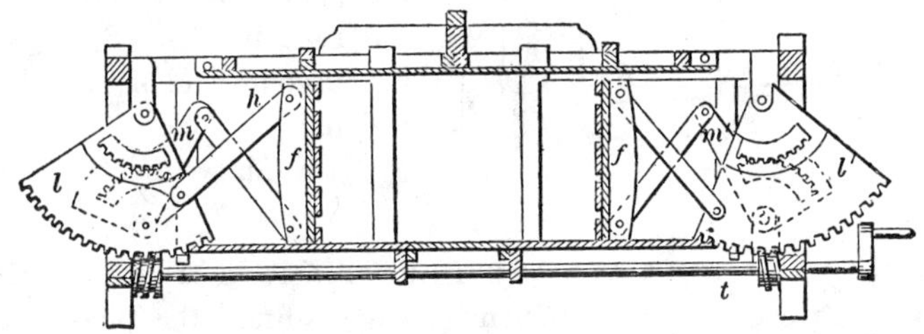

1560. SECTOR WHEEL BALING PRESS.—The large sectors are operated by the long shaft and worm gears. The double toggle joints and small sector gears extend the toggle bars with increasing power.

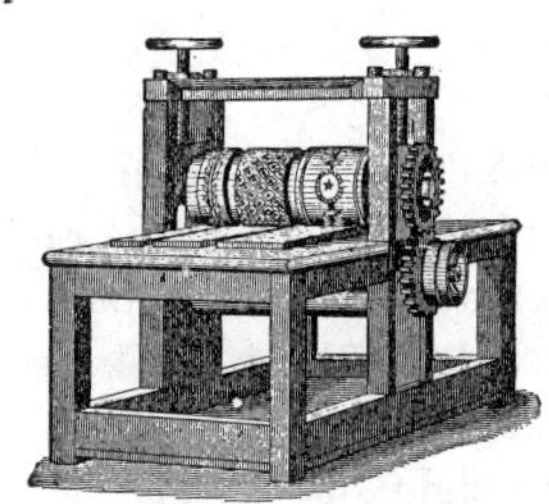

1561. WOOD COMPRESSION CARVING MACHINE.—The carved patterns are iron rings placed on a hollow iron cylinder which is heated by steam through the trunnion. The wood is steamed and passes under the roller with great pressure.

1562. BELT-DRIVEN ELEVATOR.—Worm gear and friction stop. The belt is shifted by a cam driven by the link chain from the drum shaft. The end of the drum shaft has a screw with two clamp nuts, one on each side of the chain wheel, the hub of which acts as a nut to carry the wheel against the clamp nut when it revolves and throws over the cam shipper.

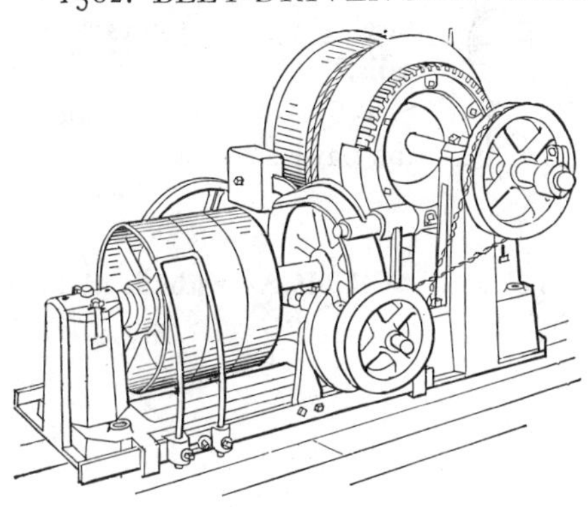

1563. SAFETY CATCH FOR ELEVATORS.—The eccentric sector levers are connected at their pivots to friction slides behind the guide rails by links. The front slides are ratchet bars on the face of the guide rails. The balance weight intensifies the action of the grips when the rope breaks. Springs are also used instead of the balance weights.

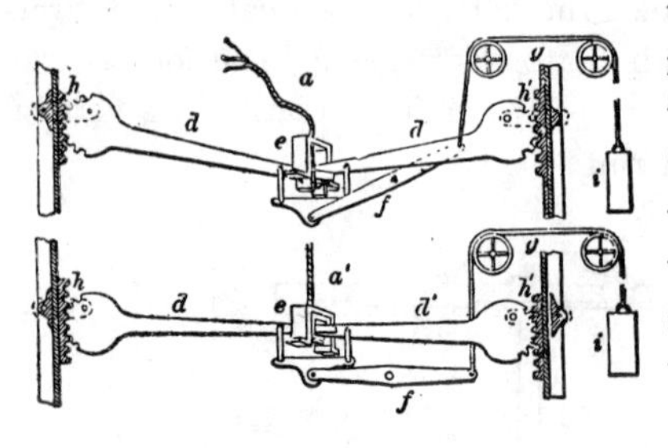

1564. Shows the grip closed.

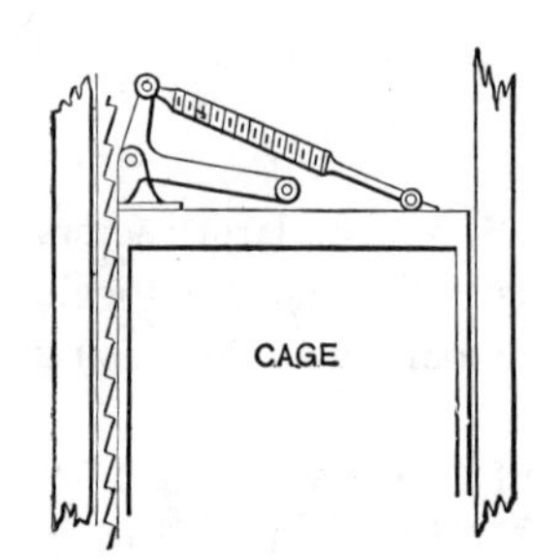

1565. ELEVATOR SAFETY GEAR.—When the cage is lifted the pivoted arm pulls the pawl clear of the rack. A breakage of the rope lets go the pawl arm, and the spring throws the pawl into the rack.

1566. SAFETY CATCH FOR ELEVATORS.—A lever pawl pivoted to each side of the elevator cage is kept clear from the rack guides by the upward pull of the cable. When the cable breaks or gives way, the balance weight or a spring intensifies the action of the pawls in closing with the rack guide rails.

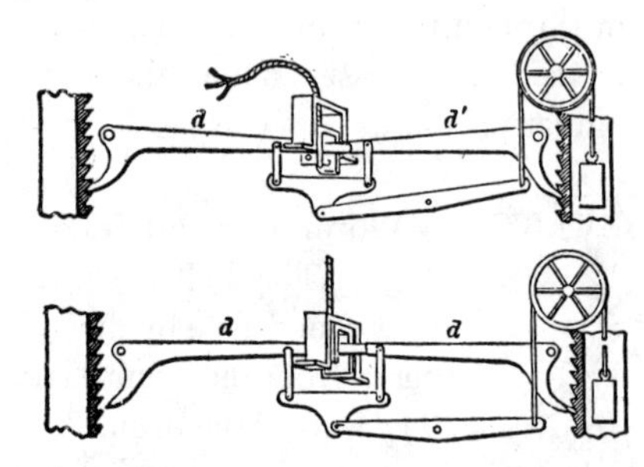

1567. Normal position of the pawls.

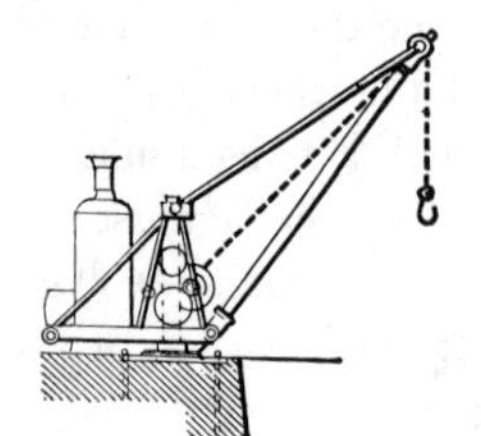

1568. SWING DERRICK, with fixed boom. Steam hoist.

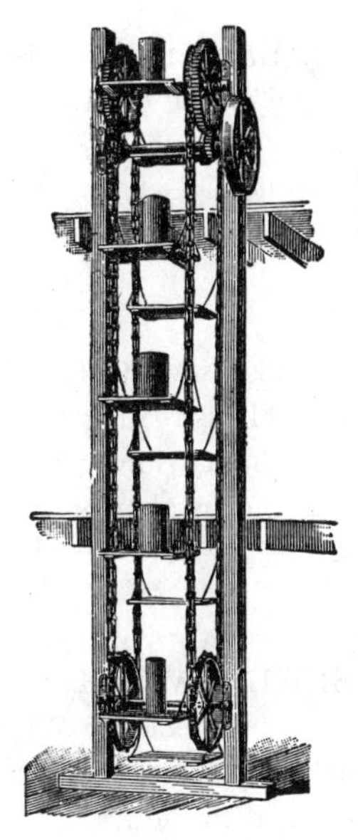

1569. PACKAGE ELEVATOR, for continuous service up or down without reversing. May be arranged for self-dumping both ways.

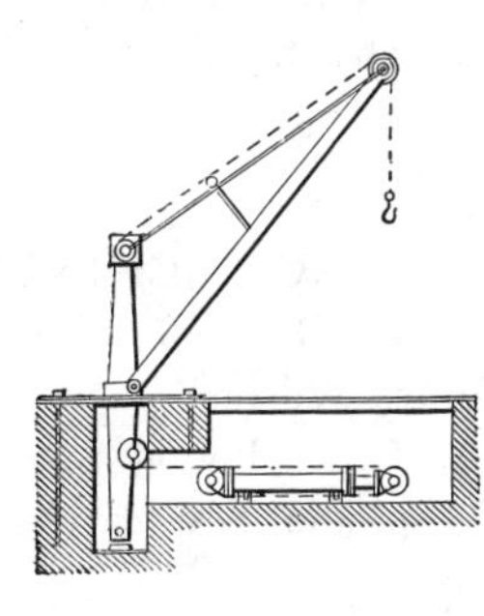

1570. POST CRANE.—Driven by hydraulic lift under the platform. The boom swings on the post. The rope is carried up the hollow post.

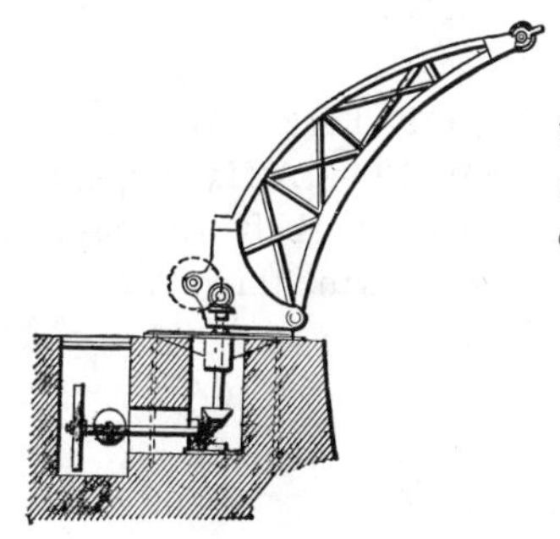

1571. WHARF CRANE, with trussed arch jib. Pivoted to turn in any direction. Power shaft turns in crane pivot.

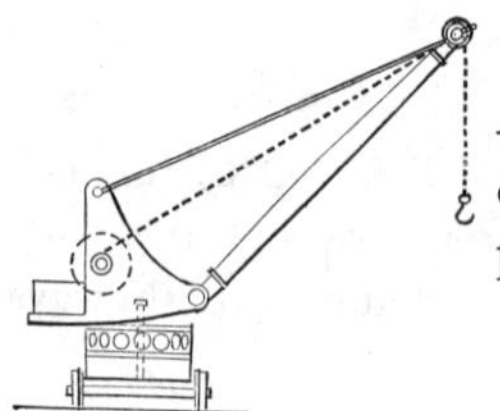

1572. AUTOMATIC BALANCE CRANE.—The rocking base shifts the centre of gravity of load and balance weight. The crane and platform revolve on radial rollers.

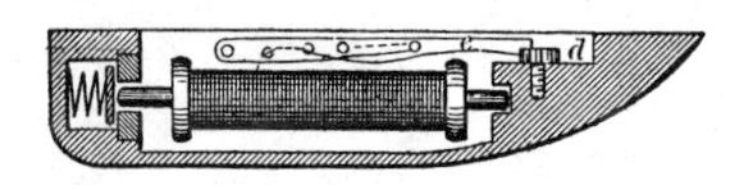

1573. SEWING-MACHINE SHUTTLE.—The thread is rove in the holes in the tension spring, which is made adjustable by the notch cam *d*.

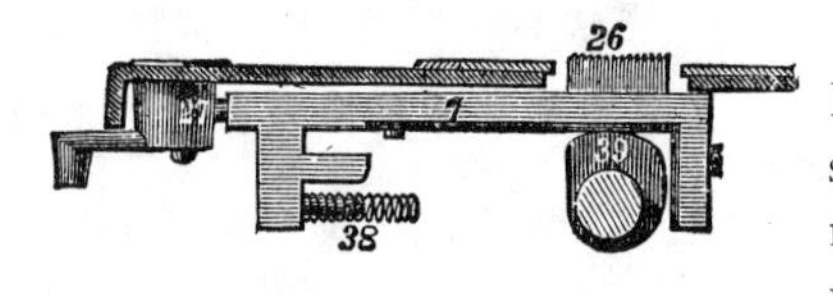

1574. SEWING-MACHINE FEED BAR, "Wheeler & Wilson" model. The toothed feed-rack 26 is fixed to the frame 7, which is lifted, moved forward, and dropped by the cam 39, and is drawn back by the spring 38. The cam stop 27 regulates the length of the stitch.

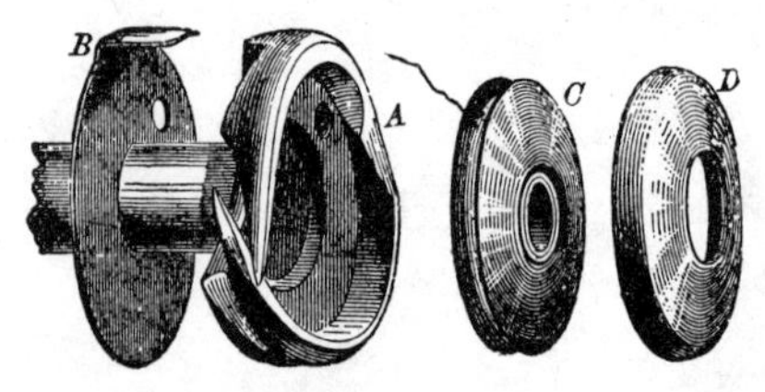

1575. SEWING-MACHINE HOOK AND BOBBIN, "Wheeler & Wilson" model. A, the hook; C, bobbin; D, case; B, spindle and carrier hook.

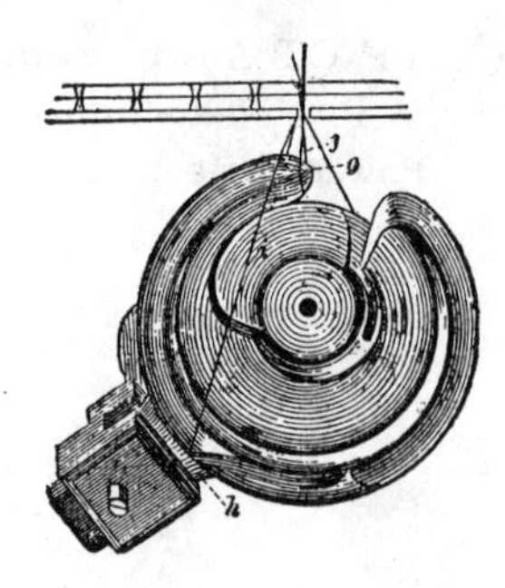

1576. HOOK OF THE "WHEELER & WILSON" SEWING-MACHINE.—The hook is rotated by the shaft, catches the needle loop, and carries the thread around a disc bobbin.

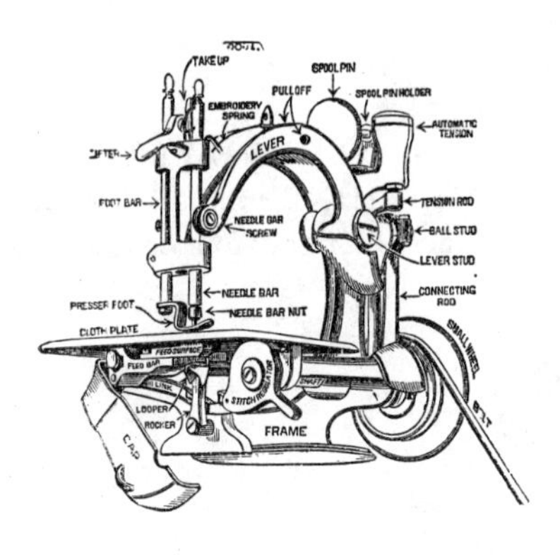

1577. SEWING-MACHINE. "Wilcox & Gibbs" model, showing the designation of parts.

1578. SPRING MOTOR, for sewing-machine. A strong coiled spring and a gear train, like a clock train on a larger scale, geared to the driving shaft. The pedal is changed and arranged as a friction stop and speed regulator.

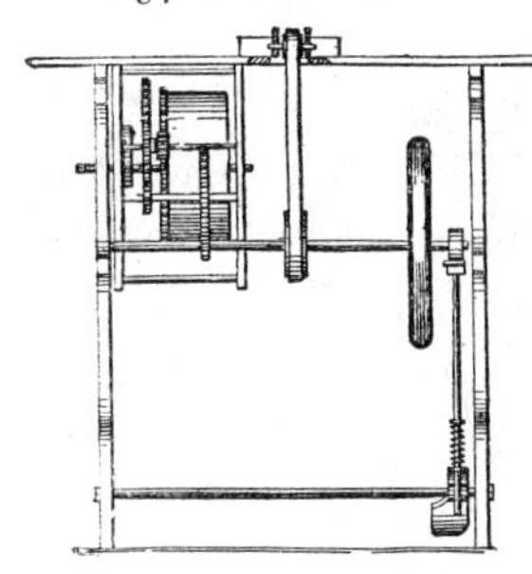

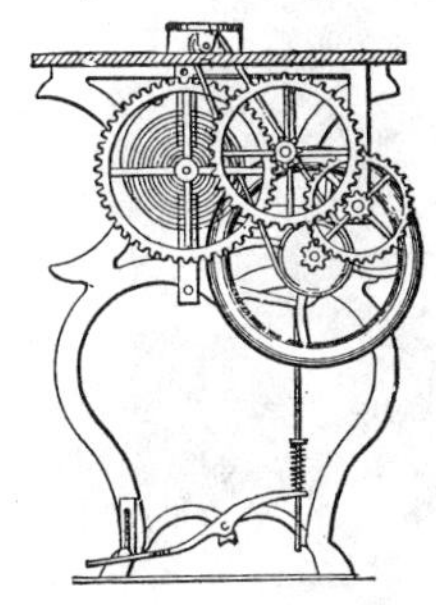

1579. End view.

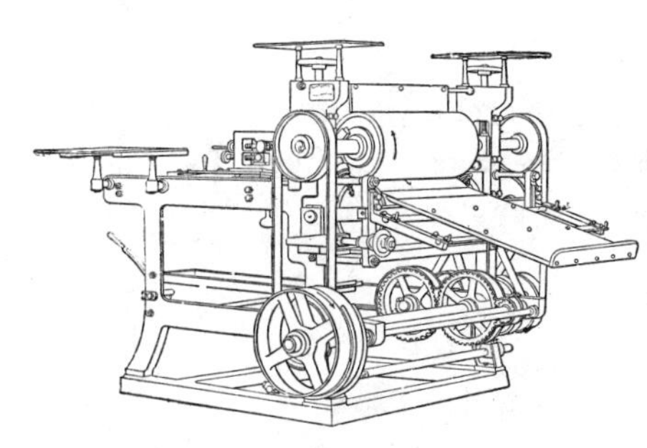

1580. TINPLATE LACQUERING MACHINE.—The roller is elastic. The lacquer is fed to the roller by small rollers and equalized by scrapers.

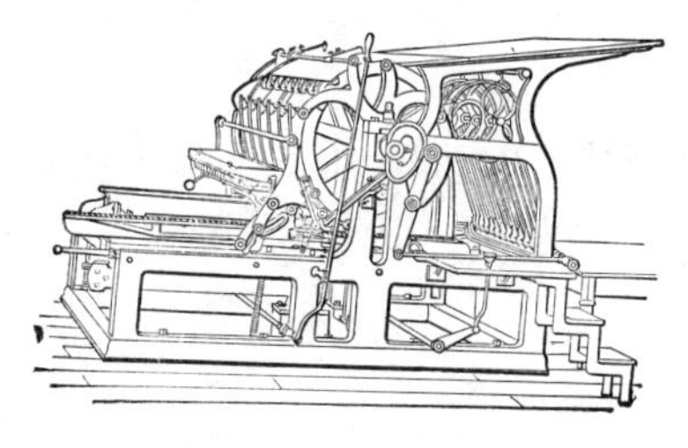

1581. SINGLE-CYLINDER PRINTING PRESS.—A type of the use of cams, levers, shafts, gearing, etc., in combination with rotary and rectilinear motion.

1582. TYPEWRITING MACHINE, "Smith" Premier model. Eighty-four characters.

1583. TYPEWRITING MACHINE, " Remington " model. Eighty-four characters.

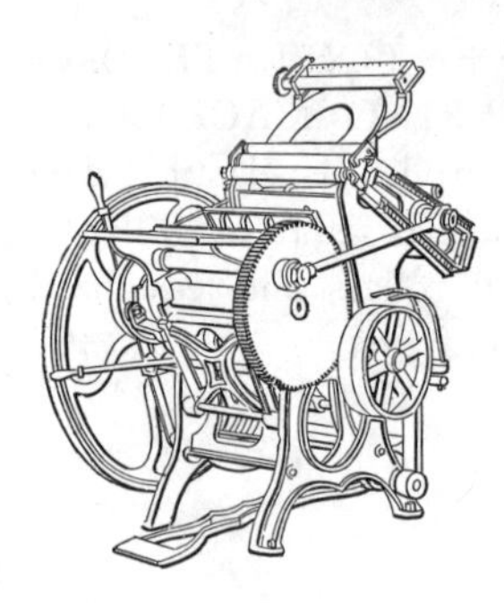

1584. "GORDON" PRINTING PRESS.—Single cylinder, for bill and letter press-work.

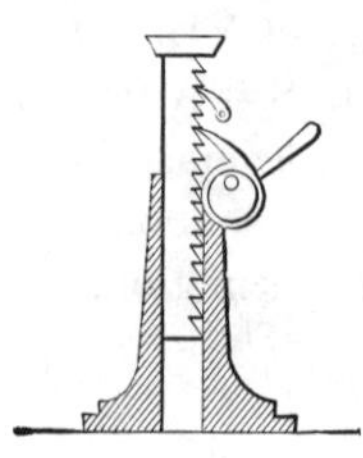

1585. RACK AND PAWL wheel lifting-jack. Lower pawl is operated by a lever or crank.

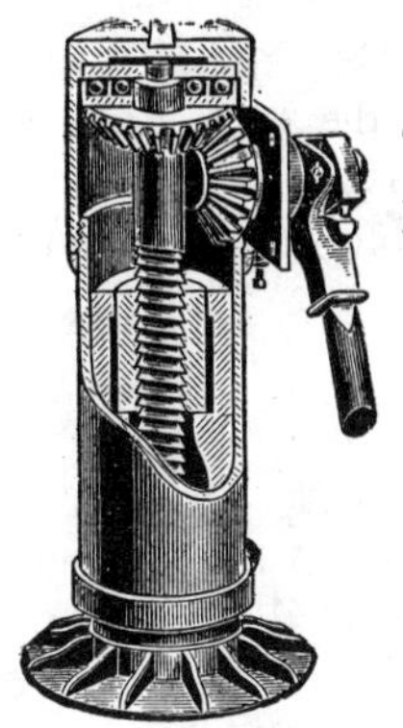

1586. BALL-BEARING SCREW JACK.—The balls run in grooves between the bearing plates.

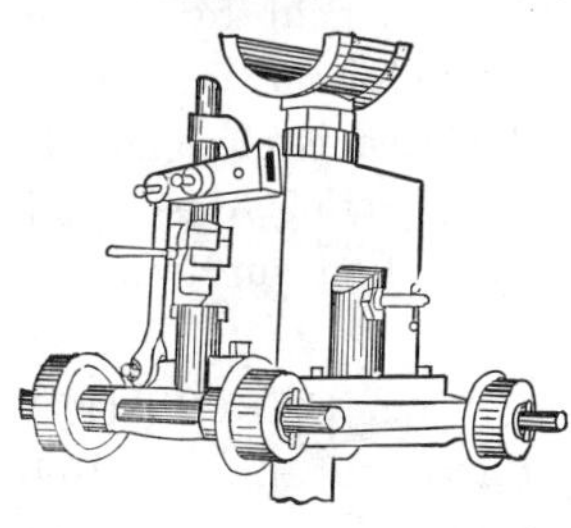

1587. HYDRAULIC TRANSFER JACK.—For lifting cars or transferring over temporary rails. The extension of the truck axles allows for adjustment to any gauge railroad.

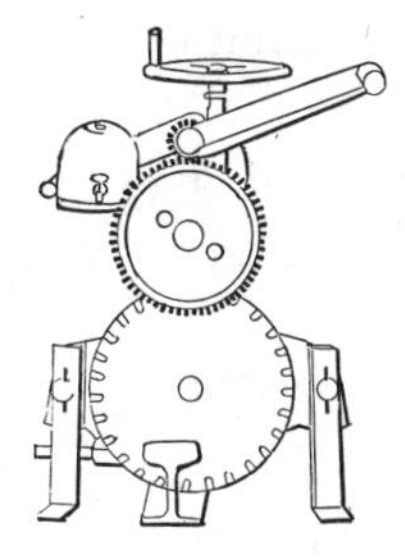

1588. RAIL-CUTTING SAW.—The saw is driven by a trundle pinnion meshing in the teeth of the saw and geared up to the crank. The saw is fed by a screw moving the gear frame down on the rail.

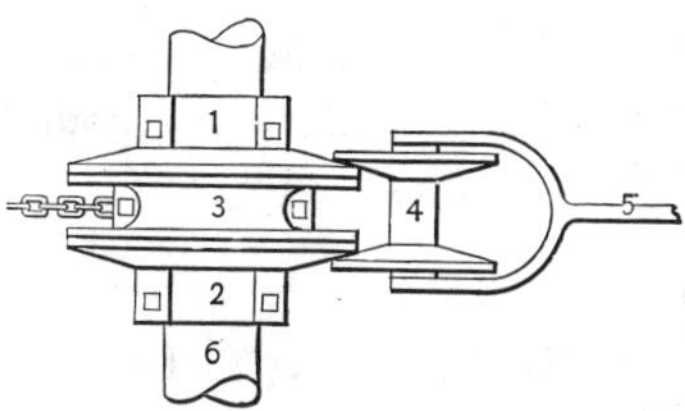

1589. PROUTY-NOBLE AUTOMATIC, OR SELF-WINDING BRAKE.—The central chain spool 3 runs loose on the car axle and between two friction flanges, one of which is fast to the axle and the other slides on a feather. The contact of the inside cones of the brake spool 4 with the outside cones of the friction flanges 1 and 2 causes the chain spool to wind up the brake chain and hold it by friction.

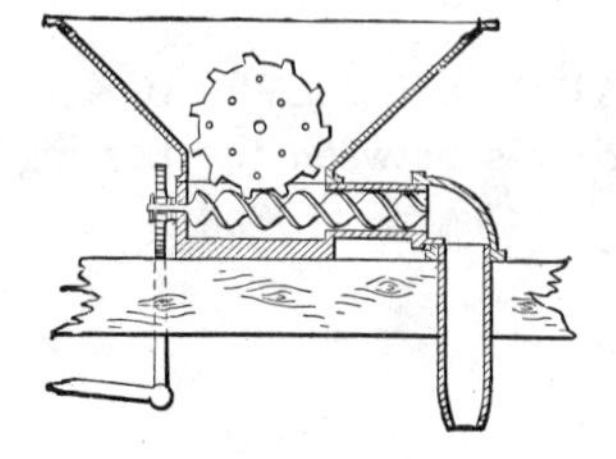

1590. STREET-CAR SAND BOX.—The operation of the lever, pawl, and rachet wheel turns the twisted carrier and at the same time revolves the toothed feed wheel.

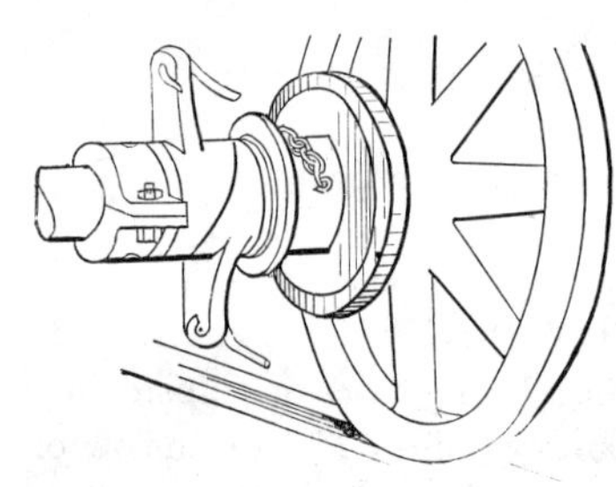

1591. FRICTION BRAKE for street-railway cars. A leather washer between the flange of the brake spool and axle flange is the friction surface. The spool is held by a short wind of the chain either way. The diagonally cut sleeve is elongated by a pull on the connecting rods, which compresses the friction surfaces.

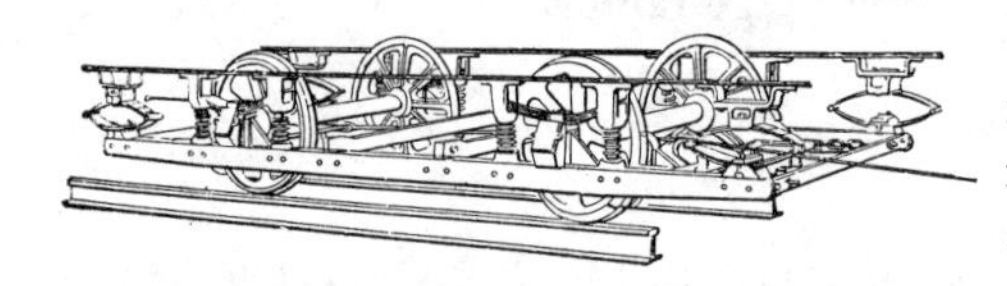

1592. CAR TRUCK for street railways. Sub-frame and compound system of springs.

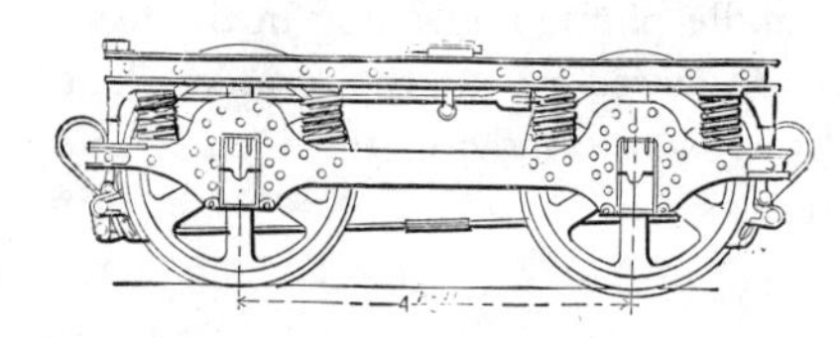

1593. STREET-CAR TRUCK with spring frame and brake connections.

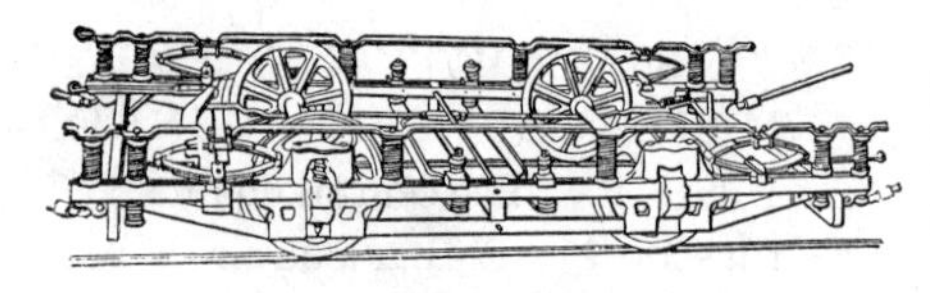

1594. CAR TRUCK for street railway. "Peckham" model. Compound system of springs.

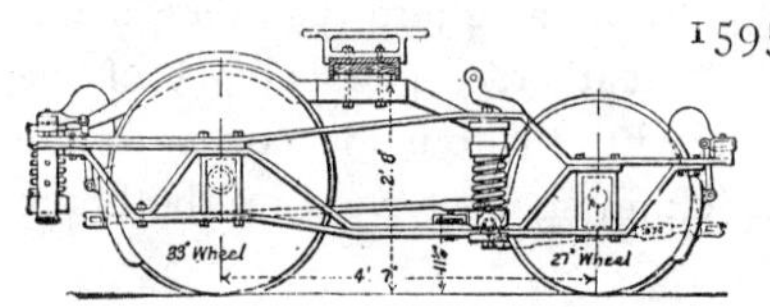

1595. TROLLEY CAR TRUCK.—The larger wheel is geared to the motor. The small wheel is the trailer.

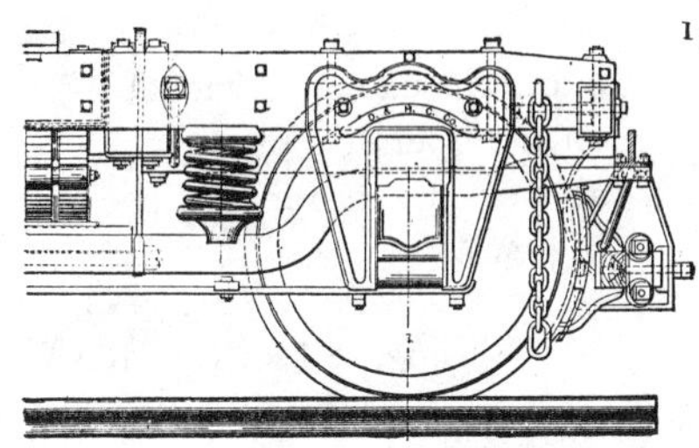

1596. FREIGHT-CAR TRUCK, forward half, with brake, beam, and safety chain; spring and bearing bar.

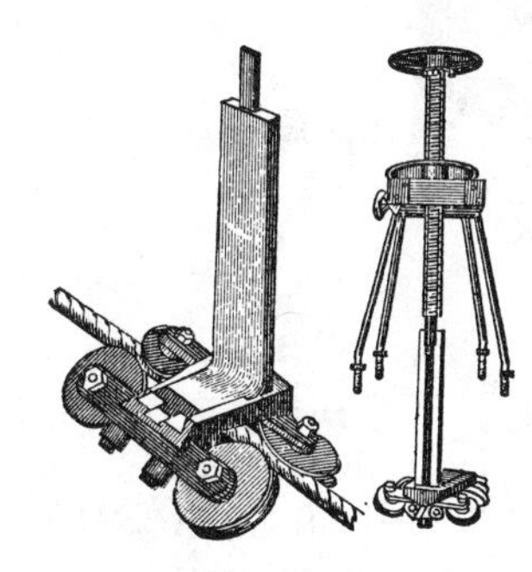

1597. CABLE RAILWAY GRIP.—Friction sheaves are drawn tightly on the cable by a vertical bar in the frame plate. Friction is increased by further tightening the grip wheel.

1598. Showing wheel connection with grip.

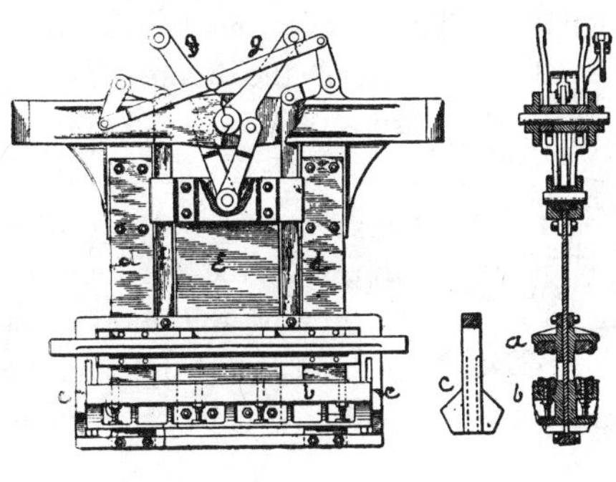

1599. CABLE GRIP FOR STREET RAILWAYS.—*a*, *b*, grip jaws and blocks; *c*, pull-up to throw the cable out of the jaws; *d*, *d*, frame plates; *e*, grip plate connected to *b*, and operated by the bell crank levers *g*, *g*; *f*, *f*, pull-up attached to frame and disengaging pieces *c*, *c*.

1600. End view of grip.

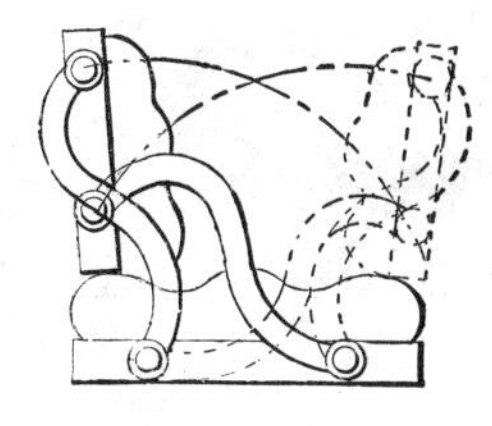

1601. LINKED HINGES for reversing car seat backs.

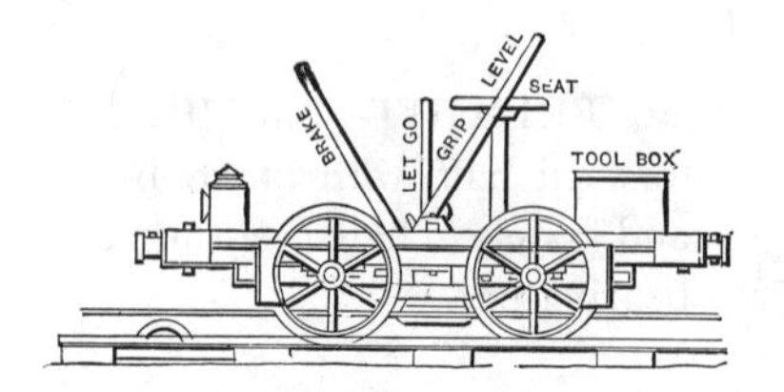

1602. ENDLESS CABLE GRIP CAR.—A stationary jaw under the cable. A movable jaw on top operated by a grip lever above. Used for towing mining cars.

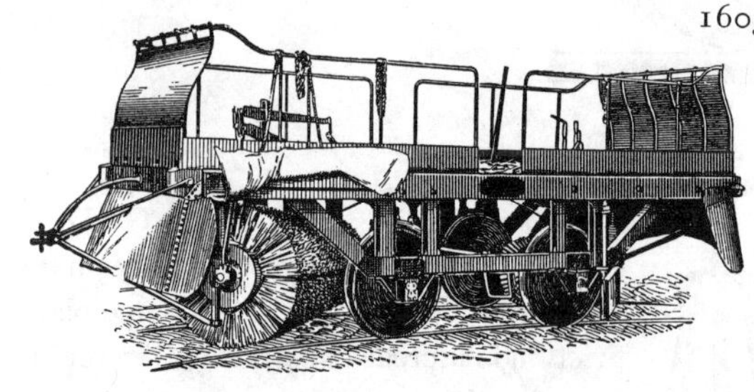

1603. STREET RAILWAY SWEEPING CAR. — The cylindrical sweeper is driven from the axle by bevel gear.

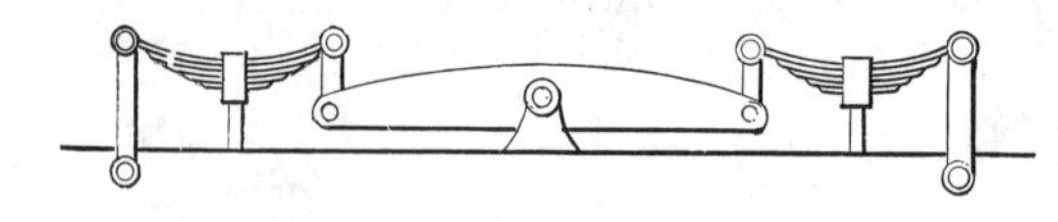

1604. EQUALIZING LEVER for distributing the load on car springs.

1605. NOVEL CAR BRAKE.—The connecting bar between the brakes is adjustable for a small movement of the brake lever to bring the brakes into operation. When the brake is put on from the front platform, with the car running either way, the motion of the front wheel tightens the brake by its friction on the brake shoe, lessening the labor of handling the brake.

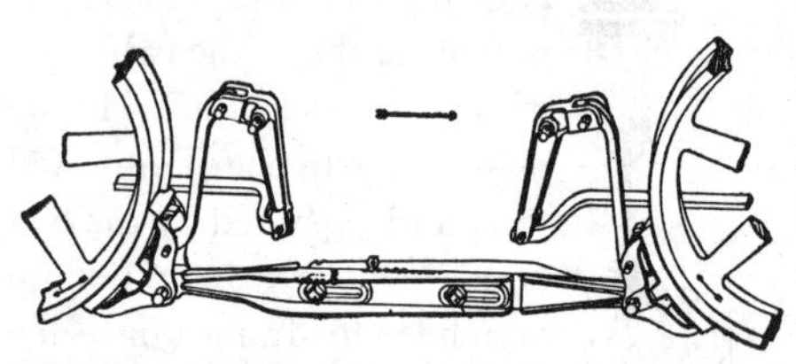

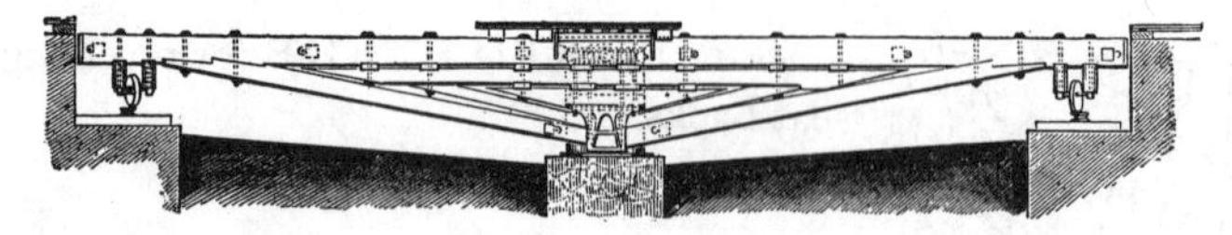

1606. WOODEN FRAME TURN-TABLE, showing method of framing.

1607. IRON FRAME TURN-TABLE, showing design of cast-iron panels. Wrought-iron top chord.

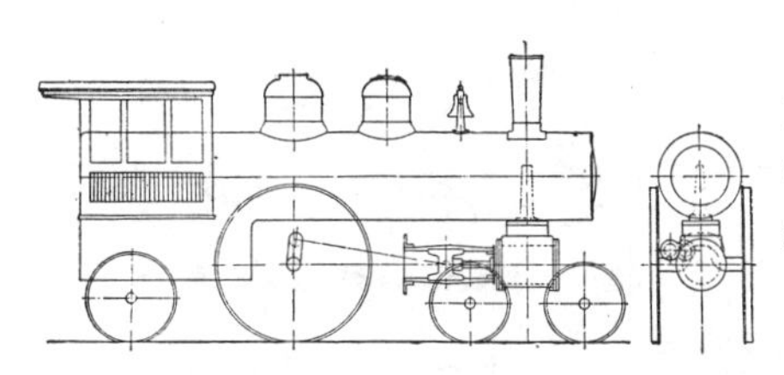

1608. SINGLE-CYLINDER LOCOMOTIVE.—Centre crank, for narrow-gauge roads.

1608*a*. End view.

1609. MODERN LOCOMOTIVE and tender.

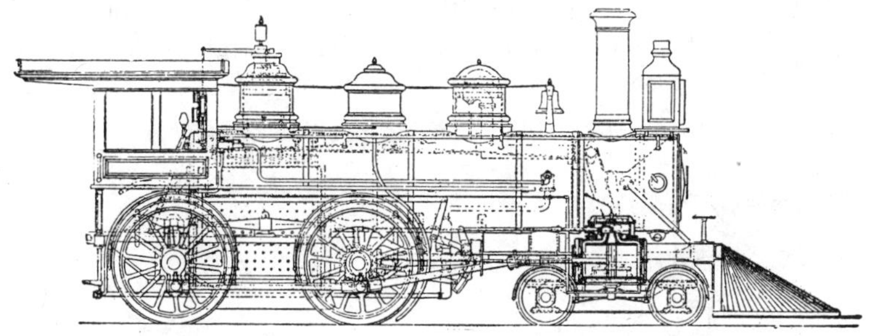

1610. PASSENGER LOCOMOTIVE.—Eight-wheel model.

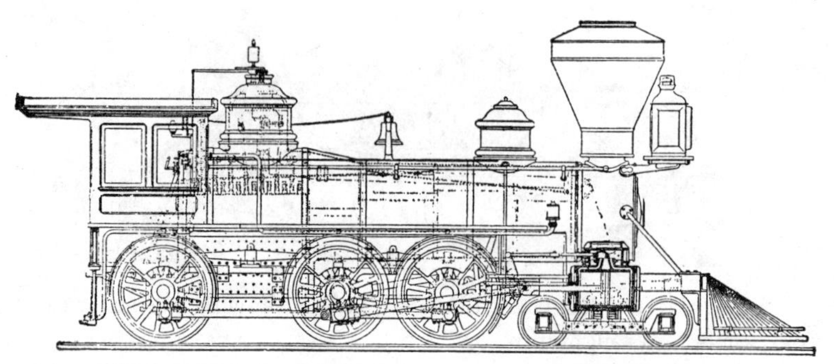

1611. TEN-WHEEL FREIGHT LOCOMOTIVE.—Recent type.

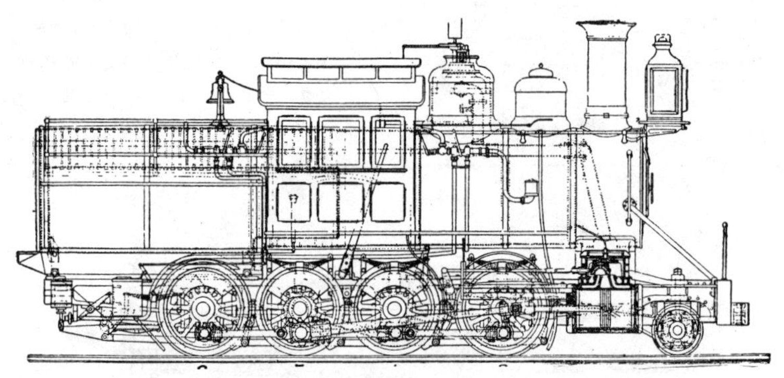

1612. FREIGHT LOCOMOTIVE.—Consolidation type.

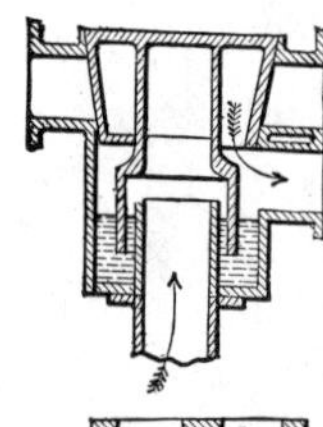

1613. CENTRE VALVE, for a gas house. A four-part valve for a purifier. Arranged to cut out any one of four purifier pans.

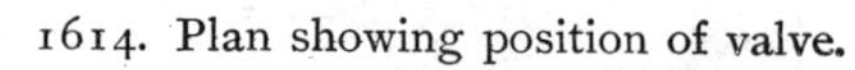

1614. Plan showing position of valve.

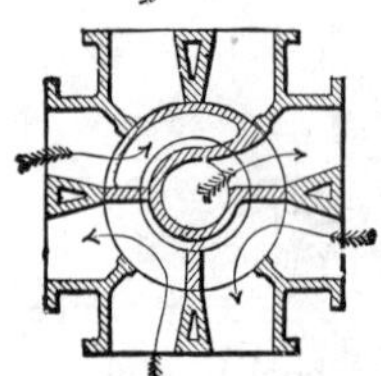

1615. DISC VALVE, for large gas pipes. The disc is revolved by a pinion meshed in a sector gear on the disc.

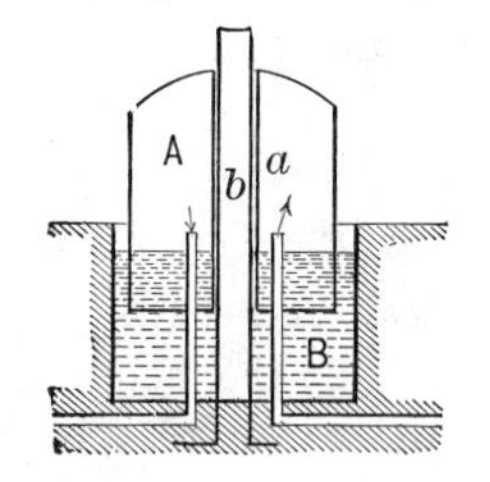

1616. CENTRE GUIDE GAS HOLDER.—

A, the holder.
b, centre guide.
a, tube sliding on centre guide.
B, tank.

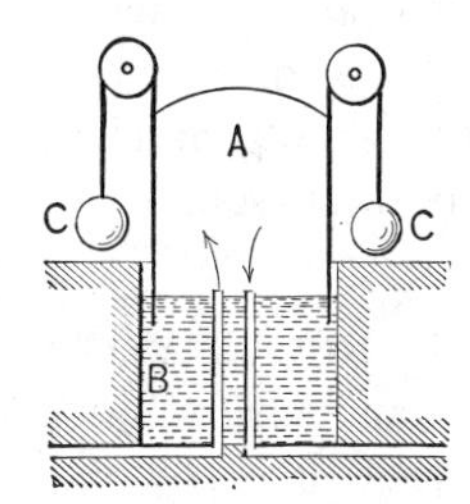

1617. COUNTER-WEIGHTED GAS HOLDER.

A, the holder.
B, the water seal.
C, the counter weights.

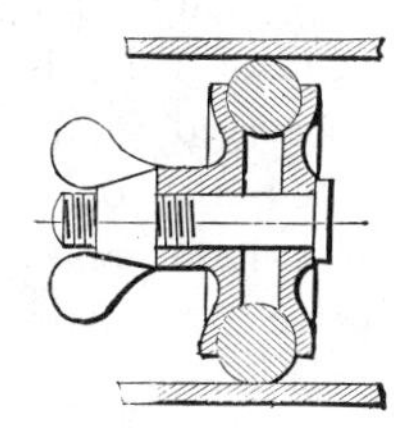

1618. EXPANDING PIPE STOPPER. – A rubber ring compressed between two flanges by a bolt and thumb screw.

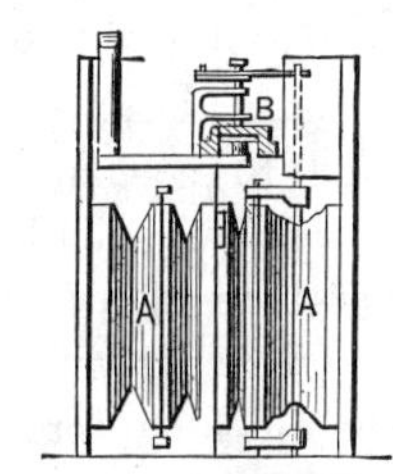

1619. LANTERN BELLOWS DRY GAS METER.—The two pair of bellow chambers, A, A, are alternately filled with gas under the service pressure, by which the movement of the central diaphragm (to which are attached pivots that move the arms of a rock shaft for each pair of bellows) is made. From the top of the rock shaft an arm revolves a spindle that operates the valve by sliding it over the different ports to the two pairs of bellows, and also revolves the gear train of the dials.

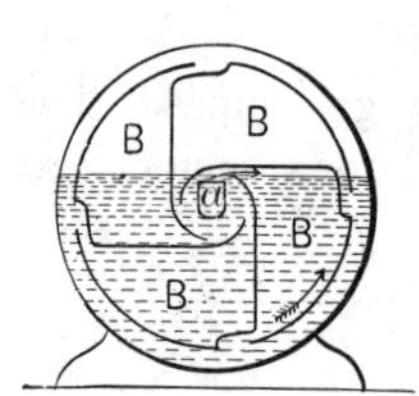

1620. WET GAS METER.—Gas enters through the hollow axis of the four compartments of the drum in a pipe, which turns up just above the water level and fills each compartment successively, and by its pressure causes the drum to revolve in the direction of the arrow and registers on a set of dials. The motion is transmitted through a counter train adapted to separate dial readings.

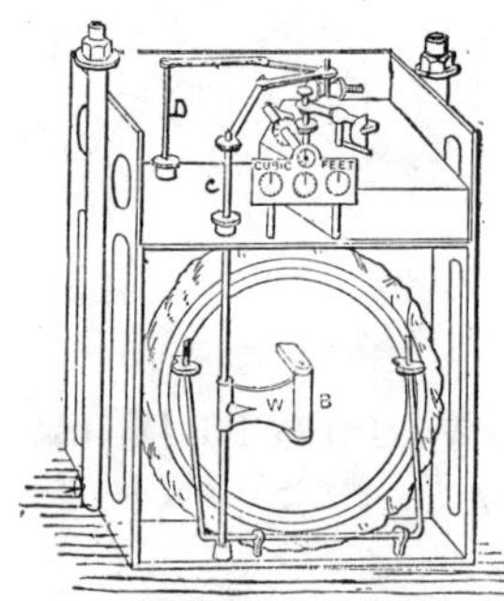

1621. DRY GAS METER and registering train. Two vertical rock shafts, C, D, are vibrated alternately by the bellows B, through the connecting arm W. By this movement the toggle arm pivoted to the rock-shaft cranks is made to swing the arm of the vertical screw-gear shaft, and to set the dial train in motion.

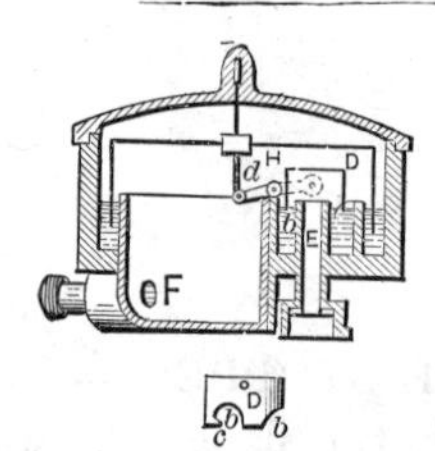

1622. GAS PRESSURE REGULATOR, "Powers" patent. The small annular recepticle around the end of the inlet pipe E is partly filled with mercury, over which the inverted cup valve is suspended to a lever, the other end of which is attached to a larger inverted cup sealed in an annular trough of mercury. F is the outlet to the lighting pipes. Any excess of pressure in the lighting system raises the large float and, through the lever, closes the cup valve to regulate the flow of gas from the service pipe.

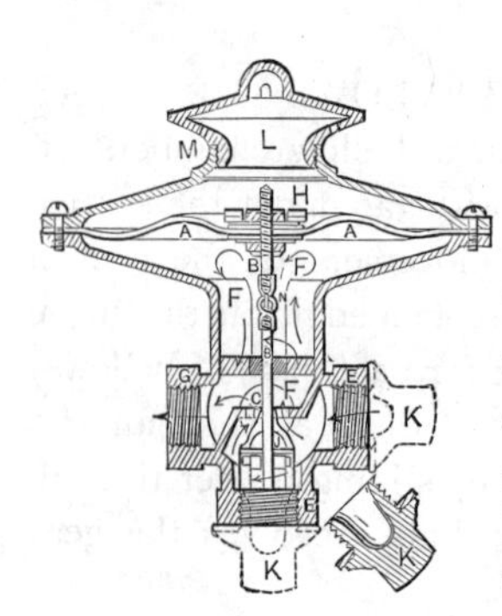

1623. GAS PRESSURE REGULATOR.—An elastic diaphragm is fastened between dished discs and connected to a conical valve disc by a light adjustable spindle. The pressure for the burners is regulated by ring weights at H, and the proper position of the valve by the nuts on the long screw at the top of the spindle. The screw cap K may be placed on either inlet as convenient.

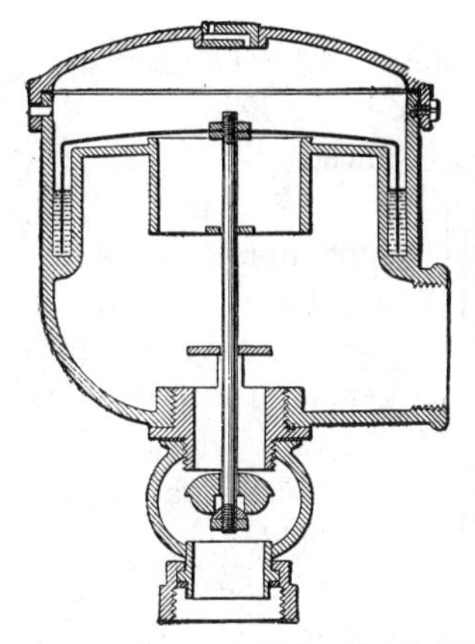

1624. GAS PRESSURE REGULATOR.—The gas flows in at the bottom and out at the side. The inverted float or basin is sealed in an annular cavity by mercury and free to rise under excessive pressure and partially to close the valve in the inlet.

1625. FUEL GAS BURNER, for stoves. Made to push into a cook stove through the side door. The fuel or natural gas enters the Bunsen tube at the right and is further mixed with air under the caps, which are also revolving dampers for shuting off the gas from one or two of the three burners.

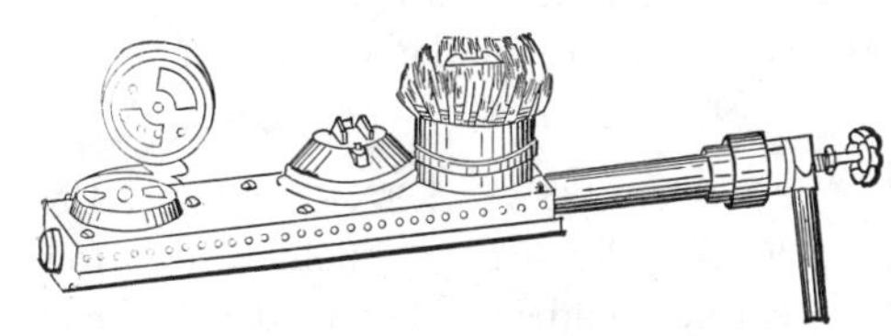

1626. GAS FURNACE.—The air injector draws the gas into an annular nozzle and mixes with it, forcing the mixed gas and air through the tube to the furnace.

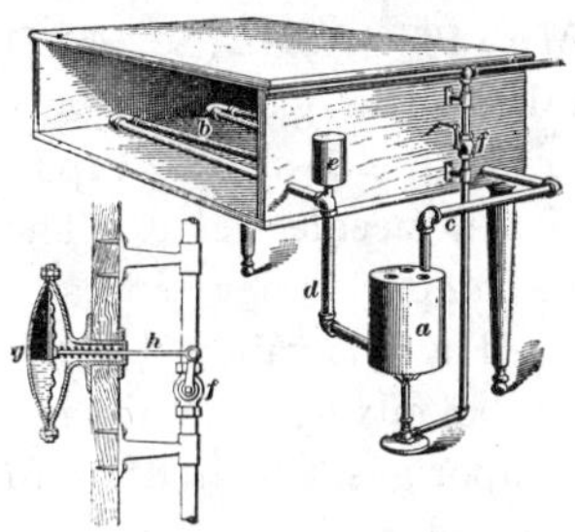

1627. GAS-HEATED INCUBATOR. — A hot-water tank heated by a small Bunsen burner or lamp. *b*, *c*, *d*, Circulating pipes; *f*, regulating cock; *e*, expansion cup.

1628. Thermostat regulator. *g*, Thermostat, consisting of a corrugated metal diaphragm between two cupped plates and connected to the lever of the wick gear or gas cock with a spring to balance the pressure of a volatile fluid on the opposite side of the diaphragm, which may be ether, which boils at 100° F.

1629. ACETYLENE GAS GENERATOR, "Troubetzkoy" model. Has a water flow governed by the rise in the holder. J, Balanced gas holder; G, water seal tank; B, B, generators, two or four; A, small water tank; *a*, pipe to convey water to generators; *b*, governing valve, operated by the rise and fall of the gas holder.

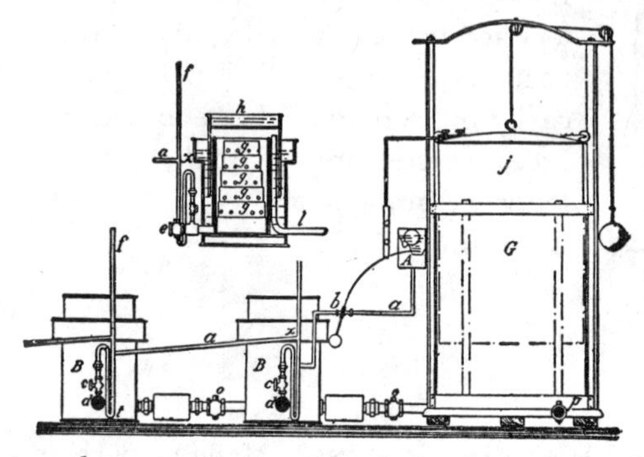

1630. Section of generator. The water enters the generators successively through the inverted siphon; *g*, *g*, *g*, pans of carbide sealed by the cap *h* in the annular water tank.

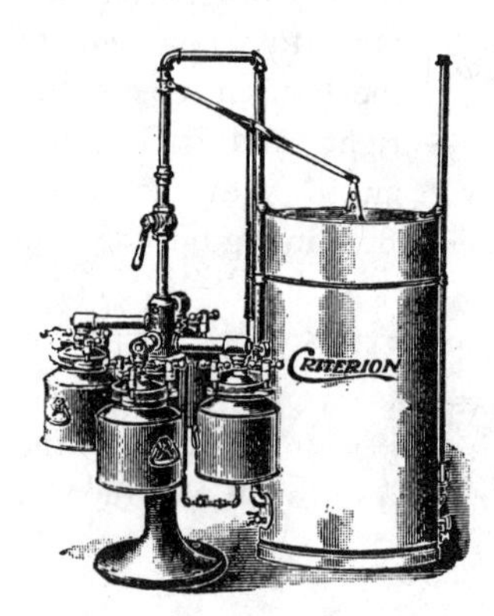

1631. ACETYLENE GAS GENERATOR.—A gas holder and four carbide holders. The holders are connected to a vertical pipe at varying heights, so that only one at a time is fed with water. The water-flow is regulated by the rise and fall of the gas holder.

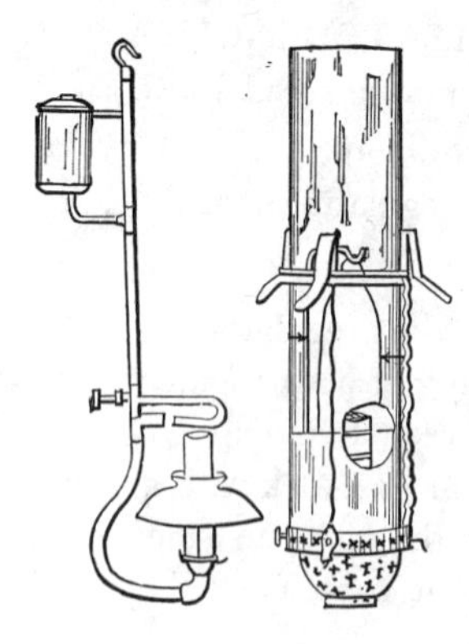

1632. AUTOMATIC GASOLINE AND MANTLE LAMP.—The gasoline flows from the reservoir to the ⊃-shaped vaporizer, regulated by a needle valve. The vaporizer over the lamp chimney generates a vapor pressure. The gas vapor is jetted into the opposite tube, mixing with air and producing an air vapor gas, which flows to the mantle burner below.

1633. Mantle and chimney.

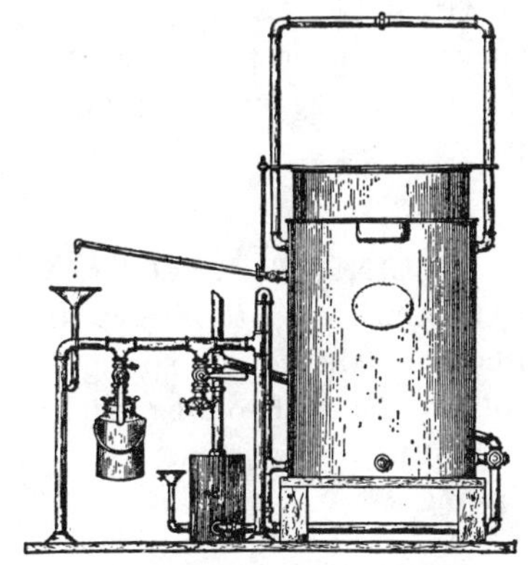

1634. ACETYLENE GENERATOR AND GAS HOLDER.—The carbide is charged into the small vessel suspended from the cross pipe, with a stopcock above. A connection for a second carbide vessel is also seen. Water from the holder runs through a jointed pipe and drips into the sealed funnel. The water nozzle is lifted by the rise of the holder and stops the flow of water. The small vessel at the bottom is a sealed washer and drip catch.

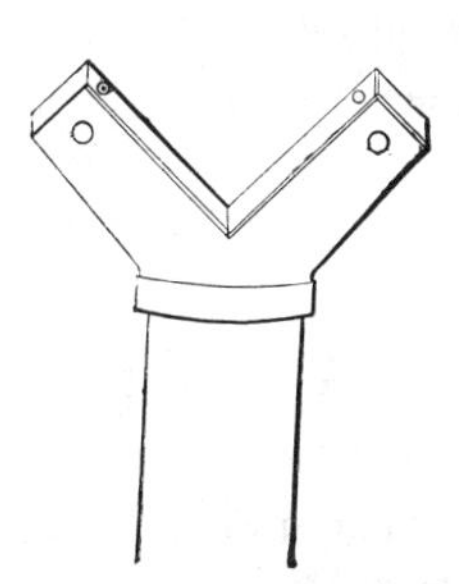

1635. ACETYLENE BURNER, made of lava. The burner holes are at an angle of 90° on inner face of the arms. The air-mixing holes are on each side of the arms. German. Gleason Manufacturing Company.

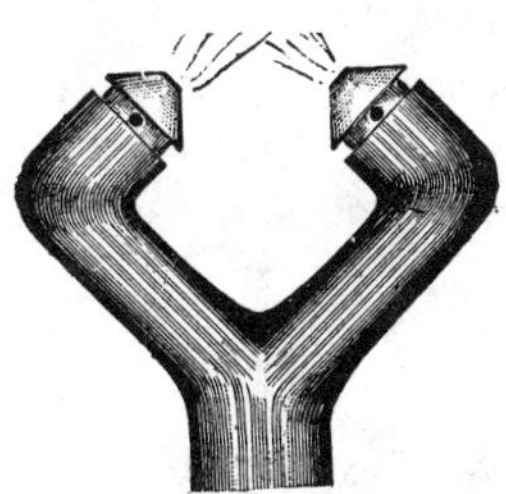

1636. ACETYLENE BURNER.—A double flame burner at right angles. The small holes in the sides of the tips allow air to enter and mingle with the acetylene gas before it is ignited, thus making a mixture of gas and air that makes a clear flame and a safe burner.

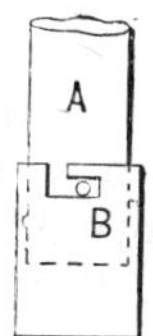

1637. BAYONET JOINT.—The pin fixed in the part A slips into the L-shaped slot of the piece B, and by turning is locked.

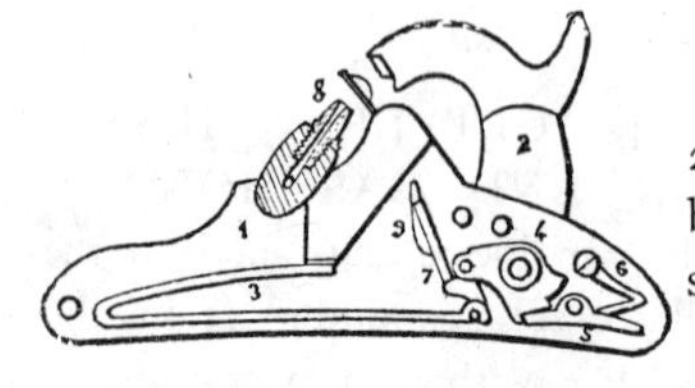

1638. GUN LOCK.—1, lock plate; 2, hammer; 3, mainspring; 4, tumbler; 5, sear or trigger lever; 6, sear spring.

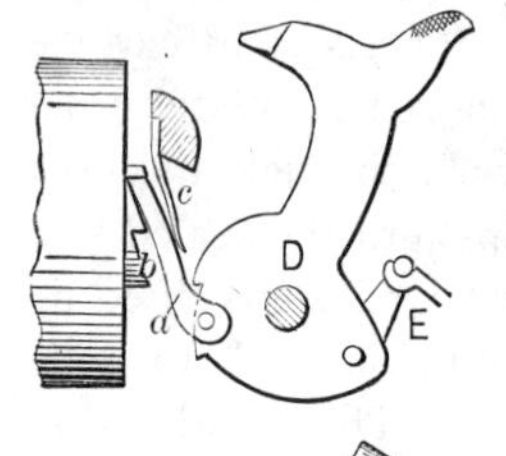

1639. COLT CYLINDER REVOLVING DEVICE for firearms. *a*, the pawl that catches the circular ratchet *b*; *c*, a spring that pushes the pawl into the teeth of the ratchet; D, the hammer butt to which is pinioned the pawl and the spring E.

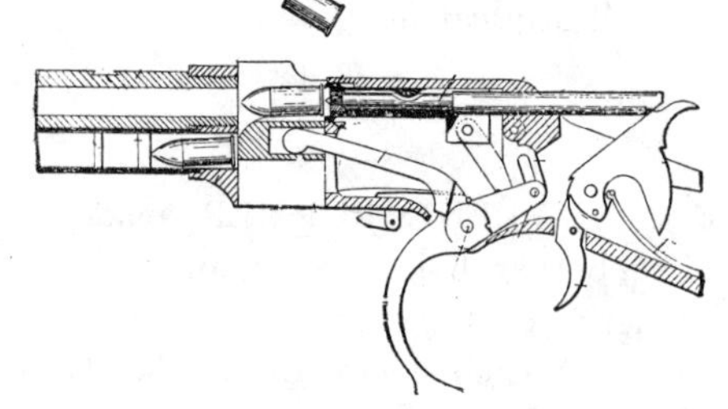

1640. MAGAZINE RIFLE, "Lee-Metford" model. Magazine in the barrel stock.

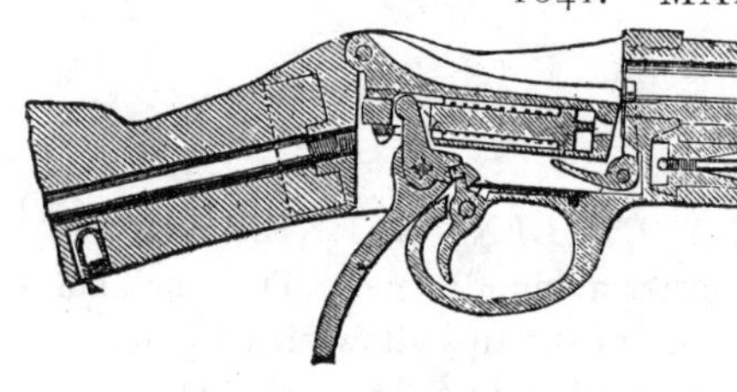

1641. "MARTINI-HENRY" RIFLE.—The breech block is pivoted at the rear end and is thrown up or down by the lever at the rear of the trigger guard. A spring plunger in the breech block, let go by the trigger, explodes the charge.

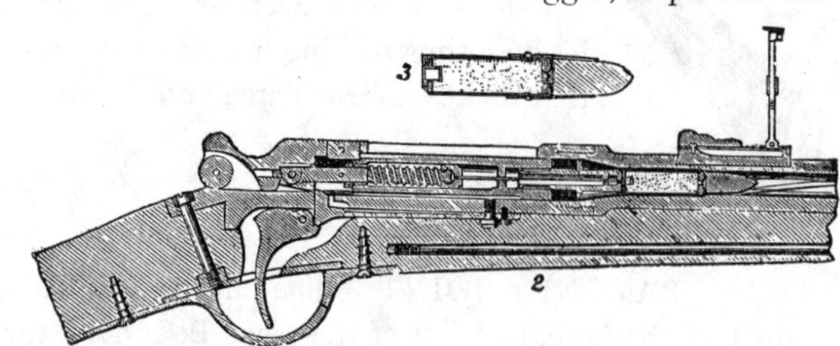

1642. CHASSEPOT GUN.—A needle gun. The cartridge is inserted by hand; the plunger runs forward and is locked by turning into a notch. Centre fire.

1643. REMINGTON RIFLE.—A breech block, operated by a handle, is pulled back to allow the cartridge to be charged by hand, when the breech and the block are locked. The hammer strikes a firing pin within the breech block.

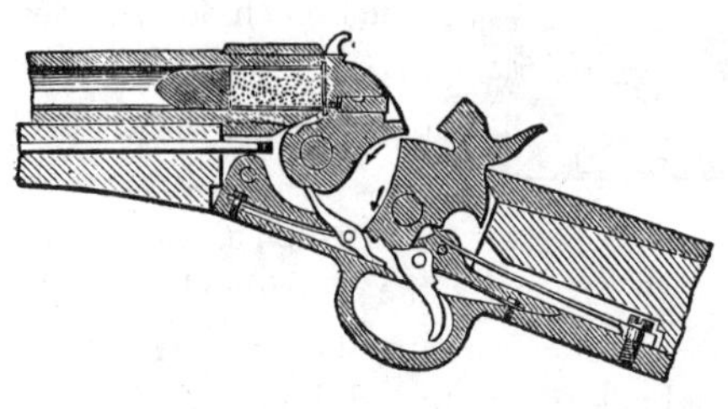

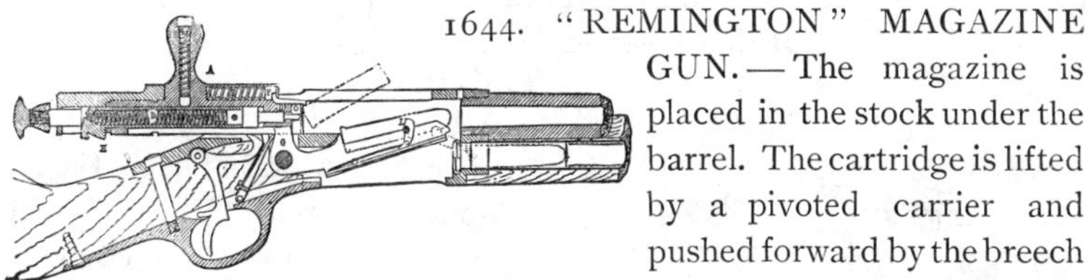

1644. "REMINGTON" MAGAZINE GUN.—The magazine is placed in the stock under the barrel. The cartridge is lifted by a pivoted carrier and pushed forward by the breech block. Central spring plunger hammer.

1645. "HOTCHKISS" MAGAZINE GUN.—The reserve cartridges are carried in the gun stock and forwarded by a light spring. The breech bolt draws back by the handle, when the cartridge is raised and pushed forward into the barrel. Centre pin spring hammer.

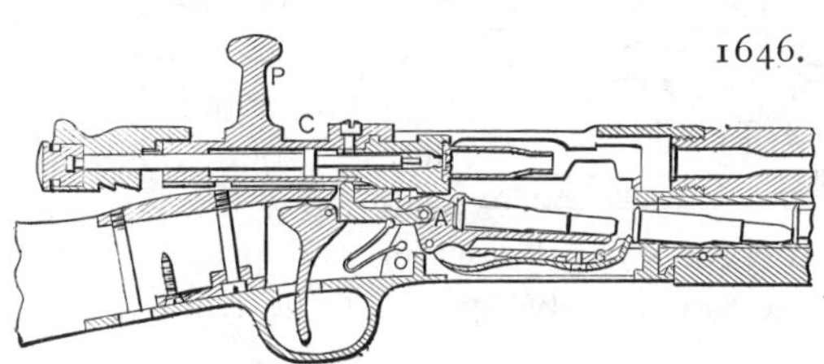

1646. "LEBEL" RIFLE.—Magazine under the barrel in the extension of the stock. A sliding breech block and piston hammer.

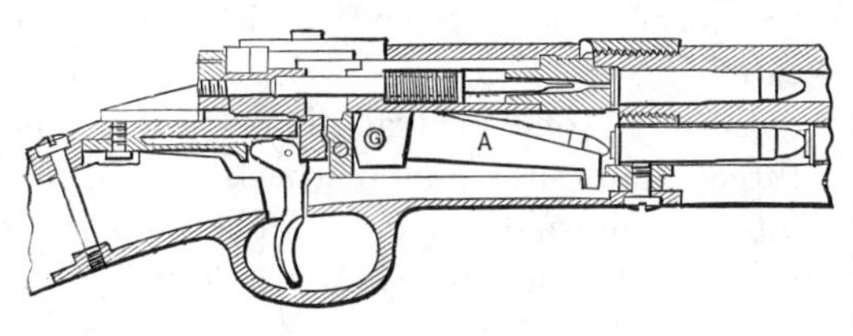

1647 "MAUSER" RIFLE.—Magazine in the forestock. The sliding breech block encloses the firing spring plunger and raises the cartridge lever A.

1648. "WINCHESTER" MAGAZINE RIFLE.—

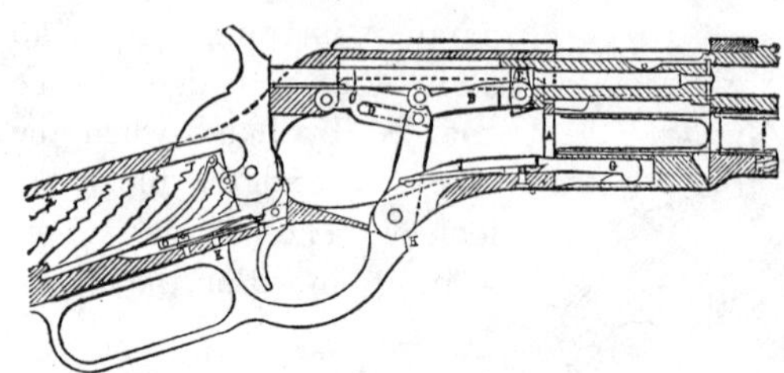

The breech lock slides in line with the barrel by a toggle link, operated by the breech lever, which also operates the cartridge lever, raising the cartridge to its position for charging. Drawing back of the breech block carries the hammer back to its firing position.

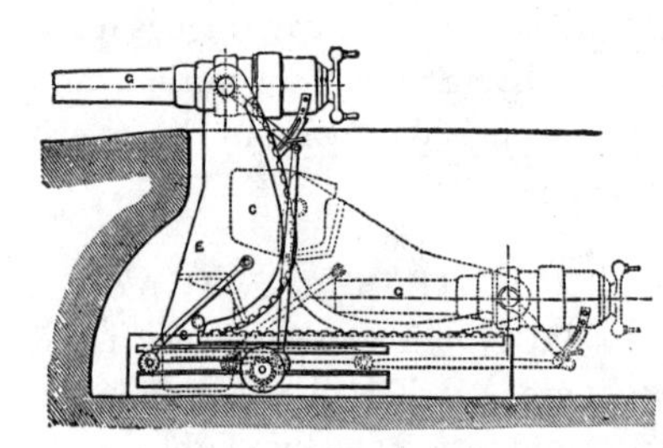

1649. DISAPPEARING GUN, "Moncrief" model. The cycloidal curved rack arm E is counterweighted, which balances the recoil of the gun by its increased leverage. The small connecting rod, rack, and pinion adjust the gun's alignment.

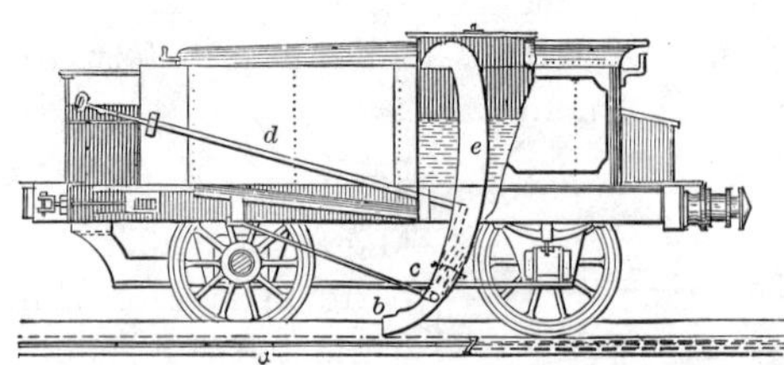

1649*a*. RAILWAY WATER LIFT. — A long water trough between the tracks. A movable spout in the tender is dropped into the water trough at an angle to scoop up the water and propel it into the tank by the speed of the train.

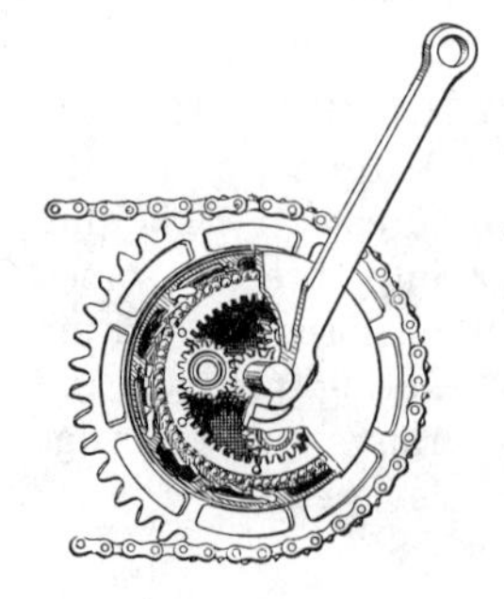

1649*b*. TWO SPEED GEAR. — The "Sunbeam" Two Speed mechanism is located inside the crank chain wheel. It consists of one central pinion wheel, and three small outer pinion wheels which gear into an internal cut ring. The change of gear is obtained by simply holding or releasing the central pinion.

INDEX.

D

E

H

I

J

K

L

M